Signaling and Communication in Plants

Series Editor

František Baluška, IZMB, Department of Plant Cell Biology, University of Bonn, Bonn, Nordrhein-Westfalen, Germany

This book series will be devoted to diverse aspects of signaling and communication at all levels of plant organization, starting from single molecules and ending at ecological communities. The individual volumes will interlink molecular biology with plant physiology and the behavior of individual organisms, right up to the system analysis of whole plant communities and ecosystems.

Plants have developed a robust signaling apparatus with both chemical and physical communication pathways. The chemical communication is based either on vesicular trafficking pathways or accomplished directly through cell-cell channels known as plasmodesmata. Moreover, there are numerous signal molecules generated within cell walls and also diffusible signals, such as nitric oxide, reactive oxygen species and ethylene, penetrating cells from exocellular space. Physical communication on the other hand is based on electrical, hydraulic, and mechanical signals.

The integrative view of this book series will foster our understanding of plant communication throughout the individual plant and with other communicative systems such as fungi, nematodes, bacteria, viruses, insects, other plants, and predatory animals.

Nabil Semmar

Secondary Metabolites in Plant Stress Adaptation

Analytic Space of Secondary Metabolites

Springer

Nabil Semmar
BIMS Laboratory
Pasteur Institute of Tunis
University of Tunis El Manar
Tunis, Tunisia

ISSN 1867-9048 ISSN 1867-9056 (electronic)
Signaling and Communication in Plants
ISBN 978-3-031-52594-0 ISBN 978-3-031-52595-7 (eBook)
https://doi.org/10.1007/978-3-031-52595-7

This Springer imprint is published by the registered company Springer Nature Switzerland AG
The registered company address is: Gewerbestrasse 11, 6330 Cham, Switzerland

Paper in this product is recyclable.

Introduction

Environment constitutes a complex space characterized by a multidimensional (multifactorial) structure varying dynamically under the effect of continuously changing physical–chemical and biological factors leading to subsequent adaptive responses of biological populations at different system scales (molecular, cellular, tissue, physiological, intraspecific, interspecific scales). Biotic and abiotic factors continuously exert stresses on living organisms and populations leading to the need of adaptive (defensive, communicative) strategies for survival and co-development. Consequently, biological compositions and biochemical patterns of ecological populations are closely associated with different abiotic and biotic stressing factors occurring at different manifestation levels, frequencies (dominant, systemic, sudden, sporadic) and scales (local/global) of environment.

Identifying the types and quantifying the levels of changing/stressing factors in environment usefully help for evaluation of living states, variation trends and working ways of biosystems (organisms, populations, communities) favoring subsequently the management of natural resources at different spatial-temporal scales.

Identification and quantification of environment stresses require reliable recording ways of dynamical and cumulative variations of abiotic factors with subsequent (associated) responses of biological organisms living in such conditions. As sessile organisms, plants are continuously exposed to local environment stresses and consequently, their metabolism tends to be highly (reliably) associated with abiotic and biotic variations at different spatial, temporal and biological scales. Correlations between plant metabolism and environment conditions are efficiently highlighted by qualitative and quantitative variations of secondary metabolites (SMs) that are widely distributed in the plant world.

SMs represent a wide field of specialized small molecules showing high structural diversity, many biological activities and not uniform botanical distributions. They are produced at low amounts because of their reactivity and not body constitutive aspects (by opposition to primary metabolites: amino acids, fatty acids, saccharides). About 200,000 SMs have been elucidated from plants, whereas the total number of plant species exceeds 350,000. This makes to conclude about the relatively small number of known SMs compared with the potential number of not discovered yet in the plant world. SMs are mainly classified into three wide metabolic classes including phenolic compounds (hydroxylated benzene structures), terpenes (C5-multiple hydrocarbon chains or cyclic structures) and nitrogen-containing compounds represented by alkaloids, cyanogenic glycosides and glucosinolates (containing sulfur in addition to nitrogen). They play key roles in defense/protection of plants against harmful abiotic and biotic factors.

Abiotic stresses can be categorized into two main types including oxidative and osmotic processes which can be induced by different environment factors. Oxidative stresses involve the formation of reactive oxygenated species (ROS) that can be caused by high radiation levels, extreme temperatures, atmospheric ozone, as well as other oxidant compounds (e.g. heavy metals). Osmotic stress is essentially linked to drought (water deficit, saline middle) and cold (freezing) conditions. These stresses occur both in air and soil strata. This book focuses essentially on air-occurring abiotic stresses due to their direct involvement in the climate change (a current worry at planetary scale).

Biotic stress includes essentially pathogen infections and herbivore attacks of plants in addition to interspecific competitions (allelopathy) between spatially neighbor plants.

Plants respond to different environment stresses by the production and regulation of different types of SMs having appropriate structural and functional characteristics. Oxidative stresses are overcome by antioxidative SMs including essentially phenolic compounds (phenol acids, flavonoids, coumarins, tannins, etc.). Antioxidative properties of these SMs are due to the structural occurrence of hydroxyl groups and conjugated double bonds favoring electronic changes and transitions required for stopping radical chain reactions. Also, phenols show original antioxidative properties through ortho-di-hydroxylated structures able to chelate metals through electrostatic interactions. SMs with conjugated double bonds are also found in terpenes (sesquiterpenes, tetraterpenes) which represent electron sources for stopping oxidative chain reactions.

Osmotic stresses, however, are managed by osmoregulatory metabolites (osmolytes) helping for water retention in plant cells living in water-deficient environments (drought) or with unavailable form of water (freezing conditions). Osmoregulatory SMs include glycosylated compounds making the aqueous intra-cell space to be more concentrated resulting in the decrease of its osmotic potential and, subsequently, the reduction (avoidance) of water loss.

Oxidative and osmotic stresses tend to disturb and fragilize cell membranes by affecting its fluidity and integrity. Due to their lipophilic character, terpenes can interact with cell membrane contributing to the regulation of its fluidity and subsequent functional stabilization.

Oxidative stress can be systematically induced by radiative stress essentially caused by UV-light range. Harmful radiations and excess of non-photosynthetic light are treated by different specialized photoreceptors (chromophores made by conjugated nitrogen heterocycles and benzene) leading to attenuate and canalize the radiative energy for next use. Under radiative stress, antioxidants (e.g. phenolic compounds, carotenoids) are produced at different relative levels depending on radiation type, intensity and ratio.

Apart from abiotic stresses, plants should overcome biotic pressures due to pathogens and herbivores. For that, different SMs-based protective strategies were developed leading to immobilization, toxication or repellence of plant enemies. Pathogens are confronted by production of phytoanticipins and phytoalexins which can interact with proteins or cell membranes to alter their functional or structural aspects. These plant defensive compounds are concerned with various metabolic classes including phenolics, terpenes and nitrogen-containing SMs (alkaloids, cyanogenic glycosides, glucosinolates). Herbivore attacks are confronted by plants through production of individual or mixed SMs causing toxication or repellence of herbivores. Toxication can occur at different biological scales including (i) cytotoxicity via mitosis-blocking SMs or cell membrane attack and (ii) physiological disorder (e.g. paralysis, adverse nervous behaviors) via neural signal perturbation. Repellence of herbivore can occur through mixtures of volatiles (terpenes) attributing specific repulsive odors to the host plant and making the herbivore to avoid it. Astringent and antifeeding compounds provide another anticipative deterring way for plants against enemies. Interestingly, when plants are attacked by specialized herbivores able to detoxify, sequester or eliminate plant toxins, they can manifest indirect defense strategy by releasing mixtures of volatile compounds attracting the natural enemy of herbivore. Sesquiterpenes and monoterpenes were revealed to play key communicative roles in this indirect defense strategy.

Apart from their functional aspects contributing in plant protection against abiotic and biotic stresses, SMs provide highly useful metabolic patterns helping for (i) chemical characterization/classification of productive plants, and (ii) assessment of plant living conditions and environment states. This attributes to phytochemical patterns a great advantage for biodiversity and environment management. Phytochemical patterns can be characterized (interpreted) under different qualitative and quantitative aspects of SMs including occurrence–absence of metabolite(s), co-occurrence of different chemical compounds with variable regulation ratios. Further pattern analysis is made by correlations or variation trends between different metabolites leading to highlight their sensitivity levels and responsiveness ways in relation with different inductive abiotic factors and biological taxa. In biodiversity field, phytochemical pattern analysis provides a strong tool for chemotaxonomic classification of plant populations representing key resiliency components of ecosystems. Environment management via phytochemical pattern analysis can be deepened by considering interaction effects of different co-manifesting environment factors on SM production profiles in different plant taxa.

A synthesis picture on involvement ways of SMs in environment management is given in Fig. 1.

Phytochemical management of environment represents an integrative field associating variable plant metabolic patterns with many structural states and functional aspects of plant growing conditions by means of multifactorial and multiscale computational analyses (Fig. 2). Phytochemical data can be of different types including identified molecular structures (chemical compositions) by spectroscopic ways,

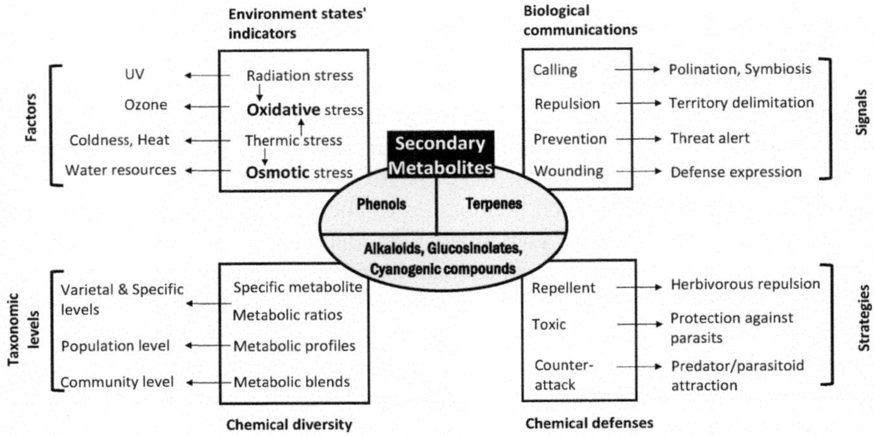

Fig. 1 Synthesis presentation on the book content focused on the interests of secondary metabolites and their patterns (production levels, regulation ratios) in the chemical assessment, monitoring and management of biological populations and natural spaces

quantified amounts by chromatographic ways (in different biological or environment matrices), *in silico* calculated molecular interaction energies, *in vitro* evaluated biological activities, etc. (Fig. 2a). Statistical analysis of these different types of data requires different appropriate methods including correlation or trend analysis between metabolites, population classification of phytochemical patterns, link analysis between biological systems and phytochemical patterns (discrimination or pattern recognition), causal–effect relationship analyses between metabolic structures/compositions and activities or functional characteristics of biological systems (Fig. 2b). Biological systems are highly diversified by their structural and functional aspects (Fig. 2c, d): under structural aspects, biosystems can be phytochemically studied for biodiversity evaluation and chemical polymorphism analysis by treating them at different scales and with different conceptions leading to hierarchical, embedded, mixture and chained systems (Fig. 2c). Under functional aspects, quantitative and qualitative analyses of SMs help for highlighting different variation trends governing biosynthesis (increase), distribution, regulation, elimination (decrease), defensive and adaptive reactions (interaction), chemical and biological activities and signaling (communicative) roles of phytocompounds in different ecological niches of plant species (Fig. 2d). Integrative combination between these three fields (phytochemistry-biology-biostatistics) provide flexible and efficient way for assessment, monitoring, control and understanding environmental systems (Fig. 2e). These steps are fundamental for protection and optimal use of natural resources helping for management strategies of complex environment in favor of sustainable development.

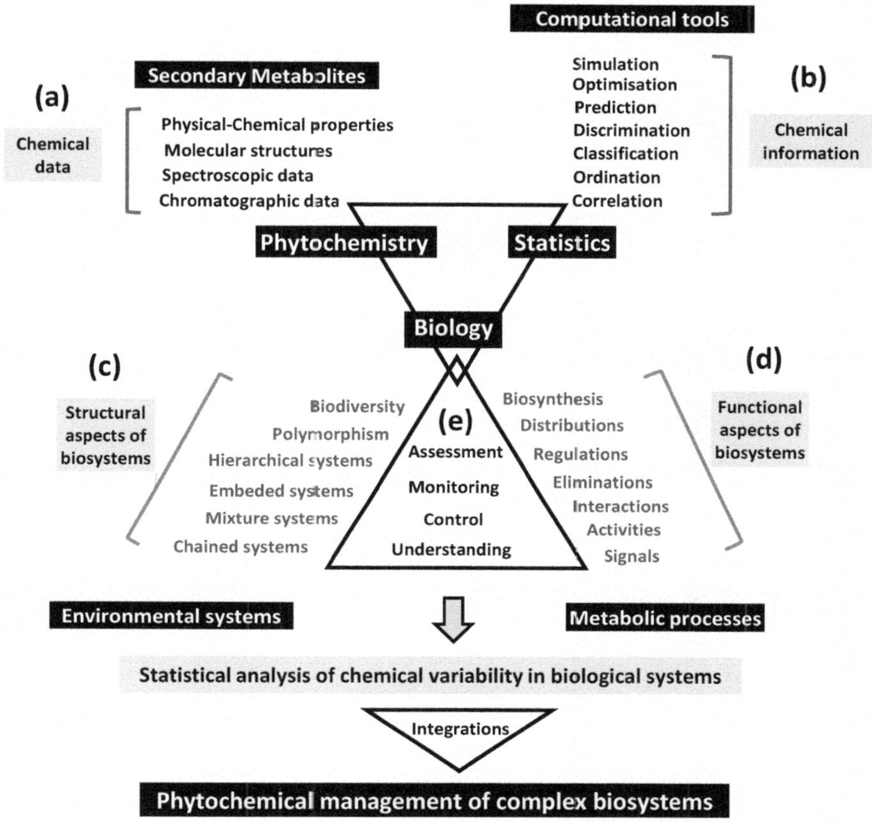

Fig. 2 Overview on the integrative aspect of variability analysis of different phytochemical data by means of different statistical tools aiming for multifactorial and multiscale assessment and monitoring of structural and functional aspects of biosystems and helping for optimal management of environment

This book is organized into four parts presenting respectively:

(1) An overview on the interest of phytochemical diversity under multiscale and multifactorial aspects. Phytochemical diversity represents the backbone of this book and is used as basis to highlight causal–effect trends between complex environment conditions and chemical composition (chemical responses) of plants.
(2) Biosynthesis pathways, structural variability and distribution of secondary metabolites in plant world. This section focuses on functional aspects of metabolic pathways and their flexible variability between plant taxa leading to specific characteristic metabolic patterns (metabotypes or chemotypes) that usefully help for assessment and monitoring of biodiversity by chemical way.

(3) Phytochemical responses of plants to abiotic stresses including high and low temperatures, water deficit, excessive or harmful radiations (e.g. UV) and ozone. These factors are mainly involved in assessment and monitoring of climate changes. The book gives several illustrative study cases on associations (correlations) between upstream abiotic variations and downstream phytochemical responses including biosynthesis of specific (signaling) molecules, variations of regulation ratios between metabolites, variable shapes of response (e.g. kinetic) curves, etc.
(4) Phytochemical responses of plants to biotic stresses due to pathogen infections and herbivore attacks. This book part provides an overview on different communicative and defensive strategies in plants through either anticipative or curative ways. These strategies involve the use of different toxic, repelling or attracting phytocompounds including phytoanticipins, phytoalexins, astringent or deterrent molecules, olfactory volatile mixtures, etc.

Better and differential understanding of the effects of different abiotic and abiotic factors on phytochemical patterns is provided by many complementary experiments based on separated or combined factors concerning plants growing under controlled (e.g. in greenhouse) or in field conditions. These multidisciplinary researches joining phytochemistry to ecology via statistical data analyses provide integrative field helping for flexible management of natural resources and sustainable development.

Contents

Part I
Overview on Governing Factors of Phytochemical Diversity

Chapter 1
Variational Aspects of Secondary Metabolites

1.1 Overview on Multifactorial, Multiphasic and Multiscale Variations of Secondary Metabolites

Phytochemistry refers to the quantitative and qualitative analysis of plant metabolites including more particularly secondary metabolites (SMs). SMs provide specialized tasks to productive plants in terms of defense and communication. They are characterized by their sensitive variations in responses to external (environmental) factors and internal (physiological, genetic) states of producer organisms. This sensitivity involves the synthesis of specific metabolites, regulation ratios of metabolites and particular sets of co-occurring metabolites (metabolic profiles) as adaptive chemical responses or protective strategies [Fig. 1.1(1)]. Environmental (biotic, abiotic) stresses exert continuous, regular or sudden disturbances to plants; consequently, plants react by defense/adaptation ways based on quantitative and qualitative variations of SMs [Fig. 1.1(1)]. Such defensive and adaptive chemical reactions are modulated by varying regulation ratios between different metabolites [Fig. 1.1(2)]. The set of all the metabolites participating in chemical responses to stress form a metabolic profile that provides an integrative picture (phytochemical pattern) on the effects of a given stress on a considered plant (taxon) [Fig. 1.1(3)].

Phytochemical patterns are associated with adaptive strategies that can be better understood by analyzing the relative variations of different metabolites involved in plant-environment interactions [Fig. 1.1(4)]. The shapes and variation ranges of such profiles provide key information on plant physiological states, chemical polymorphisms within and between populations, adaptive trends or interactive relationships at community or ecosystem scale helping for better environment management and agricultural production improvement [Fig. 1.1(5)]. At lower scale, different plant tissues show different phytochemical patterns in relation to plant states, organs' roles and living environment (soil, air; floor; light vs. shade space; etc.). At high scale, phytochemical profiles are correlated with environmental conditions (variations) leading

N. Semmar, *Secondary Metabolites in Plant Stress Adaptation*, Signaling and Communication in Plants, https://doi.org/10.1007/978-3-031-52595-7_1

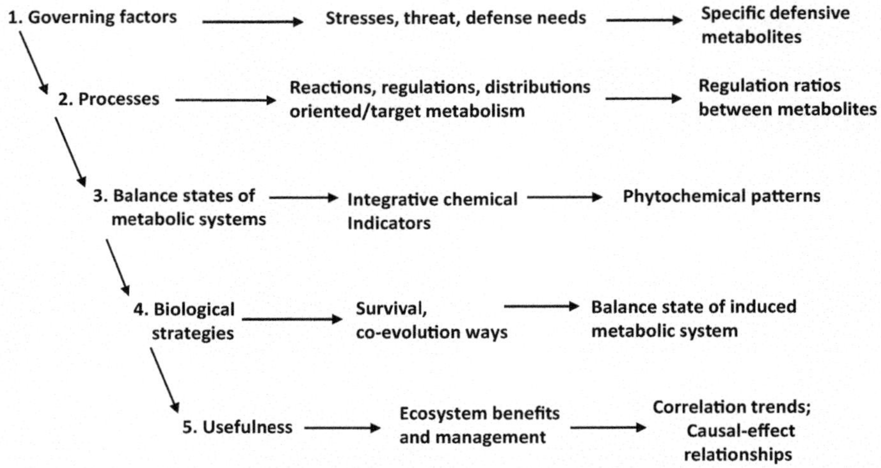

Fig. 1.1 Overview on different functional spaces governing the variability and the use of secondary metabolites in biological world

to highlight causal-effect trends governing structural and functional variability of SMs.

1.2 Overview on Stress Concept

Stress refers to adverse environmental conditions affecting growth, reproduction and yields in biological organisms (including plants). It was also referred as process or symptom associated with a critical or suffering situation (Heiser and Elstner 1998). Inductive stress and induced SMs are often connected through oxygen activation resulting in reactive oxygen species (ROS) that should be controlled by antioxidative compounds including essentially SMs. Stress factors including drought, temperature changes, light variations, biological infections, air pollution, and soil contamination lead to metabolic disturbances due to the generation of ROS including free radicals. To face oxidative stress, plants must counteract these harmful reactions by increasing antioxidative compounds or processes. SMs play key role in these defensive processes. However, if stress impacts continue, they cause the loss of control followed by harmful radical processes leading to cellular decompartmentalizations and lytic/necrotic reactions. Harmful and defense processes vary with stress type, plant species and tissue.

1.3 General Structural and Functional Characteristics of SMs

SMs differ from primary metabolites (PMs) by quantitative and qualitative characteristics leading to complementary roles under both structural and functional aspects [Fig. 1.1(2)]: PMs are constitutive of biological bodies and include saccharides, amino acids and fatty acids. They are structurally limited but quantitatively abundant leading to uniform distribution in biological world [Fig. 1.2(1)]. By their constitutive aspects, PMs are deeply involved in growth systems, but they are less sensitive to environment variations compared with SMs.

However, SMs are not constitutive of biological bodies making they are quantitatively minor [Fig. 1.2(2a)]. They are occasionally synthesized under need conditions for protection or communication signals [Fig. 1.2(2b)]. Occasional or conditional syntheses of SMs are favored by their excitable or inducible character [Fig. 1.2(2c)]. The low amounts of SMs (compared with PMs) make their small variations to be efficiently detected as chemical signals in biosystems.

The needs and ways of plant protection against many stress conditions are species- and environment-dependent resulting in high structural diversity of SMs in the plant world [Fig. 1.2(2d)]. Moreover, minor levels, occasional synthesis and high structural diversity of SMs result in their not uniform distributions with many specific or specialized roles [Fig. 1.2(2e)]. These characteristics attributes to SMs a strong usefulness for chemical characterizations of different plant taxa at different taxonomic levels [Fig. 1.2(2f)]. This refers to the chemotaxonomy field and is important for chemical identification of plant varieties associated with different production trends and potentials (types, levels, rhythms) [Fig. 1.2(2g)]. Chemical profiles of SMs are associated with different tissue and cell compartments in individual plants [Fig. 1.2(2h)]. Compartmentation results in multiple differentiation ways and subsequent specific phytochemical patterns within and between plants. Also, compartmentation represents an essential process to control activities of SMs leading to self-protection in plants.

SMs show sensitive variations with plant age, physiology and development states [Fig. 1.2(2i)]. Beyond the time-dependent variations of SMs associated with natural plant growth, SMs are involved in many interactions between plants and environment including defense, adaptation and communication signals [Fig. 1.2(2i)]. Wide chemical diversity of SMs implies flexible metabolic regulation ways resulting in different response profiles which satisfy different biological roles and signal trends [Fig. 1.2(2e, f)]. Their low amounts attribute sensitive aspects to their variations; this makes small variations to provide reliable information on plant-environment system [Fig. 1.2(2c, k)]. Also, micro-molecular nature of SMs makes their quantitative variations to be reliably converted into signals associating reactional plants and stimulant environment conditions.

By this way, SMs are implied in homeostasis processes including stabilization, regulation and survival states of biosystems [Fig. 1.2(2i, j)].

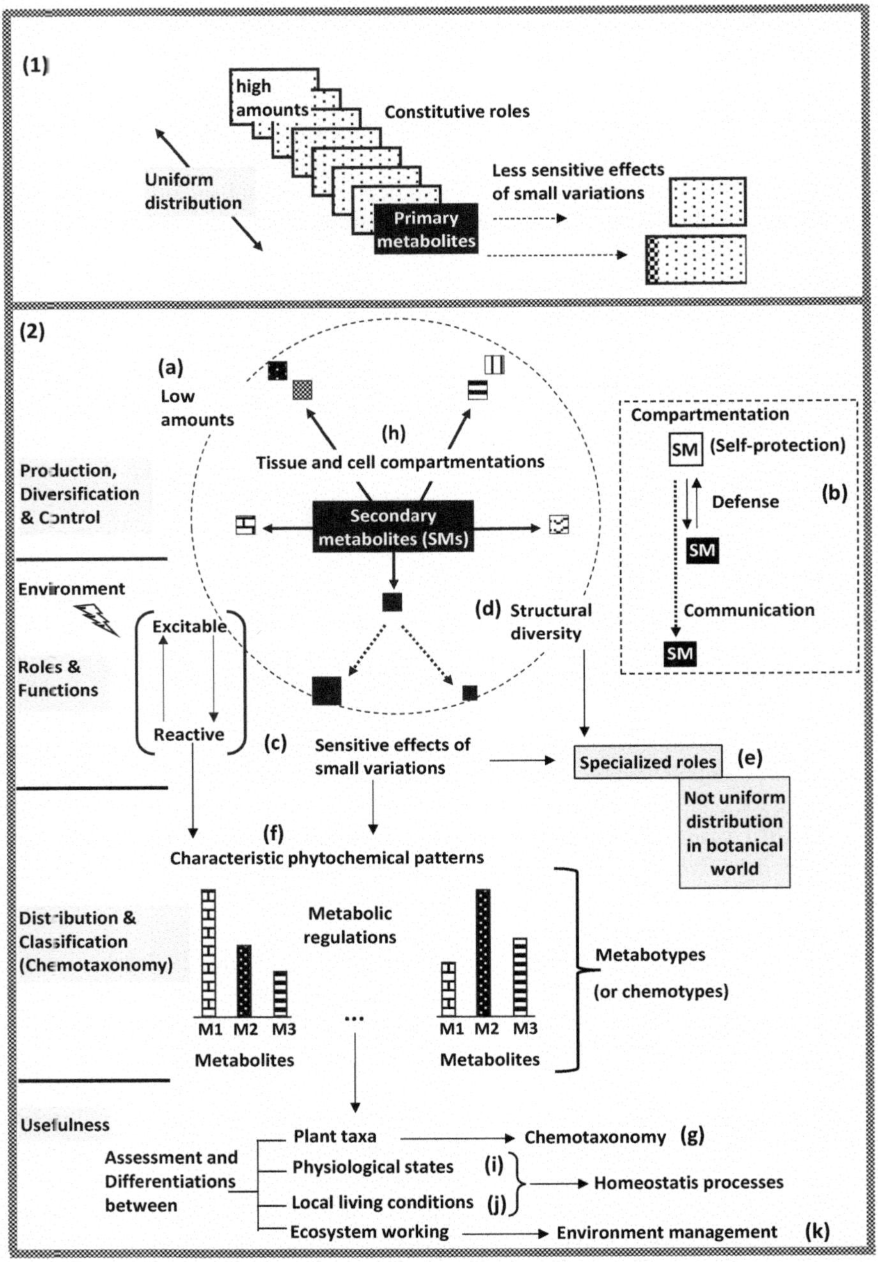

Fig. 1.2 Different quantitative and qualitative characteristics of secondary metabolites (compared with primary metabolites) helping for assessment and monitoring of biotic and abiotic states in ecosystems

Multifactorial and multidirectional variations of SMs attribute them strong usefulness for detection, correlation and classification analyses of plants responses and adaptive profiles in relation to different types and levels of inductive factors (living conditions) [Fig. 1.2 (2g, i-k)].

1.4 Overview on Variational Usefulness of SMs

Beyond development states and living conditions, SMs' patterns are helpfully used for chemical classifications and identifications of plants leading to chemotaxonomy [Fig. 1.2(2f, g)]. This usefulness is linked to the non-uniform botanical distributions of SMs at multi-taxonomical scales (class, family, genus, species, variety) [Fig. 1.2(2e)]. At the lowest taxonomical scale, individuals of a same plant species can be chemically differentiated into metabotypes defined by several metabolites produced at different regulation ratios [Fig. 1.2(2f)].

Sensitivity of SMs to abiotic factors provides an efficient way for climatic changes monitoring. For instance, protective role of polyphenolic compounds (a SM family) against UV and drought stress was highlighted in Mediterranean endemic plants through diurnal and monthly variations of their concentrations with peaks occurring in summer and at midday where UV radiations, temperature and drought reach maximal levels (Gori et al. 2019, 2020). Also, compositions of defensive compounds in plants tend to significantly vary with latitude and altitude (Moreira et al. 2014, 2018). Different Mediterranean populations of *Cistus ladanifer* living in different locations (different climatic conditions) showed different flavonoid profiles (compositions and contents) (Sosa et al. 2005).

Elevated temperatures result in higher emissions of volatile terpenes as consequence of more active synthesis in leaves and needles. Monitoring studies revealed increased emission rates of isoprene and monoterpenes (Penuelas and Staudt 2010; Holopainen et al. 2013). Subsequently to elevated night temperatures, isoprene, monoterpenes, sesquiterpenes and a homoterpene showed higher daytime emissions rates on a leaf-area basis in deciduous trees. This process associated increase in plant growth rate (influencing that of total leaf area) with increased terpenoid emissions under warmer living environment.

Environmental stresses tend to increase chemical resistances of plants. Stress-tolerant plants tend to produce more variable SMs than nontolerant ones (Sharkey et al. 2008; Loreto and Velikova 2001). In consequence, resistant or adapted plants provide a potential source of active metabolites favoring drug discovery (Macías et al. 2007).

Finally, biotic stresses in plant include pathogenic infections and herbivore attacks. These harmful situations are managed by plants via the production of SMs with both preventive and curative aspects including repelling, astringent, toxic and signalling volatile compounds.

References

Gori A, Tattini M, Centritto M, Ferrini F, Marino G, Mori J, Guidi L, Brunetti C (2019) Seasonal and daily variations in primary and secondary metabolism of three maquis shrubs unveil different adaptive responses to mediterranean climate. Conserv Physiol 7:coz070

Gori A, Nascimento LB, Ferrini F, Centritto M, Brunetti C (2020) Seasonal and diurnal variation in leaf phenolics of three medicinal mediterranean wild species: what is the best harvesting moment to obtain the richest and the most antioxidant extracts? Molecules 25:956

Heiser I, Elstner EF (1998) The biochemistry of plant stress and disease—oxygen activation as a basic principle. Ann N Y Acad Sci 851:224–232

Holopainen JK, Himanen SJ, Yuan JS, Chen F, Stewart CN (2013) Ecological functions of terpenoids in changing climates. In: Ramawat KG, Mérillon JM (eds) Natural products. Springer, Berlin

Loreto F, Velikova V (2001) Isoprene produced by leaves protects the photosynthetic apparatus against ozone damage, quenches ozone products, and reduces lipid peroxidation of cellular membranes. Plant Physiol 127:1781–1787

Macías FA, Galindo JLG, Galindo JCG (2007) Evolution and current status of ecological phytochemistry. Phytochemistry 68(22–24):2917–2936

Moreira X, Mooney KA, Rasmann S, Petry WK, Carrillo-Gavilán A, Zas R, Sampedro L (2014) Trade-offs between constitutive and induced defences drive geographical and climatic clines in pine chemical defences. Ecol Lett 17:537–546

Moreira X, Castagneyrol B, Abdala-Roberts L, Berny-Mier y Teran JC, Timmermans BGH, Bruun HH, Covelo F, Glauser G, Rasmann S, Tack AJM (2018) Latitudinal variation in plant chemical defences drives latitudinal patterns of leaf herbivory. Ecography 41:1124–1134

Penuelas J, Staudt M (2010) BVOCs and global change. Trends Plant Sci 15:133–144

Sharkey TD, Wiberley AE, Donohue AR (2008) Isoprene emission from plants: why and how. Ann Bot 101:5–18

Sosa T, Alías JC, Escudero JC, Chaves N (2005) Interpopulational variation in the flavonoid composition of *Cistus ladanifer* L. exudate. Biochem Syst Ecol 33:353–364

Chapter 2
Variational Factors and Regulation Processes of Secondary Metabolites

2.1 Multiple Induction Origins of Secondary Metabolites

Secondary metabolites are induced and vary in response to changes in environmental conditions or internal states of plants (Fig. 2.1(1, 2)). Plant sensitivity to quantitative and qualitative variations of external/internal conditions can be monitored through different phytochemical patterns characterizing different environmental conditions and biological states. By this way, downstream-formed phytochemical patterns are correlated with upstream-inducing factors associated with extrinsic or intrinsic origins. These causal-effect trends attribute high interest to secondary metabolites for understanding complex adaptive behaviors of plants on one hand, and for subsequent management of ecological and biological systems on other hand.

Inducing external conditions are associated with either abiotic or biotic stresses which have different and multiple effects on plants (Fig. 2.1(1a, b)):

- Abiotic stresses include air and soil physico-chemical factors exerting constraints through relatively high or low levels. Air-linked factors include temperature (heat, cold), water availability (humidity, rainfall, water deficit), light (sunlight, shade), UV-radiations (UV-A, UV-B) and ozone (O_3). These factors are closely involved in climatic changes representing a main current worry at planetary scale. Their effects on plants and subsequent phytochemical responses are particularly detailed and illustrated in this book. Soil-linked factors include drought, soil salinity and heavy metal attacks (Fig. 2.1(1a)).
- Biotic stresses imply cooperative or defensive interactions between plants and biological partners or enemies, respectively (Fig. 2.1(1b)). Cooperative interactions include symbiose and pollination requiring attraction processes using specific SMs or particular mixtures of SMs associated with recognizable color or odor by the expected biological partner. Some biological partners are attracted by specific odors to assist plants in their defense against herbivore attacks and parasit infections. These two biotic stresses are at the origin of production of SMs

N. Semmar, *Secondary Metabolites in Plant Stress Adaptation*, Signaling and Communication in Plants, https://doi.org/10.1007/978-3-031-52595-7_2

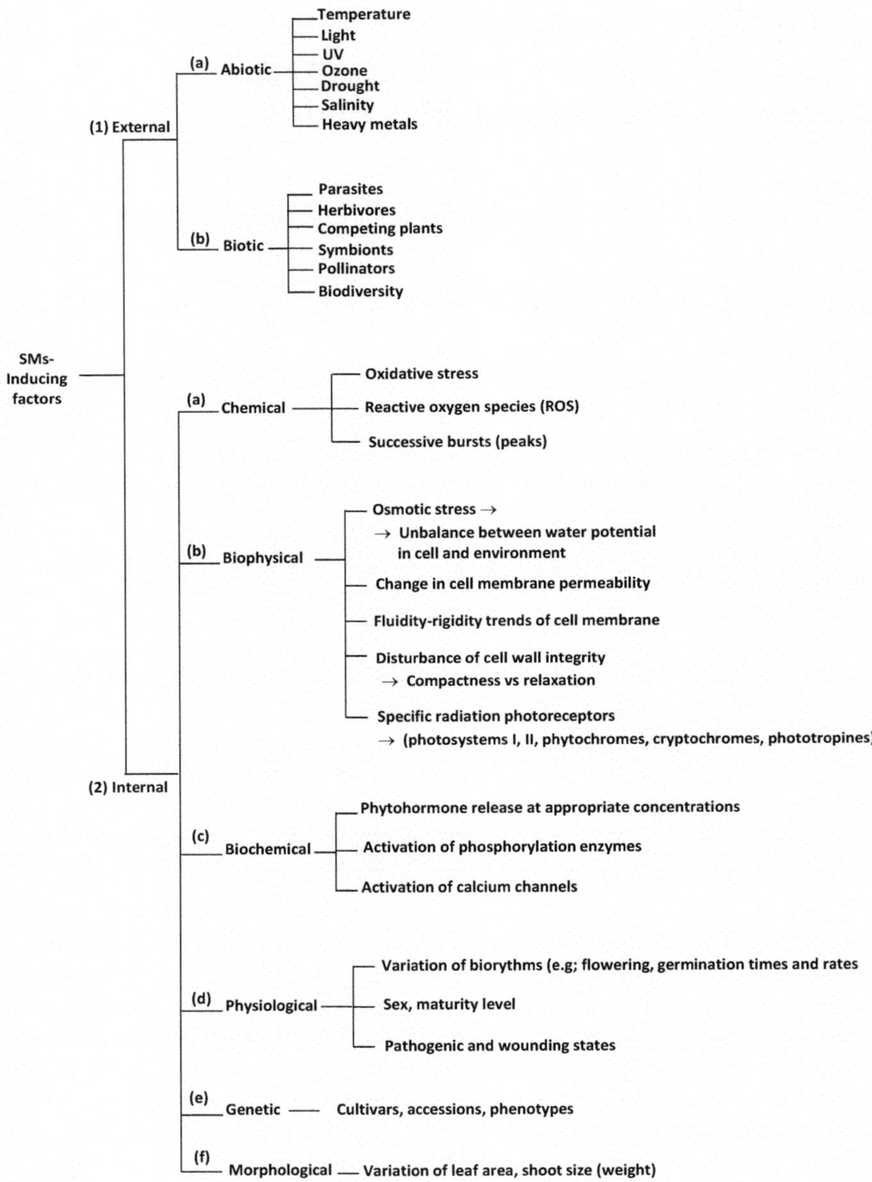

Fig. 2.1 Different types of external and internal factors involved in the production, diversification and distribution of secondary metabolites in plants

with toxic and repellent effects in plants. Futher structural diversity and patterns of produced SMs tend to be observed under more biological interactions. By this way, biodiversity in ecosystem represents a strong factor governing quantitative and qualitative productions of SMs in plants.

Plant internal factors involved in SM induction are of different types including chemical, biophysical biochemical, physiological, genetic and morphological factors/systems (Fig. 2.1(2)). SM induction occurs in downstream of stress management systems consisting of sensitive constitutional components involved in detection, transmission, transduction and conversion of stress signals from physical–chemical aspects to metabolic forms. Thus, SMs are synthesized as specialized metabolites through signals coupling harmful or alarming variations (in upstream) with protective or adaptive phytochemical patterns (in downstream). Intrinsic inducing factors of SMs include:

- Chemical factors consisting of oxidative stress due to the formation of reactive oxygen species (ROS) (Fig. 2.1(2a)). ROS and other oxidants (including radical and non-radial compounds) induce the formation of antioxidant SMs. The Inducing oxidative signal can be amplified and/or extended through double burst leading to higher efficiency of SMs' induction signal.
- Biophysical factors originated from osmotic balance disturbance and radiative excitation (Fig. 2.1(2b)). Osmotic stress is detected from unbalance between water potential of cell and that of environment. It results in different variation trends between fluidity and rigidity states of cell membrane. Osmotic stress can be amplified by some changes in cell membrane composition. Moreover, water deficit stress can result in disturbance of cell wall integrity leading to reinforce the inducing biophysical ways of SMs. Under radiative stress, SMs are induced through photoreceptors specialized in detection of different radiation ranges including photosystems I and II, phytochromes, cryptochromes and phototropines.
- SMs-inductions are also governed by biochemical processes involving phytohormones which are associated with cascades of transduction and gene expression signals in upstream or downstream steps (Fig. 2.1(2c)) (Murcia et al. 2017). Apart from their naturally endogenous aspect, phytohormones can be applied to plants as external factors (e.g. by spraying) to stimulate several protection or tolerance responses. SMs-inducing biochemical ways include also activation of trans-membrane calcium channels and phosphorylation enzymes (Fig. 2.1 (2c)).
- SMs-inducing physiological factors include changes in seasonal processes such as delayed or accelerated flowering and germination (Fig. 2.1(2d)). Inducing factors include also sex, maturity level, pathogenic and wounding states (e.g. following herbivore attacks).
- Genetic factors responsible for induction of SMs involve cultivars, accessions and/or phenotypes able to produce appropriate SMs for occurring stress adaptation (Fig. 2.1(2e)).

- Morphological factors are associated with plant growth and include leaf area and shoot size that are affected by leaf turgor pressure and mitosis impairment (Fig. 2.1(2f)).

2.2 Chemical and Cellular Factors Influencing Diversification, Distribution and Specialization Roles of Secondary Metabolites

Protective and communicative specialized roles of SMs in plants are fundamentally governed by their highly variable chemical structures involving several other aspects (Punetha et al. 2022; Yeshi et al. 2022; Hadacek 2002; Singsaas 2000):

- Hydrocarbon skeleton (called aglycone or genin) is responsible for the global shape (scaffold), elementary size and solubility properties of molecule (Fig. 2.2(1a)). Aglycones represent the initial steps in SMs' biosynthesis and basic identifiers of corresponding metabolic pathways. Metabolic pathways showing different biosynthesis rates and regulation ratios of SMs are correlated with different expression levels of key enzymes (Fig. 2.2(1b)). Biosynthesis enzymes work by promiscuity way, i.e. they are able to use different substrates to form a same product, or give different products from a same substrate (Fig. 2.3). Promiscuity system favors the development of metabolic diversity in plants leading to specialized metabolism (wide SMs' pool) in downstream of stimulating biotic and abiotic stress factors. Biosynthesis of aglycone is followed by intramolecular transformations leading to structural diversification of SMs within a same metabolic pathway (biochemical family) (Fig. 2.4a, b).
- Moreover, biosynthesis enzymes of SMs are subject to diversification through mutation processes occurring on encoding genes (Fig. 2.2(1c)). Such upstream processes favored the development of promiscuity enzymes (Leong and Last 2017; Kreis and Munkert 2019; Waki et al. 2021). Moreover, environmental constraints continuously exert selective pressures leading to further mutations which result in more phytochemical diversity where new SMs are subscribed (retained) because of their efficient roles against occurring stresses and changing conditions.
- In downstream of metabolism, SMs' diversification is favored by transformations of aglycones through chemical substitutions occurring at different carbons (positions) of molecules (Figs. 2.2 (2) and 2.4a). Aglycone can attach various chemical groups leading to the formation of new molecules with gradually varied sizes, physical–chemical properties and biological activities. This enlarges the phytochemical pools and results in more pathways and sub-pathways (Fig. 2.4b). Moreover, chemical substitutions result in steric variation of SMs leading to multiple interaction ways with other molecules.
- Substituted chemical groups can have hydrophilic or lipophilic characters resulting in different solubility properties and compartmentation ways of SMs (Fig. 2.2(3)). Compartmentation process concerns different biological structures

including cells, organs and tissues (Figs. 2.4c, 2.5 and 2.6) (Lipko et al. 2023). At cell scale, different organelles provide storage compartments of SMs according to their hydrophilic or lipophilic aspects. Hydrophilic compounds are stored in vacuole, laticifers and apoplast space of cell wall; lipophilic compounds are stored in cell membranes, cuticle, trichome, resin ducts, laticifers and oil cells (Fig. 2.6) (Cai et al. 2012). Hydrophily is favored by hydroxylation and glycosylation whereas lipophilic character is associated with hydrocarbon aspects. Further details on structures-solubility trends are presented in Sect. 3.2 of the current chapter. Cell compartmentation represents an efficient process protecting plants against self-toxication by their reactive metabolites. At higher scale, particular occurrence or accumulation of SMs in some tissues is linked to their local roles and specific needs for plant. Distribution of SMs extend beyond the plants through their emissions in environment for communicative strategies. This makes cell and tissue compartmentations to play key roles in the control of biological signals and reactivity linked to SMs (Figs. 2.2(3) and 2.5). For instance, terpenes are stored in resin ducts in conifers vs oil glands and leaf hairs in broadleaves (Singsaas 2000). Targeted communicative roles of SMs in plant-animal interactions requires their anticipated (preliminary) storages. Volatile compounds are emitted by plants from storage matrices under the effect of tissue temperature (Guenther et al. 1991). Also, SMs as phenylpropanoids tend to be relatively more distributed in leaves exposed to oxidative stress (e.g. ozone-linked) vs decreased storage in wood tissue (Richet et al. 2012).

- Metabolic diversification processes act in favor of stronger abilities of plants to face multiple types of disturbances. Initially, needed SMs are stimulated through cascade processes involving stress detection followed by specific gene induction, transduction and metabolism (Fig. 2.2(4)). In downstream, formed and stored SMs can be activated by enzymatic hydrolysis leading to release of active (toxic) components.
- Structural variability of SMs is responsible for a diversification of their activities. Different activities are associated with different target roles and fates of SMs (Fig. 2.2(5)). According to their structures, solubility properties and molecular weights, SMs will be dispatched to protect plants against specific harmful conditions or to communicate with well target biotic environment. Protective roles of SMs against abiotic stresses involve (Fig. 2.2(5)) (Kivimäenpää et al. 2022):

  (i) antioxidant activities by interactions with oxygen reactive species or heavy metals (oxidants),
  (ii) membrane protection via its fluidification favored by lipophilic SMs,
  (iii) strengthening of wall structures,
  (iv) water flux control by osmoregulation based on hydrophilic metabolites (osmolytes),
  (v) thermotolerance regulations by emission of volatile SMs that can form cooling (protecting) aerosol micro-cloud above plant tissues.

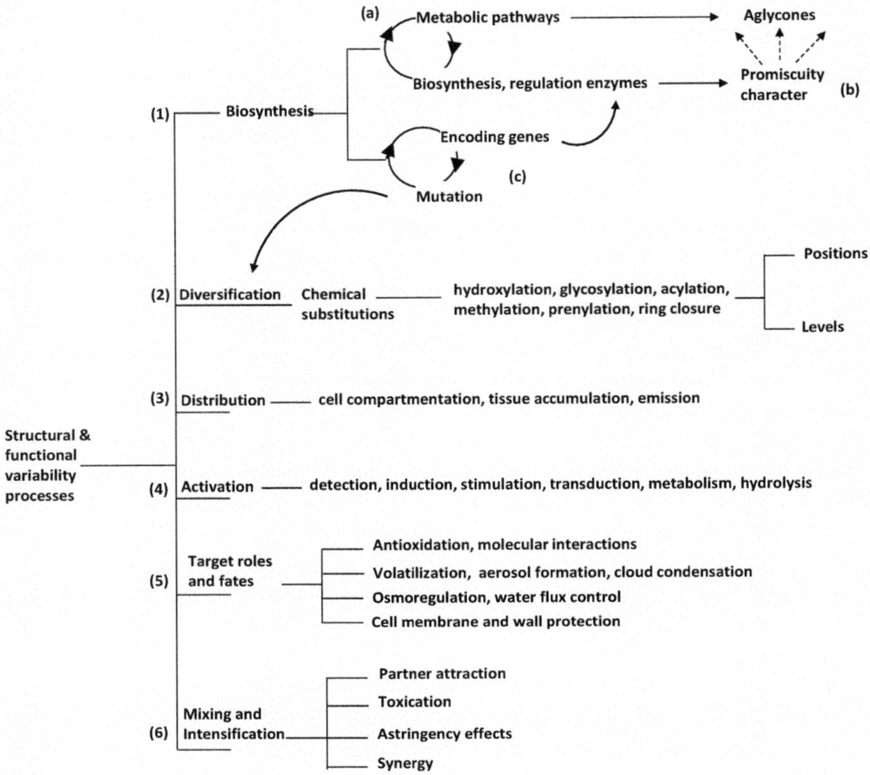

Fig. 2.2 Chemical and cellular factors influencing structural diversification, distribution and target roles of secondary metabolites against different biotic and abiotic stresses

- In biotic systems, plant SMs can be combined to provide well-defined mixtures playing key roles in (i) attraction of specific biological partners (symbiotic microorganisms, pollinators, parasitoids of herbivores) or (ii) in intensified (synergetic) repelling/toxic effects against herbivores and pathogens (Fig. 2.2(6)).

2.3 Roles of Phytochemical Variations/Patterns in Plant Adaptive Strategies and Environment Management

The interest of SMs in environment management strongly derives from their correlations or significant variation trends with different external (abiotic, biotic) conditions and internal plant states (Figs. 2.7 and 2.8). This results in more or less specific phytochemical patterns associated with different adaptive plant responses to different external conditions or internal states:

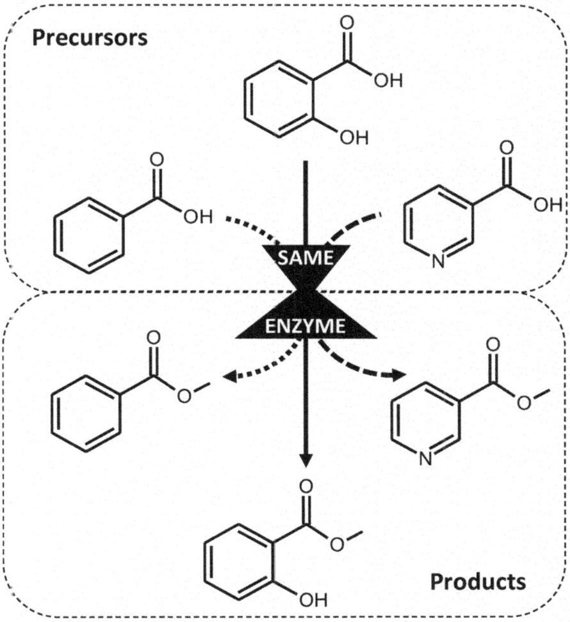

Fig. 2.3 Illustration of enzyme showing substrate promiscuity

Abiotic factors include temperature, radiations, ozone, drought, salinity and heavy metals (Fig. 2.7(1)). They are detailed in Part III. Biotic factors can be of offensive or cooperative types resulting in phytochemical patterns associated with defensive, alerting (wounding) or communicative messages (Figs. 2.7(2) and 2.8). In well-adapted plants, defensive SM-patterns are of anticipative type and can be highlighted by pre-existing SMs initially synthesized and stored as preventive strategy (Fig. 2.8(1)). Also, under unexpected or sudden stress, curative responses are manifested through repair and wound cue metabolites (Fig. 2.8(1, 2)). Defensive SMs can play toxication, repelling or deterring roles against biological enemies (e.g. herbivores, pathogens) (Fig. 2.8(1)).

Biological relationships between plants and other organisms involve specific SMs playing (i) attractive roles in symbioses, pollinations, calling of herbivores' enemies, or (ii) repulsive effects against competing neighbor plants (allelopathy) or animal or microbial harmful (Fig. 2.8(3)). Specific SMs and more generally phytochemical patterns vary with plant physiology and taxonomy (genetics) (Fig. 2.7(3, 4)). Physiologic states of plants vary with their reproduction phases and fitness, growth and maturation levels, circadian rhythms, health states and disease types (Fig. 2.7(3)).

Phytochemical patterns associated with different abiotic and biotic conditions provide key information on multiple interactions between plants and living conditions helping for environment protection and management. Also, environment management calls for optimal use of bioresources by identifying phytochemical patterns associated with high plant fitness. Such patterns could be potential candidates for agriculture development.

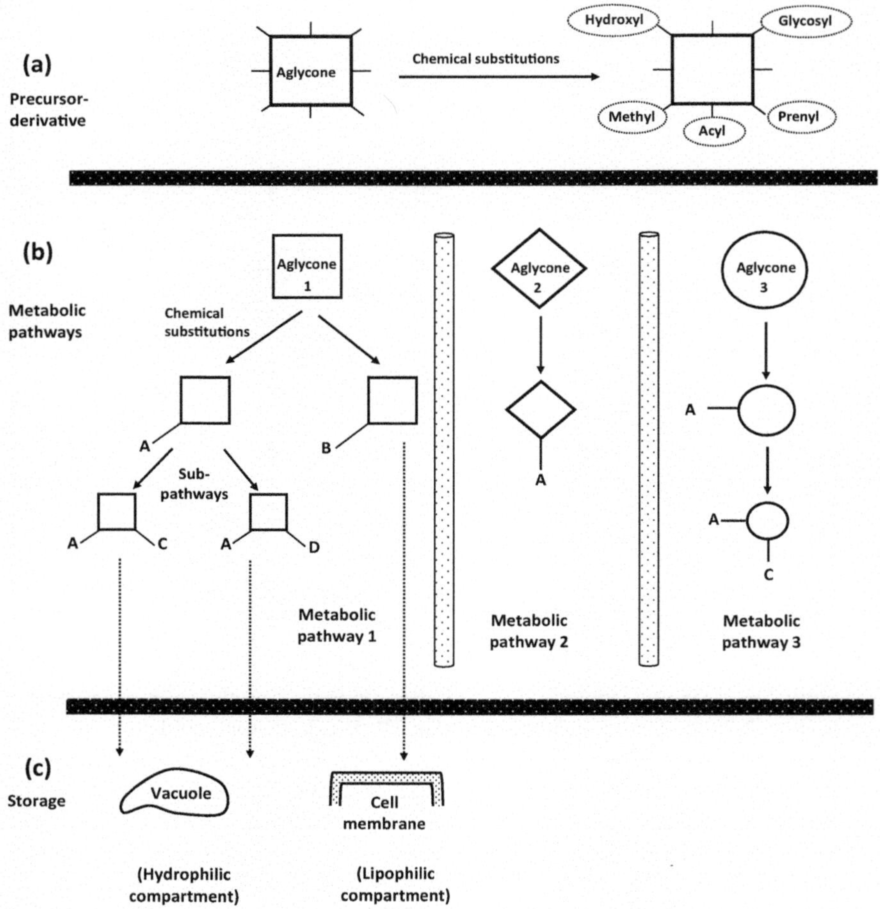

Fig. 2.4 Overview on biosynthesis (**a**), metabolic regulation and diversification (**b**) and distribution/compartmentation (**c**) of secondary metabolites in plants

Phytochemical patterns provide also reliable fingerprints for classification/identification of plants at different taxonomic levels: from class, family to species and variety levels (Fig. 2.7(4)). Such patterns are generally made by metabolites which are mainly or specifically produced by the considered plant taxon. Beyond their occurrences, metabolites can be functionally characterized by their regulation ratios leading to fine analyses at lower taxonomic levels (e.g. phytochemical varieties of a same species). This usefully helps for identification of different metabolic or chemical phenotypes that can be associated with potentially adapted plant varieties in natural or disturbed conditions as well as original accessions or cultivars for agriculture.

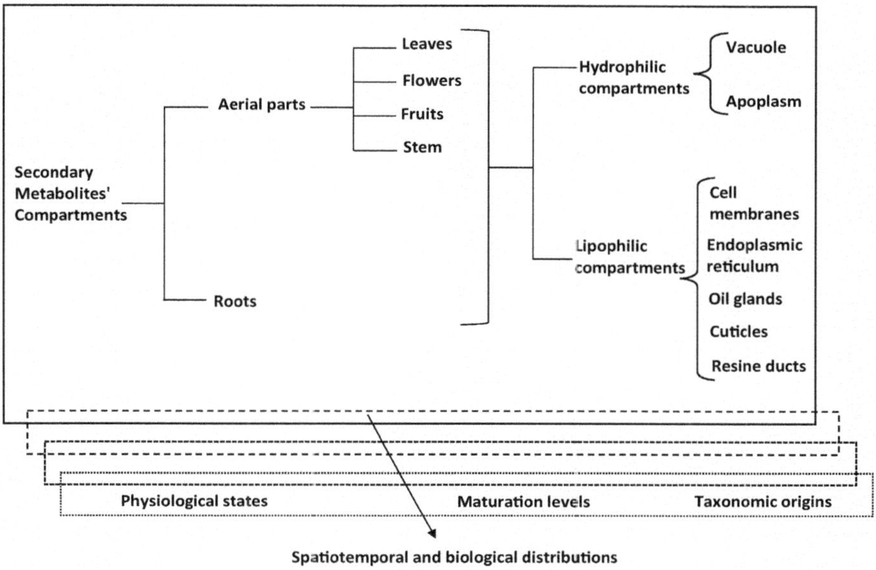

Fig. 2.5 Cell and tissue compartmentation ways of secondary metabolites influenced by chemical properties, spatiotemporal variations and different biological factors (e.g. taxonomy, physiology)

Secondary metabolites are at the origin of adaptive biological strategies favoring survival of biological species and resilience of ecosystems. Plant species or varietal species showing wider production pools of secondary metabolites tend to resist more to threats than those with more restricted phytochemical equipment. Biosystems with high phytochemical diversity (e.g. wild natural spaces) showed stronger ecological stability than agronomic spaces (e.g. monocultures or some cultivated species).

Improvement of stability or resilience of cultivated systems (agrosystems) find favorable solution through the replication of natural ecosystem processes and patterns (Altieri et al. 1995; Gliessman 1998). In agroecosystem management, phytochemistry of crop individuals or cultivars is investigated to identify effective patterns against pest and pathogens.

SMs show different quantitative and qualitative variation patterns associated with different extrinsic and intrinsic variations and manifesting at different molecular scales (Fig. 2.9):

- At intramolecular scale, SMs' occurrences depend on plant need and induction factors (Fig. 2.9a). Occurring SMs show variable amounts that can be correlated with favorable biosynthesis conditions (Fig. 2.9a, b).
- At intermolecular scale, metabolites are relatively considered through their regulation ratios. Different regulation ratios between same metabolites can be associated with different genetic forms (plant varieties) and can be governed by appropriate environmental conditions inducing the expressions of enzymes responsible for

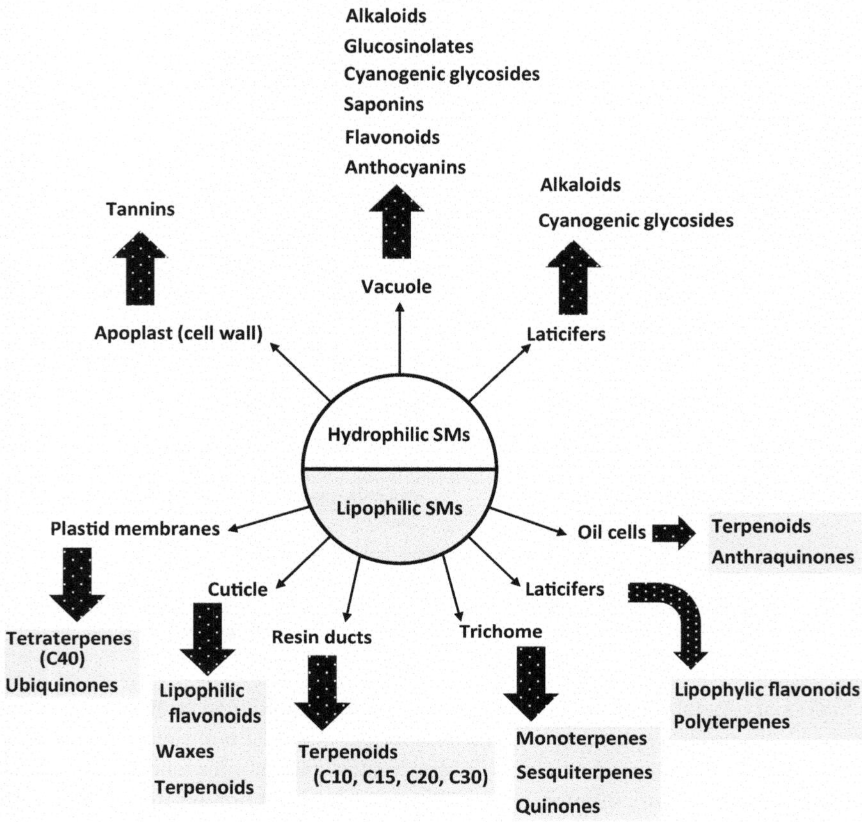

Fig. 2.6 Different cell compartments associated with different solubility characteristics of secondary metabolites

biosynthesis of needed metabolites at the expense of other ones (Fig. 2.9c). In illustrative example of Fig. 2.9c, metabolite M2 shows higher regulation in case B than A, although its amount is lower in B than A.

- At meta-molecular scale, global environmental conditions and multiple genetic expressions lead to several shapes of phytochemical profiles containing several metabolites with different regulation ratios called metabotypes or chemotypes (Fig. 2.9d). Metabotypes or chemotypes can be correlated with both structural and functional factors. Structural factors in plant world consist of different taxonomical levels going from plant classes to varieties or cultivars. Functional factors are linked to the effects of dominant or periodic environmental conditions. Understanding the meaning of metabotypes combining several SMs with different regulation ratios provide useful keys for environment management.

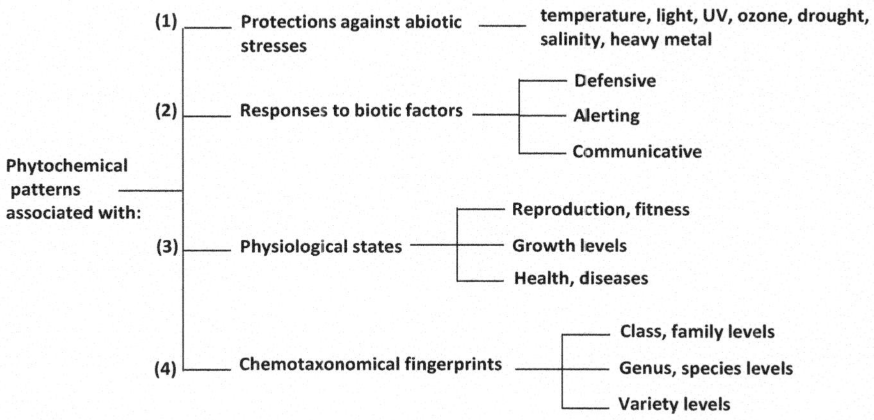

Fig. 2.7 Different variational sources of secondary metabolites resulting in different phytochemical patterns

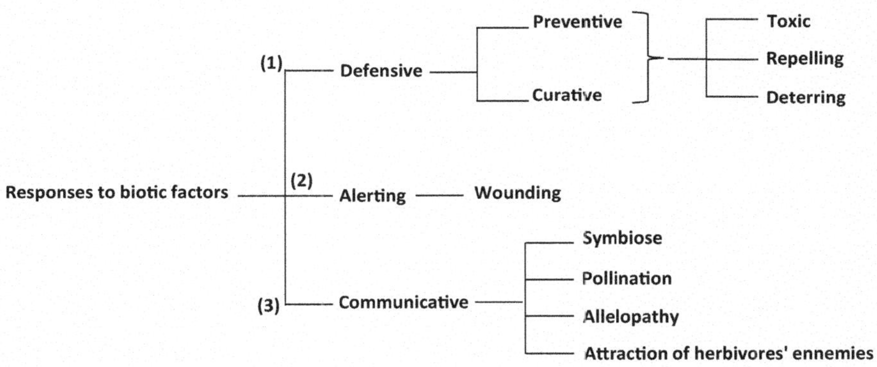

Fig. 2.8 Different types of plant responses to biotic factors implying secondary metabolites

At environmental scale, natural or undisturbed habitats are characterized by high plant diversity resulting in high phytochemical pool which provides a strong functional basis for stability and protection of ecosystem. Illustrative case is given by plant-microbe interaction: threatening effects of pathogen microbes on plants tend to be limited by high diversity of competing microorganisms (Rodriguez and Redman 1997). Thus, competing micro-organism community favored by high plant (phytochemical) diversity tends to exert strong constraints against pathotype evolution.

Phytochemical diversity is also applied as strength for better agroecosystem management at individual, population or community scales. This aim is further reached and undertaken by applying more biodiversity (more phytochemical diversity) in situ and with time (Letourneau et al. 2011; Kremen et al. 2012; Oliver et al. 2015a, b).

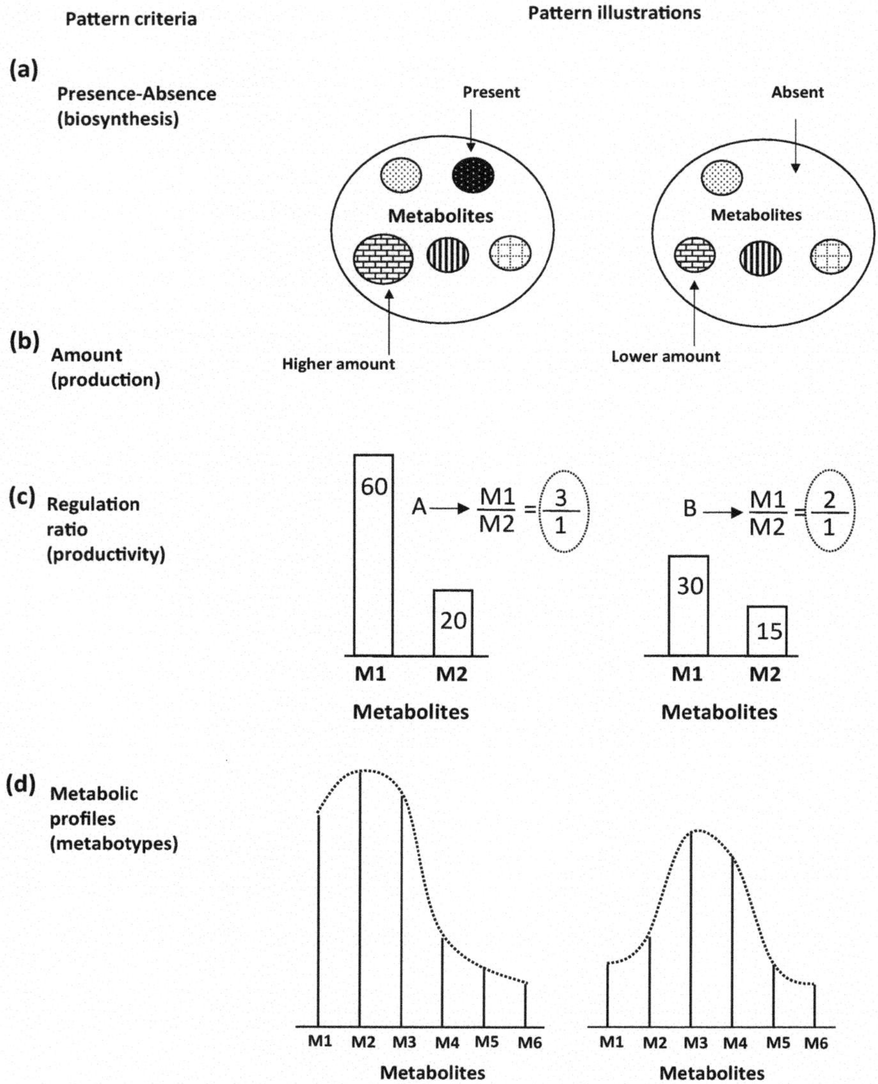

Fig. 2.9 Different types of phytochemical patterns based on metabolic occurrences (biosynthesis) (**a**), production levels (amounts) (**b**), regulation ratios (relative levels) (**c**) and their combinations into variable shape patterns (metabotypes or chemotypes) (**d**)

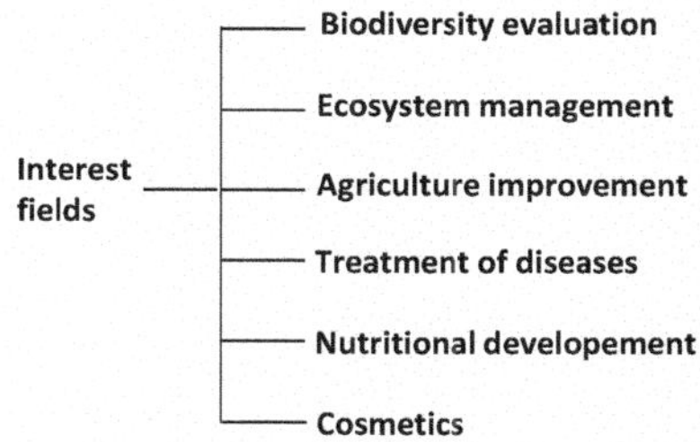

Fig. 2.10 Different research and application fields of SMs in human life

2.4 Overview on multiple Interest Fields of Secondary Metabolites

SMs' analyses provide useful ways helping for deep understanding, management and optimal use of natural systems. They help for (i) chemical characterization of biodiversity, (ii) agriculture production improvements through the reproduction in agrosystems of phytochemical aspects of wild biosystems, (iii) health protection through drug development and (iv) nutritional valorization through preparations of labeled food matrices (Fig. 2.10).

In pharmacology, SMs are subject to structure-activity analyses and molecular modeling (docking) helping for drug designs. In nutrition, SMs are responsible for organoleptic properties and are used as food additives. In cosmetic, essential oils are sources of complex mixtures for preparation of fragrances. In agriculture, structural variability of SMs provide a biotechnological platform for bioinsecticide development (Hadacek 2002). In environmental science, analysis of structure-function trends of SMs helps for better understanding their protective roles against stresses in favor of biodiversity balance and protection. Moreover, exposition of plants to different kinds and levels of stresses provides discovering way of optimal productions of SMs associated with different constraints.

References

Altieri MA, Farrell JG, Hecht SB, Liebman M, Magdoff F, Murphy B, Norgaard RB, Sikor TO (1995) Toward sustainable agriculture. In: Altieri MA (ed) Agroecology, 2nd edn. CRC Press, Boca Raton. ISBN: 978-0429495465

Cai Z, Anja K, Dietrich K, Iryna S (2012) Exudation: an expanding technique for continuous production and release of secondary metabolites from plant cell suspension and hairy root cultures. Plant Cell Rep 31:461–477

Gliessman SR, Engles E, Krieger R (1998) Agroecology: ecological processes in sustainable agriculture. CRC Press, Boca Raton

Guenther AB, Monson RK, Fall R (1991) Isoprene and monoterpene emission rate variability : observations with Eucalyptus and emission rate algorithm development. J Geophys Res 96:10799–10808

Hadacek F (2002) Secondary metabolites as plant traits: current assessment and future perspectives. Crit Rev Plant Sci 21(4):273–322

Kivimäenpää M, Riikonen J, Valolahti H, Elina H, Holopainen JK, Holopainen T (2022) Effects of elevated ozone and warming on terpenoid emissions XE "terpenoid emissions" and concentrations of Norway spruce depend on needle phenology and age. Tree Physiol 42:1570–1586

Kreis W, Munkert J (2019) Exploiting enzyme promiscuity to shape plant specialized metabolism. J Exp Bot 70(5):1435–1445

Kremen C, Miles A (2012) Ecosystem services in biologically diversified versus conventional farming systems: benefits, externalities, and trade-offs. Ecol Soc 17:40

Leong BJ, Last RL (2017) Promiscuity, impersonation and accommodation: evolution of plant specialized metabolism. Curr Opin Struct Biol 47:105–112

Letourneau DK, Armbrecht I, Salguero Rivera B, Montoya Lerma J, Jiménez Carmona E, Daza MC, Escobar S, Galindo V, Gutiérrez C, López, SD, López Mejía J, Acosta Rangel AM, Herrera Rangel J, Rivera L, Saavedra CA, Torres AM, Reyes Trujillo A (2011) Does plant diversity benefit agroecosystems? A synthetic review. Ecol Appl 21:9–21

Lipko A, Paczkowski C, Perez-Fons L, Fraser PD, Kania M, Hoffman-Sommer M, Danikiewicz W, Rohmer M, Poznanski J, Swiezewska E (2023) Divergent contribution of the MVA and MEP pathways to the formation of polyprenols and dolichols in *Arabidopsis.* Biochem J 480:495–520

Murcia G, Fontana A, Pontin M, Baraldi R, Bertazza G, Piccoli PN (2017) ABA and GA3 regulate the synthesis of primary and secondary metabolites related to alleviation from biotic and abiotic stresses in grapevine. Phytochemistry 135:34–52

Oliver TH, Heard MS, Isaac NJB, Roy DB, Procter D, Eigenbrod F, Freckleton R, Hector A, Orme CDL, Petchey OL, Proença V, Raffaelli D, Suttle KB, Mace GM, Martín-López B, Woodcock BA, Bullock JM (2015a) Biodiversity and resilience of ecosystem functions. Trends Ecol Evol 30:673–684

Oliver TH, Isaac NJB, August TA, Woodcock BA, Roy DB, Bullock JM (2015b) Declining resilience of ecosystem functions under biodiversity loss. Nat Commun 6:10122

Punetha A, Kumar D, Suryavanshi P, Padalia RC, KatanaPalya Thimmaiah V (2022) Environmental abiotic stress and secondary metabolites production in medicinal plants: a review. J Agr Sci-Tarim Bili 28(3):351-262

Richet N, Tozo K, Afif D, Banvoy J, Legay S, Dizengremel P, Cabane M (2012) The response to daylight or continuous ozone of phenylpropanoid and lignin biosynthesis pathways in poplar differs between leaves and wood. Planta 236:727–737

Rodriguez RJ, Redman RS (1997) Fungal life-styles and ecosystem dynamics: biological aspects of plant pathogens, plant endophytes and saprophytes. Adv Bot Res 24:169–195

Singsaas EL (2000) Terpenes and the thermotolerance of photosynthesis. New Phytol 146:1–4

Waki T, Takahashi S, Nakayama T (2021) Managing enzyme promiscuity in plant specialized metabolism: a lesson from flavonoid biosynthesis Mission of a "body double" protein clarified. BioEssays 43:2000164

Yeshi K, Crayn D, Ritmejeryte E, Wangchuk P (2022) Plant secondarymetabolites produced in response to abiotic stresses has potential application in pharmaceutical product development. Molecules 27:313

Chapter 3
Overview on Structural and Functional Diversity of Secondary Metabolites

3.1 General Classifications of Secondary Metabolites

Secondary metabolites (SMs) are structurally classified into three main classes including **(a)** phenolic compounds, **(b)** terpenes and **(c)** nitrogen-containing compounds (Fig. 3.1).

Phenolic compounds are characterized by the presence of phenolic cycle(s) consisting of benzene containing hydroxyl (OH) group (Fig. 3.1a). According to their number and type of cyclization, phenolic compounds are categorized into different structural families including phenylpropanoid and derivatives, flavonoids, coumarins and tannins:

- Phenylpropanoids contain one aromatic cycle (C6) with acid lateral chain with three carbons leading to carbon-skeleton formula of C6C3 (Fig. 3.2a). They are represented by coumaric acid and derivatives issued from cinnamic acid.
- Phenylpropanoids (C6C3) can provide C6C1 phenolic compounds by β-oxidation (Fig. 3.2b). Thus, β-oxidation of cinnamic acid gives benzoic acid, and that of *p*-OH-coumaric acid gives *p*-OH-benzoic acid; etc.
- Coumarins derive from cyclization of C3-lateral chain of phenylpropanoids leading to bicyclic C6C3-compounds (Fig. 3.2c). Cyclization occurs after hydroxylation of phenylpropanoid at ortho-position of lateral chain (Shimizu 2014).
- Flavonoids are made by two phenolic cycles (C6, C6) linked by a central heterocycle of three carbons (C3) (Fig. 3.2d). This results in carbon-skeleton formula of C6C3C6. They are issued from condensation of phenylpropanoid with three malonyl-CoA leading to chalcone-type flavonoid (precursor of flavonoids) (Rammohan et al. 2020).
- Stilbenes are C6C2C6-compounds containing two benzenes linked by a central chain of two carbons (Fig. 3.2e). They are relatively less frequent compared with the other phenolic families (Liu et al. 2021).

N. Semmar, *Secondary Metabolites in Plant Stress Adaptation*, Signaling and Communication in Plants, https://doi.org/10.1007/978-3-031-52595-7_3

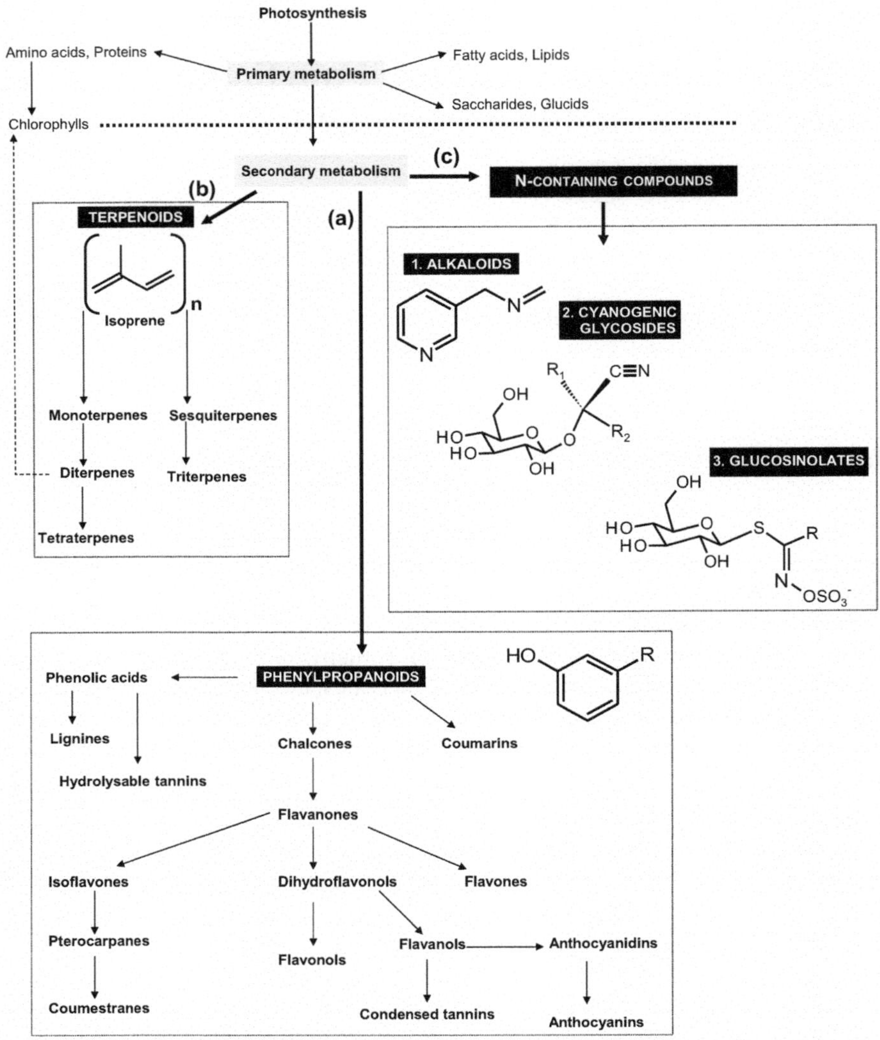

Fig. 3.1 Backbone structures of phenolic (**a**), terpene (**b**) and nitrogen-containing (**c**) compounds representing three highly structurally diversified secondary metabolite classes

- Tannins are polyphenolic polymers made by repeated motifs of particular flavonoids (flavanols) or particular benzoic acid (gallic acid). These two polymerization types provide condensed (Fig. 3.2f) and hydrolysable (Fig. 3.2g) tannins, respectively.

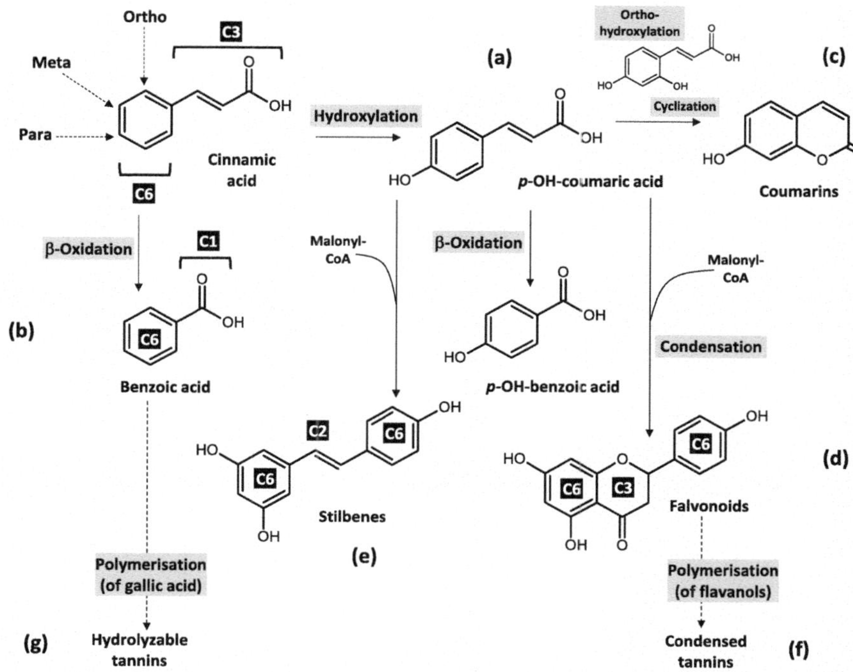

Fig. 3.2 Different structural families of phenolic compounds with their metabolic links

Terpenes are essentially hydrocarbon SMs issued from sequential polymerizations of a C5-unit called isoprene (Fig. 3.1b). These polymerizations result in different terpene families with multiple C5 numbers including (Fig. 3.3): hemiterpenes (C5), monoterpenes (C10), sesquiterpenes (C15), diterpenes (C20), sesterpenes (C25), triterpenes (C30), tetraterpenes (C40) and polyterpenes (polymers).

Nitrogen-containing compounds are classified into three families including alkaloids, cyanogenic glycosides and glucosinolates (Fig. 3.1c). Alkaloids are structurally characterized by the occurrence of nitrogen in heterocycle and/or aliphatic chain structures (Fig. 3.3(1c1)). They have generally alkaline character. Cyanogenic glycosides are structurally characterized by an aglycone attaching a sugar moiety on one hand and a cyanide group (C≡N) on other hand (Fig. 3.3(1c2)). Glucosinolates are nitrogenous metabolites containing a sulfate and a thioglucose moiety (Fig. 3.3(1c3)). The R-group shows high structural diversity including several types of side chains (alkenyl, alkyl, thioalkyl), aromatic and indolic structures (Blažević et al. 2020). Hydrolysis of glucosinolates commonly produces isothiocyanate (S=C=N–R) which is toxic to many organisms. Glucosinolates are abundant in Brassicaceae.

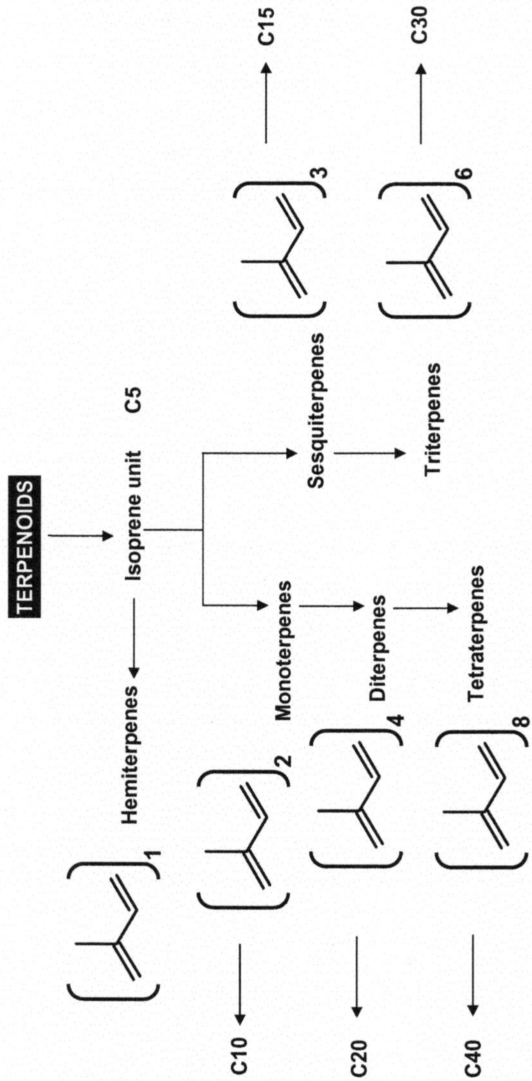

Fig. 3.3 Differentiation of terpenoid families by the number of carbons constituting the molecular scaffold

3.2 Metabolic Processes Governing Structural Variability at Different Molecular Scales

Biosynthesis of SMs is characterized by a multitude of sequential metabolic processes occurring during and after aglycone synthesis, and consisting of gradual additions of new structural components leading to molecular diversification (Figs. 3.4 and 3.5). At aglycone level, structural diversification occurs through cyclization, formation of new double bond(s) and variation of molecular configuration or conformation (Fig. 3.4). Some cases are presented for illustration:

- In flavonoids, chalcone is distinguished by the absence of heterocycle (Fig. 3.4a). Its cyclization results in the formation of flavanone, precursor of different flavonoid families containing a central oxygenated cycle (Fig. 3.4b).
- Among flavonoids, isoflavone (Fig. 3.4c) can submit a cyclization between its ketone and lateral benzene leading to the formation of pterocarpan scaffold containing two oxygenated heterocycles (Fig. 3.4d).
- In sesquiterpene, cyclization is a very common process resulting in high structural diversification. For instance, germacrane (Fig. 3.4e) shows monocyclic scaffold which gives eudesmane after internal cyclization (Fig. 3.4f).

The formation of double bonds in aglycone represents a common mechanism contributing to structural variability of SMs:

- In flavonoids, central heterocycle is characterized by either absence or occurrence of double bond between carbons C2 and C3 (Fig. 3.4g, h). Absence of such a double bond concerns flavane, flavanol, flavanone and flavanonol scaffolds. However, this double bond is present in flavone, flavonol, isoflavone families.
- In sesquiterpenes, the occurrence of double bond(s) shows variable positions between different carbons of aglycon. For instance, germacrene A has 11,12-double bond (Fig. 3.4i), whereas in germacrene B, the double bond is localized between C7 and C11 (Fig. 3.4j). Other germacrene types are characterized by other π-bond positions including germacrenes C, D, E (Semmar 2010).

Close (similar) aglycone scaffolds can differ by configuration aspects. This is commonly found in cyanogenic glycosides (nitrogen-containing compounds). Illustrative case is given for (*S*)-heterodendrin and (*R*)-epiheterodendrin (Fig. 3.4k, l). Apart from configuration, molecular conformation brings additional way for structural diversification. This is common in sesquiterpenes as illustrated by *E*,*E*-germacrene and *Z*-*E*-germacrene scaffolds (Fig. 3.4m, n) (Semmar 2010).

After aglycone formation, structural variability pursues through a serial of several possible chemical substitutions occurring at different molecular positions and with different levels (Fig. 3.5). Substitution positions are associated with different carbons that have standard numbers (Fig. 3.5a). Substitutions can be concerned with hydrophilic chemical groups (Fig. 3.5b, c) or hydrophobic ones (Fig. 3.5d–f).

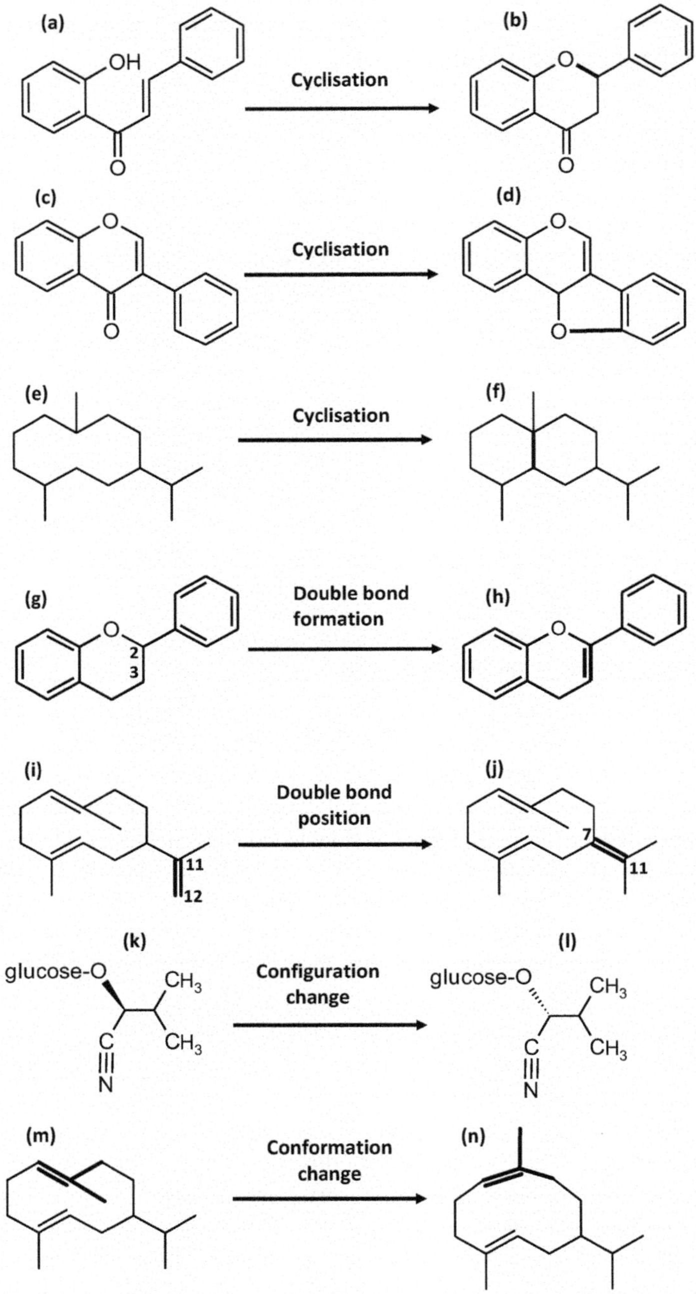

Fig. 3.4 Different mechanisms involved in aglycone structure changes of secondary metabolites

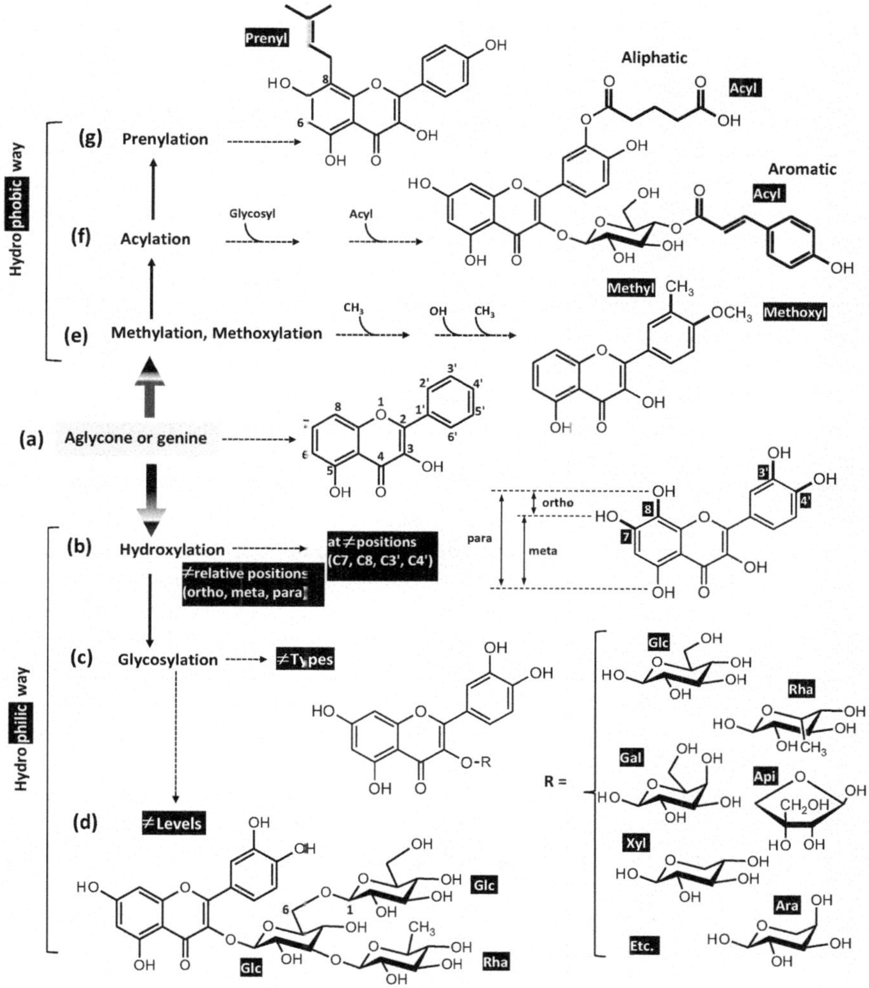

Fig. 3.5 Different chemical substitution processes resulting in structural diversification of secondary metabolites from a given aglycone

Hydrophilic substitutions include hydroxyl and glycosyl groups (Fig. 3.5b, c). Hydroxyl groups are initially attached at different carbons of aglycone (Fig. 3.5b). In cyclic aglycones, substituted carbons can be characterized by their relative positions including ortho, meta and para types (Fig. 3.5b). Ortho position concerns two immediately neighbor substituted carbons; meta position occurs when two substituted carbons are separated by another intermediate carbon on aglycone cycle; para position concerns two opposite substituted carbons on the cycle. Ortho, meta and para positions are generally used to describe the localization of a substituted chemical group by reference to a main substituent on the aglycone cycle: common case is given by phenolic acids in which the main substituent on aglycone consists of lateral carboxylic chain serving as reference to localize other substitutions occurring later on benzen cycle (Fig. 3.2a).

Hydroxylation serves as a main door for glycosylation and other substituents. In this case, one talks about O-glycosylation which occurs after the proton of hydroxyl is lost and substituted by a sugar unit (or chain) (Fig. 3.5c). In less common case, sugar can be directly attached on carbon of aglycon leading to C-glycosyl. This is enzymatically limited to some carbons (positions) of aglycon. In flavonoid (phenolic SMs), C–C substitutions are concerned with positions 6 and 8 of aglycone. Substituted sugar units are various including hexoses (e.g. glucose, galactose, rhamnose, fucose, mannose, quinovose) and pentoses (e.g. xylose, arabinose, apiose) (Fig. 3.5c). Saccharide units can be sequentially (or simultaneously) substituted leading to glycosylated chains with different lengths (glycosylation levels) (Fig. 3.5d). For instance, among disaccharides, 6-rhamnosyl-glucose and 6-glucosyl-glucose correspond to rutinose and gentiobiose, respectively.

Beside hydroxyl and glycosyl groups which increase molecular hydrophily, other chemical groups increase the hydrophobic aspect of molecule. They include methyl, acyl and prenyl groups (Fig. 3.5e–g). When methylation occur on hydroxyl group, it results in the formation of methoxyl group (OCH_3) (Fig. 3.5e). Structural diversification is highly increased by various acyl groups that can be of aromatic (cyclic) or aliphatic types (Fig. 3.5f). Hydrophobic chemical groups include also prenyl side-chain (Fig. 3.5g) (Yazaki et al. 2009). There are different prenyl chain structures including (i) 3,3-dimethylallyl substituent, (ii) geranyl or E-3,7-dimethyl-2,6-octadienyl, (iii) 1,1-dimethylallyl, (iv) lavandulyl or 5-methyl-2-isoprophenyl-hex-4-enyl, etc. (Mukne et al. 2011).

Hydrophily and hydrophoby have governing roles in activities and storage ways of SMs (Chen et al. 2014). Hydrophilic groups are responsible for the storage of SMs in aqueous compartments such as vacuole and apoplast. However, hydrophobic substituents contribute to the storage of SMs into lipidic compartments such as cell membranes. They significantly facilitate interactions and binding processes of SMs with cell membranes and target proteins (Yang et al. 2015).

3.3 Functional Involvement of Phytochemical Compounds in Protective Processes Against Adverse Oxidative Effects

Oxidation represents the most common adverse effect resulting from many environmental stresses. Oxidative processes result in the production of reactive oxygen species (ROS) which are highly toxic against nucleic acids, proteins, lipids and chlorophyll pigments (Fig. 3.7) (Chen et al. 2022). ROS include radicals (superoxide, hydroxyl radical, peroxyl, alkoxyl) and non-radical forms (singlet oxygen, hydrogen

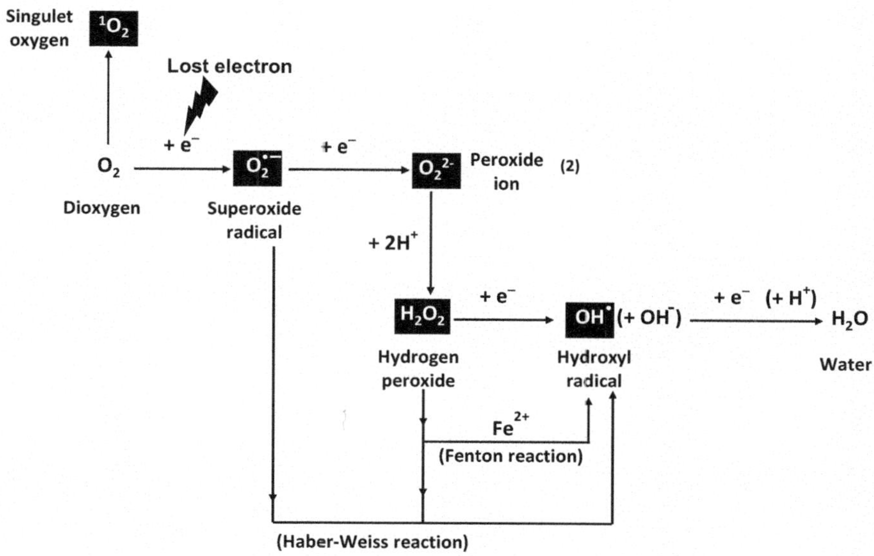

Fig. 3.6 Different reactive oxygen species (black squares) issued from serial of chemical reactions where single (lost) electron moves. Neutralization process consists in adding successively four electrons to dioxygen to form water

peroxide) (Fig. 3.6). Radicals are chemical species having a single (nonpaired) electron resulting in high reactivity for balancing peripheral electronic layer. Increase of ROS in cells results in oxidative stress in biological organisms. Oxidative stress level tends to increase under unbalance states between ROS production and defense systems. In consequence, antioxidative mechanisms are needed for biological survival. In addition to oxidative stresses, non-oxidative stresses can be amplified by ROS effects.

ROS appear by two successive processes:

(i) Electron leak from mitochondrial or chloroplast electron transport chain, followed by
(ii) Reaction of the lost electron with molecular oxygen (O_2) without other electron acceptors.

Formation and transformation of ROS occur in different plant cell compartments including mitochondria, chloroplast, peroxisome and cytosol which contain different catalytic enzymes (Fig. 3.7). When some electrons leak from electronic transport chains in mitochondria or chloroplast, some oxygen molecules (O_2) are reduced by single electrons leading to the formation of superoxide radical:

$$O_2 + 1e^- \rightarrow O_2$$

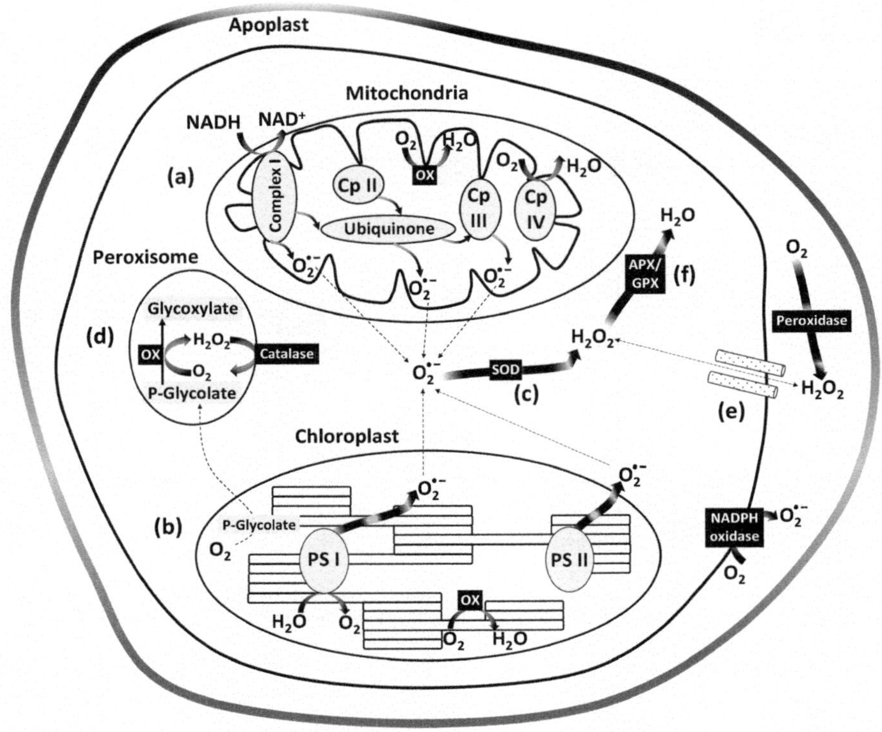

Fig. 3.7 Processes and enzymes catalyzing the formation and transformation of different reactive oxygen species in mitochondria, chloroplast and peroxisome of plant cell under environmental stress. Legend: Cp, complex; PS I, PS II, photosystems I and II; SOD, superoxide dismutase; APX, ascorbate peroxidase; GPX, glutathione peroxidase; OX, oxidase

This reaction occurs in several biochemical systems including ubiquinone (or coenzyme Q), NADH-deshydrogenase of mitochondria membrane and chloroplast (photosystems I and II) (Fig. 3.7a, b).

Oxidative effects of $O_2^{\cdot -}$ on organic compounds can be direct or indirect via its reaction with hydrogen peroxide (H_2O_2) leading to the formation of hydroxyl radical or other radicals (e.g. peroxynitrite radicals from NO) (Fig. 3.6, Haber-Weiss reaction).

Superoxide radical has a life duration of some seconds favoring both its diffusion and scavenging. It is eliminated by its transformation into hydrogen peroxide (H_2O_2) which has longer life duration (of some minutes). This transformation occurs by the dismutation effect of superoxide dismutases (SODs) (Fig. 3.7c):

$$O_2^{\cdot -} + O_2^{\cdot -} \xrightarrow{\text{SOD, } 2H^+} H_2O_2 + O_2$$

SODs are metallo-enzymes located in different cell compartments including Mn-SOD (mitochondrion), Fe–, Cu–Zn-SODs (chloroplast), Mn, Cu–Zn-SODs (peroxisome), Cu–Zn-SOD (cytosol) (Alscher et al. 2002).

Hydrogen peroxide (H_2O_2) can be also produced by bielectronic reduction of O_2 in presence of oxidases (e.g. glycolate oxidase, amine oxidase) (Fig. 3.7d). It is not radical because it has paired peripheric electrons. Like $O_2^{\cdot -}$, it can diffuse far from its production site; due to its chemical similarity with H_2O, it can cross cell membrane via aqueous channels (aquaporins) (Fig. 3.7e). Intracellular concentration of H_2O_2 is regulated by different enzymes including ascorbate peroxidase (APX), catalase (in peroxysomes) or glutathion peroxidase (GPX) (cytosolic essentially) (Fig. 3.7d, f). In peroxysomes, H_2O_2 production occurs by oxidation of glycolate to glycoxylate during photorespiration (Mittler et al. 2004) (Fig. 3.7d). Catalase accelerates dismutation reaction of H_2O_2 which produces O_2 and H_2O:

$$H_2O_2 + H_2O_2 \rightarrow 2H_2O + O_2$$

Toxic effect of hydrogen peroxide is linked to its transformation into hydroxyl radical ($^{\cdot}OH$) in presence of metallic cations like Cu^+ or Fe^{2+} (Fenton or Haber Weiss reactions) (Fig. 3.6):

$$H_2O_2 + Fe_2^{2+} \rightarrow OH + Fe^{3+} + OH^- \text{ (Fenton reaction)}$$

$$H_2O_2 + O_2^{\cdot -}\left(+Fe^{3+}\right) \rightarrow {}^{\cdot}OH + O_2 + OH^-\text{(Haber – Weiss reaction)}$$

Hydroxyl radical has very short life duration (microsecond) compared with the other ROS. This results in higher reactivity. Harmful reactions are of three types:

- Removal of one electron: $^{\cdot}OH + Fe^{2+} \rightarrow Fe^{3+} + OH^-$
- Removal of a H-atom from organic substrate (RH): $OH + RH \rightarrow R^{\cdot} + H_2O$
- Addition to a double bond: $^{\cdot}OH$ + >C=C< → $^{\cdot}$C–C(OH)–

In this last case, hydrogen peroxide can react with DNA by substituting on double bonds of nucleic bases (Fig. 3.8). It results in energetic excess which initializes degrading chain reactions of nucleic acids. Also, $^{\cdot}OH$ can react with polyunsaturated fatty acids of membrane phospholipids and lipoproteins leading to lipidic peroxidation chains. Moreover, constitutive amino-acids of proteins reveal high sensitivity to attacks by $^{\cdot}OH$. Also, $^{\cdot}OH$ radicals are able to inactivate active sites of enzymes.

Harms of hydroxyl radicals are limited by their low diffusion (low molecules' movements). However, due to their low diffusion and high reactivity, hydroxyl radicals are extremely harmful locally. In counter part, superoxide radicals have lower reaction rate constants making they are locally more controllable. In return, superoxide radicals can cause more effects on more distant tissue targets because of their diffusion ability linked to their higher life duration. More generally, the toxicity of $O_2^{\cdot -}$ tends to have indirect aspect through reaction with H_2O_2 to generate $^{\cdot}OH$.

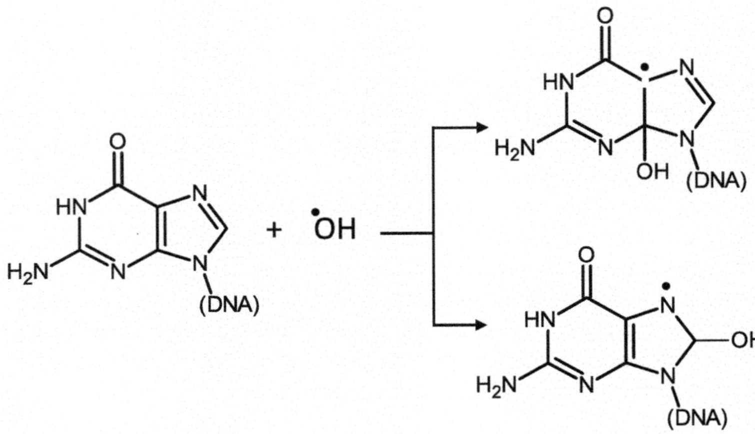

Fig. 3.8 Radical reactions between guanine and hydroxyl radical resulting in different energy excess forms causing degrading chain reactions of nucleic acids (and nucleotides)

Apart from their toxic effects, ROS play key roles in signal transduction events. Under optimal growth conditions, low levels of ROS are produced in different cell organelles (mitochondria, chloroplasts, peroxisomes). Recent studies highlighted that the oxidative burst associated with rapid increase in ROS production play essential role in mediating responses against pathogen infections and abiotic stresses (Mittler et al. 2004; Torres and Dangl 2005). Thus, ROS show both negative (toxic) and positive (signal transduction) effects leading to the need to control their intracell steady states for optimal living conditions of plants.

3.3.1 Protective Lipoxygenase Pathway Products

Lipoxygenases (LOX) are ubiquitous enzymes in the biological world (animals, plants, microbes). In plants, they play key roles in the productions of regulatory and signaling metabolites in both physiological processes and defense against various environment stresses (Fig. 3.9a). They transform polyunsaturated fatty acids (PUFA) into PUFA hyperoxides which are highly reactive products (Fig. 3.9c). Phospholipids and galactolipids of cell membrane represent preliminary source of PUFA which are initially released by phospholipases and/or galactosidases (Fig. 3.9b) to give hyperoxides by LOX actions (Fig. 3.9c). Due to their high reactivity, hyperoxides are ready to be converted into signal and defense products called phyto-oxylipins (Fig. 3.9d). Various phyto-oxylipins are obtained from hyperoxides by the actions of various enzymes including (Fig. 3.9e):

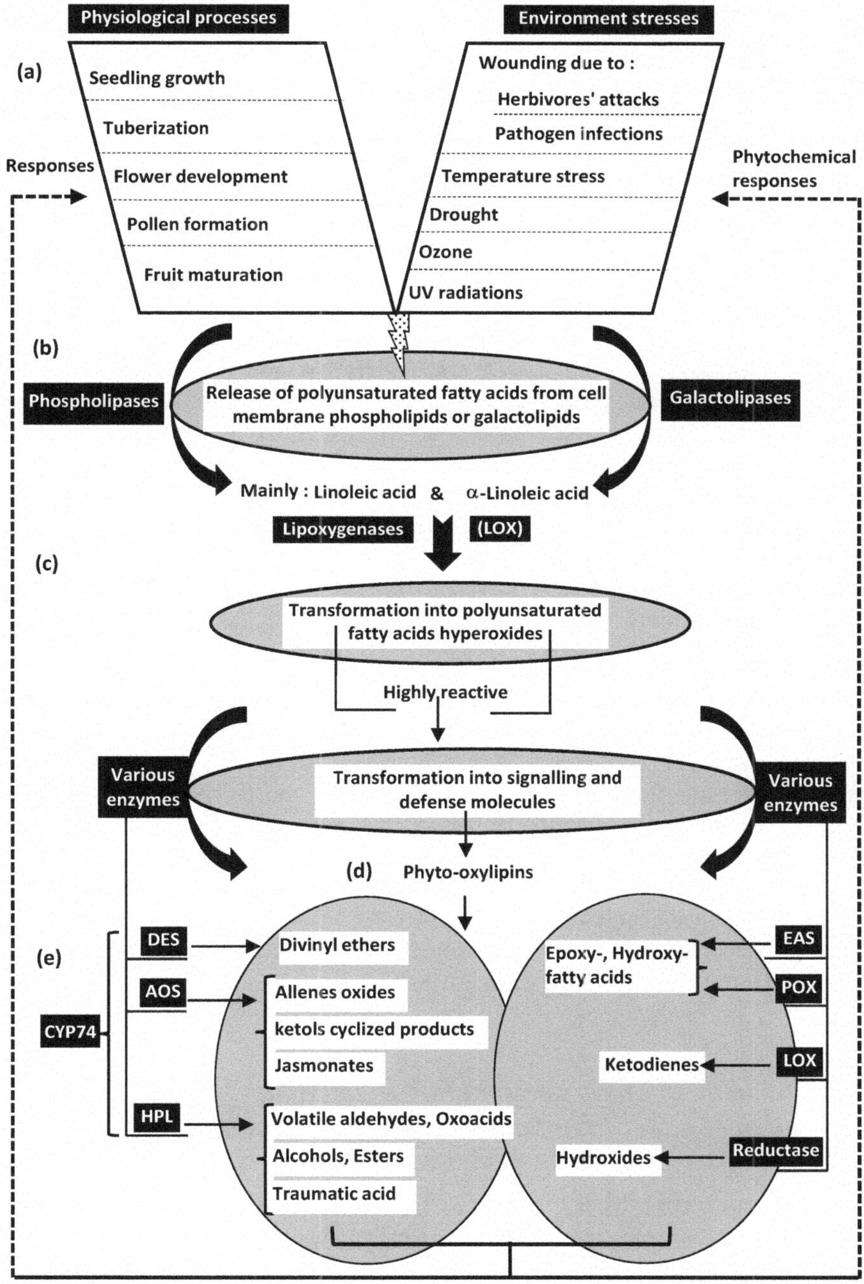

Fig. 3.9 General presentation of lipoxygenase pathway in plants as response-system to physiological needs and environment stresses. LOX, lipoxygenase; DES, divinyl ether synthase; AOS, Allene oxide synthase; EAS, epoxy alcohol synthase; POX, peroxygenase; HPL, hydroperoxide lyase; CYP74, cythochrome P450 family 74

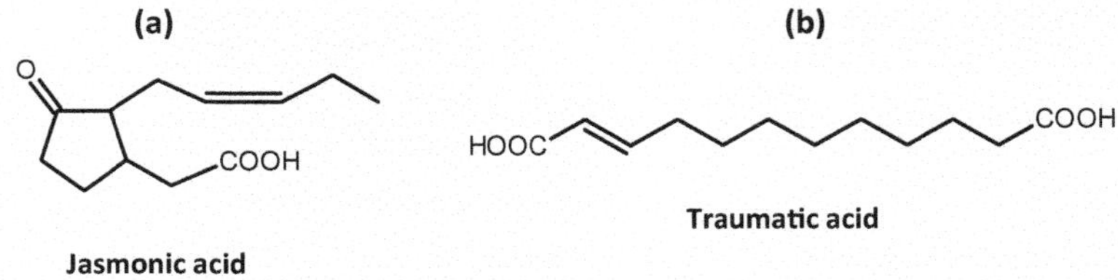

Fig. 3.10 Structures of some C12 fatty acids involved in wounding signals

- Divinyl ether synthase (DES) giving divinyl ethers.
- Allene oxide synthase (AOS) giving allene oxides, α- and γ-ketols cyclized products followed by jasmonic acid (JA) and its derivatives (Fig. 3.10a).
- Hydroperoxide lyase (HPL) giving volatile aldehydes, alcohols, esters and traumatic acid (Fig. 3.10b).
- Peroxygenase (POX) or epoxy alcohol synthase (EAS) giving epoxy-hydroxy fatty acids.
- LOX catalyzing the cleavage of O–O bond to give alkoxy radicals followed by their self-rearrangement in ketodienes.
- Reductase reducing the PUFA hydroperoxides to give less reactive products, the fatty acid hydroxides.

Plant LOX-responses to stresses include emissions of biogenic volatile organic compounds (BVOCs). Such particular BVOCs are called green leaf volatiles (GLVs) and consist of six carbon (C6) compounds including alcohols, aldehydes and esters (Figs. 3.9e and 3.11) (Loreto et al. 2006; Zhou et al. 2021). GLVs generally consist among the first plant responses to stress (Copolovici and Niinemets 2010). Their emission rates tend to be quantitatively associated with stress severity (Beauchamp et al. 2005; Copolovici and Niinemets 2010; Fall et al. 1999; Niinemets 2010).

Other BVOCs include low molecular weight SMs (isoprene, monoterpenes, sesquiterpene, phenol acids) which are emitted during and after stress to bring large signaling roles or specific metabolic responses. The blends of emitted volatile SMs depends on the stress type (Niinemets 2010).

3.3.2 *Structural and Functional Characteristics of Secondary Metabolites Against Oxidation Stress*

Apart from enzymatic defense reactions (Fig. 3.7), important production of SMs provides a strong non enzymatic defense system against ROS due to their antioxidative properties (Masamba and Kappo 2021). SMs play key chemical protective roles by preventing ROS to reach their biological targets. Beyond their ability to react with ROS, antioxidant SMs are able to regenerate *in vivo* for further uses and chemical protections.

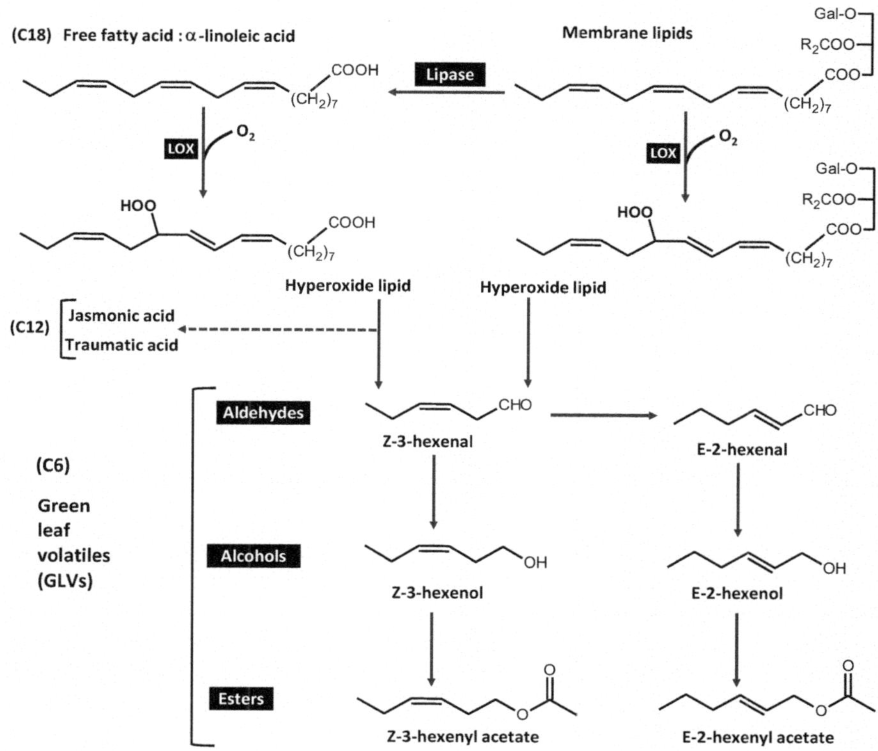

Fig. 3.11 LOX metabolic pathways leading to the formation of volatile C6-compounds known as green leaf volatiles (GLVs)

Antioxidation mechanisms show different aspects:

- Neutralization of free radical by transfer of H-atom from the antioxidant (A-OH):

$$R^{\cdot} + \text{A-OH} \rightarrow \text{RH} + \text{AO}^{\cdot} \rightarrow\rightarrow \text{ Stabilization : AO}^{\cdot} + R^{\cdot} \rightarrow \text{ AOR(Stable)}$$

- Transformation of free radical into anion by acquisition of one electron from the antioxidant (A-OH):

$$R^{\cdot} + \text{A-OH} \rightarrow R^{-} + \text{AOH}^{\cdot} \rightarrow\rightarrow \text{ Stabilization}$$

- Chelation of transition metal ions (high oxidative effects) by antioxidants leading to the formation of stable complexes where the metals are entrapped and prevented from generation of free radicals. Chelation of iron by flavonoids provides antioxidative (anti-ROS) way via Fenton reaction (Fig. 3.6).

Fig. 3.12 Different chemical structures favoring antioxidative activity

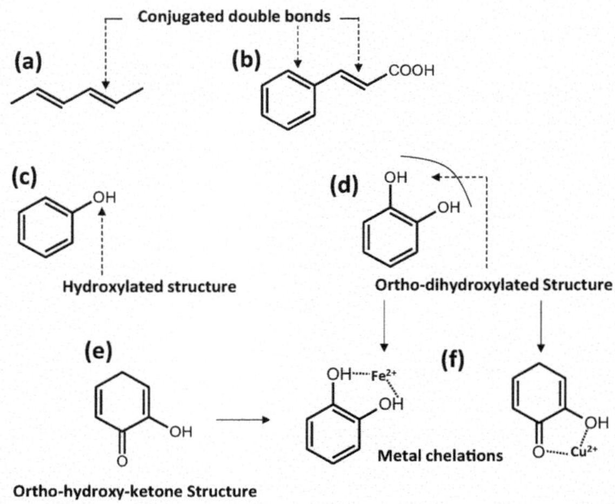

Antioxidative ability in SMs is favored by different chemical structures including conjugated double bonds, hydroxylation and ortho-dihydroxyl states (Fig. 3.12). Conjugated double bounds (Figs. 3.12 a, b) favor electronic transfer; hydroxylation favors entire proton transfer containing electron (Fig. 3.12c, d); ortho-di-OH state offers appropriate chemical structure for chelation (Fig. 3.12d, f). Chelation is also favored by ortho-hydroxy-ketone structures (Fig. 3.12e, f). These structural characteristics are found in phenolic compounds, fat-soluble vitamin E (-tocopherol) and water-soluble vitamin C (L-ascorbic acid).

Antioxidant activity measured in different flavonoids (flavonols, flavone, flavanol, flavane) revealed increasing trend with hydroxylation level (Fig. 3.13) (Melidou et al. 2005): antioxidant activity decreased according to the following trend: quercetin (flavonol; five OH with two ortho-di-OH) > myricetin (six OH) > taxifolin (flavanol; five OH) > kaempferol (flavonol; four OH) > chrysin (Flavonol; three OH), apigenin (flavone; three OH), naringenin (flavane; three OH). An exception concerned the quercetin (five OH) which showed higher antioxidant activity than myricetin (six OH); this indicated more efficient antioxidation associated with one ortho-di-OH system than two ortho-di-OH on a same aromatic ring. Also, double bond of heterocycle has key role in antioxidant activity as quercetin (higher activity) and taxifolin (lower activity) structurally differed by this double bond (Fig. 3.13).

Beyond unimolecular aspect, antioxidant capability can be originated from synergetic effects between antioxidant metabolites and enzymes.

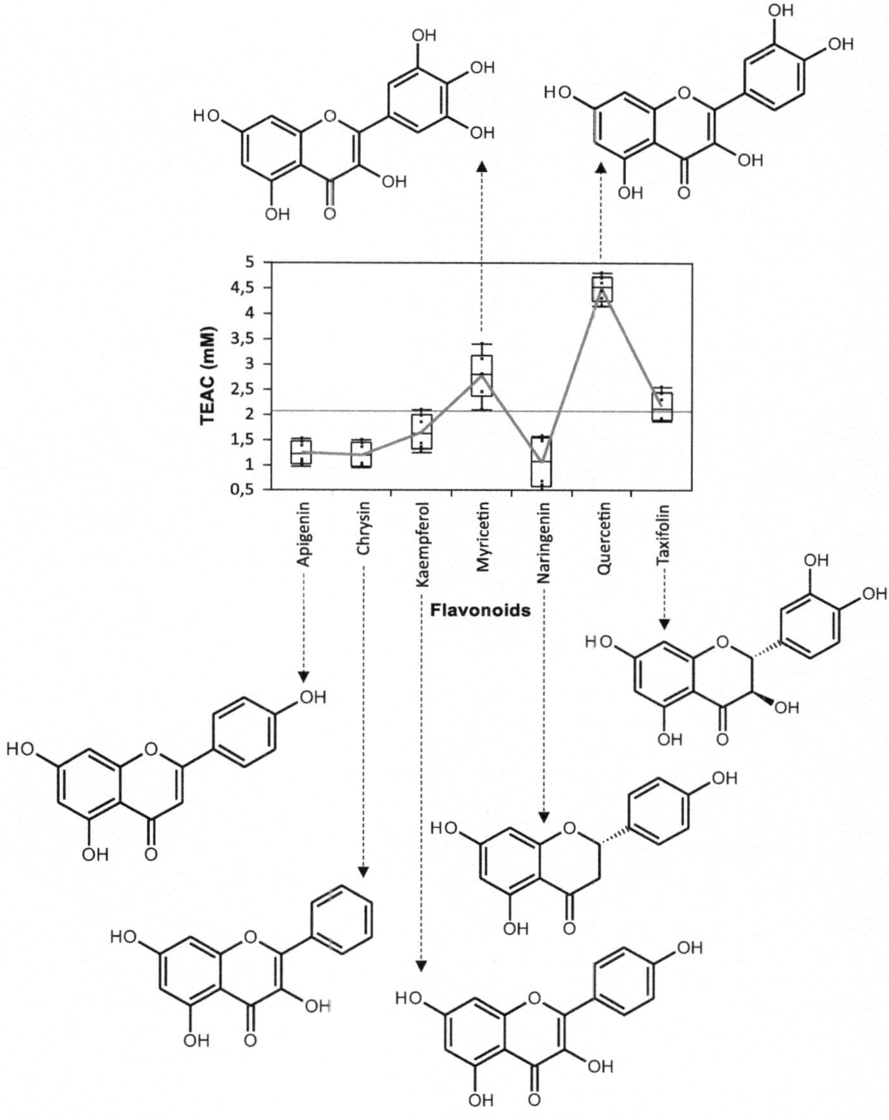

Fig. 3.13 Variation of antioxidative activity in different flavonoids showing different levels and relative positions of hydroxylation. TEAC, Trolox Equivalent Antioxidant Capacity

References

Alscher RG, Erturk N, Heath LS (2002) Role of superoxide dismutases (SODs) in controlling oxidative stress in plants. J Exp Bot 53(372):1331–1341

Beauchamp J, Wisthaler A, Hansel A, Kleist E, Miebach M, Niinemets U et al (2005) Ozone induced emissions of biogenic VOC from tobacco: relationships between ozone uptake and emission of LOX products. Plant Cell Environ 28:1334–1343

Blažević I, Montaut S, Burčul F, Olsen CE, Burow M, Rollin P, Agerbirk N (2020) Glucosinolate structural diversity, identification, chemical synthesis and metabolism in plants. Phytochemistry 169:112100

Chen D, Mubeen B, Hasnain A, Rizwan M, Adrees M, Naqvi SAH, Iqbal S, Kamran M, El-Sabrout AM, Elansary HO, Mahmoud EA, Alaklabi A, Sathish M, Din GMU (2022) Role of promising secondary metabolites to confer resistance against environmental stresses in crop plants: current scenario and future perspectives. Front Plant Sci 13:881032

Chen X, Mukwaya E, Wong MS, Zhang Y (2014) A systematic review on biological activities of prenylated flavonoids. Pharm Biol 52(5):655–660

Copolovici L, Niinemets Ü (2010) Flooding induced emissions of volatile signalling compounds in three tree species with differing waterlogging tolerance. Plant Cell Environ 33:1582–1594

Fall R, Karl T, Hansel A, Jordan A, Lindinger W (1999) Volatile organic compounds emitted after leaf wounding: on-line analysis by proton-transfer-reaction mass spectrometry. J Geophys Res 104:15963–15974

Liu W, Feng Y, Yu S, Fan Z, Li X, Li J, Yin H (2021) The flavonoid biosynthesis network in plants. Int J Mol Sci 22(12824):1–18

Loreto F, Barta C, Brilli F, Nogues I (2006) On the induction of volatile organic compound emissions by plants as consequence of wounding or fluctuations of light and temperature. Plant Cell Environ 29:1820–1828

Masamba P, Kappo AP (2021) Immunological and biochemical interplay between cytokines, oxidative stress and schistosomiasis. Int J Mol Sci 22:7216

Melidou M, Riganakos K, Galaris D (2005) Protection against nuclear DNA damage offered by flavonoids in cells exposed to hydrogen peroxide: the role of iron chelation. Free Radic Biol Med 39:1591–1600

Mittler R, Vanderauwera S, Gollery M, Van Breusegem F (2004) Reactive oxygen gene network of plants. Trends Plant Sci 9:490–498

Mukne AP, Viswanathan V, Phadatare AG (2011) Structure prerequisites for isoflavones as effective antibacterial agents. Pharmacog Rev 5:13–18

Niinemets Ü (2010) Mild versus severe stress and BVOCs: thresholds, priming and consequences. Trends Plant Sci 15:145–153

Rammohan A, Reddy JS, Sravya G, Rao CN, Zyryanov GV (2020) Chalcone synthesis, properties and medicinal applications: a review. Environ Chem Lett 18(1):433–458

Semmar N (2010) Chemotaxonomical analyses of herbaceous plants based on phenolic and terpenic patterns. In: Flexible tools to survey biodiversity in grasslands. Nova Science Publishers, New York, p 257.

Shimizu BI (2014) 2-Oxoglutarate-dependent dioxygenases in the biosynthesis of simple coumarins. Front Plant Sci 5(549):1–7

Torres MA, Dangl JL (2005) Functions of the respiratory burst oxidase in biotic interactions, abiotic stress and development. Curr Opin Plant Biol 8:397–403

Yang X, Jiang Y, Yang J, He J, Sun J, Chen F, Zhang M, Yang B (2015) Prenylated flavonoids, promising nutraceuticals with impressive biological activities. Trends Food Sci Technol 44:93–104

Yazaki K, Sasaki K, Tsurumaru Y (2009) Prenylation of aromatic compounds, a key diversification of plant secondary metabolites. Phytochemistry 70:1739–1745

Zhou ZW, Wu QY, Ni ZX, Hu QC, Yang Y, Zheng YC, Bi WJ, Deng HL, Liu ZZ, Ye NX, Lai ZX, Sun Y (2021) Metabolic flow of C6 volatile compounds from LOX-HPL pathway based on airflow during the post-harvest process of Oolong Tea. Front Plant Sci 12:738445

Part II
Biosynthetic Pathways, Structural Diversification Processes and Roles of Different Secondary Metabolite Families

Chapter 4
Phenolic Compounds

4.1 Biosynthesis Pathways

Phenolic SMs include several structural families which are sequentially issued the ones from the others through different processes including condensation, cyclization, conjugation, ionization, substitution and polymerization (Fig. 4.1):

- Condensation reaction is responsible for the formation of molecular skeletons called aglycones or genins representing different metabolic classes. It occurs between different precursor molecules leading to the formation of a more complex structure representing the scaffold of a given structural class from which several derivatives can be synthesized through different processes (described in the next paragraphs) leading to different metabolic families. In phenolic metabolism, condensation can initially occur between aromatic structures including C6C3 (cinnamic acids) and C6 (benzoic acids) types leading to the formation of C15 (C6C3C6) or C14 (C6C2C6) compounds containing two benzenes linked by a C3- or C2-chain (Fig. 4.1a). Condensation process is initially responsible for the formation of chalcones and stilbenes (Fig. 4.2f, g). Chalcones are precursors of different flavonoid families.
- Cyclization results in the closing of C3-chain linking the two benzene structures leading to the apparition of oxygenated heterocycle within the C6C3C6 scaffold (Fig. 4.1b). This process is involved in the synthesis of flavonoids, coumarins and aurones (Fig. 4.2d, h, i).
- Conjugation or desaturation will concern heterocycle through the apparition of double bond(s) at some locations associated with some carbon pairs (Fig. 4.1c, d). The C2–C3 pair is the most concerned with conjugation process (Fig. 4.1c); it is a characteristic of several flavonoid families including flavones, flavonols and isoflavones (Fig. 4.2j, k). Other carbon pairs can be concerned with conjugation according to the phenolic family (Fig. 4.1d).
- Ionisation of heterocyclic oxygen is at the origin of anthocyanidin formation (Figs. 4.1d and 4.2k).

N. Semmar, *Secondary Metabolites in Plant Stress Adaptation*, Signaling and Communication in Plants, https://doi.org/10.1007/978-3-031-52595-7_4

- Hydroxylation represents the first chemical substitution occurring on the aglycone making the molecule to be ready to substitute the proton of OH by different chemical groups (Fig. 4.1 e1, e2). This favors structural diversification within a same metabolic family. Hydroxylation of an aglycone represents a chemical way for next glycosylation and acylation.
- Dimerization or polymerization occurs through repeated condensation of a same molecular scaffold leading to the apparition of a new metabolic family (Fig. 4.1f). This process is responsible for the synthesis of different tannin families (condensed and hydrolysable tannins) and phlobaphenes (Figs. 4.2m and 4.4t, w).

The wide structural diversity of phenylpropanoids results from several biosynthesis pathways issued from the *p*-coumaroyl-CoA pathway (common precursor) which is synthesized from the deamination of phenylalanine (Fig. 4.2) (Liu et al. 2021).

Aromatic cycle of phenols is initially provided by phenylalanine which is an aromatic amino acid synthesized via the shikimate pathway (Fig. 4.2a) (Wang et al. 2018a). The general phenylpropanoid pathway begins when phenylalanine ammonia-lyase (PAL) catalyzes the deamination of phenylalanine to trans-cinnamic acid

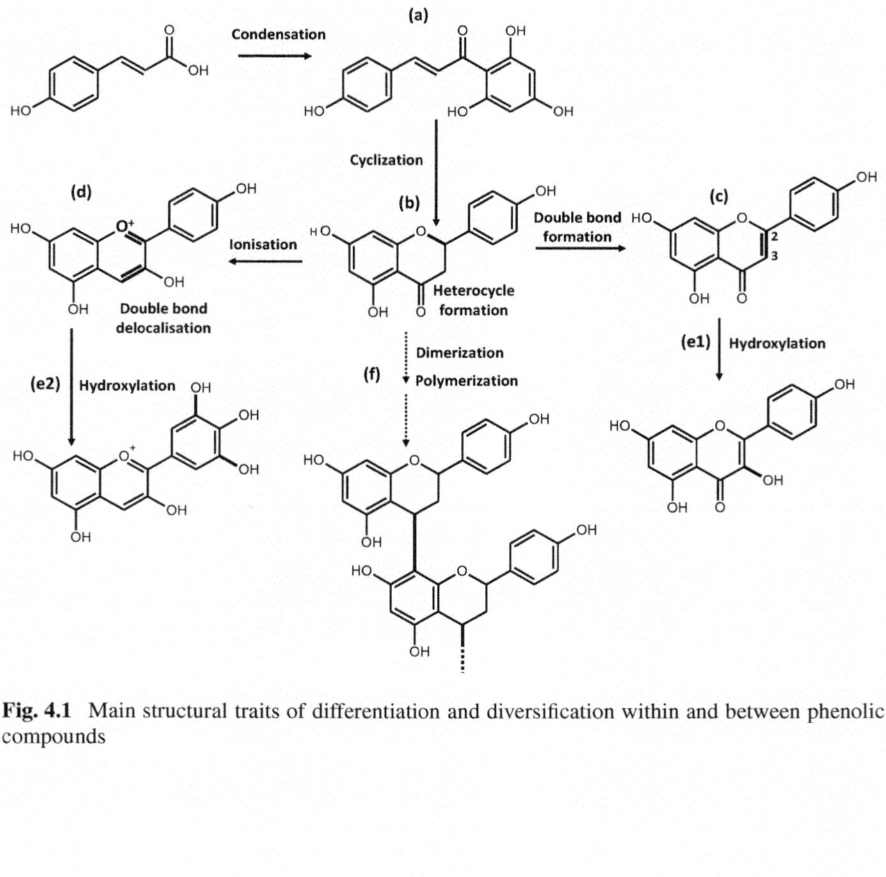

Fig. 4.1 Main structural traits of differentiation and diversification within and between phenolic compounds

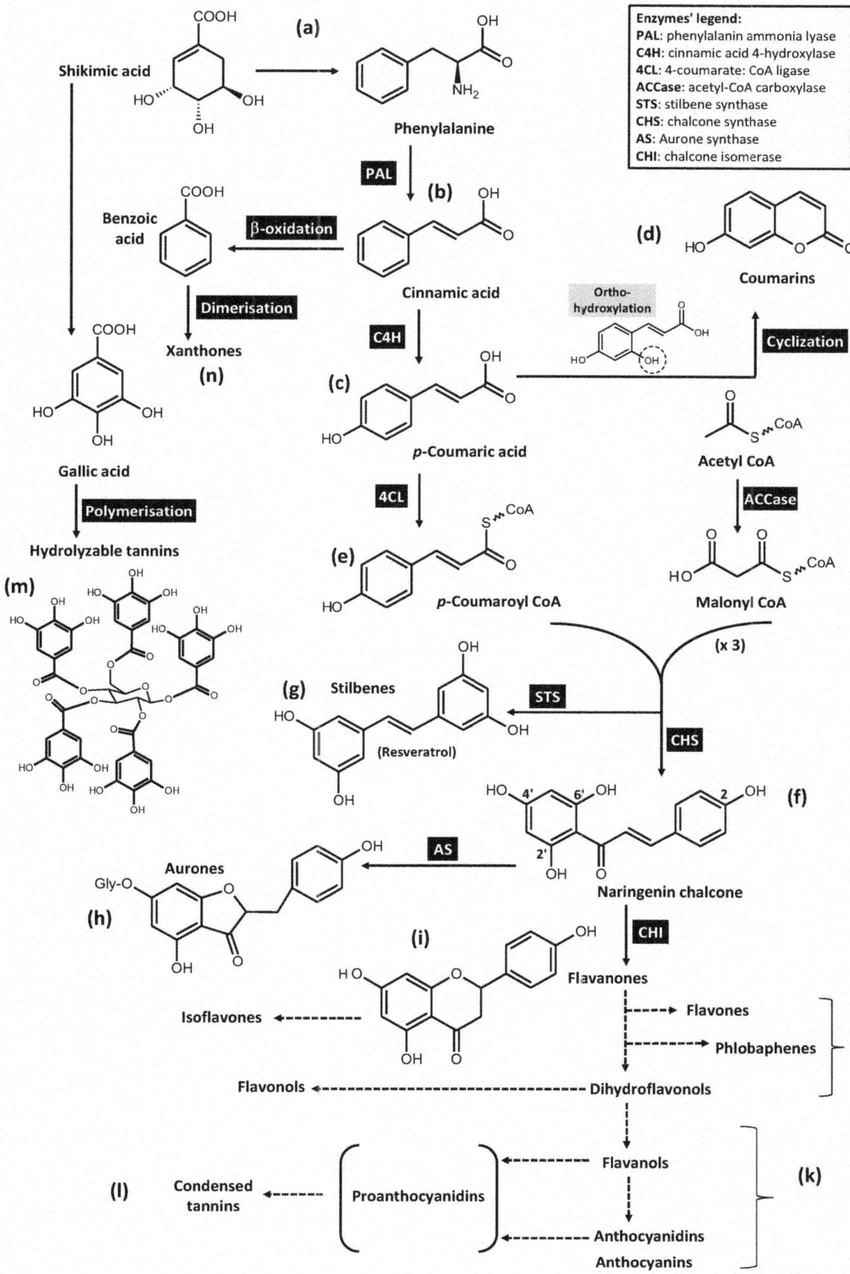

Fig. 4.2 Overview on metabolic links between different families of phenolic compounds

(Fig. 4.2b). Moreover, PAL is a key enzyme responsible for carbon flux conversion from primary to secondary metabolism in plants (Barros and Dixon 2020). Cinnamic acid is hydroxylated to *p*-coumaric acid by cinnamic acid 4-hydroxylase (C_4H), a cytochrome P450 monooxygenase (Fig. 4.2c) (Wohl and Petersen 2020). Cyclization of *p*-coumaric acid after its ortho-hydroxylation leads to the formation of coumarins (Fig. 4.2d). *p*-Coumaric acid plays crucial role in the synthesis of different phenolic classes: co-enzyme A (CoA) is added to *p*-coumaric acid by 4-coumarate:CoA ligase (4CL) leading to the formation of *p*-coumaroyl-CoA (Fig. 4.2e). *p*-Coumaroyl-CoA will submit condensation with three molecules of malonyl-CoA to start flavonoid biosynthesis with initial apparition of chalcones and stilbenes (Fig. 4.2f, g).

Flavonoid biosynthesis starts from the condensation of one *p*-coumaroyl-CoA with three malonyl-CoA (derived from acetyl-CoA) to provide either chalcones or stilbenes (Fig. 4.2f, g). From this condensation, naringenin chalcone appears by the action of chalcone synthase (CHS) (Fig. 4.2f). Naringenin chalcone, a tetrahydroxylated chalcone (4, 2′, 4′, 6′-OH chalcone), represents the first appeared flavonoid with C15-skeleton (C6C3C6) and serves as precursor for the synthesis of downstream flavonoids. Chalcones differ from the other flavonoids by the absence of heterocycle.

Apart from chalcones, stilbenes are also directly issued from condensation of *p*-coumaroyl-CoA with malonyl-CoA by the action of stilbene synthase (STS) (Fig. 4.2g). Resveratrol is the precursor of most of plant stilbenes via various metabolic processes including isomerization, glycosylation, methylation, prenylation, etc. (Chong et al. 2009). However, this flavonoid family is relatively rare because of its 14 carbons-content rather than 15C. Stilbenes show limited distribution in the plant world; they were found in grapevine, pine, sorghum and peanut (Chong et al. 2009; Peng et al. 2013).

Chalcones are subject to intra-molecular cyclization leading to the formation of aurones and flavanones (Fig. 4.2h, i). Flavanones are formed by the action of chalcone isomerase (CHI); they represent precursors of various flavonoid families including isoflavones, dihydroflavonols, flavones and phlobaphenes (Figs. 4.2j, 4.3 and 4.4). Dihydroflavonols are further transformed by different catalyzing enzymes to give flavonols and flavanols which are precursors of anthocyanidins and proanthocyanidins (Figs. 4.2k and 4.4). At the downstream of phenylpropanoid metabolic network, gallic acid (benzoic acid type) and flavanols can be subject to polymerization leading to the formation of hydrolysable and condensed tannins, respectively (Figs. 4.2l, m and 4.4). Also, condensation of two benzoic acids provides a metabolic way for the synthesis of xanthones (Fig. 4.2n).

Further details on enzymatic steps and metabolic diversity of different flavonoid families are presented in Figs. 4.3 and 4.4.

Within chalcone family, naringenin chalcone can be transformed to isoliquiritigenin (4, 2′, 4′-trihydroxychalcone) by the action of chalcone reductase (CHR) (Fig. 4.3a). Both naringenin chalcone and isoliquiritigenin can be converted to the flavanones: naringenin and liquiritigenin (Fig. 4.3b, c), respectively, via the cyclization effect of chalcone isomerase (CHI) leading to the formation of heterocyclic ring C in the flavonoid pathway (Nabavi et al. 2018; Lin et al. 2021). CHI and chalcone

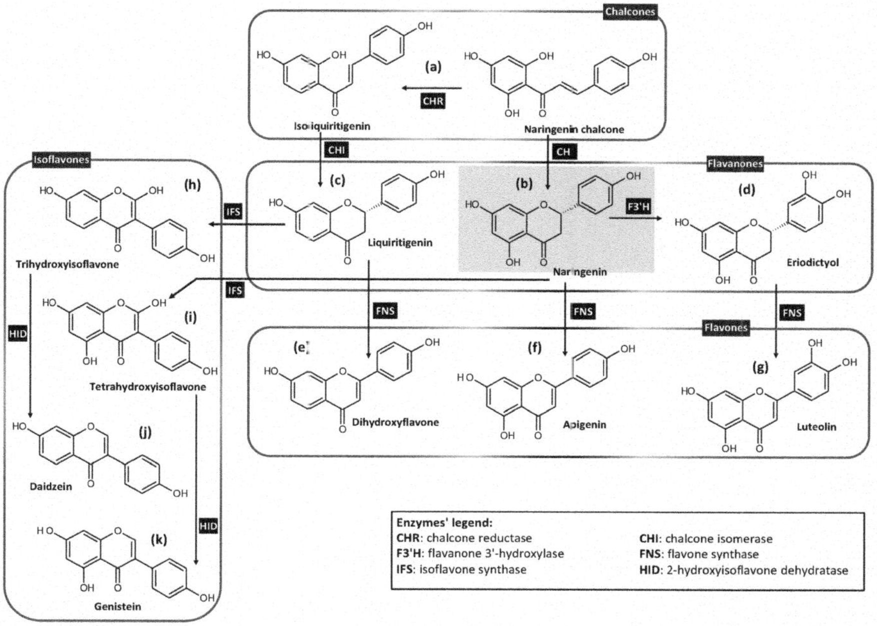

Fig. 4.3 Sequential biosynthesis pathways of isoflavones and flavones from flavanones issued from chalcones

synthase (CHS) represent key rate-limiting enzymes in the flavonoid biosynthesis pathway (Deng et al. 2013; Zhang et al. 2017; Yin et al. 2019).

Within flavanone family, naringenin can be converted to eriodictyol by hydroxylation of carbon 3′ (Fig. 4.3d). Acquisition of double bond between C2 and C3 of heterocycle leads to the formation of flavones (Fig. 4.3e–g). This conjugation occurs by the action of flavone synthase (FNS) leading to the formation of dihydroxyflavone, apigenin and luteolin from liquiritigenin, naringenin and eriodictyol, respectively.

Moreover, flavanones represent precursors of isoflavones via the enzyme action of isoflavone synthase (IFS) (Fig. 4.3h, i). Liquiritigenin and naringenin are converted to tri- and tetra-hydroxyisoflavones, respectively (Fig. 4.3h, i). Within isoflavones, these two preliminarily formed metabolites can be dehydroxylated in carbon 2 by 2-OH-isoflavone dehydratase (HID) to give daidzein and genistein, respectively (Fig. 4.3j, k).

C3-hydroxylation of flavanones results in the formation of dihydroflavonols (Fig. 4.4a–c) which serve as precursors for flavonols (Fig. 4.4d–f). During this sequential metabolic process, mono-, di- and tri-hydroxylated flavanones in B-ring are converted to analogous dihydroflavonols (by flavanone 3-hydroxylase, F3H) then flavonols (by flavonol synthase, FLS). Thus, naringenin, eriodictyol and pentahydroxyflavanone are respectively converted to dihydrokaempferol, dihydroquercetin

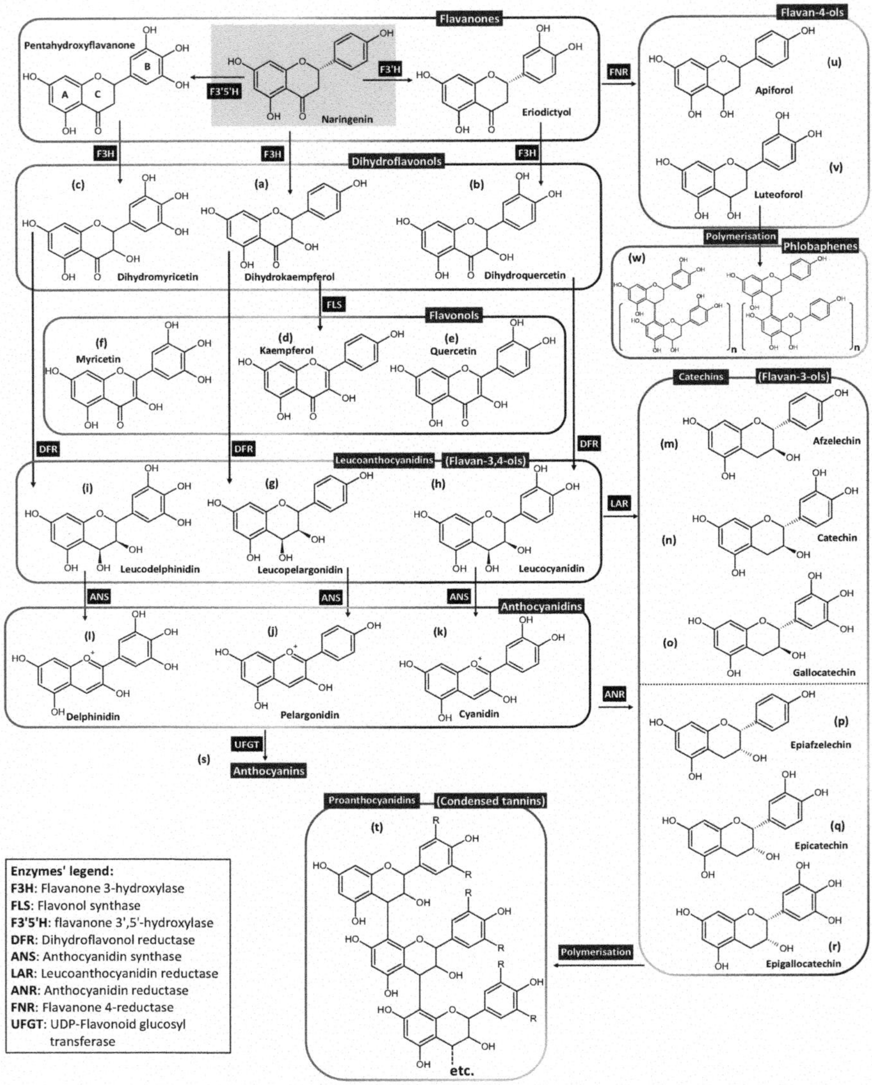

Fig. 4.4 Biosynthesis pathways of different flavonoid families from flavanones including dihydroflavonols (**a**–**c**), flavonols (**d**–**f**), flavan-3,4-ols (**g**–**i**), anthocyanidins (**j**–**l**), flavan-3-ols (**m**–**r**), anthocyanins (**s**), proanthocyanidins (**t**), flavan-4-ols (**u**, **v**), phlobaphenes (**w**)

and dihydromyricetin (dihydroflavonols, Fig. 4.4a–c), and these later are respectively converted to kaempferol, quercetin and myricetin (flavonols, Fig. 4.4d–f).

Moreover, dihydroflavonols can be converted to leucoanthocyanidins by the enzyme action of dihydroflavonol-4-reductase (DFR) which catalyzes the formation of a OH-group at carbon 4 of heterocycle (ring C) (Fig. 4.4g–i). Leucoanthocyanidins have the backbone structure of flavan-3,4-ols and are precursors of anthocyanidins (Fig. 4.4j–l) and catechins (Fig. 4.4m–r). Conversion of leucoanthocyanidins to anthocyanidins occurs by anthocyanidin synthase (ANS) giving pelargonidin, cyanidin and delphinidin (Fig. 4.4j–l) from leucopelargonidin, leucocyanidin and leucodelphinidin, respectively (Fig. 4.4g–i). Therefore, glycosylation of anthocyanidins by glycosyl-transferase results in the formation of anthocyanins (Fig. 4.4s).

Conversion of leucoanthocyanidins to catechins occurs by the enzyme action of leucoanthocyanidin reductase (LAR) which drives the dehydroxylation of C-4 in C-ring (Wang et al. 2018b). Leucoanthocyanidin with mono-, di- and tri-hydroxylated B-ring (leucopelargonidin, leucocyanidin, leucodelphinidin) (Fig. 4.4g–i) are converted to their analogous catechins (afzelechin, catechin, gallocatechin) (Fig. 4.4m–o). Catechins have a backbone structure of flavan-3-ols and are characterized by the presence of two chiral centers (carbons C2 and C3). Isomerization based on *cis*- and *trans*-configurations exists leading to *cis*-flavan-3-ols and *trans*-flavan-3-ols. *Trans*-flavan-3-ols include afzelechin, catechin, gallocatechin (Fig. 4.4m–o). *Cis*-flavan-3-ols include epiafzelechin, epicatechin and epigallocatechin (Fig. 4.4p–r). *Cis*-flavan-3-ols can be directly formed from anthocyanidins by the action of anthocyanidin reductase (ANR)(Fig. 4.4p–r). This enzyme catalyzes the removal of double bond in heterocycle leading to convert pelargonidin, cyanidin and delphinidin (Fig. 4.4j–l) into epiafzelechin, epicatechin and epigallocatechin, respectively (Fig. 4.4p–r). LAR and ANR are NADPH-dependent reductases and represent the key and rate-limiting enzymes in the biosynthesis of catechins and proanthocyanidins (Fig. 4.4t).

Proanthocyanidins, also called condensed tannins, derive from the polymerization of catechins (flavan-3-ols) (Fig. 4.4t). Also, polymerization can concern flavan-4-ols which are directly derived from flavonone via the enzyme action of flavanone 4-reductase (Fig. 4.4u, v). Such a polymerization leads to the formation of phlobaphenes (Fig. 4.4w).

4.2 Structural Diversification Processes and Chemotaxonomical Trends

Biosynthesis pathways of phenolic compounds initially involve structural diversification of aglycones leading to different metabolic families. Aglycone pattern provides wide chemical differentiation traits between plants essentially at high taxonomical levels (characteristic patterns of plant classes, families) (Fig. 4.5a).

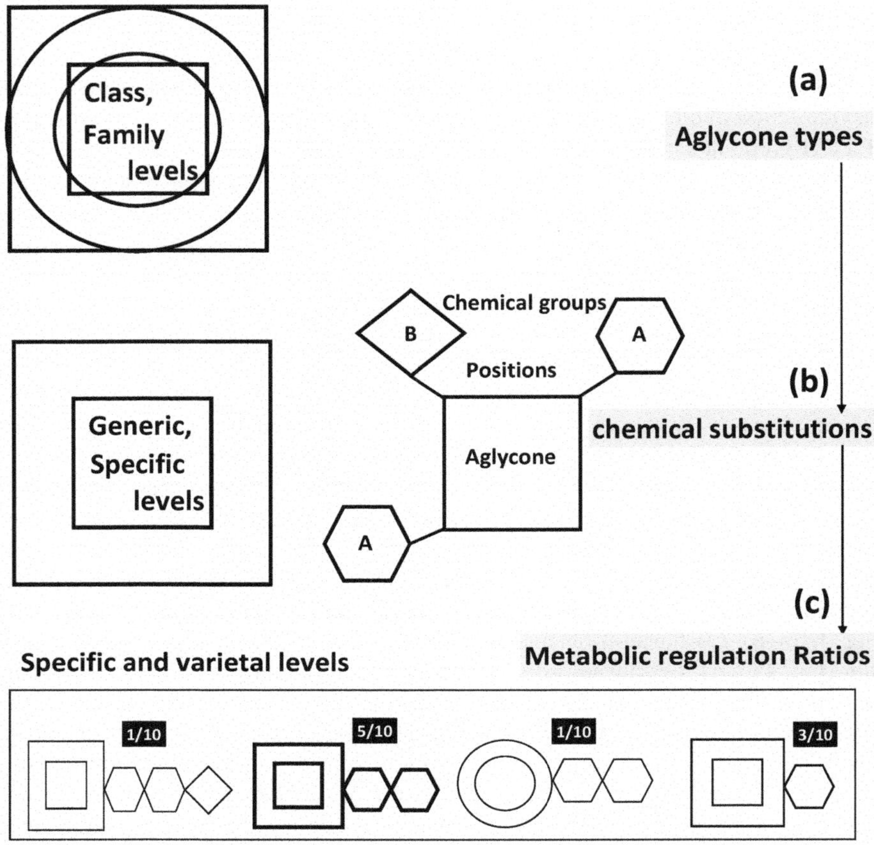

Fig. 4.5 Overview on different metabolic criteria observed in phenolic compounds and serving for chemical characterization of plants at different taxonomical levels (chemotaxonomical interests)

Beyond aglycone synthesis governing different phenolic classes, metabolic variability is amplified by several chemical substitutions occurring on a same aglycone at different locations (carbons) and with different levels leading to the formation of different molecules deriving from a same aglycone (Fig. 4.5b). Phytochemical patterns based on (combining) different substituted metabolites provide original chemical fingerprints associated with different plant genera and species. At lower scale, different plant varieties can be metabolically characterized by different regulation ratios of same synthesized phenolic compounds (Fig. 4.5c). Different regulation metabolic patterns (called metabotypes or chemotypes) are made by sets of metabolites co-occurring at different relative levels (ratios).

Chemical substitutions of phenolic aglycones include (Semmar 2010):

- Hydroxylation and glycosylation resulting in more hydrophilic molecules.
- Methylation, acylation and prenylation resulting in more hydrophobic (less hydrophilic) compounds.

4.2.1 Phenolic Differentiations at Plant Family Level

This book is not intended for chemotaxonomy but it provides wide lines on the interest of the phytochemistry in plant classification and recognition. Several illustrative cases are given on the interest of phytocompounds in attributing chemical fingerprint to different plant taxa.

At the aglycone level, flavanones, flavones, flavonols and isoflavones show differential distributions in several plant families (Fig. 4.6):

- Poaceae (monocotyledons) are characterized by the production of flavones which can be glycosylated at C8 (and C6) via a C–C bond (rather than C–O bond) (Fig. 4.6c). The C8 and C6 are the only carbons of flavonoids to attach sugars via C–C bond, i.e. without preliminary hydroxylation. Among the non-glycosylated flavones, tricine (Fig. 4.6b) is highly characteristic of Poaceae; it derives from the 3′, 5′-methylation of tricetin (Fig. 4.6a).
- Caryophyllaceae (dicotyledons) were also characterized by production of 6-C- and 8-C-glycosyl flavones (Jakimiuk et al. 2022). Such a phenolic pattern is also found in Poaceae (monocotyledons) and attributes to these two families high chemotaxonomical values within the plant world (Fig. 4.6a). Chemotaxonomical differentiation between the two plant families involves further metabolic classes (e.g terpenes): Caryophyllaceae were strongly characterized by original triterpene patterns based on gypsogenic acid, 16-OH-gypsogenic acid, quillaic acid and gypsogenin (Fig. 5.19) (Cheikh-Ali et al. 2019).
- Rutaceae (dicotyledons) include plants of *Citrus* genus. They are chemically characterized by the production of flavanones with different glycosylation forms (Fig. 4.6d–f).
- Lamiaceae (dicotoyledons) include aromatic plants with various fragrances. They show the production of flavones with methoxylation and 7-glycosylation aspects (Fig. 4.6g, h) (Frezza et al. 2021); the 7-O-glycosylation can be followed by acylation which provides an additional way of structural diversification (Fig. 4.6h) (Frezza et al. 2019).
- Asteraceae (dictotyledons) represents a wide family of flowering plants. These plants essentially produce flavonols and flavones (both 87% of flavonoid pool) with relatively higher trend for flavonols (52%) (Fig. 4.7a) (Emerenciano et al. 2001). The remaining (minor) part of flavonoid pool (13%) is concerned with different aglycone types including aurones, chalcones and flavanones. Aurones and chalcones were revealed to occur relatively more in Heliantheae tribe (Fig. 4.7a). Flavones and flavonols showed positive correlation indicating global positive trend in their biosynthesis at Asteraceae family scale (Fig. 4.7b). However, some tribes of Asteraceae tend to produce more flavones than flavonols, including Lactuceae and Cardueae (Fig. 4.7b, c). This was indicated by a ratio of produced flavones to flavonols > 1 (Fig. 4.7c).

Flavonols have often methoxylated aspects due to multiple initial hydroxylations (Fig. 4.6j). Quercetagetin is a highly hydroxylated flavonol and was found in different genera of Asteraceae including *Tagetes* (particularly) (Fig. 4.6i) (Abid and Qaiser 2003; Wollenweber et al. 1997).

- Fabaceae (dicotyledons) are also characterized by production of flavonols but with relatively high glycosylation levels (Fig. 4.8a). Moreover, Fabaceae are characterized by the production of isoflavonoids which show relatively restricted distribution in the plant world. By this way, isoflavonoids provide strong chemotaxonomical fingerprints to the Fabaceae family. Isoflavonoids of Fabaceae show highly diversified structures including isoflavones (Fig. 4.8b), isoflavanones (Fig. 4.8c), rotenoids (Fig. 4.8d), pterocarpans (Fig. 4.8e), isoflavans (Fig. 4.8f), isoflavanquinones (Fig. 4.8g), isoflav-3enes (Fig. 4.8h) and coumestans (Fig. 4.8i). Structural diversity within the isoflavonoid class is increased by several metabolic processes including prenylation, glycosylation, dimerization, etc. (Fig. 4.8j–l).

When a given flavonoid aglycone shows a wide botanical distribution, its multiple glycosylation ways provide structural diversification making plants of different families to be distinguished. For instance, quercetin (flavonol) shows different characteristic glycosylation pathways in different plant species affiliated to some plant families (Fig. 4.9). Thus, in onion (*Allium cepa*, Liliaceae) (monocotyledon), quercetin shows characteristic monoglucosylation structure at C4′ in addition to the diglucosides at 3,4′ and 7,4′ positions (Fig. 4.9a–c) (Kwak et al. 2017). However, other quercetin

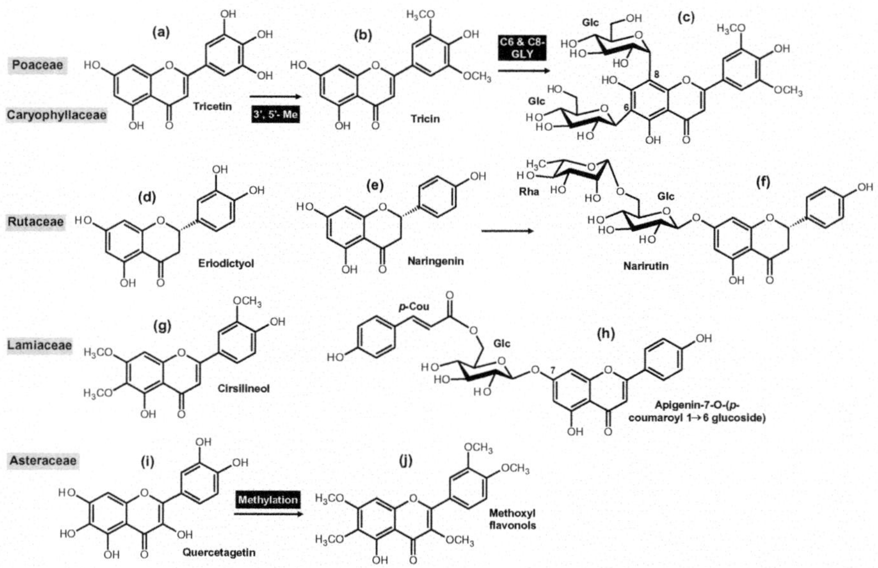

Fig. 4.6 Overview on associations between different flavonoid types and plant families

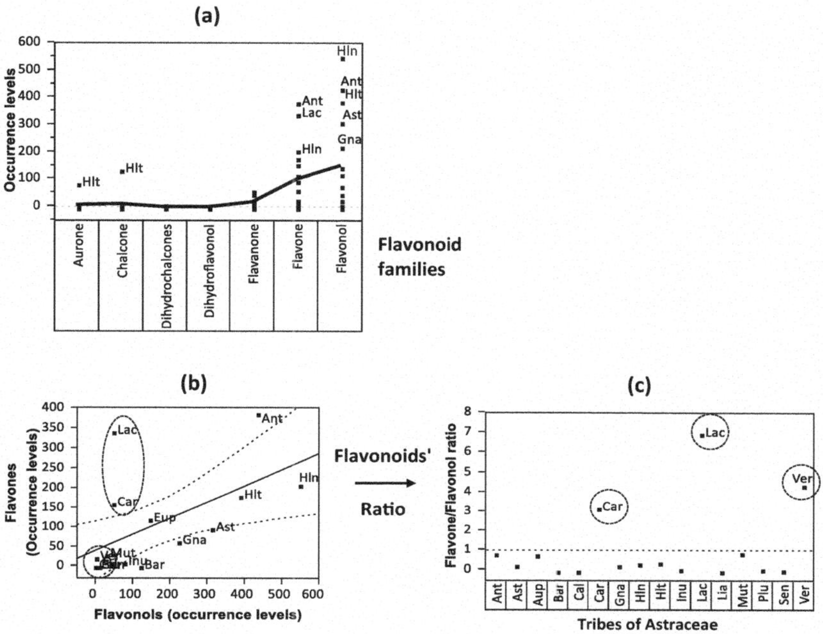

Fig. 4.7 Overview on the flavonoid pool in different tribes of Asteraceae family. (**a**) Occurrence levels of different flavonoid families; (**b**) occurrence levels of flavones; (**c**) regulation ratios between flavone and flavonol families (the most synthesized flavonoids). Legend: Anthemideae (Ant), Astereae (Ast), Barnadesieae (Bar), Calenduleae (Cal), Cardueae (Car), Eupatorieae (Eup), Gnaphalieae (Gna), Helenieae (Hln), Heliantheae (Hlt), Inuleae (Inu), Lactuceae (Lac), Liabeae (Lia), Mutiseae (Mut), Plucheeae (Plu), Senecioneae (Sen), Vernonieae (Ver)

glycosides such as quercitrin, rutin, hyperoside or avicularin are more common in the plant world (Fig. 4.9d–g).

4.2.2 Phenolic Differentiations at Plant Generic and Specific Levels

Within a same plant genus, different plant species can be chemically distinguished by some specific metabolites or more generally by different production and regulation levels of same metabolites; Production is quantified by amount which is expressed as micro-quantity of SM per dry mass of plant ($\mu g\ g^{-1}$). Two illustrative examples on phenolic patterns are given for the genera *Vaccinium* and *Aronia* (Ericaceae), and *Citrus* (Rutaceae) (Figs. 4.10, 4.11, 4.12, 4.13, 4.14, 4.15, 4.16 and 4.17).

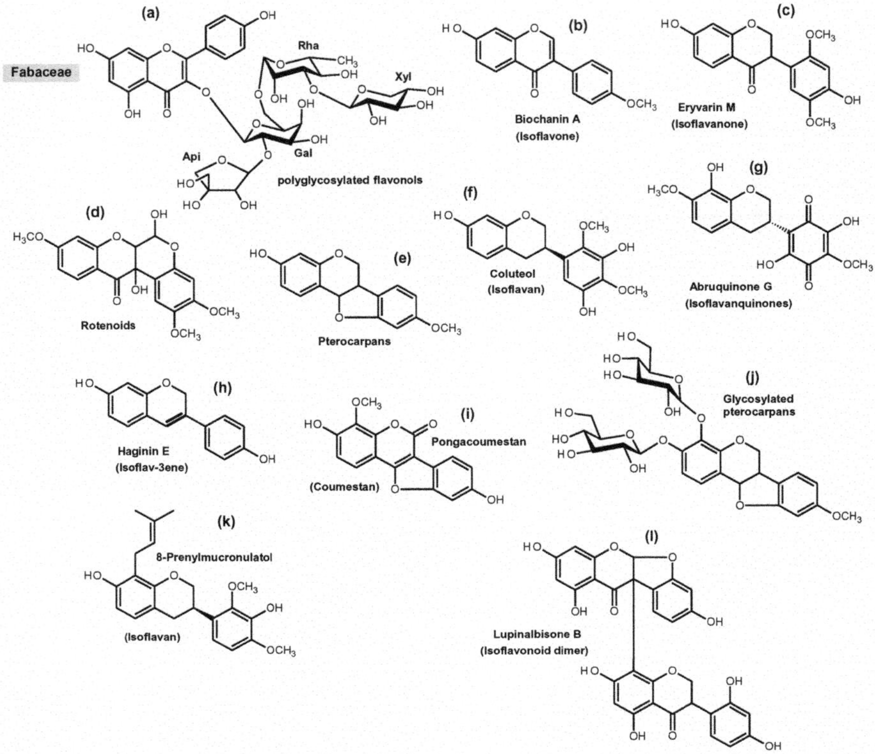

Fig. 4.8 Flavonoid structures representing phenolic patterns of Fabaceae family

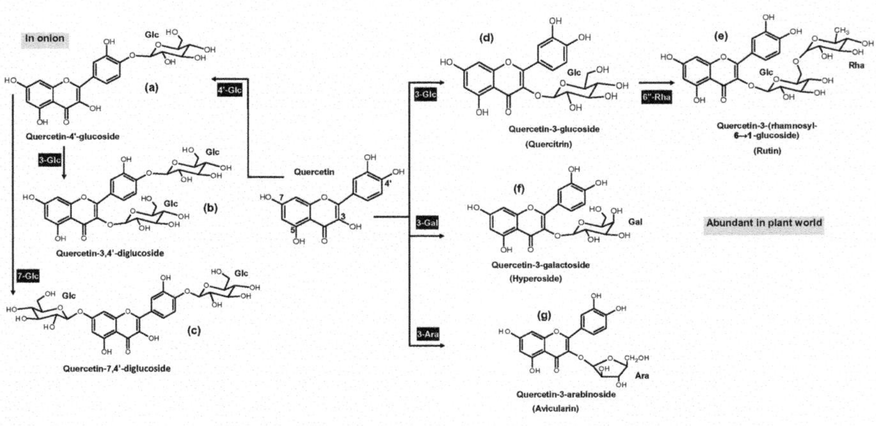

Fig. 4.9 Different glycosylation forms of quercetin showing more distributed or more target occurrences in plant world

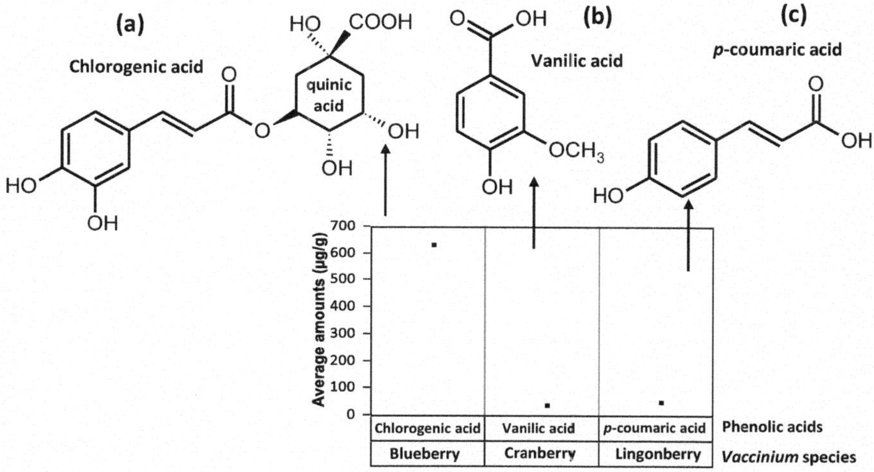

Fig. 4.10 Some phenolic acids specifically found in different *Vaccinium* species (Ericaceae)

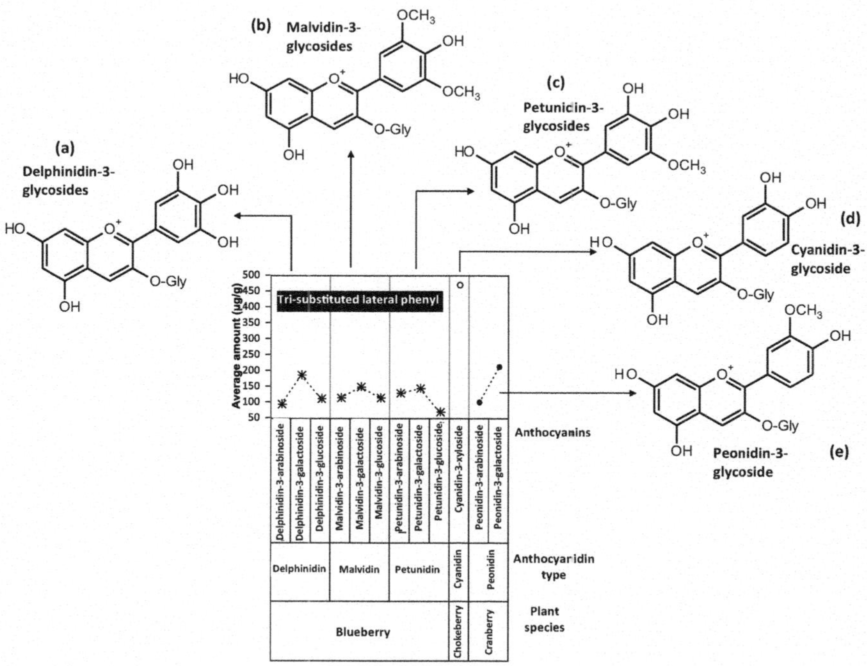

Fig. 4.11 Different anthocyanins (**a–e**) specifically found in the fruits of blueberry, cranberry (*Vaccinium*) and chokeberry (*Aronia*) species

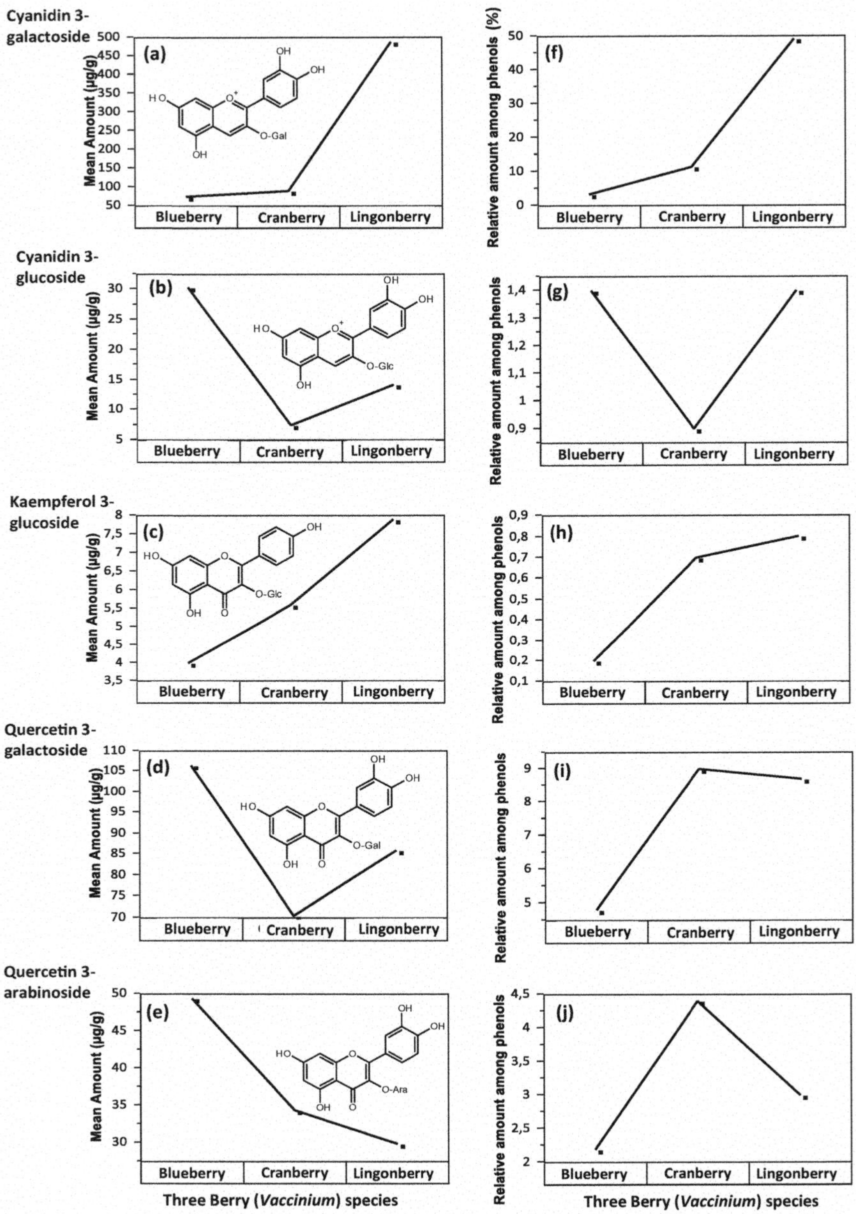

Fig. 4.12 Variations of mean amounts (**a–e**) and associated metabolic regulation levels (**f-j**) of different anthocyanins and glycoside flavonols synthesized by three *Vaccinium* species: *V. corymbosum* (blueberry), *V. macrocarpon* (cranberry) and *V. idis-idaea* (lingonberry)

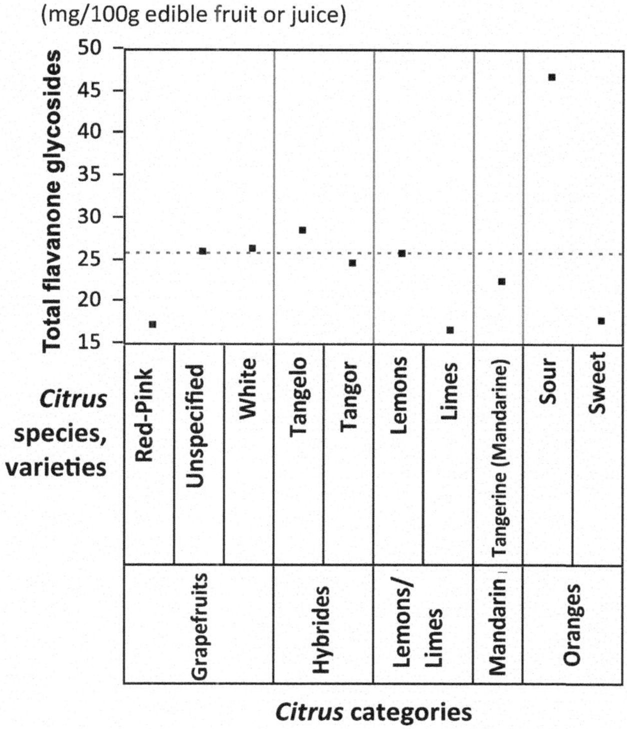

Fig. 4.13 Total contents of flavanone glycosides in different *Citrus* species and varieties

Chemical differentiation of Vaccinium species. *Vaccinium* (Ericaceae) includes plants with blue-red fruits due to high production of anthocyanins. Flavonoid glycosides were analyzed in the fruits of three *Vaccinium* species including *V. corymbosum* (blueberry), *V. macrocarpon* (cranberry) and *V. idis-idaea* (lingonberry) (Zheng and Wang 2003). A fourth species, *Aronia melanocarpa* (chokeberry), was also considered for intergeneric comparison. Phytochemical analysis of fruits revealed the production of specific anthocyanidins and phenolic acids in different species (Figs. 4.10 and 4.11). Further metabolic differentiations were quantitatively provided by several glycosides shared by different species but showing different production and regulation levels (Fig. 4.12). The three *Vaccinium* species (blueberry, cranberry and lingonberry) were distinguished by the production of chlorogenic acid, vanillic acid and *p*-coumaric acid in their fruits, respectively (Fig. 4.10). These three acids show different structural complexities with C6C1-scaffold for vanillic acid (Fig. 4.10b), C6C3 for *p*-coumaric acid (Fig. 4.10c) and esterified caffeic acid with quinic acid (giving chlorogenic acid) (Fig. 4.10a). Under quantitative aspect, blueberry species

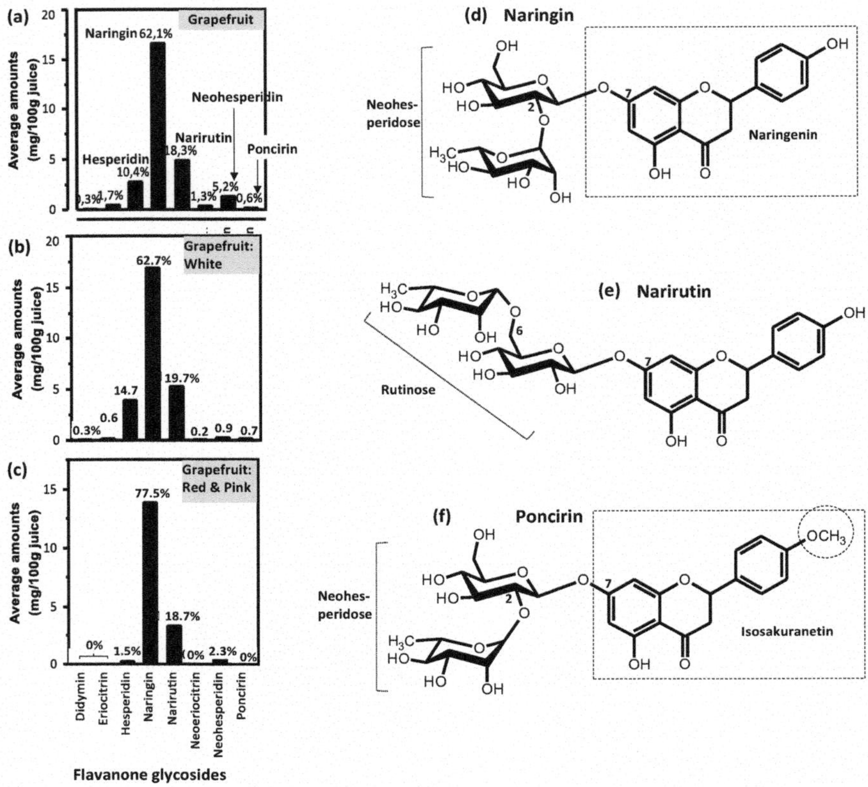

Fig. 4.14 Flavanone glycoside patterns of grapefruits (**a**–**c**) and chemical structures of characteristic compounds including naringin (**d**), narirutin (**e**) and poncirin (**f**)

was distinguished by high production of chlorogenic acid, whereas the productions of the two other specific acids in the other species were frankly lower (Fig. 4.10).

Blueberry species was also distinguished by the production of different anthocyanins based on delphinidin (Fig. 4.11a), malvidin (Fig. 4.11b) and petunidin (Fig. 4.11c) in fruits. These metabolites are based on anthocyanidin that are trisubstituted on lateral phenyl (B-ring). Each of these anthocyanidins showed 3-glycosylation by glucose, galactose or arabinose. Galactosyls showed higher amounts than glucosyls or arabinosyls. However, cranberry and chokeberry showed specific anthocyanins based on disubstituted B-ring (Fig. 4.11d, e). Chokeberry specifically produced cyanidin-3-xyloside (Fig. 4.11d); this showed well-distinct glycosylation (3-Xyl) in anthocyanins of *Aronia* species compared with *Vaccinium* species.

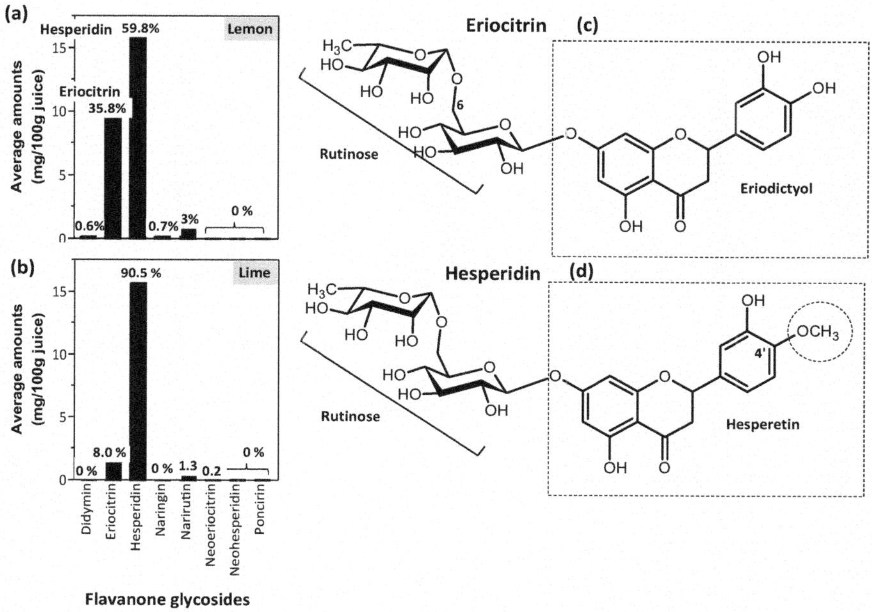

Fig. 4.15 Flavanone glycoside patterns of lemons (**a**, **b**) and chemical structures of characteristic compounds including eriocitrin (**c**) and hesperidin (**d**)

However, cranberry was distinguished by the synthesis of 3-galactosyl and 3-arabinosyl of peonidin (Fig. 4.11e) with higher production level of peonidin-3-galactoside. Finally, although lingonberry was not characterized by specific anthocyanin, it was distinguished by high production and/or regulation levels for some flavonoid glycosides found in the three species (Fig. 4.12).

Further chemical differentiation between the three *Vaccinium* species was provided by production levels and regulation ratios of other glycosides synthesized by all the species including cyanidin, kaempferol and quercetin glycosides (Fig. 4.12): cyanidin-3-galactoside showed high amounts in lingonberry compared with the two other berry species (Fig. 4.12a). These amounts (> 450 $\mu g\ g^{-1}$) exceeded those of all the metabolites analyzed in the three plant species. Moreover, this production level associated with this single glycoside represented 50% of the analyzed phenolic pool (Fig. 4.12f). These results showed that *V. idis-idaea* tends to be specialized in the production of cyanidin-3-galactoside. Moreover, this species showed relatively higher production of kaempferol-3-glucoside compared with the two other species (Fig. 4.12c). Although the amounts of this metabolite were generally low, their associated ratios revealed a metabolic regulation four times higher in lingonberry than blueberry (Fig. 4.12h). Blueberry was characterized by relatively high levels of cyanidin-3-glucoside at both synthesis and regulation aspects (Fig. 4.12b, g). However, the three species showed contrasted

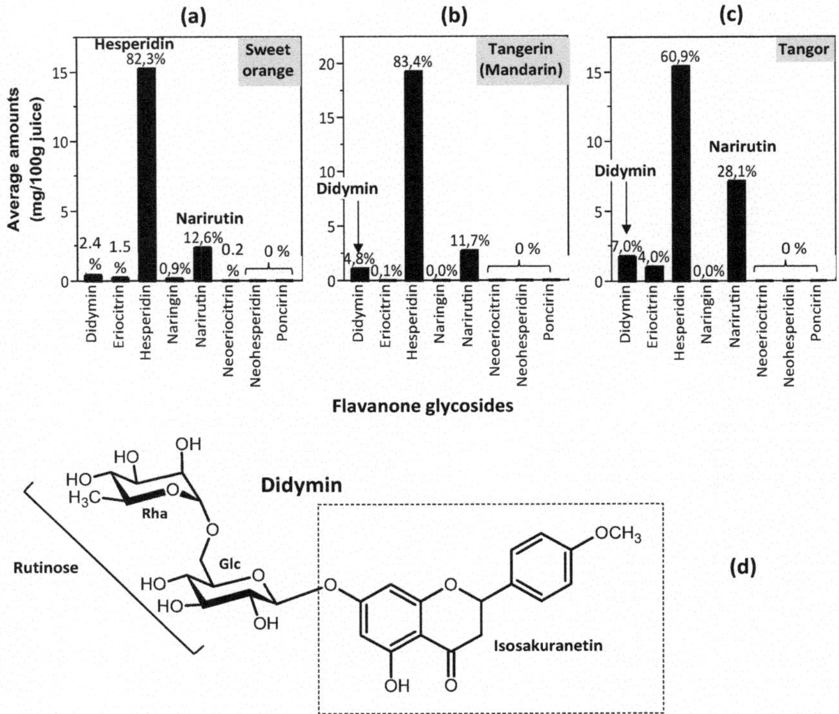

Fig. 4.16 Flavanone glycoside patterns of sweet orange (**a**), mandarin (or tangerine) (**b**) and Tangor (sweet orange × mandarin hybrid) (**c**) with chemical structures of didymin (**d**), a rare compound in *Citrus* genus found in these three species

metabolic trends concerning quercetin-3-galactoside (Fig. 4.12d, i): while cranberry showed minimal amounts of quercetin-3-galactoside (70.4 $\mu g\ g^{-1}$) (Fig. 4.12d), such a production level was associated with relatively high metabolic regulation ratio (relatively to the total amount of all the metabolites: 70.4 $\mu g\ g^{-1}$/783.3 $\mu g\ g^{-1} = 0.09$) compared with blueberry (Fig. 4.12i). However, blueberry showed a maximal production value (106.4 $\mu g\ g^{-1}$) which corresponded to a minimal regulation ratio (106.4/2210 = 0.05) when it is relativized to the whole metabolic pool (2210 $\mu g\ g^{-1}$). This contrasted aspect between production and regulation was also observed for quercetin-3-arabinoside (especially for blueberry) (Fig. 4.12e, j). These results showed the importance of regulation ratios for better evaluating the status of different metabolite in the metabolic machinery (pattern) of the considered plant species.

Chemical differentiation of Citrus species. Several *Citrus* (Rutaceae) species were chemically differentiated by flavanone glycosides (FGs) in the fruits (Peterson et al. 2006a, b). Grapefruits, oranges, lemons, mandarins and hybrids showed different flavanone patterns (Figs. 4.13, 4.14, 4.15, 4.16 and 4.17). Eight FGs were analyzed in

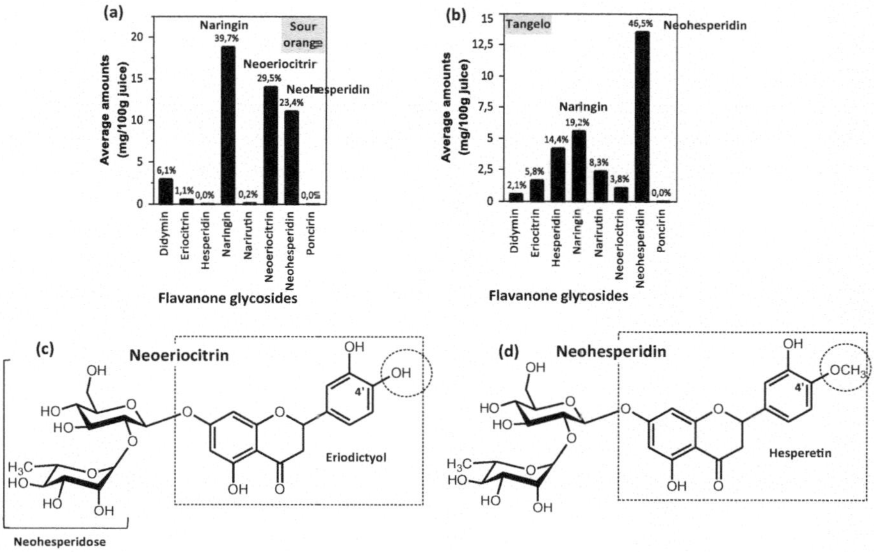

Fig. 4.17 Flavanone glycoside patterns of sour orange (**a**) and Tangelo (grapefruit × mandarin hybrid) (**c**) with chemical structures of characteristic compounds including neoeriocitrin (**c**) and neohesperidin (**d**)

the fruits consisting of naringenin, eriodictyol, hesperetin and isosakuranetin substituted by rutinose or neohesperidose at C7 (Fig. 4.14, 4.15, 4.16 and 4.17). The different *Citrus* showed strong chemical differentiations based on different average production levels and regulation ratios of the eight FGs. Under global production aspect, total flavanone contents were the highest in the sour orange (*Citrus aurantium*) (47.4 mg/100 g edible fruit) (without rind, pith and seeds) (Fig. 4.13). More generally, total flavanone contents (TFC) provided well-contrasted separations between *Citrus* species of same categories (Fig. 4.13): thus, sour orange and sweet orange (*C. sinensis*) showed high and low (47.4 vs 17.9 mg/100 g) TFC, respectively; moreover, lemon (yellow lemon, *C. limon*) and lime (green lemon, *C. aurantifolia*) were well differentiated by moderate (26.4 mg/100 g) and low (17.3 mg/100 g) TFC, respectively; also, the white and red-pink varieties of grapefruit (*C. X paradisi*) showed moderate and low TFC, respectively.

Beyond TFC, individual FGs were more or less highly synthesized and regulated leading to specific phytochemical patterns of different *Citrus* species (Figs. 4.14, 4.15, 4.16 and 4.17): different varieties of grapefruits (*C. X paradisi*) were strongly characterized by high regulations of naringin (> 60% TFC) combined with moderate ratios of narirutin (> 18%) (Fig. 4.14). Interestingly, poncirin was quasi-absent in the studied citrus species except the grapefruits where it was detected at negligible (non null) levels associated with regulation ratios < 1% (Fig. 4.14a, b). At grapefruit varietal level, this minor FG showed occurrence in white grapefruit versus absence in red-pink variety. White and red-pink grapefruits were further chemically differentiated by hesperidin which showed moderate regulation levels in white (14.7% of TFC) versus low levels in red-pink variety (1.5% TFC).

Lemon (*Citrus limon*) and lime (*C. aurantifolia*) species were chemically differentiated by FG patterns combining highly regulated hesperidin with moderately or lowly regulated eriocitrin (Fig. 4.15c, d): Lime showed the simplest pattern which was dominated by the hesperidin (Fig. 4.15d) representing more than 90% of TFC (Fig. 4.15b). The remaining FGs (10%) were mainly concerned with eriocitrin (Fig. 4.15c) which showed a regulation ratio of 8% (Fig. 4.15b). However, although hesperidin and eriocitrin were dominant in FG pattern of *C. limon* (> 95% TFC), their metabolic ratios were less contrasted (more balanced) than in *C. aurantifolia* (Fig. 4.15a vs. Fig. 4.15b): hesperidin and eriocitrin showed regulation ratios of 59.8% and 35.8% (balanced pattern), respectively in *C. limon* (vs. 90.5% and 8% in *C. aurantifolia*) (contrasted pattern).

Sweet orange (*Citrus sinensis*), mandarin (*C. reticulata*) and their hybrid (Tangor) were distinguished from the other citrus by high regulation peak of hesperidin combined with moderate level of narirutin (Fig. 4.16a–c). Tangor was differentiated from its two mother species by higher narirutin regulation (28.1% of TFC) at the expense of hesperidin (60.9%) (Fig. 4.16c). However, sweet orange and mandarin whish showed similar ratios of hesperidin (82–84%) and narirutin (11–13%) tend to be well differentiated by didymin despite its low regulation ratios: relatively low metabolic ratio of didymin in mandarin (4.8% TFC) (Fig. 4.16b) was revealed to be double of that in sweet orange (2.4%) (Fig. 4.16a). Moreover, despite its minor level, didymin provided additional differentiation aspect to Tangor because of its higher regulation ratio (7% TFC) (three once higher than that of sweet orange).

Finally, the three most contributive FG (narirutin, hesperidin, didymin) to phytochemical patterns of sweet orange, mandarin and Tangor (hybrid) were concerned with 7-rutinosyl substitution (Figs. 4.14e, 4.15d, 4.16d).

Sour orange (*Citrus aurantium*) was revealed to have frankly different FG pattern (Fig. 4.17a) from that of sweet orange (*C. sinensis*) (Fig. 4.16a). FG pattern of sour orange combined moderate regulation ratios of neoeriocitrin and neohesperidin (Fig. 4.17a, c, d) in presence of relatively higher regulation of naringin. By opposition to sweet orange (and mandarin and Tangor), the three characteristic compounds of sour orange were associated with 7-glycosylated flavanones by neohesperidose (rather than rutinose). However, didymin synthesis occurred in sour orange with regulation ratio of 6.1%. By its occurrence/absence, this minor compound tends to chemically separate two sets of citrus species including orange (sour and sweet), mandarin and Tangor (sensitive synthesis of didymin) on one hand versus lemons and grapefruits (negligible or absent synthesis of didymin) on other hand.

Finally, Tangelo (hybrid citrus between mandarin and grapefruit) showed the most complex FG pattern because of gradual and moderate regulations of quasi-all the compounds except poncirin (Fig. 4.17b). These multiple co-occurrences of FGs attribute to Tangelo a specific pattern based on higher phytochemical diversity. Moreover, this pattern was differentiated by the highest regulation ratio of neohesperidin

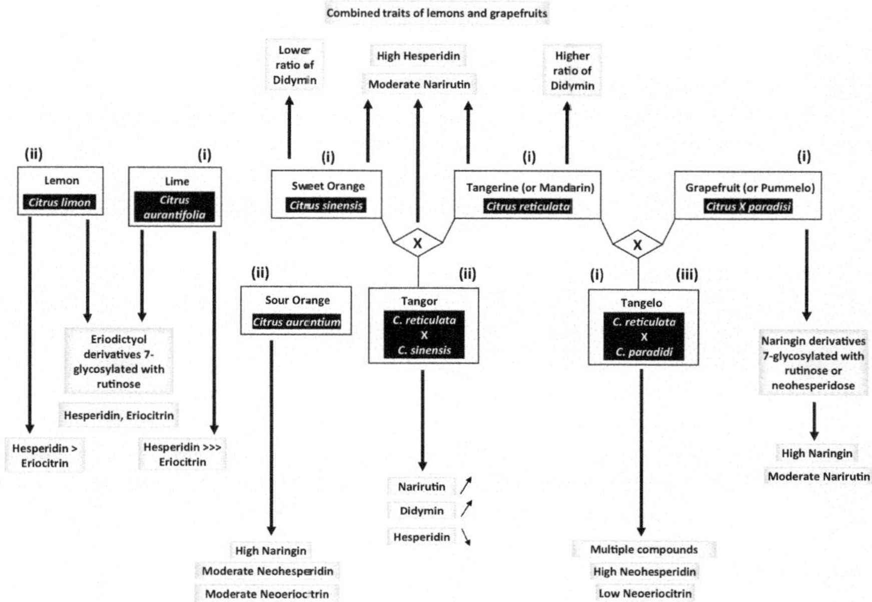

Fig. 4.18 Synthesis diagram showing characteristic flavanone glycosides in different *Citrus* species and varieties. Characteristic patterns involve relatively high (low) regulation ratios of some compounds leading to (**i**) a dominant compound, (**ii**) co-occurrence of some specific compounds, (**iii**) multiple compounds with differential regulation ratios

(46.5% TFC); however, neoeriocitrin showed low regulation levels by opposition to its moderate levels in sour orange (Fig. 4.17b vs. Fig. 4.17a).

A synthesis diagram of differentiation metabolic ways of different *Citrus* species and varieties is given in Fig. 4.18.

4.2.3 Phenolic Differentiations at Intra-specific Level

Within a same plant species, different varieties can be chemically characterized by different amounts and regulation levels of same metabolites. Further information can be provided by analyzing metabolites in different plant parts (organs, tissues, layers, etc.). An illustrative case is provided by analysis of quercetin glycosides in four Californian varieties of onion (*Alium cepa*): (i) *Chief*, (ii) *Denali*, (iii) *Sequoia* and (iv) *Cowboy* (Lee and Mitchell 2011). Flavonol compositions of onion bulb were analyzed by considering four bulb layers: external paper or skin (layer 0), outer layer (labeled 1), second layer (2) and inner layer. Five structural forms of quercetin were identified in onion bulb: free quercetin, quercetin-3-glucoside, 4′-glucoside, 3,4′-glucoside (Fig. 4.9a, b, d) and isorhamnetin 4′-glucoside (Fig. 4.20); isorhamnetin corresponds to 3′-methyl quercetin.

Flavonol amounts showed decreasing trends with the depth of onion bulb; maximal production levels were observed in the outer layer (labeled 1) versus minimal levels in inner layer (Fig. 4.19a–e). Higher flavonol contents in the outer layer could be linked to its direct exposition to environment conditions (oxidative stresses). This decreasing trend was relatively shared by the thin external layer (bulb-covering paper, labeled 0) concerning some flavonols including quercetin aglycon, quercetin 4′-O-glucoside and isorhamnetin 4′-O-glucoside (Fig. 4.19a, d, e).

Chief variety showed higher production levels (amounts) of flavonols than the three other varieties concerning the three first bulbe layers (0, 1, 2) (Fig. 4.19a–e). On this basis, it can be qualified as the most productive onion variety among the four studied ones. Among the five flavonols in *Chief* variety, quercetin 4′-O-glucoside (Q-4′-*Glc*) showed the highest levels of both production and regulation in the outer layer (2500 mg/100 g DW) (Fig. 4.19d, i): the high average amount (Fig. 4.19d) corresponded to a regulation ratio of 0.65 (65% of the flavonol pool) in the outer layer (bulb layer 1) (Fig. 4.19i). Regulation ratios of Q-4′-*Glc* in the other bulb layers were also higher in *Chief* variety than the three other onion varieties. For the other flavonols, although they had amount peaks in *Chief* variety, they were not the best regulated in this variety compared with other onion varieties:

Denali variety was characterized by the highest metabolic regulation of isorhamnetin 4′-*O*-glucoside (Iso-4′-*Glc*) (Fig. 4.19f), quercetin 3,4′-*O*-diglucoside (Q-3,4′-*Glc*) (Fig. 4.19g) and quercetin-3-O-glucoside (Q-3-*Glc*) (Fig. 4.19h). These regulation peaks in *Denali* variety were revealed by maximal or high relative amounts (regulation ratios) (Fig. 4.19f–h) whereas their associated amounts were intermediate with less distinguished aspects in the studied population (Fig. 4.19a–c). This shows the importance of working with regulation ratios for identifying and characterizing metabolic patterns of different plant varieties (tissues). Other metabolic trends characterizing onion varieties were highlighted by Iso-4′-*Glc* and Q-3,4′-*Glc* (Fig. 4.19f, g): in *Denali* and *Chief*, (i) regulation ratios of Iso-4′-*Glc* decreased from onion paper (layer 0) to inner bulb layer (Fig. 4.19f), whereas (ii) regulation peak of Q-3,4′-*Glc* showed opposite trend by increasing with the bulb depth (Fig. 4.19g). Such a negative trend between Q-3,4′-*Glc* and Iso-4′-*Glc* in different onion layers (Fig. 4.20a, b) could be indicative of metabolic competition between quercetin 3-*O*-glucoside and isorhamnetin aglycone for 4′-glucosylation. Apart from the metabolic regulation peaks of Q-3,4′-*Glc* and Iso-4′-*Glc* (Fig. 4.19f, g), *Denali* variety was also characterized by minimal regulations for quercetin 4′-*O*-glucoside (Q-4′-*Glc*) (Fig. 4.19i). Moreover, Q-3,4′-*Glc* and Q-4′-*Glc* regulations showed opposite variation through the four bulb layers in the four onion varieties, with more extreme state in *Denali* variety (Fig. 4.19g vs. Fig. 4.19i) (Fig. 4.20c). This negative trend could be indicative of two competing metabolic (sub)pathways where glucose tends to be fixed on the C4′ of either quercetin (giving Q-4′-*Glc*) or its derivative Q-3-*Glc* (giving Q-3,4′-*Glc*). The later

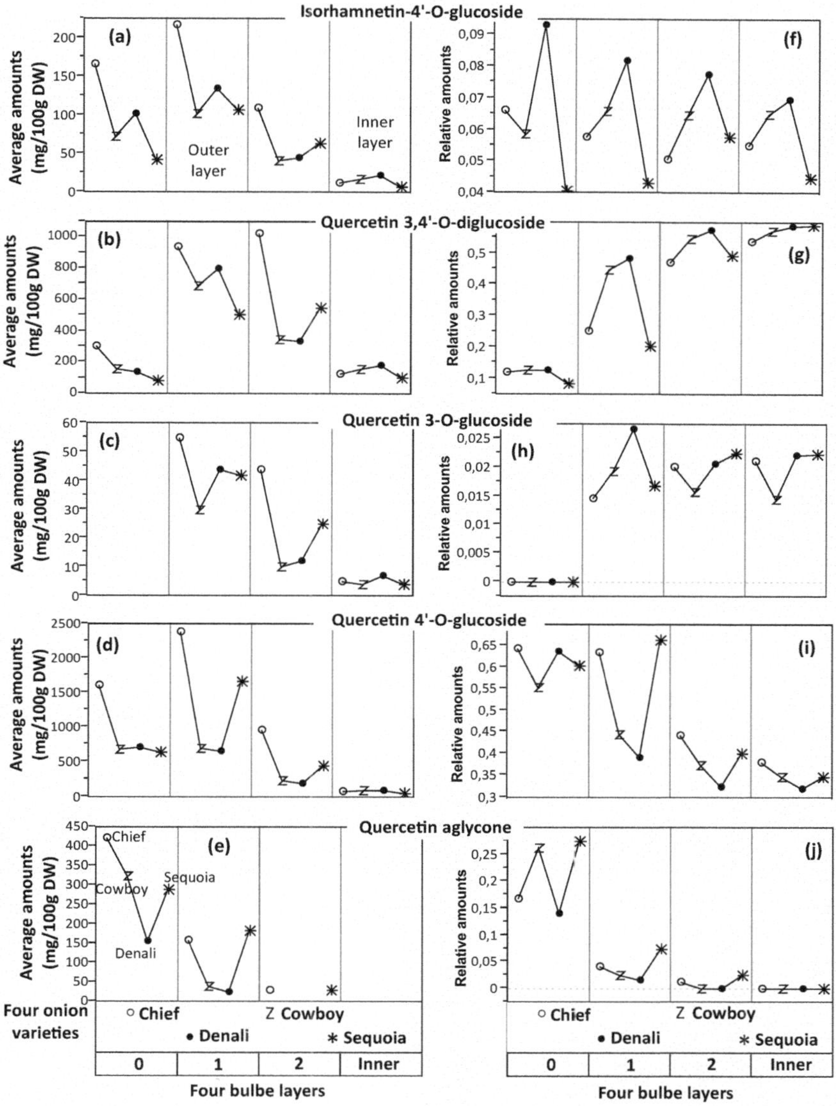

Fig. 4.19 Variations of amounts and associated regulation ratios of quercetin and derivative glucosides in different bulb layers of four Californian onion varieties of the species *Alium cepa*. Legend: 0, thin external layer; 1, outer layer; 2, mediane layer; inner, inner layer

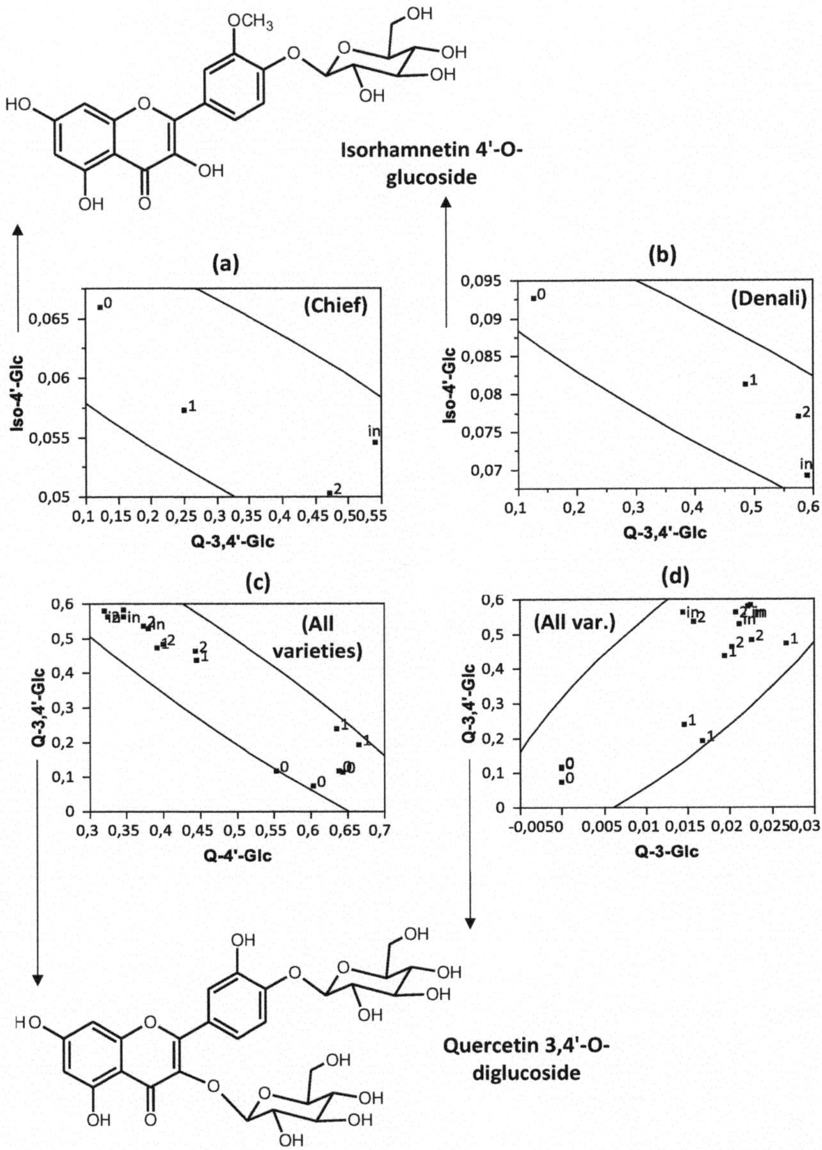

Fig. 4.20 Variation trends between regulation ratios of quercetin (Q) and derivative glucosides (Glc) showing positive or negative correlations in Californian *Allium cepa* (onion). Labelled points corresponded to different analyzed onion bulb layers: 0, external or paper layer; 1 outer layer; 2, second layer; in, inner or deep layer

process tends to be confirmed by positive correlation between Q-3,4′-*Glc* and Q-3-*Glc* at population scale (Fig. 4.20d). Finally, *Denali* was also characterized by minimal regulation ratios of quercetin aglycone in different bulb layers by opposition to *Sequoia* (Fig. 4.19e, j).

Sequoia variety showed relatively high amounts of quercetin aglycone in paper, outer and second layers (Fig. 4.19e); these amounts were associated with maximal regulation ratios in *Sequoia* compared with the three other onion varieties (Fig. 4.19j). Also, Sequoia showed minimal production and regulation levels of Iso-4′-*Glc* compared with the other onion varieties (Fig. 4.19a, f).

Finally, Cowboy variety showed intermediate flavonol pattern among the four studied onion varieties. It can be relatively characterized by minimal amounts and regulation ratios of Q-3-*Glc* (Fig. 4.19c, h).

References

Abid R, Qaiser M (2003) Chemotaxonomics study of *Inula* L. (S. STR.) and its allied genera (Inuleae-Compositae) from Pakistan and Kashmir. Pak J Bot 35(2):127–140

Barros J, Dixon RA (2020) Plant phenylalanine/tyrosine ammonia-lyases. Trends Plant Sci 25:66–79

Cheikh-Ali S, Farman M, Lacaille-Dubois MA, Semmar N (2019) Structural organization of saponins in caryophyllaceae. Phytochem Rev 18:405–441

Chong J, Poutaraud A, Hugueney P (2009) Metabolism and roles of stilbenes in plants. Plant Sci 177:143–155

Deng X, Bashandy H, Ainasoja M, Kontturi J, Pietiainen M, Laitinen RA, Albert VA, Valkonen JP, Elomaa P, Teeri TH (2013) Functional diversification of duplicated chalcone synthase genes in anthocyanin biosynthesis of *Gerbera hybrida*. New Phytol 201:1469–1483

Emerenciano VP, Militão JSLT, Campos CC, Romoff P, Kaplan MAC, Zambon M, Brant AJC (2001) Flavonoids as chemotaxonomic markers for Asteraceae. Biochem Syst Ecol 29(6):947–957

Frezza C, Venditti A, Serafini M, Bianco A (2019) Phytochemistry, chemotaxonomy, ethnopharmacology, and nutraceutics of Lamiaceae. In: Atta-ur-Rahman (ed) Studies in natural products chemistry, vol 62, pp 125–178

Frezza C, Venditti A, Giuliani C, Foddai S, Cianfaglione K, Maggi F, Fico G, Guiso M, Nicoletti M, Bianco A, Serafini M (2021) Occurrence of flavonoids in different Lamiaceae taxa for a preliminary study on their evolution based on phytochemistry. Biochem Syst Ecol 96:1–8

Jakimiuk K, Wink M, Tomczyk M (2022) Flavonoids of the caryophyllaceae. Phytochem Rev 21:179–218

Kwak JH, Seo JM, Kim NH, Arasu MV, Kim S, Yoon MK, Kim SJ (2017) Variation of quercetin glycoside derivatives in three onion (*Allium cepa* L.) varieties. Saudi J Biolog Sci 24:1387–1391

Lee J, Mitchell AE (2011) Quercetin and isorhamnetin glycosides in onion (*Allium cepa* L.): varietal comparison, physical distribution, coproduct evaluation, and long-term storage stability. J Agric Food Chem 59(3):857–863

Lin S, Singh RK, DA Moehninsi N (2021) R2R3-MYB transcription factors, StmiR858 and sucrose mediate potato flavonol biosynthesis. Hortic Res 8:25

Liu W, Feng Y, Yu S, Fan Z, Li X, Li J, Yin H (2021) The flavonoid biosynthesis network in plants. Int J Mol Sci 22(12824):1–18

Nabavi SM, Šamec D, Tomczyk M, Milella L, Russo D, Habtemariam S, Suntar I, Rastrelli L, Daglia M, Xiao J et al (2018) Flavonoid biosynthetic pathways in plants: versatile targets for metabolic engineering. Biotechnol Adv 38:107316

Peng W, Qin R, Li X, Zhou H (2013) Botany, phytochemistry, pharmacology, and potential application of *Polygonum cuspidatum* Sieb.et Zucc.: a review. J Ethnopharmacol 148:729–745

Peterson JJ, Dwyer JT, Beecher GR, Bhagwat SA, Gebhardt SE, Haytowitz DB, Holden JM (2006a) Flavanones in oranges, tangerines (mandarins), tangors, and tangelos: a compilation and review of the data from the analytical literature. J Food Compos Anal 19:S66–S73

Peterson JJ, Beecher GR, Bhagwat SA, Dwyer JT, Gebhardt SE, Haytowitz DB, Holden JM (2006b) Flavanones in grapefruit, lemons, and limes: a compilation and review of the data from the analytical literature. J Food Compos Anal 19:S74–S80

Semmar N (2010) Chemotaxonomical analyses of herbaceous plants based on phenolic and terpenic patterns. Flexible tools to survey biodiversity in grasslands. Nova Science Publishers, New York, p 257

Wang ZL, Wang S, Kuang Y, Hu ZM, Qiao X, Ye M (2018a) A comprehensive review on phytochemistry, pharmacology, and flavonoid biosynthesis of *Scutellaria baicalensis*. Pharm Biol 56:465–484

Wang P, Zhang L, Jiang X, Dai X, Xu L, Li T, Xing D, Li Y, Li M, Gao L et al (2018b) Evolutionary and functional characterization of leucoanthocyanidin reductases from *Camellia sinensis*. Planta 247:139–154

Wohl J, Petersen M (2020) Functional expression and characterization of cinnamic acid 4-hydroxylase from the hornwort *Anthoceros agrestis* in physcomitrella patens. Plant Cell Rep 39:597–607

Wollenweber E, Dörr M, Fritz H, Papendieck S, Yatskievych G, Roitman JN (1997) Exudate flavonoids in asteraceae from Arizona, California and Mexico. Z Naturforsch 52c:301–307

Yin YC, Zhang XD, Gao ZQ, Hu T, Liu Y (2019) The research progress of chalcone isomerase (CHI) in plants. Mol Biotechnol 61:32–52

Zhang X, Abrahan C, Colquhoun TA, Liu CJ (2017) A proteolytic regulator controlling chalcone synthase stability and flavonoid biosynthesis in *Arabidopsis*. Plant Cell 29:1157–1174

Zheng W, Wang SY (2003) Oxygen radical absorbing capacity of phenolics in blueberries, cranberries, chokeberries, and lingonberries. J Agric Food Chem 51:502–509

Chapter 5
Terpenes

5.1 General Structural Aspects

Terpenes are C5-multiple compounds synthesized by head-to-tail condensations of C5-units leading to C5 (hemiterprenes), C10 (monoterpenes), C15 (sesquiterpenes), C20 (diterpenes), C25 (sesterpenes), C30 (triterpenes), C40 (tetraterpenes) and higher sizes compounds (polyterpenes) (Fig. 5.1). The initial C5-unit precursor has 2-methylbutane structure (Fig. 5.1a); its conjugation results in the formation of 2-methyl-1,3-butadiene or isoprene representing the hemiterpene family (Fig. 5.1b). C10- to C25-terpenes are synthesized by head-to-tail condensation between lower C5-size units (Fig. 5.1c–f), whereas C30- and C40-terpenes involve tail-to-tail condensation ways (Fig. 5.1g, h). Beyond C40, polymerization of C5-units leads to the formation of polyterpenes (Fig. 5.1i) which represent plant natural rubbers.

Terpene metabolism starts with phosphorylated 2-methyl-1,3-butadienes under the form of isopentenyl diphosphate (IPP) and its allylic isomer dimethylallyl diphosphate (DMAPP) (Figs. 5.2f and 5.4a, b). IPP and DMAPP submit head-to-tail condensations by trans-prenyltransferases (TPTs) leading to the formation of several prenyl diphosphates with various chain length $(C5)_n$ serving as precursors for different terpene families (Fig. 5.4).

Initially, IPP and DMAPP are synthesized through two separated metabolic pathways (Figs. 5.2 and 5.3):

- Cytosolic mevalonic acid (MVA) pathway consisting of six steps (Fig. 5.2).
- Plastidial methylerythriol phosphate (MEP) pathway consisting of seven steps (Fig. 5.3) (Banerjee and Sharkey 2014).

The existence of two biosynthesis pathways provides strong indication on the importance of terpenoids in biological world (Holopainen et al. 2013). (i) Archae, animals and fungi have MVA pathway, (ii) green algae have MEP pathway, whereas plants and bacteria possess both pathways (Allamand et al. 2023; Lipko et al. 2023).

N. Semmar, *Secondary Metabolites in Plant Stress Adaptation*, Signaling and Communication in Plants, https://doi.org/10.1007/978-3-031-52595-7_5

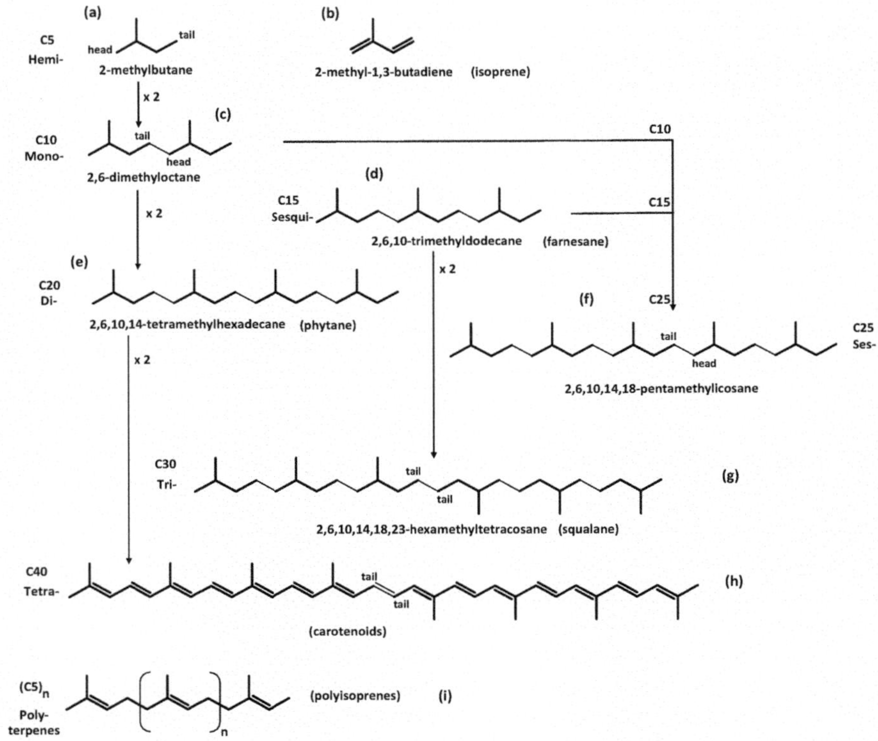

Fig. 5.1 Scaffold structures of different terpene families issued from head-to-tail (or tail-to-tail) condensation between C5-unit-based compounds

5.2 MVA and MEP Biosynthesis Pathways

MEP represents the main metabolic pathway for biosynthesis of monoterpenes (C10, diterpenes (C20 and tetraterpenes (C40) whereas MVA represents the main metabolic pathway of sesquiterpenes (C15) and triterpenes (C30) (Sawai and Saito 2011; Thimmappa et al. 2014). The genes of both pathways were revealed to have opposite sensitivity to light/dark conditions (Tholl 2015; Vranova et al. 2013; Ghassemian et al. 2006): Light tends to stimulate transcript accumulation of MEP pathway genes versus downregulation of MVA pathway genes.

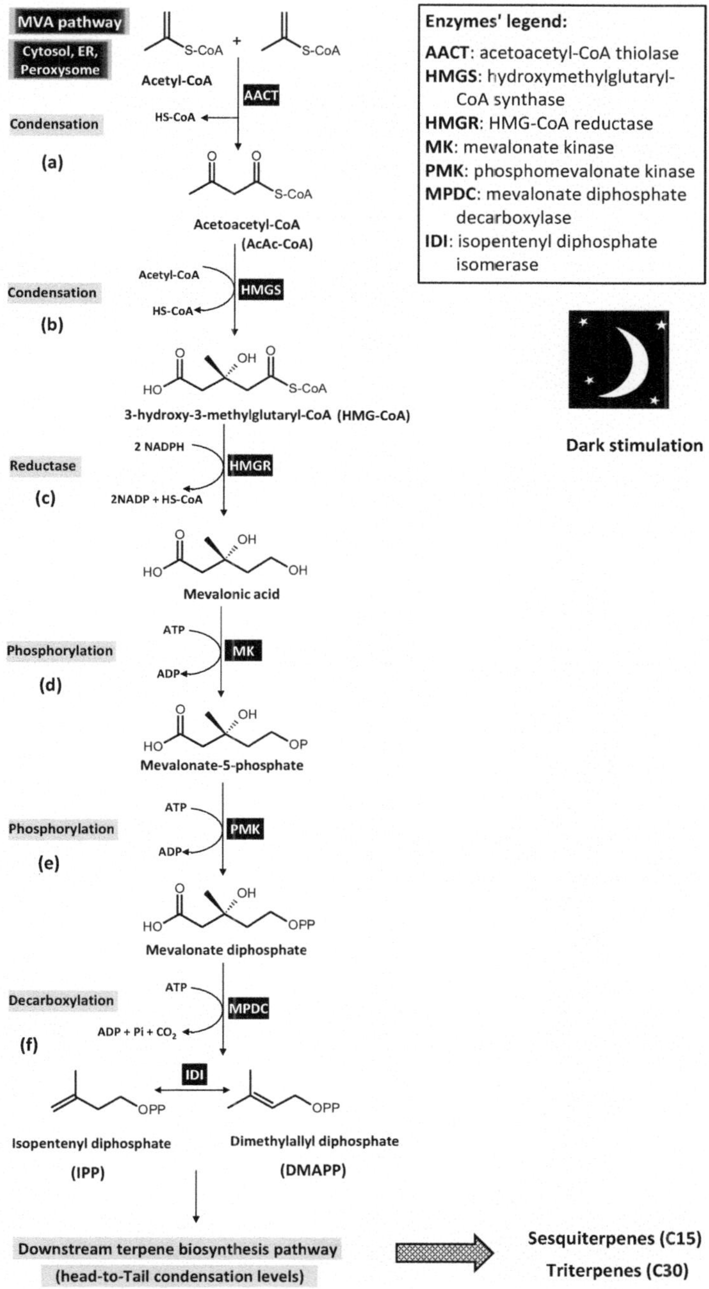

Fig. 5.2 Cytosolic mevalonate (MVA) pathway leading to the formation of isopentenyl diphosphate (IPP) and its isomer dimethylallyl diphosphate (DAMPP) (two terpene precursors)

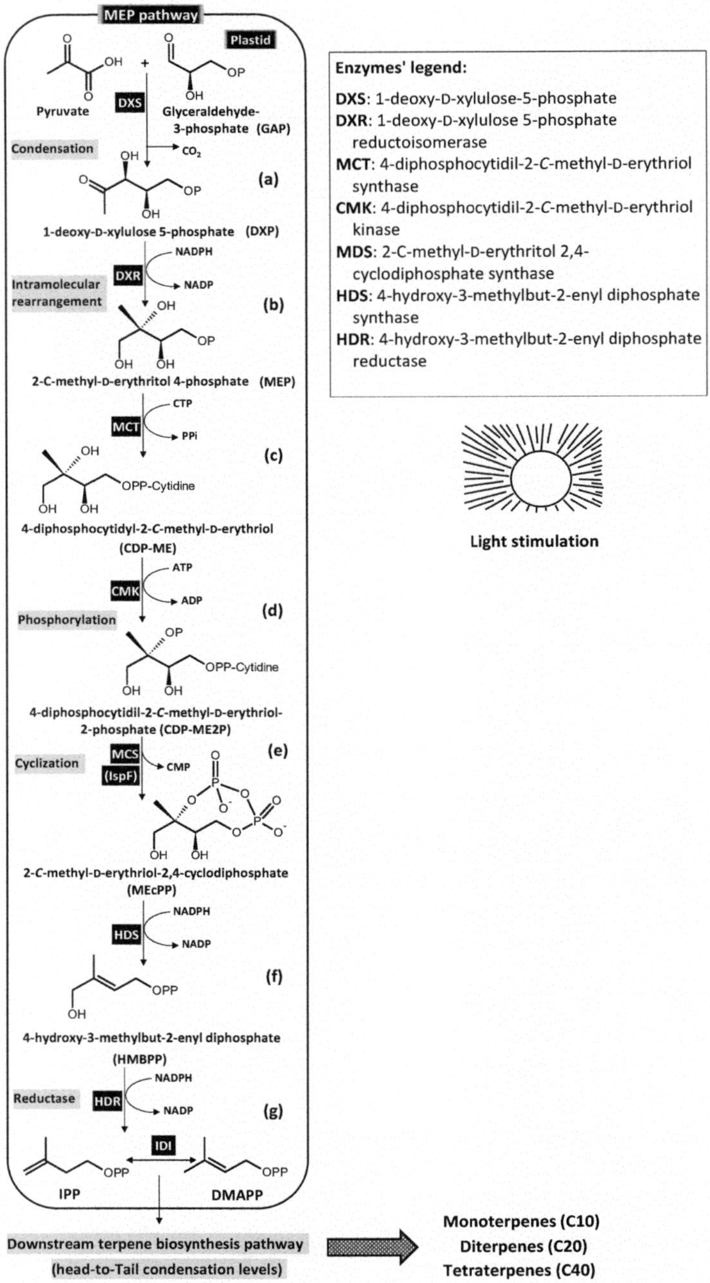

Fig. 5.3 Plastidic methylerythriol phosphate (MEP) pathways leading to the formation of isopentenyl diphosphate (IPP) and its isomer dimethylallyl diphosphate (DMAPP)

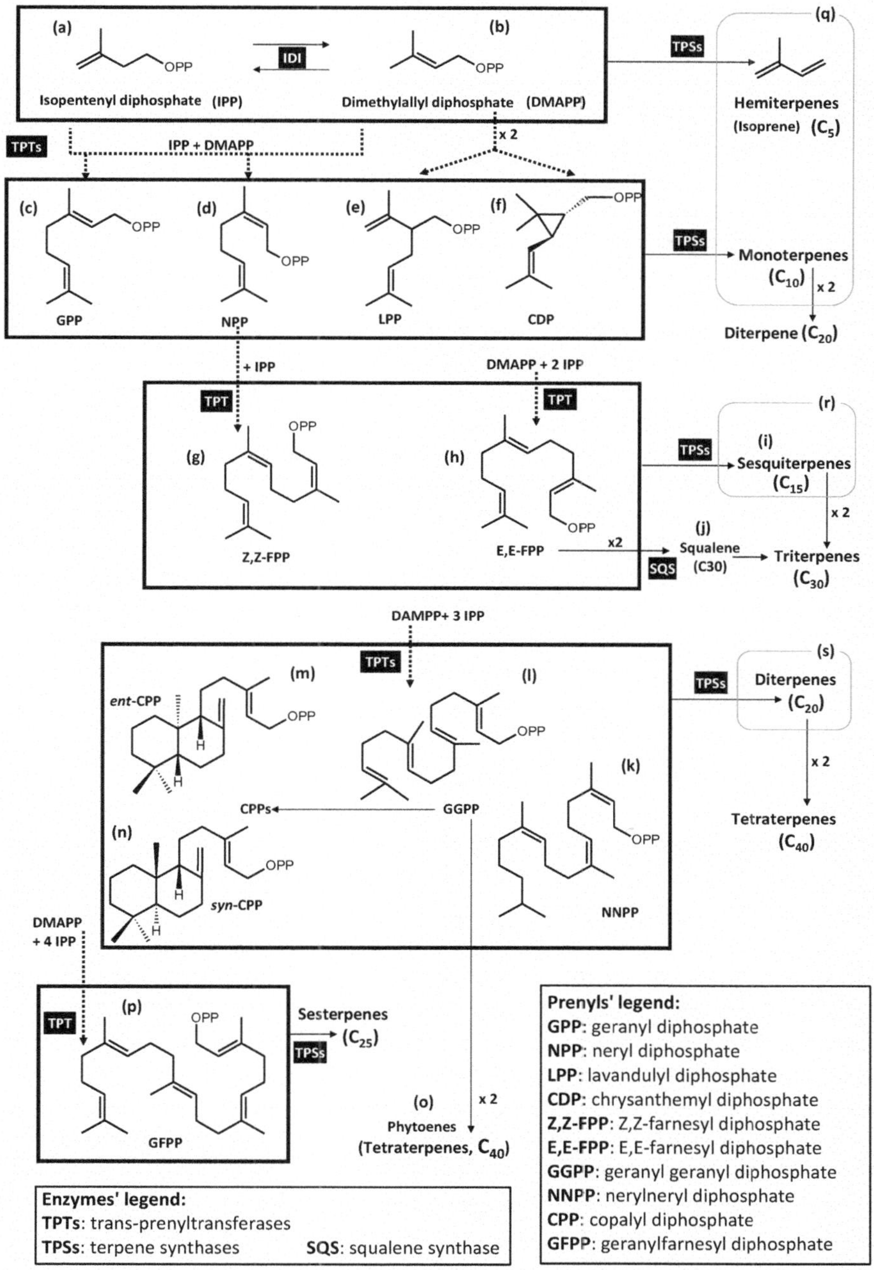

Fig. 5.4 Sequential prenylations based on tail-to-head condensation process leading to the formation of various $(C5)_n$ precursors of different terpene families (hemiterpenes C5, monoterpenes C10, sesquiterpenes C15, diterpenes C20, sesterpenes C25, triterpenes C30, tetraterpenes C40)

5.2.1 MVA pathway *(Fig. 5.2)*

MVA pathway is initiated by the condensation of two acetyl-CoA giving acetylacetyl-CoA (AcAc-CoA) (Fig. 5.2a). This process is catalyzed by a thiolase-type enzyme acetyacetyl-CoA thiolase, AACT). (2) Further condensation occurs by the introduction of a third acetyl CoA leading to the formation of 6 carbons-molecule, hydroxymethylglutaryl-CoA (HMG-CoA) (Fig. 5.2b). This step is catalyzed by hydroxy-3-methylglutaryl-CoA synthase (HMGS). (3) Therefore, HMG-CoA is converted to *R*-mevalonate by HMG-CoA reductase (HMGR) involving the reduction of two NADPH (Fig. 5.2c). HMGR is a key enzyme playing critical regulatory role in the biosynthesis of isoprenoid compounds (Re et al. 1995). HMGR activity showed flexible functional aspect associated with its topological orientations and transmembrane location favoring compartmentation of different portions of the protein. Moreover, HMGR activity is sensitive to various external stimuli leading to its important role in organism-environment interactions (Brooker and Russell 1975, 1979; Wong et al. 1982; Tholl 2015). (4) Mevalonate is subject to phosphorylation- and decarboxylation-linked transformation leading to its conversion into isopentenyl diphosphate (IPP) (Fig. 5.2d–f): the two successive phosphorylation processes occur by the enzyme effects of mevalonate kinase (MK) and phosphomevalonate kinase (PMK) leading to the formation of mevalonate-5-phosphate (Fig. 5.2d) and mevalonate diphosphate (Fig. 5.2e), respectively. Finally, mevalonate diphosphate is decarboxylated through another ATP-dependent phosphorylation step to be converted into IPP (and its isomer DMAPP). This step is catalyzed by mevalonate diphosphate decarboxylase (MPDC) (Fig. 5.2f).

5.2.2 MEP pathway *(Fig. 5.3)*

MEP pathway is initiated by the formation of 1-deoxy-D-xylulose 5-phosphate (DXP) from condensation of pyruvate with glyceraldehyde-3-phosphate (GAP) with CO_2 elimination (Fig. 5.3a). This step is catalyzed by DXP synthase (DXS) which is a thiamin diphosphate-dependent enzyme. (2) Then, DXP submits an intramolecular rearrangement to be converted into 2-C-methyl-D-erythritol 4-phosphate (MEP) (Fig. 5.3b). MEP formation is catalyzed by DXP reductoisomerase (DXR) occurring with an NADPH-dependent reduction. (3) MEP is converted to 4-diphosphocytidyl-2-C-methyl-D-erythritol (CDP-ME) by catalytic effect of 4-diphosphocytidyl-2-C-methyl-D-erythritol synthase (MCT) (Fig. 5.3c). (4) CDP-ME is phosphorylated by CDP-ME kinase (CMK) giving CDP-ME 2-phosphate (CDP-ME2P) (Fig. 5.3d) (Rohdich et al. 2000). (5) Therefore, CDP-ME2P is cyclized by 2-C-methyl-D-erythriol 2,4-cyclodiphosphate synthase (MCS) (or 2-C-methyl-D-erythritol 2,4-cyclodiphosphate synthase, IspF) to give 2-C-methyl-D-erythritol 2,4-cyclodiphosphate (MEcPP) (Fig. 5.3e). This step occurs with elimination of one cytidine monophosphate (CMP). (6) In the sixth step, MEcPP is converted into hydroxymethylbutenyl diphosphate (HMBPP) (Fig. 5.3f). This process is catalyzed by HMBPP synthase (HDS) and involves two electron-reduction using NADPH. The MEP pathway is finalized by another reduction process through which HMBPP is transformed to a mixture of IPP and DMAPP by HMBPP reductase (HDR) (Fig. 5.3g). At this step, IPP and DMAPP are isomerized by isopentenyl diphosphate isomerase (IDI).

5.3 Structural Diversification of $(C5)_n$-Terpene Precursors from Prenyl Condensation Processes

IPP and DMAPP represent the fundamental C5 units for terpene synthesis (Fig. 5.4a, b). Terpene synthesis occurs in two steps consisting of IPP and/or DMAPP condensations by transferase enzymes followed by dephosphorylation by terpene synthases (TPSs). Head-to-tail condensations between IPP and DMAPP are catalyzed by *trans*-prenyl transferases (TPTs) leading to a serial of multiple C5-structure precursors of terpenes (Fig. 5.4c–m) (Tholl 2015; Zhou and Pichersky 2020):

Several stereoisomeric C10-units are obtained by addition of two DMAPP units or one DMAPP with one IPP (Fig. 5.4c–f). These C10-compounds are precursors of monoterpenes and include geranyl diphosphate (GPP), neryl diphosphate (NPP), lavandulyl diphosphate (LPP), chrysanthemyl diphosphate (CDP) (Fig. 5.4c–f).

C15 compounds are obtained from (C10 + C5)-processes such as NPP + IPP or DMAPP + 2IPP leading to *Z*,*Z*-farnesyl diphosphate (*Z*,*Z*-FPP) and *E*,*E*-farnesyl diphosphate (*E*,*E*-FPP), respectively (Fig. 5.4g, h). These compounds are precursors of sesquiterpenes (Fig. 5.4i). Head-to-head dimerization of *E*,*E*-FPP by squalene synthase (SQS) leads to the synthesis of squalene, a C30 compound playing a key precursor role in triterpene (C30) metabolism (Fig. 5.4j).

C20-compounds with multiple configurations can be obtained from DAMPP + 3IPP (C5 + 3xC5) processes catalyzed by various TPTs. Illustrative examples include nerylneryl diphosphate (NNPP), geranylgeranyl diphosphate (GGPP), copalyl diphosphate (CPPs) (Fig. 5.4k–n). Head-to-head dimerization of GGPP by phytoene synthase leads to the synthesis of phytoene, a C40-compound playing a key precursor role in tetraterpene (C40) metabolism (Fig. 5.4o).

C25-compounds are relatively rare compared with the other $(C5)_n$ compounds. Among C25-precursors, geranylfarnesyl diphosphate (GFPP) can be formed from the condensation of one DAMPP with four IPP (Fig. 5.4p). GFPP serves as precursor for biosynthesis of sesterpenes.

Terpenes issued from *trans*- (or *E*-) prenyltransferases (TPTs) generally concern the most of plant world. However, terpenes based on *cis*- (or *Z*-) prenyl diphosphates were revealed to be particularly found in Solanaceae family leading to a strong chemotaxonomical aspect.

These different prenyl phosphate compounds serve as substrates for different terpene synthases (TPSs). TPSs (i) remove the diphosphate group and (ii) shape specific aglycones (molecular carbon skeleton) associated with different terpene families (monoterpenes, sesquiterpenes, diterpenes, triterpenes) (Fig. 5.4q–s) (Chen et al. 2011). Dissociation of the pyrophosphate anion from prenylated structures by TPSs provide carbonium (or carbenium) ions which are ready for cyclization leading to many diversified structures within a same terpene family. Aglycone sculpting by TPSs provides a first metabolic process for structural diversity of terpenes. Structural characterization of terpene aglycones are governed by several processes including cyclization (number and shape of cycles associated with different stereoisomeric aspects) combined with variable numbers and positions of double links. These structural aspects within different terpene families are illustrated in Figs. 5.5, 5.6, 5.7, 5.8, 5.9, 5.10, 5.11 and 5.12.

5.3.1 Monoterpenes

Monoterpenes include acyclic, monocyclic and bicyclic structures (Fig. 5.5); monoterpenes with a same cyclization level are differentiated by their conjugation aspect due to the number and positions of double bonds. Illustrative acyclic structures include lavandulol and geraniol which are found in *Lavandula* and *Rosa* plant genera, respectively (Fig. 5.5a, b). These hydroxylated acyclic monoterpenes have two double bonds that are concerned with different carbons. In upstream, they are derived from LPP and GPP, respectively. Monocyclic monoterpenes can be illustrated by limonene (in *Citrus*, Rutaceae) and β-phellandrene (in *Solanum*, Solanaceae) which are issued from GPP and NPP, respectively (Fig. 5.5c, d). The locations of the intra-cyclic and lateral double bonds make easy the distinction between these two structurally close aglycones. However, chrysanthemol (in Asteraceae), issued from CDP, showed original monocyclisation due to the presence of cyclopropane with lateral hydroxylation formed after the removal of diphosphate group (Fig. 5.5e). Bicyclic monoterpenes can be illustrated by δ-2-carene (in *Solanum*) and 1,8-cineol (in *Eucalyptus*, Myrtaceae) (Fig. 5.5f, g) characterized by the occurrence of cyclopropane and cyclic ether, respectively. Close bicyclic structures of monoterpenes can be illustrated by α-pinene and β-pinene (in *Pinus*, Pinaceae) that differ by the location of double bond (intracyclic in α-pinene vs. external in β-pinene) (Fig. 5.5h, i). Oxygenation (by hydroxylation, ether link, lactone group, etc.) provides an additional downstream process for structural diversification of (mono)terpenes (Fig. 5.5j, k).

A particular family of monoterpenes, called iridoids, is characterized by a cyclopentanopyran scaffold which is synthesized through an alternative cyclization of GPP (Fig. 5.5j) (Krithika et al. 2015; Miettinen et al. 2014; Sampaio-Santos and Kaplan 2001). Illustrative example is given for nepetalactol derived from geraniol (Fig. 5.5b, j). This metabolism involves a preliminary hydroxylation of geraniol leading to hydroxygeraniol; the di-hydroxylated metabolite will be transformed to a di-aldehyde (oxogeranial) by deprotonation of its OH groups; the formed aldehyde is ready for cyclization which transforms it to iridoid type monoterpene (nepetalactol in illustrative case) (Fig. 5.5j). The name "iridoid" is originated from an ant genus, *Iridomyrmex*, in which species were characterized by secreted defense compounds called iridomyrmecin, iridolactone and iridodial (Bruneton 1995; Dinda and Debnath

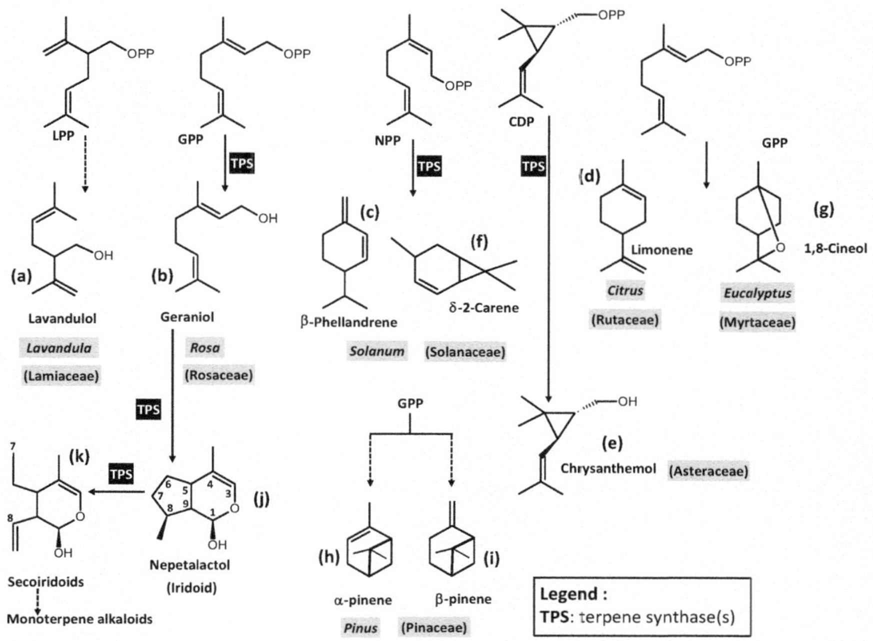

Fig. 5.5 Different chemical structures of monoterpenes illustrating metabolic diversity associated with cyclisation (type, level), double bond (number, location) and oxygenation (when occurring)

2013). In plants, iridoids were found in different families including Oleaceae, Apocyanaceae, Rubiaceae, Caprifoliaceae, Scrophulariaceae, Bigniniaceae and Lamiaceae (Dinda and Debnath 2013; Jensen et al. 2002). Oxidative cleavage at C7, C8-bond of the cyclopentane moiety of iridoids results in the formation of secoiridoids (Fig. 5.5k). Secoiridoids were found both as glycosides and as secoiridoid-indole alkaloids (O'Connor and Maresh 2005).

5.3.2 *Sesquiterpenes*

Sesquiterpenes show high structural diversity of aglycones due to their cyclisation levels and types (Fig. 5.6). Different structures are synthesized from farnesyl-derivative cations including nerolidyl, humulyl and germacradienyl cations. These different cations serve as substrates for different terpene synthases (TPSs) followed by cyclase enzymes for the biosynthesis of the great variety of sesquiterpene aglycones with different cyclization levels and conjugation aspects:

Initially, farnesyl cation (precursor of sesquiterpenes) is formed through dephosphorylation of farnesyl diphosphate (FPP) (Fig. 5.6). Farnesyl cation can be directly converted into (*E*,*E*)-β-farnesene (acyclic sesquiterpene) by deprotonation

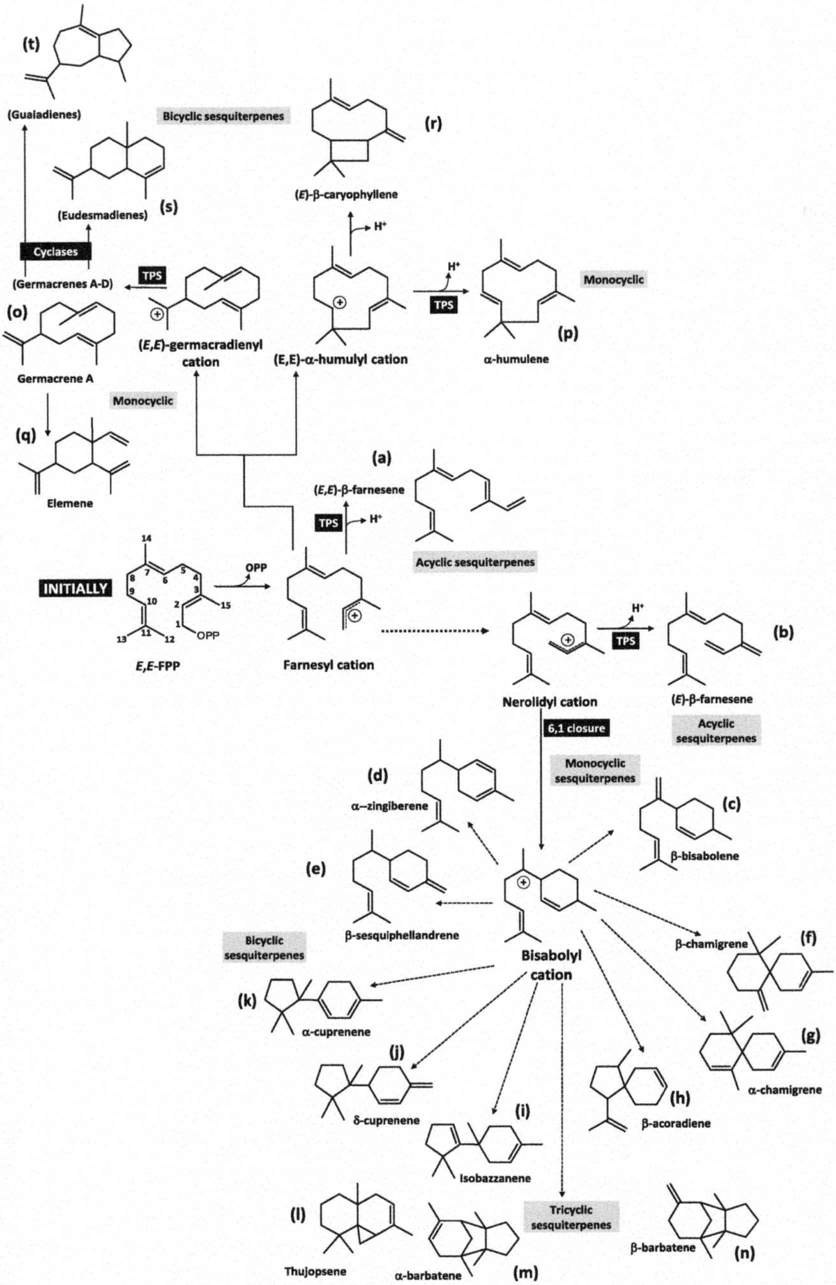

Fig. 5.6 Different aglycones of sesquiterpenes derived from farnesyl diphosphate (E,E-FPP) and showing different cyclisation types and levels. TPS, terpene synthase

and double bond rearrangement catalyzed by TPS (Fig. 5.6a). Acyclic sesquiterpenes include also (*E*)-β-farnesene issued from deprotonation of nerolidyl cation (Fig. 5.6b). (6,1)-Cyclization of nerolidyl cation provides bisabolyl cation which serves as precursors for many sesquiterpenes showing different cyclization levels (Fig. 5.6c–n): monocyclic sesquiterpenes deriving from bisabolyl cation include β-bisabolene, α-zingiberene and β-sesquiphellandrene (Fig. 5.6c–e). These metabolites based on bisabolane aglycone differ by locations of their double bond triplets (1,2; 10,11; 7,14) in (c), (2,3; 4,5; 10,11) in (d), (1,2; 3,15; 10,11) in (e). Bicyclic sesquiterpenes deriving from bisabolyl cation are various by combining cyclohexene with cyclohexane or cyclopentane (cyclopentene) (Fig. 5.6f–k): combined cyclohexene-cyclohexane structures include β-chamigrene (Fig. 5.6f); double cyclohexene structures include α-chamigrene (Fig. 5.6g); cyclohexene, cyclopentane structures include β-acoradiene, isobazzarene, δ-cuprenene, α-cuprenene (Fig. 5.6h–k). Further cyclisation can occur in bisabolyl cation resulting in the formation of tricyclic sesquiterpenes (Fig. 5.6l–n). Illustrative structures include thujopsene containing cyclohexane, cyclohexene and cyclopropane (Fig. 5.6l), and α-barbatene and β-barbatene containing cyclohexane or cyclohexene co-occurring with two cyclopentanes (Fig. 5.6m, n). These two close compounds are differentiated by the location of double bond: intra-cyclic in (m) and external in (n).

Apart from the bisabolyl pathway, other original structures of sesquiterpenes are directly derived from farnesyl cation (Fig. 5.6o–s): monocyclic sesquiterpenes deriving from farnesyl cation include germacrenes (Fig. 5.6o) and α-humulene (Fig. 5.6p) (after deprotonation). Molecular rearrangement of germacrenes leads to elemenes (Fig. 5.6q). Cyclisation of germacrenes and humulene results in the formation of eudesmenes and caryophyllene, respectively, which are bicyclic sesquiterpenes (Fig. 5.6r, s). More structural diversity of bicyclic sesquiterpenes is provided by guaiadiene structures deriving from another cyclisation way of germacrenes (Fig. 5.6t).

Guaiane-type sesquiterpenes (Fig. 5.7a) can be further diversified by tricyclisation ways including acquisition of lactone as well as closures occurring between different carbons (Fig. 5.7). Lactone cycle can be formed by closure between C12 and C6 or C8 (Fig. 5.7b–d). The occurrence of methyl (C14) on C5 rather than C4 is less common in guaiane and results in the formation of pseudoguaianolide family (Fig. 5.7d). Further guaiane families are generated from cyclization process between C11 and other carbons including C1, C6 and C10 leading to 11–1, 11–6 and 11–10-guaianes, respectively (Fig. 5.7e–g).

Finally, a part from the widely distributed *E*,*E*-FPP pathway in plant world (Fig. 5.6), sesquiterpenes can be synthesized from *Z*,*Z*-FPP in Solanaceae (Fig. 5.8a) (Zhou and Pichersky 2020). This original pathway attributes chemotaxonomical value to this plant family. Bicyclic and tricyclic derivatives are illustrated by α-bergamotene and α-santalene, respectively (Fig. 5.8b, c).

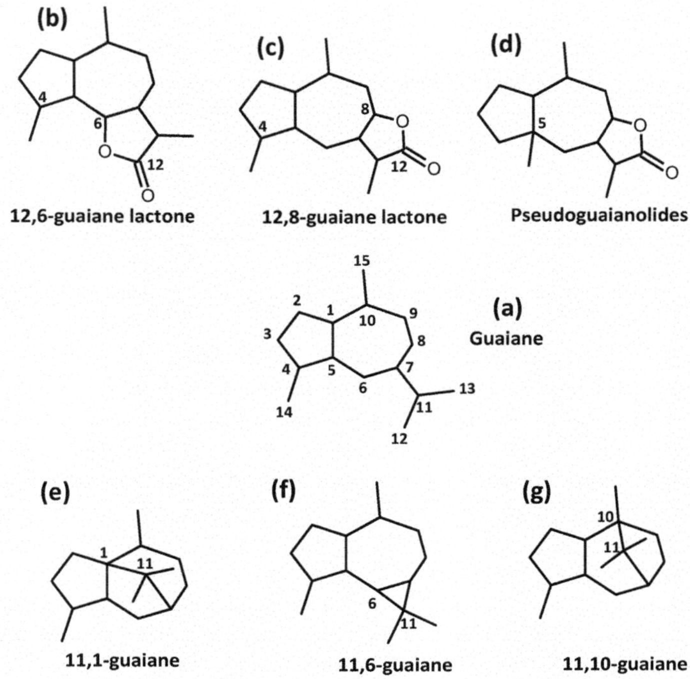

Fig. 5.7 Different aglycone structures of guaianes based on different cyclisation types and locations

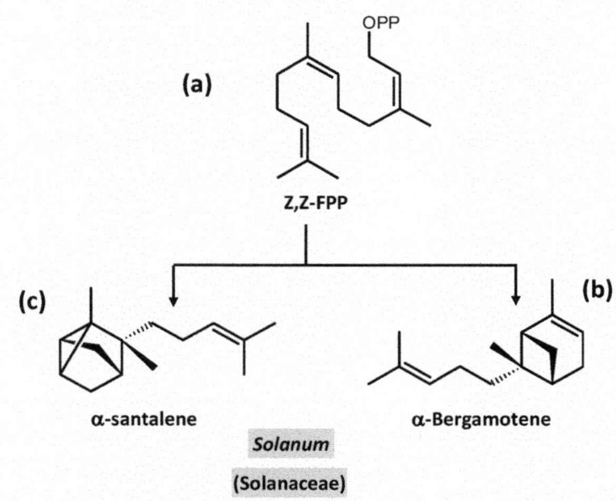

Fig. 5.8 Some sesquiterpenes deriving from the Z,Z-FPP and characterizing the Solanaceae family

5.3.3 Diterpenes

Diterpenes are mostly non-volatile compounds forming plant resins and latex. They are mainly found in Coniferae (resin plants) and Euphorbiaceae (latex plants) in addition to other plant families including Lamiaceae (Holopainen et al. 2013; Piozzi et al. 2006). Diterpenes are synthesized from GGPP from which the allylic pyrophosphate group is removed resulting in carbocation which can initiate multiple reaction paths to give a great variety of diterpene aglycones showing different cyclization levels (Fig. 5.9). Reaction paths depend on the bond conformation at the active site of enzyme and include: carbocation-addition to double bonds, chemical group moving (Wagner-Meerwein rearrangement), molecular stabilization by proton shifting (hybrid shift), deprotonations, etc. Acyclic diterpene structures include geranyllinalol (Fig. 5.9a); bicyclic structures include casbene (Fig. 5.9b); tricyclic structures include abietadiene and taxadiene (Fig. 5.9c, d). Tetracyclic structures include *ent*-kaurene (Fig. 5.9e) that is precursor of the ubiquitous gibberellic acid phytohormones. Also, tetracyclic diterpenes include *syn*-stemar-13-ene (Fig. 5.9f) that derives from *syn*-CPP and serving as precursors for oryzalexin S (a phytoalexin). Other phytolexins (plant defense diterpenes) are derived from *ent*-CPP including oryzalexins A–F.

Apart from the cyclization levels, structural diversity of diterpenes is governed by double bond locations and different oxygenation groups. Illustrative example is provided by tricyclic diterpenes (abietane, primarane and dehydroabietane) found in the conifer Sitka spruce (*Picea sitchensis*) (Hamberger et al. 2011) (Fig. 5.10): in these diterpenes, carbon 18 is subject to different substitutions including CH_3, CH_2OH, CHO or COOH leading to diterpene olefins, alcohols, aldehydes and acids, respectively. Abietane type shows four aglycone families characterized by different locations of double bonds: abietadiene (π at 7, 8 and 13, 14), palustradiene (8, 9 and 13, 14), levopimaradiene (8, 14 and 12, 13), neoabietadiene (8, 14 and 13, 15) (Fig. 5.10a–d). Pimarane type is characterized by a double bond occurring between C15 and C17 whereas a second one has variable location including 8, 14 for pimaradiene, 7,8 for isopimaradiene and 8, 14 for sandaracopimaradiene (Fig. 5.10e–g). Pimaradiene and sandaracopimaradiene differ at α and β configuration levels. Dehydroabietane type is characterized by an entirely conjugated cycle containing three double bonds at (C8, C9), (11, 12) and (13, 14) (Fig. 5.10h). Apart from the variable double bond locations, tricyclic aglycones are further diversified by oxygenation of C18: for instance, Substitution of C18 by CH_2OH, CHO and COOH gives abietadienol, abietadienal and abietic acid, respectively (Fig. 5.10a2–a4).

Beyond the tricyclic aglycones, diterpene pattern of *Picea sitchensis* is further diversified by tetracyclic structure of kaurene type (Fig. 5.10i). Substitutions of C18 by CH_3, CH_2OH, CHO and COOH result in *ent*-kaurene, *ent*-kaurenol, *ent*-kaurenal and *ent*-kaurenoic acid (Fig. 5.10i1–i4). These four aglycones serves as precursors for the synthesis of ubiquitous gibberellic acid-based phytohormones (gibberellins).

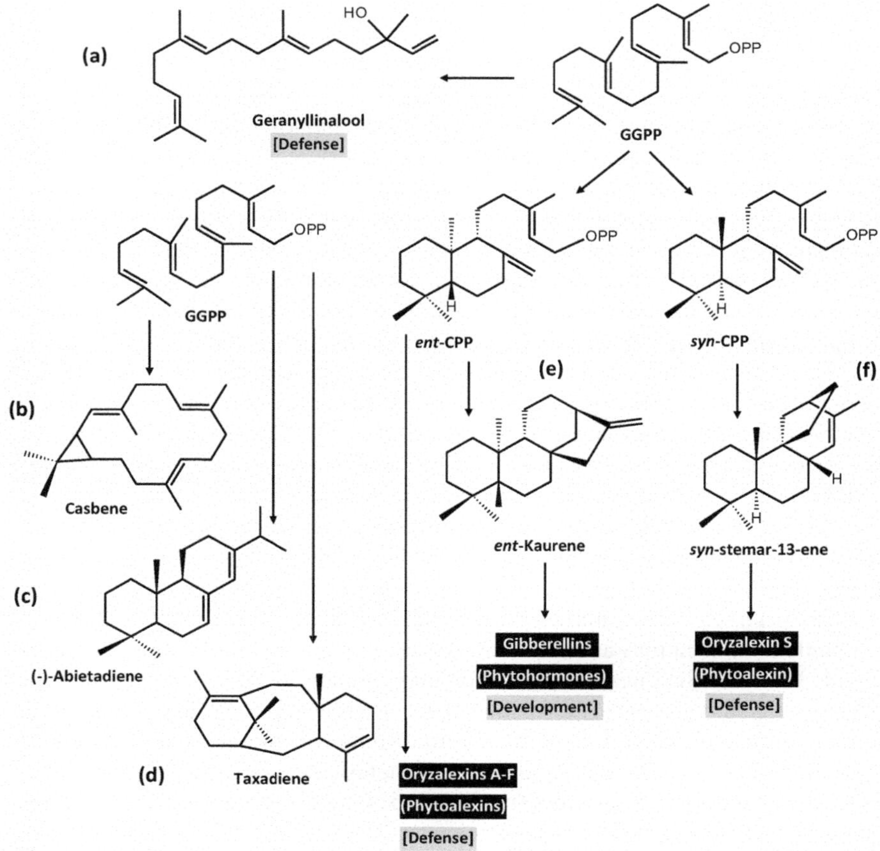

Fig. 5.9 Different diterpene structures showing different cyclization levels

This diterpene pool of *Picea sitchensis* showed strong quantitative and qualitative variations depending on plant tissue (Fig. 5.11). Diterpenes' amounts were significantly higher in bark and phloem tissue compared with wood and xylem one (Fig. 5.11a, c, e, g vs. Fig. 5.11b, d, f, h). Diterpene pattern of bark-phloem tissue was characterized by relatively high amounts of palustradiene and oxygenated derivatives (palustradiol, palustradial, palustric acid) which showed pattern peaks (Fig. 5.11a, c, e, g). However, metabolic regulation ratios showed decreasing trend from palustradiene to palustric acid because of the apparition of other peaks associated with other compounds including levopimaradienal (Fig. 5.11e) and levopimaric acid (Fig. 5.11g). Thus, double peak pattern tended to develop with increasing biosynthesis amounts that occur in the way of diterpene olefins → alcohols → aldehydes → acids. Increasing trend of biosynthesis amounts was also observed in wood-xylem tissue but with frankly lower levels (Fig. 5.11b, d, f, h). In this tissue, diterpene pattern was characterized by relatively high amount of levopimaradiene

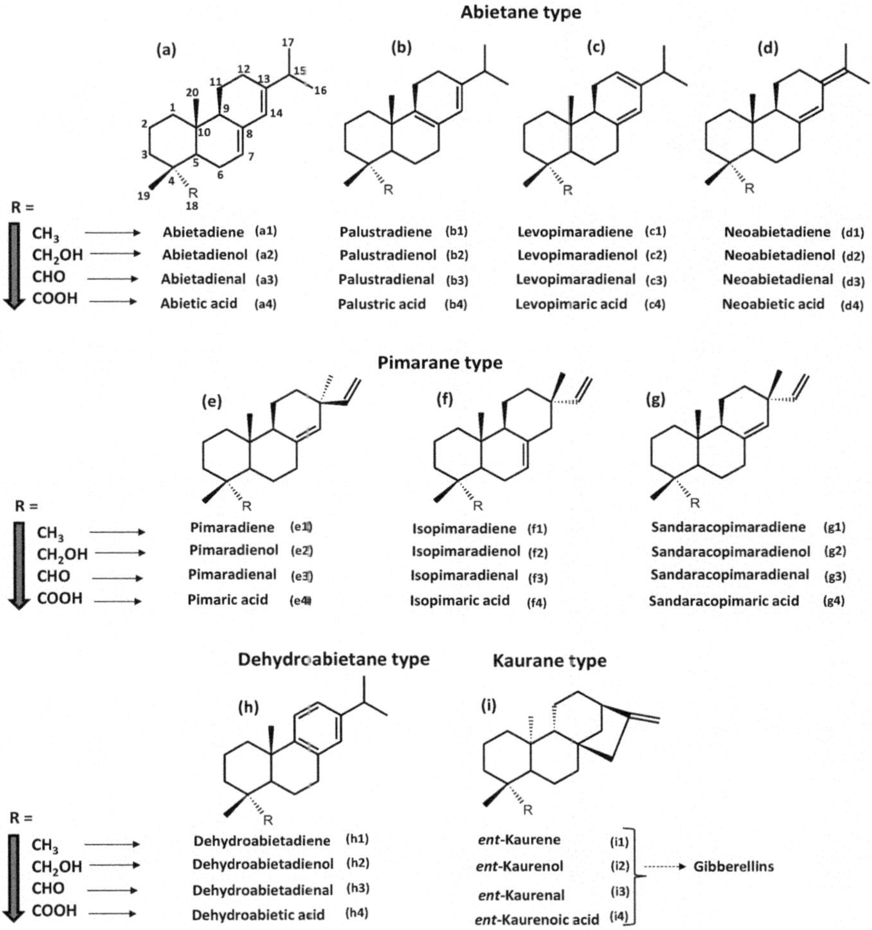

Fig. 5.10 Different aglycone structures of olefin (1), alcohol (2), aldehyde (3) and acid (4) diterpenes in *Picea sitchensis* characterized by variable locations of double bonds and different chemical substitutions of C18

and its oxygenated derivatives with increasing metabolic regulations for aldehyde and acid compared to olefin and alcohol compounds: regulation ratios of levopimaradienal and levopimaric acid in their respective diterpene patterns (Fig. 5.11f, h) were revealed to be twice higher than those of levopimaradiene and levopimaradiol (Fig. 5.11e, g). Diterpene resin acids were revealed to be specialized SMs of the oleoresin in conifers. Their biosynthesis involves enzyme actions of diterpene synthases (DiTPS) and cytochrome P450s of the CYP720B family (Hamberger et al. 2011).

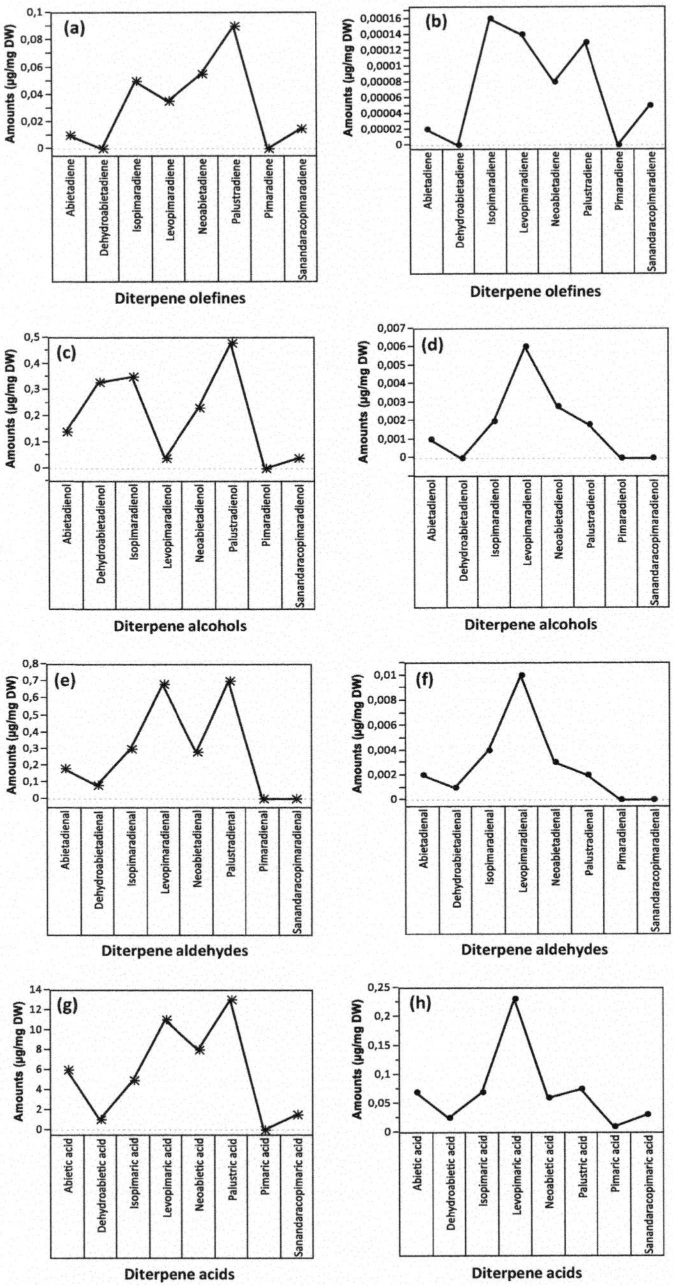

Fig. 5.11 Different diterpene patterns of *Piceae sitchensis* associated with olefin-, alcohol-, aldehyde- and acid-type compounds synthesized in different plant tissues: bark-phloem (**a**, **c**, **e**, **g**) and wood-xylem (**b**, **d**, **f**, **h**) tissues

Apart from the conifers, structural diversity of diterpenes with chemotaxonomical interests concern several other plant taxa. An illustrative case is given for Lamiaceae showing high variability of aglycones synthesized by diTPS and P450 and leading to characteristic molecules of different plant genera/species (Fig. 5.12) (Zerbe et al. 2014; Chen et al. 2017). Diterpene structures include labdane (bicyclic and tricyclic) derivatives, lactone type and tetracyclic neoclerodane type compounds: Bicyclic labdane-derivatives were found in the species *Pogostemon cablin* showing the production of Labda-8-ene type diterpenes (Fig. 5.12a). Labdane tricyclic derivatives can be represented by forskolin (or coleonol) found in *Coleus* genus (*Coleus forskohlii*, *C. barbatus*) (Fig. 5.12b). Also, tricyclic structures include grindelic acid found in *Grindelia robusta* and issued from 7,13-CPP (Copalyl diphosphate) by enzyme actions of diTPS and P450 (Fig. 5.12c). Tetracyclic diterpenes in Lamiaceae include several structures among which marrubiin is a lactone type with a furane cycle, commonly found in the genus *Marrubium* as well as in other genera (*Leonitis*, *Phlomis*) (Fig. 5.12d) (Zerbe et al. 2014). Apart from labdane type derivatives, tetracyclic diterpenes of Lamiaceae include *neo*-clerodane-type compounds such as salvinorin A (in *Salvia divinorum*) and ajugarin I (in *Ajuga reptans*) (Fig. 5.12e, f) (Ortega et al. 1982; Johnson et al. 2019). Neo-clerodanes have similar stereoisomer properties to clerodin (insect antifeedant) (Fig. 5.12 j). Clerodane class shows high stereochemistry (Fig. 5.12g, h) resulting in high usefulness in chemotaxonomical interest at both inter-generic and inter-specific levels (Merritt and Ley 1992; Coll and Tandron 2008). In addition to stereochemical variations, structural diversification of diterpene aglycones within clerodane (and other) class is further provided by several types of chemical substitutions (hydroxylation, acetylation, methylation, prenylation, etc.), conjugations and cyclization processes occurring at different carbons (Fig. 5.12i, j) (C4, C6, C11, C15–C19, etc.) (Coll and Tandron 2008).

5.3.4 Triterpenes

Triterpenes (C30) and sterols derive from MVA pathways where they are synthesized from 2,3-oxidosqualene which is directly issued from squalene after the formation of a 2,3-epoxy cycle (Fig. 5.13a, b). This heterocycle is formed by the enzyme action of squalene epoxidase (SQE) or monooxygenase. Due to its long open structure with conjugated aspect, 2,3-oxidosqualene shows high geometrical flexibility leading to folding ways from which different biosynthesis pathways can take place (Fig. 5.13c–h): 2,3-oxidosqualene can be folded according to two different conformations including chair-boat–chair (C–B–C) type and chair-chair-chair (C–C–C) one (Fig. 5.13c, d). C–B–C (Fig. 5.13 c) serves as basis for the synthesis of sterols characterized by the presence of a sterane aglycone (tetracyclic aglycone) hydroxylated at C3. C–B–C pathway provides triterpenes serving as precursors of sterols and including lanosterol, cucurbitadeniol and cycloartenol (Fig. 5.13e–g):

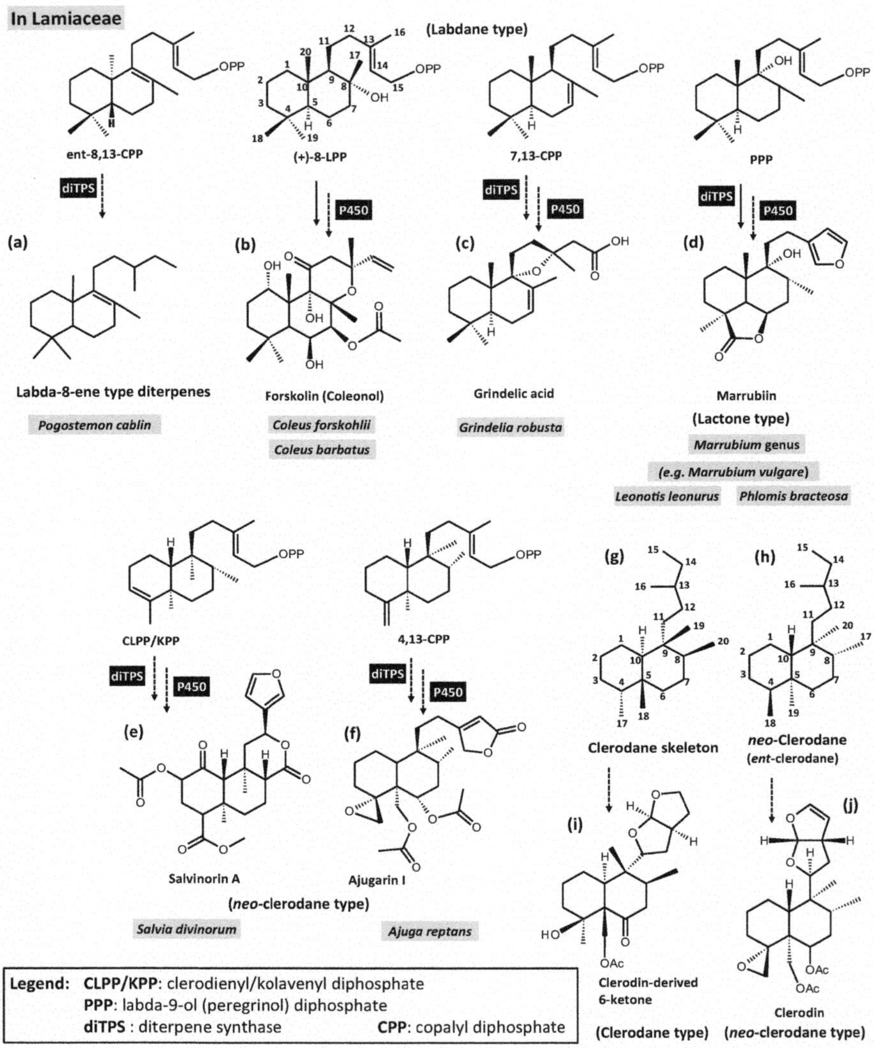

Fig. 5.12 Different structures of di-, tri- and tetracyclic diterpenes with labdane, lactone and neoclerodane structures found in Lamiaceae

Lanosterol (Fig. 5.13e) is synthesized by lanosterol synthase (LAS) and is used in fungus and animal world for the biosynthesis of ergosterol and cholesterol, respectively. LAS are also found in plants but these enzymes appear to be involved to a small extent in the synthesis of phytosterols by opposition to cycloartenol synthases (CAS) that are more common in plant world (Thimmappa et al. 2014; Ohyama et al. 2009).

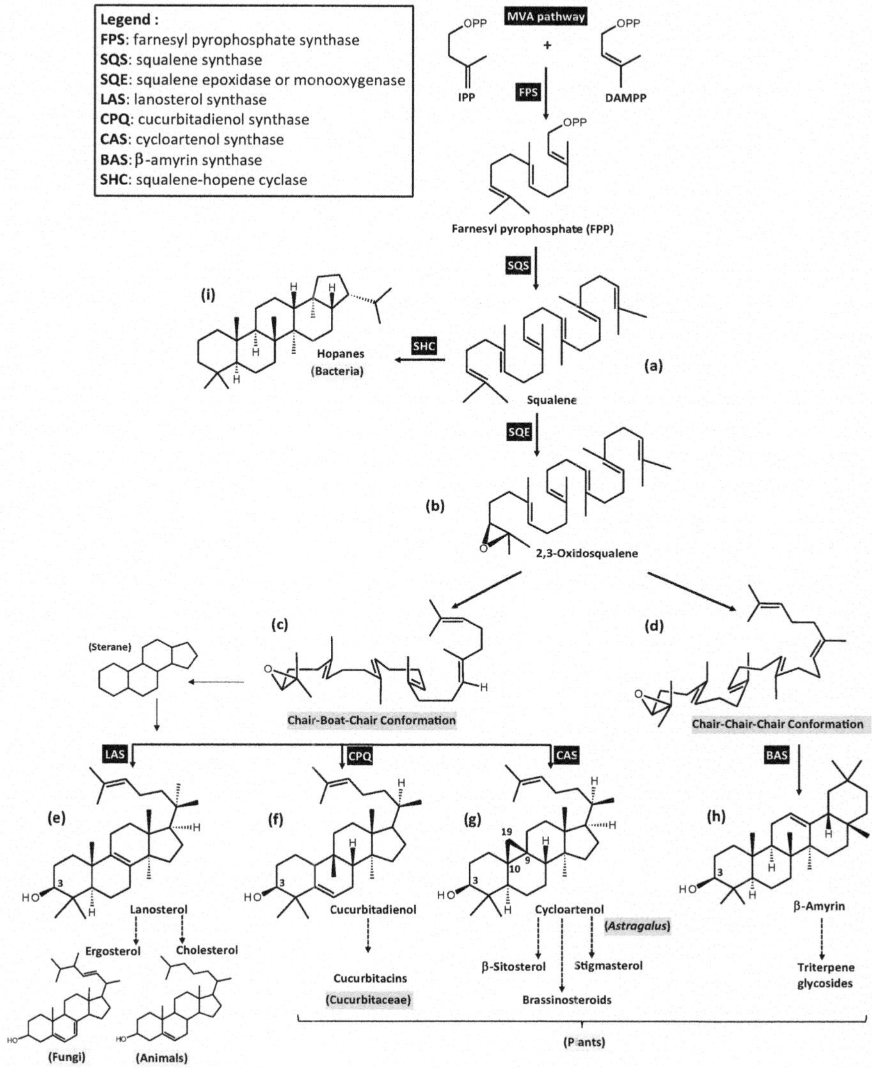

Fig. 5.13 Overview on the synthesis of different phytosterol and triterpene scaffolds in plants from the 2,3-oxidosqualene

Cucurbitadienol (Fig. 5.13f) is synthesized by cucurbitadienol synthase (CPQ) and serves as precursor for biosynthesis of cucurbitacins (Chen et al. 2005). Initially, cucurbitacins were found in Cucurbitaceae as responsible for their bitter principles (Rehm et al. 1957). Thereafter, they were revealed to have a wide botanical distribution including many plant families (Brassicaceae, Scrophulariaceae, Begoniaceae,

Elaeocarpaceae, Datiscaceae, Desfontainiaceae, Polemoniaceae, Primulaceae, Rubiaceae, Sterculiaceae, Rosaceae, Thymelaeaceae) (Schun and Cordell 1985; Miro 1995; Monoz et al. 2000). Chemical structures of cucurbitacins show multiple oxygen substitutions at different positions. Under functional aspect, they were noted to manifest large (nonspecific) cytotoxic activity (Chen et al. 2005).

Cycloartenol is synthesized by cycloartenol synthase (CAS) (Fig. 5.13g). It is structurally characterized by the occurrence of a cyclopropane between C9, C10 and C19. It serves as precursor for several phytosterol classes including membrane constituents and phytohormones: β-sitosterol, stigmasterol, brassinosteroids. Cycloartenol-based compounds showed high chemotaxonomical interests in some plant taxa such as the genus *Astragalus* (the largest botanical genus) (Sarraj-Laabidi et al. 2018).

In addition to the C-B-C pathway responsible for the synthesis of phytosterols, 2,3-oxidosqualene can be folded according to a C–C–C way to provide a wide pool of triterpenes with pentacyclic structures basically illustrated by β-amyrin (Fig. 5.13h).

By opposition to plants, fungi and animals, bacteria synthesize C30-compounds directly from squalene rather than 2,3-oxidosqualene leading to hopane-based structures via the enzyme action of squalene-hopene cyclase (SHC) (Fig. 5.13i).

Synthesis and structural diversification of C30-aglycones (phytosterols and triterpenes) in plants are based on sequential metabolic processes involving folding, cyclization, double bond (π) formation and ring expansion before further variability acquisition by various chemical substitutions (Fig. 5.14):

Different folding ways result in the formation of C–B–C and C–C–C conformations (Fig. 5.14a, b). These two conformations have a 2,3-oxirane (2,3-epoxylated) cycle which is at the origin of C3-hydroxyl structure that is initially observed after whole molecular cyclization (Fig. 5.14c, d). Both C–B–C and C–C–C conformations cyclize to protosteryl and dammarenyl cations, respectively, which have ionized C20 (Fig. 5.14c, d).

Ionization can move through different carbons leading to the formation of double bond(s) at well-defined locations: protosteryl cation (ionized at C20) gives C8-cation (Fig. 5.14e) that can be transformed to lanosterol (Fig. 5.14f). Ionization migration also gives C9-cation (Fig. 5.14g) that can be transformed either to cucurbitadeniol via 5,6-double bond (Fig. 5.14h) or cycloartenol via 9,10,19-cyclopropane formation (Fig. 5.14i).

Similarly to protosteryl cation (C–B–C derivative), dammarenyl cation (C–C–C derivative) shows a C20-ionization after which a ring expansion occurs leading to the formation of baccharenyl cation with ionized C18 (Fig. 5.14i). This ring expansion resulted in the transformation of terminal C5-cycle to C6-one (Fig. 5.14 d vs 5.14 i). Baccharenyl cation is further cyclized by acquiring an additional C5-ring leading to the formation of pentacyclic compound (lupyl cation) with ionized C20 (Fig. 5.14j). The C20-ionization of lupyl cation favors the formation of lupeol containing a 20,29-double bond (Fig. 5.14k). Further ring expansion process transforms lupyl cation to germanicyl cation showing six cyclohexane cycles with ionized C19 (Fig. 5.14l). Ionization migration to C13 gives oleanyl cation (Fig. 5.14m) which directly provides β-amyrin after the acquisition of a 12,13-double bond (Fig. 5.14n).

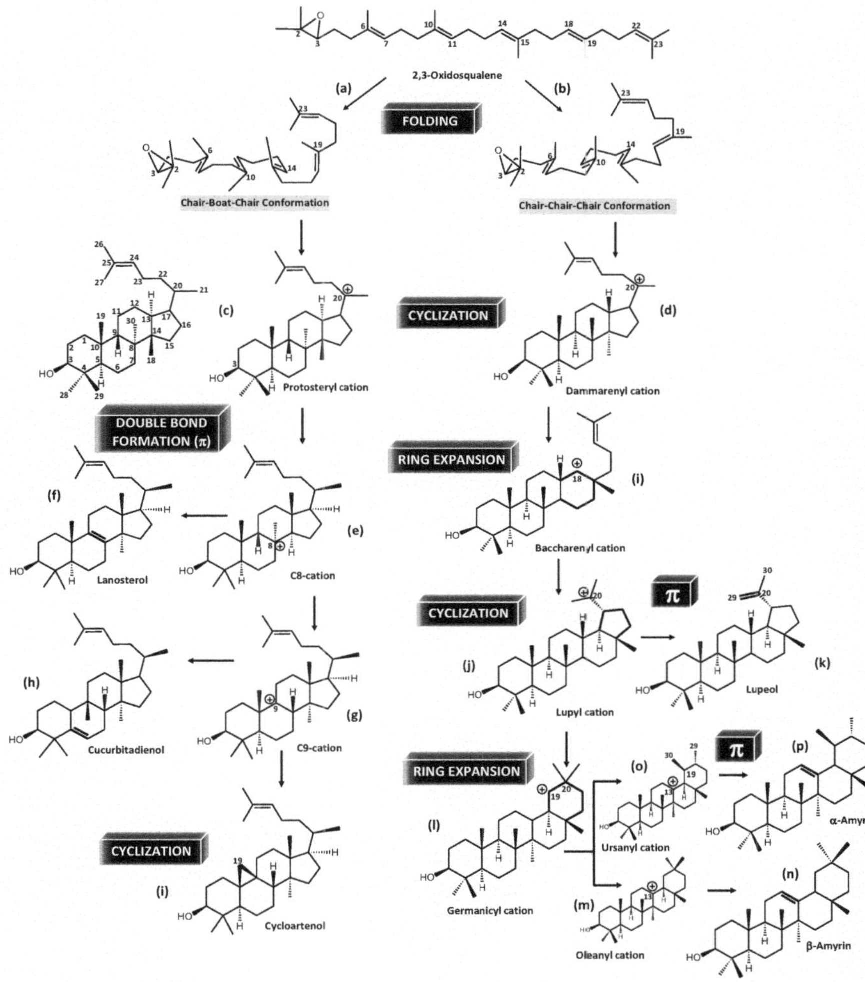

Fig. 5.14 Different sequential metabolic processes consisting of molecular folding and ionization followed by conjugation, cyclization and ring expansion leading to the formation of different phytosterol and triterpene scaffolds

Moreover, C19-ionization in germanicyl cation can result in moving of the C30 (methyl) from C20 to C19 attachment position leading to ursanyl cation with ionized C13 (Fig. 5.14o). This ionization will favor the formation of a 12,13-double bond leading to α-amyrin (Fig. 5.14p).

Beyond these upstream scaffolding processes, structural variability of triterpene aglycones is increased by hydroxylation, carboxylation and epoxylation reactions catalyzed by different cytochrome enzymes (Figs. 5.15 and 5.16). Illustrative example is given for β-amyrin (Fig. 5.15):

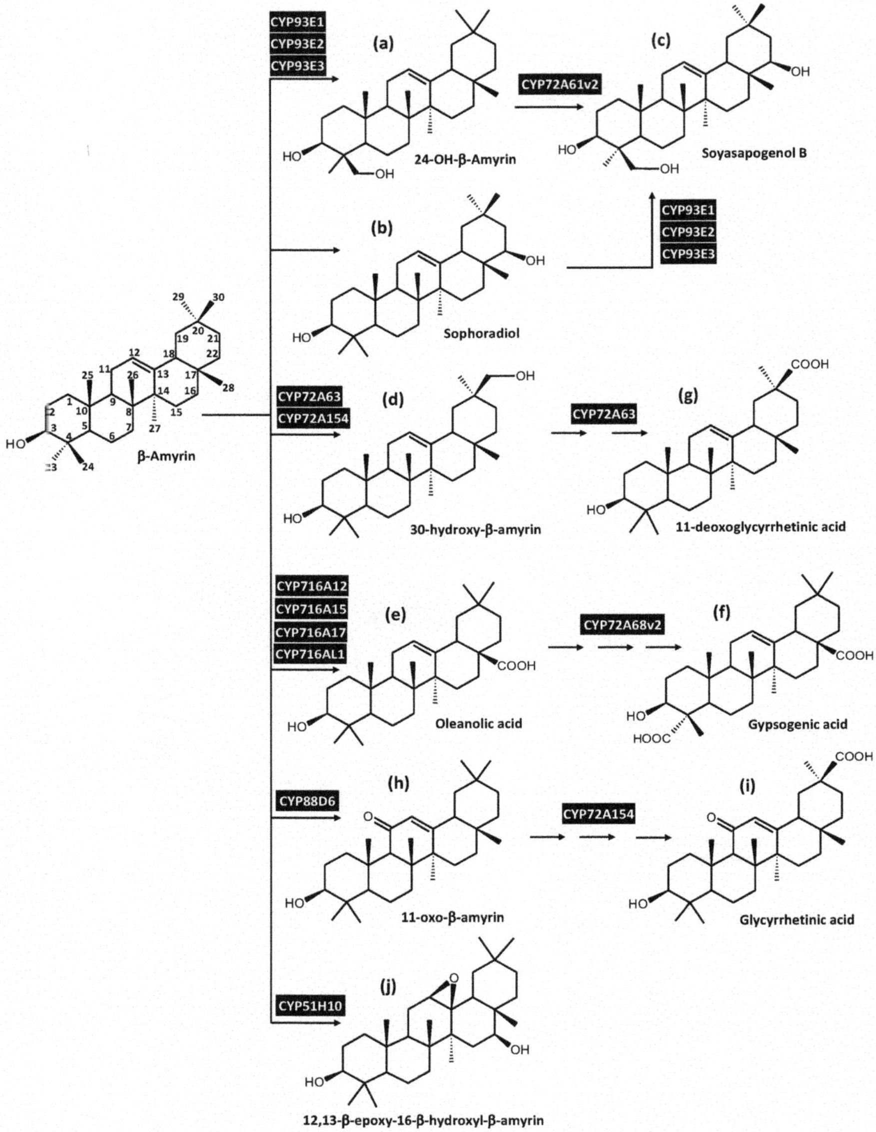

Fig. 5.15 Different oxygenation processes (hydroxylation, carboxylation, epoxidation) leading to structural diversification of triterpene aglycones. Illustrative case is given for β-amyrin-derivatives

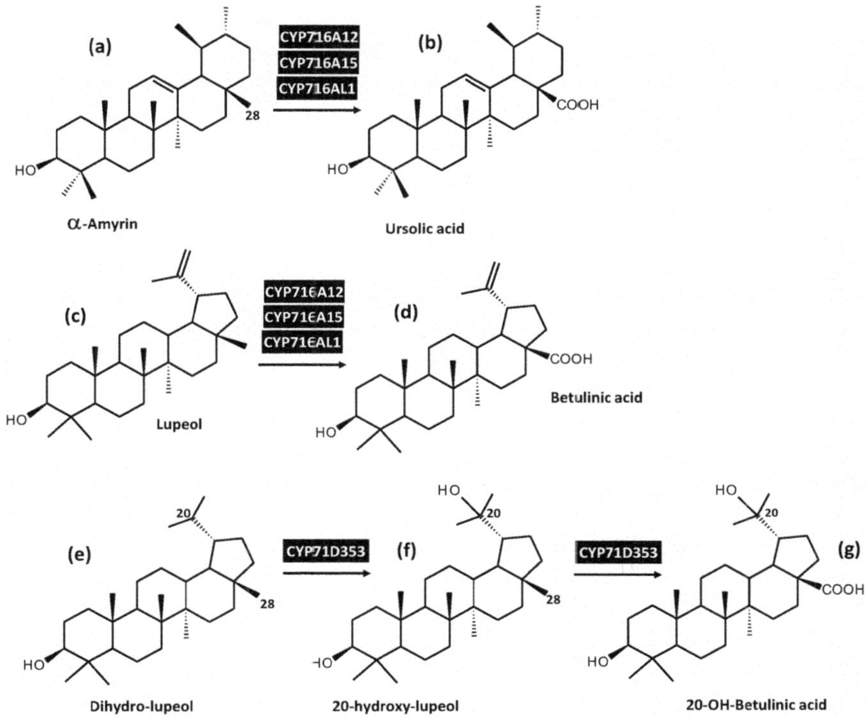

Fig. 5.16 Hydroxylation and carboxylation of triterpene aglycones leading to their structural diversification. Illustrative cases given for α-amyrin and lupeol-derivatives

Hydroxylation of β-amyrin at C24 or C22 results in the formation of 24-hydroxy-β-amyrin and sophoradiol, respectively (Fig. 5.15a, b). These two mono-hydroxylated triterpenes can be further hydroxylated to give soyasapogenol B (22, 24-diOH-β-amyrin) (Fig. 5.15c). These hydroxylation processes are catalyzed by the cytochrome enzymes of CYP93E-types (E1, E2, E3). Hydroxylation of β-amyrin can be concerned with C30 leading to the formation of 30-OH-β-amyrin (Fig. 5.15d); catalytic effect involves CYP72A-type cytochrome enzymes (CYP72A63, CYP72A154).

Apart from hydroxylation, structural diversification of triterpene aglycones can be further increased by carboxylation (Fig. 5.15e–g, i): carboxylation of β-amyrin at C28 gives oleanolic acid (Fig. 5.15e). This process is favored by cytochrome enzymes of CYP716A-types (CYP716A12, CYP716A15, CYP716A17, CYP716AL1). Furthermore, carboxylation of oleanolic acid at C23 gives gypsogenic acid (Fig. 5.15f). C23-carboxylation involves catalytic action of CYP72A68v2. Also, carboxylation of β-amyrin at C30 results in the formation of 11-deoxoglycyrrhetinic acid (Fig. 5.15g). This triterpene corresponds to 11-deoxygenation of glycyrrhetinic acid (Fig. 5.15i) which is obtained from β-amyrin through preliminary formation of C11-ketone (giving 11-oxo-β-amyrin) (Fig. 5.15h) followed by C30 carboxylation. Finally, structural diversification of β-amyrin-derived aglycones can involve epoxidation process; illustrative example is given by 12,13-β-epoxy-16-β-hydroxy-β-amyrin (Fig. 5.15j) via the catalytic action of CYP51H10.

Analogous diversification ways based on oxygenation processes of aglycones are given for α-amyrin, lupeol (Fig. 5.16) and cycloartenol (Fig. 5.17). The 28-Carboxylation of α-amyrin results in the formation of ursolic acid (Fig. 5.16a, b). Similar carboxylation of lupeol results in the formation of betulinic acid (Fig. 5.16c, d). These carboxylation processes are catalyzed by cytochrome enzymes of CYP716A (A12, A15, AL1). Also, 28-carboxylation of 20-hydroxy-lupeol (Fig. 5.16f) (derived from dihydro-lupeol; Fig. 5.16e) gives 20-OH-betulinic acid (Fig. 5.16g); this structural diversification is catalyzed by CYP71D353.

Structural diversification of triterpene aglycones is also provided by epoxidation that can be largely found in cycloartenol derivatives. Illustrative example is given for the cycloartenol-based triterpenes found in the genus *Astragalus* (Fig. 5.17) (Sarraj-Laabidi et al. 2018). Cycloartenol in *Astragalus* showed multiple epoxidation levels of lateral cycle(s) going from one to three cooccurring epoxides in lateral part of molecule (Fig. 5.17g). Cycloartenol structures with acyclic lateral part (Fig. 5.17a) as well as cyclized ones containing oxalane (or tetrahydrofurane) (Fig. 5.17b, c) were revealed to be the most common structures in *Astragalus* genus. These two heterocycles contained epoxide at C20, 24 and C20, 25, respectively. Less common tetrahydrofurane cycle was concerned with 16,23-epoxylated bond (Fig. 5.17d). Cycloartenol aglycone with higher epoxidation levels showed lower frequencies at the generic level. They chemically characterized some *Astragalus* species resulting in high chemotaxonomical values at specific level (Sarraj-Laabidi and Semmar 2017). These structures concerned cycloartenol aglycones containing two and three epoxidated cycles (Fig. 5.17e–g). Di-epoxidation of cycloartenol occurred at (C16, C23) and (C16, C24) (Fig. 5.17e) and at (C24, C16) and (C24, C20) (Fig. 5.17f). Epoxidation bonds in triepoxide structure were concerned with C16-O-C23, 24-O-25 and 23-O-26 (Fig. 5.17g).

After hydrophobic aglycone formations, triterpenes are glycosylated at different levels and at different carbons leading to the formation of saponins. Glycosylation aspects (sugar types, sequences, levels, locations) have governing roles in bioactivities of saponins (Fuchs et al. 2017). Glycosylation level in saponins is highly variable going from monoglycosides to decaglycosides (Cheikh-Ali et al. 2019). Saponins based on a same aglycone with the same glycosylation type can differ by their glycosylated substitution locations (carbons). The C3 tends to be relatively more glycosylated than the other carbons because of its preliminary hydroxylation originated from 2,3-oxidosqualene. The number of glycosylated carbons in a saponin defines its desmosylation level; thus, saponins glycosylated at one, two and three carbons correspond to mono-, bi- and tri-desmosides, respectively. Different glycosylation and desmosylation levels of saponins are illustrated in Fig. 5.18 including a triglycoside monodesmoside (at C3) of gypsogenin (Fig. 5.18a), octaglycoside bidesmoside (at C3, C28) of gypsogenin (Fig. 5.18b), pentaglucoside bidesmoside (at C23, C28) of gypsogenic acid (Fig. 5.18c, d), and triglycoside tridesmoside (at

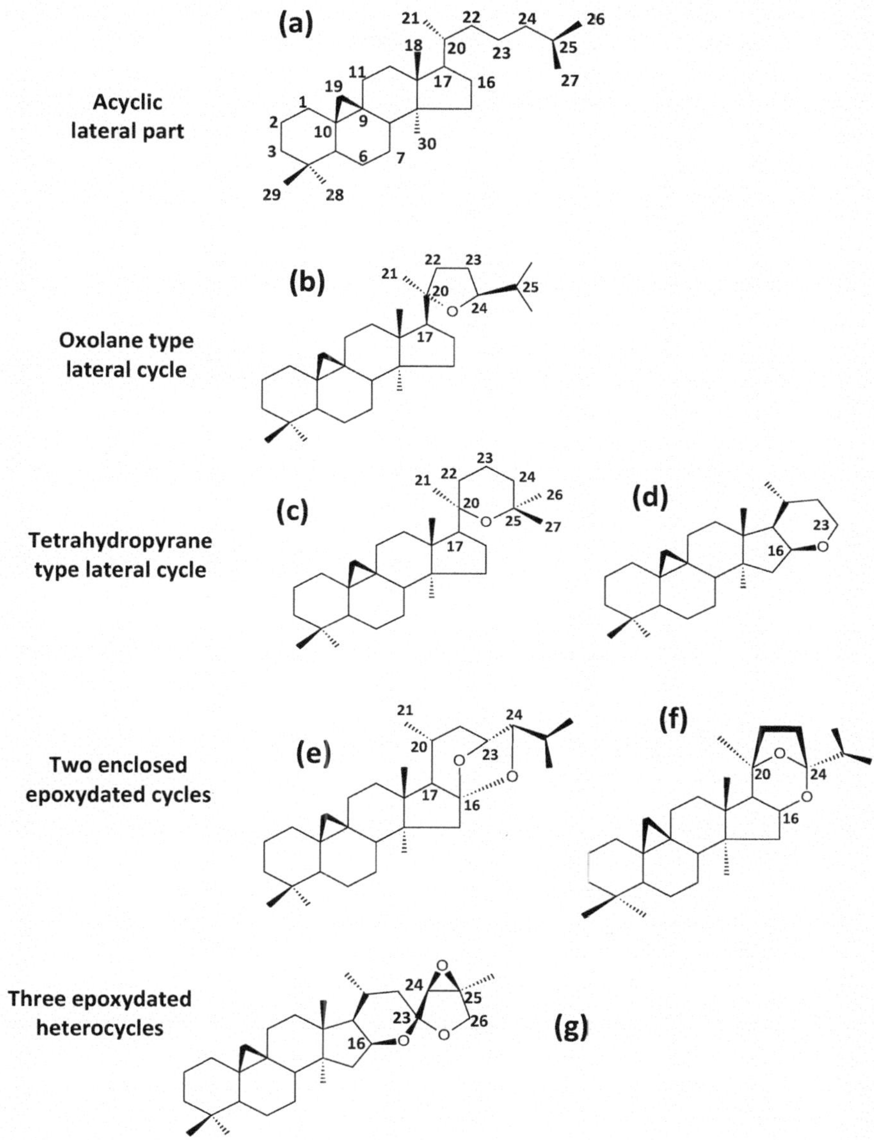

Fig. 5.17 Different cycloartane-based structures found in the genus *Astragalus* showing increasing epoxidation level of lateral cycle(s)

C3, C6, C25) of cycloartenol (Fig. 5.18e, f) (Bechkri et al. 2020; Koike et al. 1998). Glycosylated carbons are initially hydroxylated or carboxylated. Commonly glycosylated carbons vary according to the aglycone type: in cycloartenol of *Astragalus* genus, relative glycosylation level tends to decrease in the way of C3 (most glycosylated), C6 and C24 followed by C25 and C16 (Sarraj-Laabidi et al. 2017). It has been shown that cycloartenol type (with lateral chain or heterocycle) significantly influences glycosylation patterns of saponins (Sarraj-Laabidi et al. 2017).

In β-amyrin type aglycones, C3 and C28 tend to be relatively more glycosylated compared with other carbons (less commonly glycosylated) (Fig. 5.18a, b, d and 5.19) (Thimmappa et al. 2014; Cheikh-Ali et al. 2019). Relative glycosylation levels of C3 and C28 tend to be influenced by the type of β-amyrin derived triterpene (Cheikh-Ali et al. 2019; Hammami et al. 2021): for instance, in Caryophyllaceae family, regulation level of 28-glycosylation tends to increase at the expense of C3-glycosylation in gypsogenic acid and 16-OH-gypsogenic acid (Fig. 5.19a, b). However, C3-glycosylation tends to be relatively more favored in quillaic acid and gypsogenin (Fig. 5.19c, d).

5.3.5 *Tetraterpenes*

Tetraterpenes are C40-compounds made by eight isoprene units (C5) (Fig. 5.20a, b) (Britton et al. 1998, 2008). The simplest case concerns the phytoene showing eight sequential isoprene units (Fig. 5.20b) forming a polyene chain containing 9 conjugated double bonds (Fig. 5.20c).

Structural diversification is provided by several metabolic processes including:

- Desaturation leading to the apparition of additional double bonds in derived molecules (Fig. 5.21a, b).
- Different arrangements or cyclizations of terminal part leading to different types of rings at the opposite ends of polyene chain (Fig. 5.21c–e).
- Oxygenation of the molecule leading to the formation of xanthophylls (Fig. 5.22).
- Other chemical substitutions such as acylation (Fig. 5.22e).

Tetraterpenes are essentially plant pigments commonly called carotenoids (Figs. 5.21 and 5.22). According to their strict hydrocarbon or oxygenated aspects, carotenoids are divided into carotenes (Fig. 5.21) and xanthophylls (Fig. 5.22), respectively.

Carotenes are synthesized from phytoene by enzyme actions of desaturase and cyclase (Fig. 5.21). Phytoene (9 double bonds, colorless) (Fig. 5.21a) acquires additional double bonds by catalytic effect of desaturase leading to the formation of lycopene (11 double bonds, red) (Fig. 5.21b). Further structural diversification in carotene class is provided by catalytic action of cyclase enzymes leading to the formation of different types of carotenes showing different cyclic terminal groups

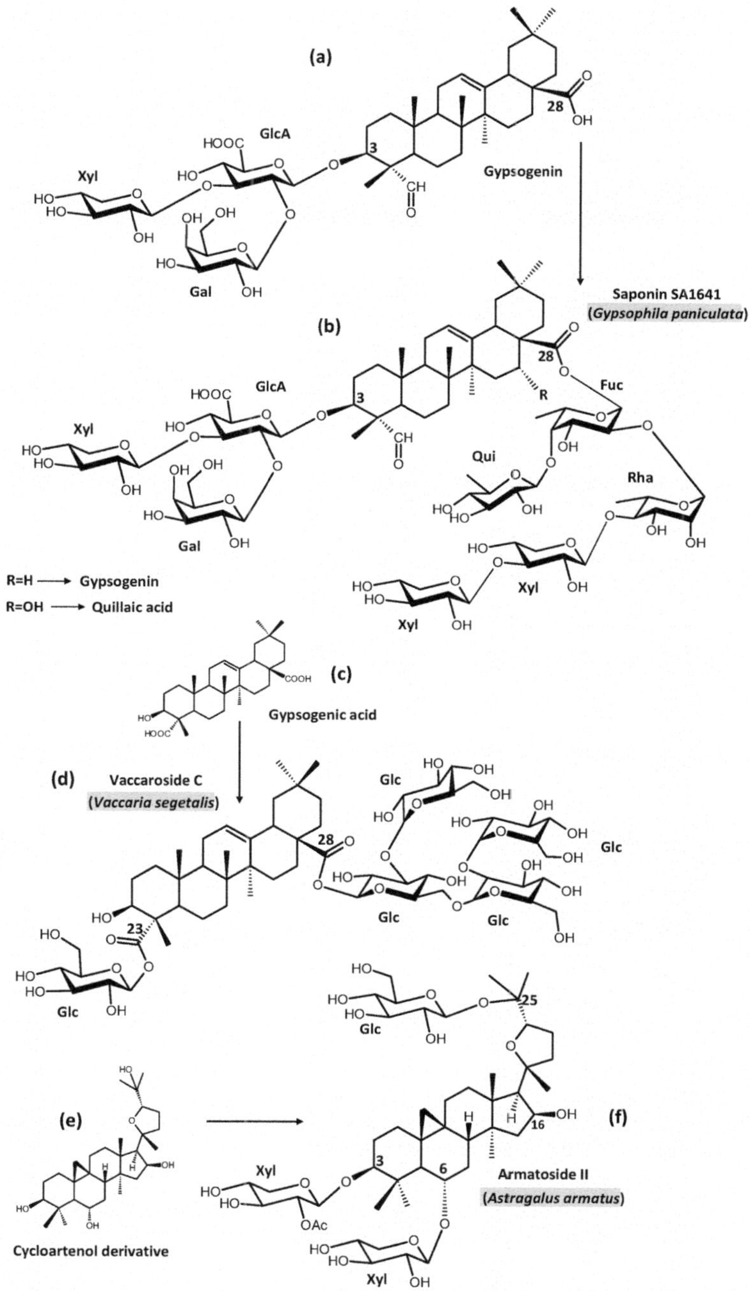

Fig. 5.18 Saponins based on different triterpene aglycones showing different glycosylation and desmosylation levels (**a**, **b**, **d**, **f**). Structures based on gypsogenin (**a**, **b**), gypsogenic acid (**c**, **d**) and cycloartenol (**e**, **f**)

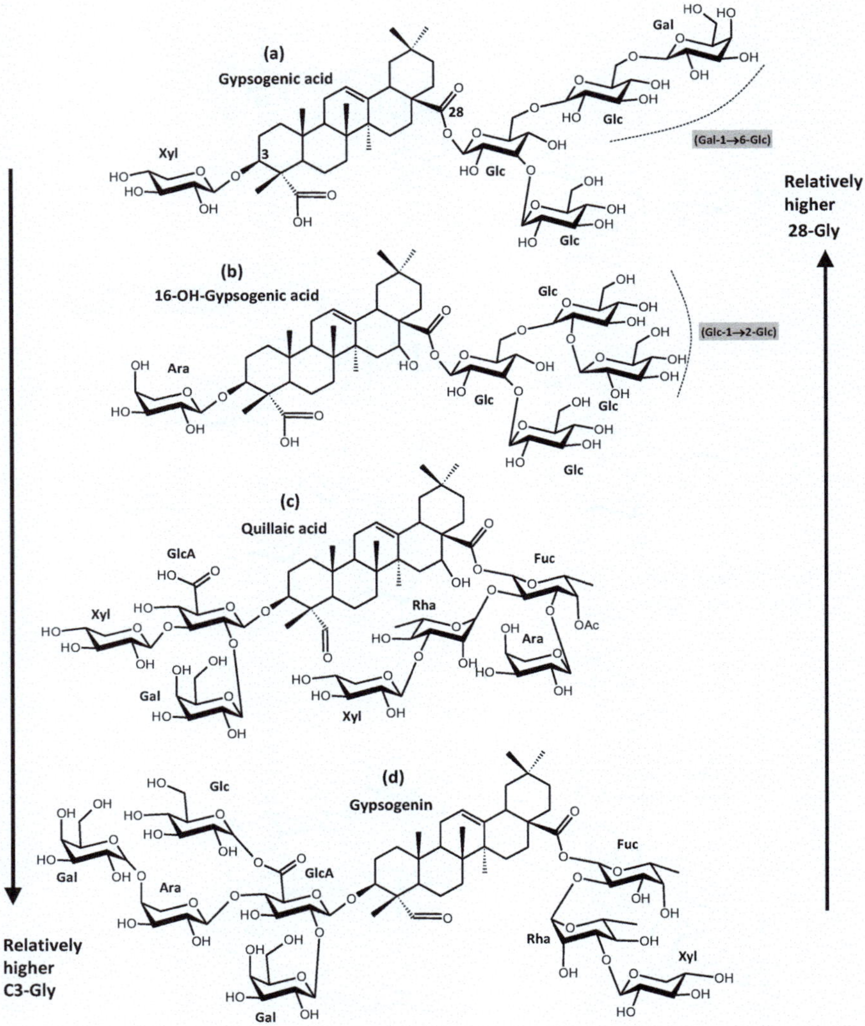

Fig. 5.19 Different saponins based on β-amyrin scaffold showing opposite glycosylation ratios between C3 and C28 according to the aglycone type. **a** Gypsogenic acid-, **b** 16-OH-gypsogenic acid-, **c** quillaic acid-, **d** gypsogenin-based saponins. The four saponins were found in Caryophyllaceae characterized by these four types of triterpene aglycones

(Fig. 5.21c–e): the most common carotene is the β-carotene (Fig. 5.21c). β-carotene (orange) differs from α-carotene (orange-yellow) by the location of double bond of the second terminal cycle: C5′,C6′-bond in β-carotene (Fig. 5.21c) versus 4′,5′-bond

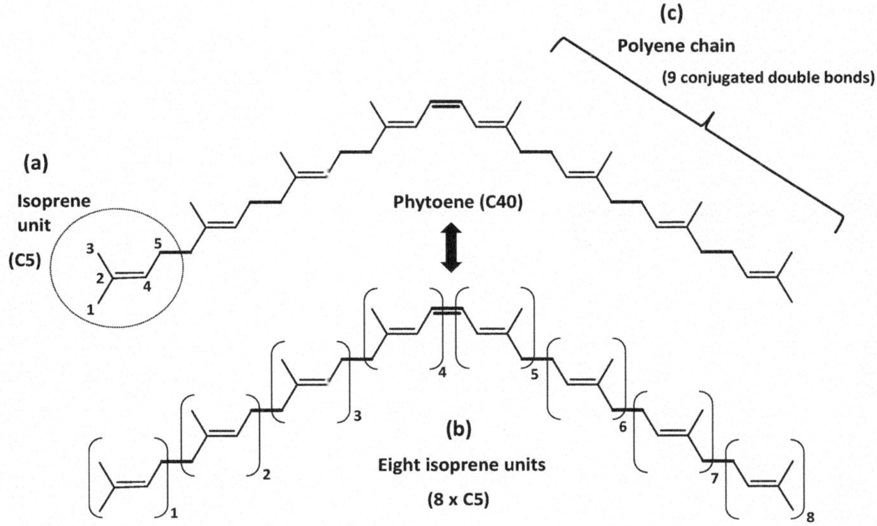

Fig. 5.20 Illustration of tetraterpene structure made by 40 carbons organized into 8 isoprene (C5) units with a polyene chain containing 9 conjugated double bounds

in α-carotene (Fig. 5.21d). Various terminal cycles with different double bond locations and levels, in addition to different cycle shapes, result in various carotene structures including α-, β-, γ-, ε-, κ-, ϕ- and χ-carotenes (Fig. 5.21e) (Maoka 2020; Garcia-Plazaola et al. 2007). The number and location of conjugated double bonds represent determining factors of maximal absorption wavelength of pigment molecules and their subsequent color (Melendez-Martinez et al., 2007).

Xanthophylls derive from carotenes by oxygenation process occurring mainly in the terminal cycles (Fig. 5.22). Oxygenations are catalyzed by carotene hydroxylases, ketolases and epoxidases. Thus, β-cryptoxanthin (Fig. 5.22a) derives from 3-hydroxylation of β-carotene (Fig. 5.21 c). Also, lutein (Fig. 5.22b) derives from 3,3′-dihydroxylation of α-carotene (Fig. 5.21 d). However, such a dihydroxylation applied to β-carotene leads to the formation of zeaxanthin (Fig. 5.22c). Ketone-substitutions of zeaxanthin at C4 and C4′ result in the formation of astaxanthin (Fig. 5.22d). Oxygenation patterns in xanthophylls can be diversified by epoxidation process, lactone formation and substitutions occurring outside of the terminal cycles; a multi-oxygenated xanthophyll by hydroxyls, ketones, epoxides and lactone is illustrated by peridinin (Fig. 5.22e).

Xanthophylls play key roles in light (excess, deficit) management in plants through a process called xanthophyll cycles (Fig. 5.23) (Britton et al. 2008; Li et al. 2000). By the xanthophyll cycles, excess of light energy is dissipated via de-epoxidation processes of xanthophylls in order to protect the photosynthetic machinery against light-induced oxidative stress (Fig. 5.23d, e) (Della Penna and Pogson 2006; Tanaka et al. 2008). Back reaction (epoxidation reaction) regenerates epoxidated xanthophyll

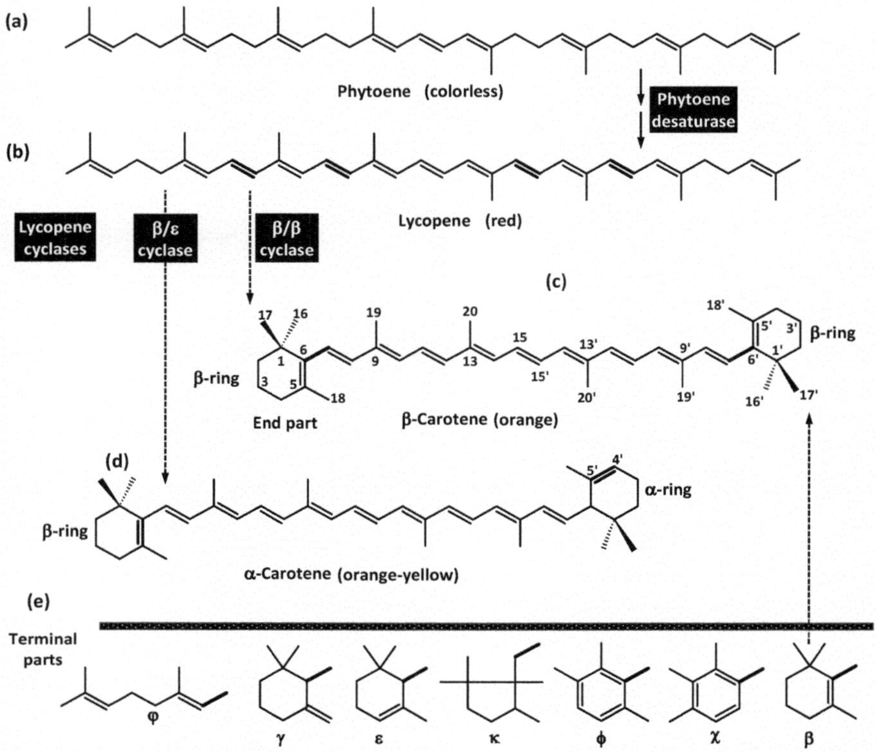

Fig. 5.21 Biosynthesis of carotenes (non-oxygenated tetraterpenes) from phytoene and lycopene (**a-d**) and their structural diversification by various types of terminal cyclisation (**e**). The bold segments (atomic links) at C6 and C6' in (**c**) indicate the locations of terminal cyclizations

leading to complete the cycle (Fig. 5.23f, g); it acts in the way of light harvesting under light deficit or darkness conditions.

This protective process has general occurrence in the plant world including higher plants, phytoplankton, algae, bryophytes (Maoka 2020). In higher plants, two xanthophyll cycles (de-epoxidation, epoxidation) were highlighted: the first cycle associates violaxanthin (V) with zeaxanthin (Z) (Fig. 5.23a, c); the second one involves lutein epoxide (Lx) and lutein (L) (Fig. 5.22b). The V-Z cycle has ubiquitous aspect whereas Lx-L is taxonomically limited. Xanthophyll cycle is initiated by light-driven de-epoxidation of violaxanthin (di-epoxide) to zeaxanthin (not epoxidated) via antheraxanthin (monoepoxide) (Fig. 5.23a–c); de-epoxidation is carried out by de-epoxidase type enzyme, the violaxanthin de-epoxidase (VDE) (Fig. 5.23d, e). Back reaction of xanthophyll cycle occurs in darkness or under low light by catalytic effect of zeaxanthin epoxidase resulting in regeneration of violaxanthin (Fig. 5.23f, g).

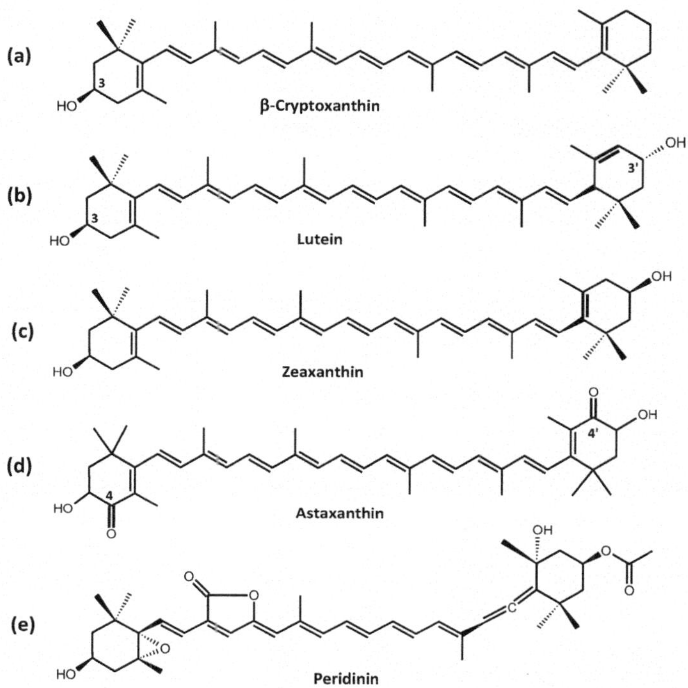

Fig. 5.22 Different structures of xanthophylls diversified by different oxygenation processes (hydroxylation, ketonation, epoxidation) occurring at different location of the molecules

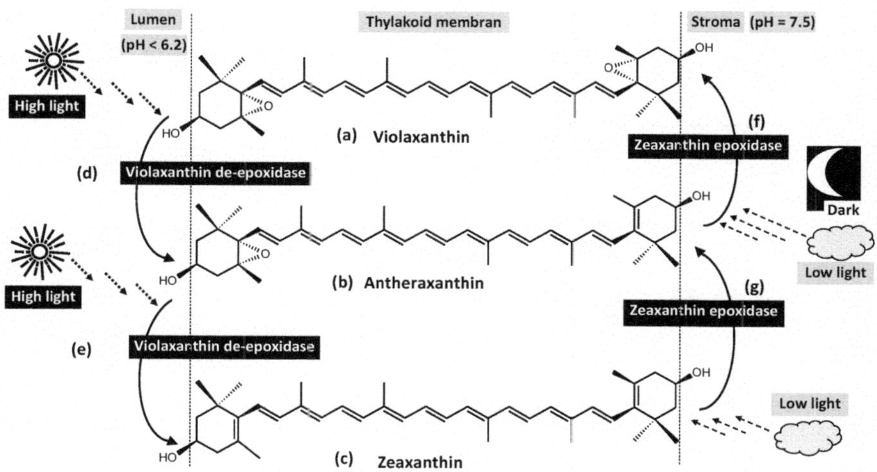

Fig. 5.23 Light-induced xanthophyll cycle involving epoxidation/de-epoxidation reactions of xanthophylls helping plants to manage excess and deficit light energy for protection and better efficiency of photosynthetic system

Light absorption exceeding plant's capacity for CO_2 fixation results in excitation energy of pigment antenna (Fig. 5.24a). Excess energy will be dissipated under thermal form (Fig. 5.24d) via xanthophyll cycle (Fig. 5.24c). Light-induced xanthophyll cycle starts by the formation of a proton gradient (ΔH^+ or ΔpH, across the thylakoid membranes) which is generated by photosynthetic proton pumping (Fig. 5.24b) (Horton et al. 1996). Protonation (Fig. 5.24b) results in lumen acidification ($pH < 6.2$) leading to activation of violaxanthin de-epoxidase (VDE) (Fig. 5.23d, e and 5.24c). De-epoxidation of violaxanthin leads to heat dissipation as protective mechanism against excess light energy (Fig. 5.24d). However, dissipation of proton gradient (Fig. 5.24e) acts in opposite way by favoring the activity of zeaxanthin epoxidase (ZE) which generates violaxanthin (Fig. 5.24f). This back reaction of xanthophyll cycle favors further conservation of light energy under low irradiance conditions leading to optimisation of photosynthetic yield. Under limiting light conditions, xanthophylls (and more generally carotenoids) harvest supplemental light energy and provide it to chlorophylls through singlet–singlet excitation transfer (Maoka 2020).

Excessive energy absorbed by chlorophylls is transferred to carotenoids which release it through polyene vibration mechanism (Fig. 5.20c) (Britton et al. 2008). Thermal energy dissipation represents a flexible diffusion way of excess light energy absorbed by chlorophylls within the photosynthetic apparatus. This results in quenching of the chlorophyll fluorescence yield leading to the so-called non-photochemical quenching (NPQ), a mechanism that can be efficiently assessed by chlorophyll fluorescence signal (measured *NPQ* parameter) (Bilger and Björkman 1990). Photo-protection against light stress conditions involves the participation of both carotenoids (lipid protective antioxidants) with light-harvesting proteins in NPQ

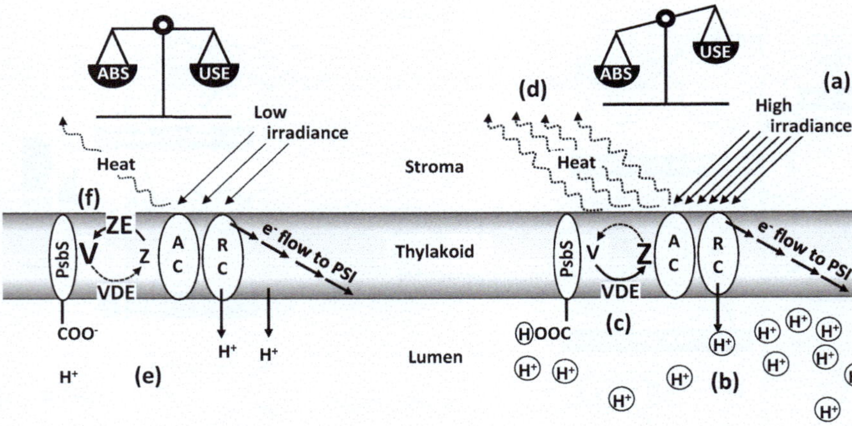

Fig. 5.24 Modulation of heat dissipation through xanthophyll cycle mechanisms leading to photo-protection **a** and light harvesting **b** in favor of photosynthesis machinery under high and low irradiance, respectively. ABS, absorption; USE, use of absorbed radiation (not converted in heat). Z, zeaxanthin; ZE, zeaxanthin epoxidase; VDE, violaxanthin deepoxidase; Psbs, chlorophyll-binding protein of photosystem II; PSI, photosystem I

mechanism. PSbS was revealed as intrinsic chlorophyll-binding protein of photosystem II (PSII) playing key role in non-photochemical quenching favoring balance between dissipation and utilization of light energy (Fig. 5.24). This dynamic balance process minimizes the generation of oxidizing molecules leading to protect plants against photo-oxidative damage (Li et al. 2000).

Apart from its role in excess light dissipation, xanthophyll de-epoxidation was reported to occur also in darkness under different environmental stresses according to biological organisms (Fernandez-Marin et al. 2021; Latowski et al. 2011). Reported stimulated stresses include freezing in angiosperms, dessication (strong water deficit) in angiosperms, bryophytes and pteridophytes, high temperatures in Pheophyceae, low oxygen in diatoms and Pheophyceae (Fernandez-Marin et al. 2021; Garcia-Plazaola et al. 2012). Light-independent xanthophyll de-epoxidation was linked to middle acidification by proton sources other than photosynthetic electron transport chain (Li et al. 2016; Girolomoni et al. 2020): alternative proton sources include chloroplastic ATP hydrolysis (i.e. reversed ATP synthase activity) resulting in lumen acidification (Gilmore and Yamamoto 1992; Gilmore and Björkman 1994, 1995; Verhoeven et al. 1999). Persisting increased ATP levels associated with retained zeaxanthin in darkness under cold conditions supported the proton source hypothesis in different plant species. Chlororespiration represents an alternative electron transport process that can play key role in acidification of thylakoid lumen (Peltier et al. 2016; Alric and Johnson 2017; Tan et al. 2020).

Apart from violaxanthin cycle which is ubiquitous in high plants, lutein epoxide (Lx) cycle shows taxonomically restricted distribution (Garcia-Plazaola et al. 2007; Esteban et al. 2007). Lx cycle involves the de-epoxidation of Lx to lutein under high light versus back reaction (L epoxidation to Lx) under low light. Many plant species lack Lx which is associated with the α-branch of carotenoid metabolism by opposition to the widely distributed zeaxanthin (de-epoxidated violaxanthin) associated with the β-branch. This lacking seems to be due to the kinetics of back reaction (Garcia-Plazaola et al. 2007): epoxidation of lutein to Lx (Lx regeneration mechanism) shows much slower kinetics than epoxidation of zeaxanthin to antheraxanthin or violaxanthin (Fig. 5.23). Some plant families showed efficient epoxidation ability of lutein to Lx. Lauraceae is among the angiosperm families showing Lx cycle with interspecific variations (Esteban et al. 2007). Previous studies on different Lauraceae species showed general positive correlation between violaxanthin pool size and growth irradiance. However, correlation between Lx pool size and growth irradiance significantly varied between species, with stronger functional aspect in *Ocotea foetens* (Aiton.) Benth compared to three other Lauraceae species, *Apollonias barbujana* (Cav.) Bornm., *Persea indica* (L.) Spreng and *Laurus azorica* (Seub.) Franco. These results suggested species-dependent dynamics of Lx cycle leading to interesting functional chemotaxonomic aspects.

Beyond their roles in photosynthetic systems, carotenoids play key roles in coloration, signal emissions and anti-oxidative protections in non-photosynthetic organs (Britton et al. 1998, 2008; Maoka 2009):

Red color appears during the ripening stage of many fruits and seeds due to the formation of anthocyanins and/or carotenoids. For instance, greenish-yellow to red color transition during ripening of tomato is governed by the conversion of phytoene (colorless) to lycopene (red pigment) in the pericarp (Fig. 5.21a, b). In addition to its pigmentation role, lycopene play strong role in singlet oxygen quenching and photoprotection (Britton et al. 1998, 2008).

During the summer-autumn transition, *Pittosporum tobira* (Tobera in Japanese) changes the color of its seeds due to influencing physical factors and subsequent biochemical processes (Maoka 2009): in summer, the seeds are covered with a capsule and have a yellow color. During this period, violaxanthin (Fig. 5.25a) represents the dominant xanthophyll form in the seeds. In autumn, the seeds are exposed to sunlight and submit a color change to red associated with the conversion of violaxanthin to tobiraxanthin A (Fig. 5.25c) by oxidative cleavage of C5–C6 and C5′–C6′ epoxy-bonds (Fig. 5.25b) (Maoka 2009, 2020).

Tobiraxanthin A shows a stronger anti-oxidative activity than violaxanthin because of more conjugated double bonds (11 vs. 9); the two additional conjugated double bonds in tobiraxanthin A are originated from the introduction of two conjugated carbonyl groups at C6 and C6′ in the polyene chain of violaxanthin. This attributes to tobiraxanthin A strong quenching ability of singlet oxygen induced by sunlight.

In addition to this protective physical process (excess energy dissipation by NPQ), carotenoids are involved in highly antioxidative chemical reactions against reactive oxygen species (ROS) by taking up oxygen by polyene or by reacting with oxidative groups (Fig. 5.26): in presence of oxidant such as ROS, the carotenoid reacts by its polyene chain which will be cut at a given double bond leading to multiple possible oxidized derivatives. Illustrative case is given for astaxanthin (Fig. 5.26a)

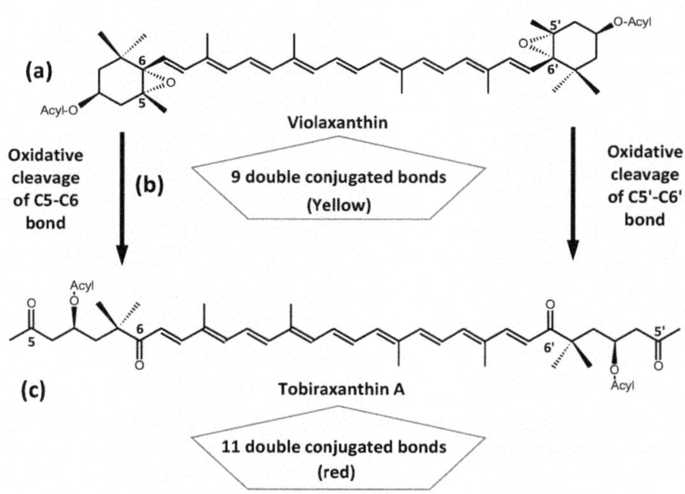

Fig. 5.25 Color change of the seeds of *Pittosporum tobira* (Tobera in Japanese) from yellow to red due to oxidative cleavage of C5–C6 and C5′–C6′ epoxide bonds in violaxanthin leading to tobiraxanthin A formation

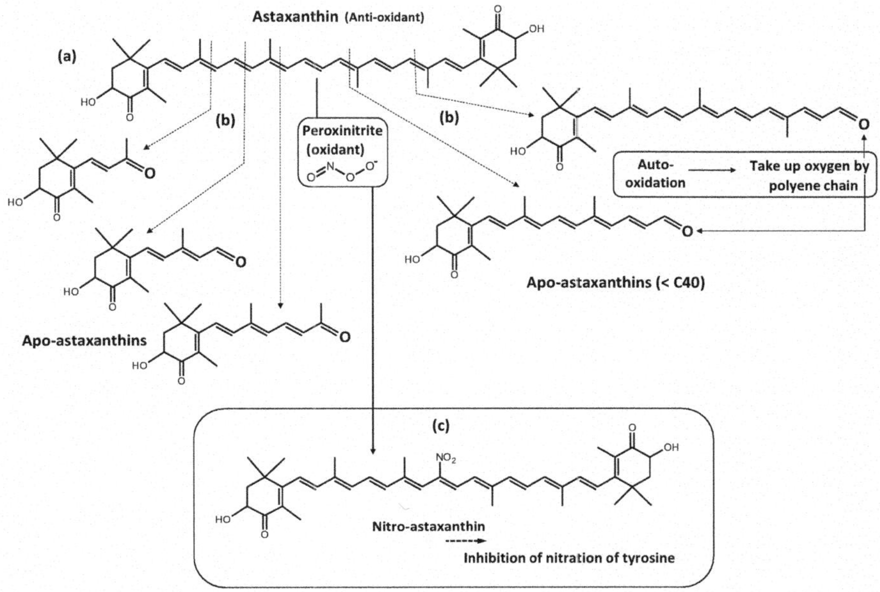

Fig. 5.26 Anti-oxidative mechanisms provided by carotenoids against oxygen and nitrogen reactive species (ROS, RNS). Illustrative case given by astaxanthin

which takes up reactive oxygen at a given double of its polyene chain (Etoh et al. 2012). This auto-oxidative process leads to the formation of several possible stable (nonreactive) derivatives due to the transformation of reactive oxygen to aldehyde (Fig. 5.26b). These derivatives are called apo-carotenoids (apo-astaxanthin) because they have less than 40 carbons. Apart from its involvement with ROS, astaxanthin showed strong interactive ability with reactive nitrogen species (RNS) such as peroxinitrite ion (Maoka and Etoh 2016): astaxanthin can form nitro-astaxanthin (stable compound) by reacting with peroxynitrite (oxidant) (Fig. 5.26c). By this way, astaxanthin captures nitro group on its polyene chain by forming nitro-astaxanthin on one hand and by inhibiting the nitration of tyrosine on other hand.

These antioxidative processes against ROS and RNS illustrated by astaxanthin were also highlighted for different carotenoids including β-carotene, lutein, zeaxanthin, capsanthin, and fucoxanthin (Maoka 2020; Tsuboi et al. 2011).

References

Allamand A, Piechowiak T, Lièvremont D, Rohmer M, Grosdemange-Billiard C (2023) The multifaceted MEP pathway: towards new therapeutic perspectives. Molecules 28:1403

Alric J, Johnson X (2017) Alternative electron transport pathways in photosynthesis: a confluence of regulation. Curr Opin Plant Biol 37:78–86

Banerjee A, Sharkey TD (2014) Methyl erythritol 4-phosphate (MEP) pathway metabolic regulation. Nat Prod Rep 31(8):1043–1055

Bechkri S, Alabdul Magid A, Sayagh C, Berrehal D, Harakat D, Voutquenne-Nazabadioko L, Kabouche Z, Kabouche A (2020) Triterpene saponins from *Silene gallica* collected in North-Eastern Algeria. Phytochemistry 172:112274

Bilger W, Björkman O (1990) Role of the xanthophyll cycle in photoprotection elucidated by measurements of light-induced absorbance changes, fluorescence and photosynthesis in leaves of *Hedera canariensis*. Photosynth Res 25:173–185

Britton G, Liaaen-Jensen S, Pfander H (1998) Carotenoids, vol. 3. Biosynthesis and metabolism. Birkhäuser, Basel

Britton G, Liaaen-Jensen S, Pfander H (2008) Carotenoids, vol. 4. Natural functions. Birkhäuser, Basel

Brooker JD, Russell DW (1975) Subcellular localization of 3-hydroxy-3-methylglutaryl coenzyme A reductase in *Pisum sativum* seedlings. Arch Biochem Biophys 167:723–729

Brooker JD, Russell DW (1979) Regulation of microsomal 3-hydroxy-3-methylglutaryl coenzyme A reductase from pea seedlings: rapid post-translational phytochromemediated decrease in activity and in vivo regulation by isoprenoid products. Arch Biochem Biophys 198:323–334

Bruneton J (1995) Pharmacognosy, phytochemistry and medicinal plants. Intercept Ltd., Hampshire, pp 475–485

Cheikh-Ali S, Farman M, Lacaille-Dubois MA, Semmar N (2019) Structural organization of saponins in caryophyllaceae. Phytochem Rev 18:405–441

Chen JC, Chiu MH, Nie RL, Cordell GA, Qiu SX (2005) Cucurbitacins and cucurbitane glycosides: structures and biological activities. Nat Prod Rep 22:386–399

Chen F, Tholl D, Bohlmann J, Pichersky E (2011) The family of terpene synthases in plants: a mid-size family of genes for specialized metabolism that is highly diversified throughout the kingdom. Plant J 66:212–229

Chen XY, Berim A, Dayan FE, Gang DR (2017) A (−)-kolavenyl diphosphate synthase catalyzes the first step of salvinorin A biosynthesis in *Salvia divinorum*. J Exp Bot 68:1109–1122

Coll J, Tandron YA (2008) Neo-Clerodane diterpenoids from *Ajuga*: structural elucidation and biological activity. Phytochem Rev 7:25–49

Della Penna D, Pogson BJ (2006) Vitamin synthesis in plants: tocopherols and carotenoids. Annu Rev Plant Biol 57:711–738

Dinda B, Debnath S (2013) Monoterpenes: Iridoids. In: Ramawat KG, Mérillon JM (eds) Natural products. Springer, Berlin

Esteban R, Jiménez E, Jiménez MS, Morales D, Hormaetxe K, Becerril JM, Garcia-Plazaola JI (2007) Dynamics of violaxanthin and lutein epoxide xanthophyll cycles in Lauraceae tree species under field conditions. Tree Physiol 27:1407–1414

Etoh H, Suhara M, Tokuyama S, Kato H, Nakahigashi R, Maejima Y, Ishikura M, Terada Y, Maoka T (2012) Auto-oxidation products of astaxanthin. J Oleo Sci 61:17–21

Fernandez-Marin B, Roach T, Verhoeven A, Garcia-Plazaola J (2021) Shedding light on the dark side of xanthophyll cycles. New Physiologist 230:1336–1344

Fuchs H, Niesler N, Trautner A, Sama S, Jerz G, Panjideh H, Weng A (2017) Glycosylated triterpenoids as endosomal escape enhancers in targeted tumor therapies. Biomedicines 5(2):1–25

Garcia-Plazaola JI, Shizue Matsubara A, Osmond CB (2007) The lutein epoxide cycle in higher plants: its relationships to other xanthophyll cycles and possible functions. Funct Plant Biol 34:759–773

Garcia-Plazaola JI, Esteban R, Fernandez-Marin B, Kranner I, Porcar-Castell A (2012) Thermal energy dissipation and xanthophyll cycles beyond the *Arabidopsis* model. Photosynth Res 113:89–103

Ghassemian M, Lutes J, Tepperman JM, Chang HS, Zhu T, Wang X, Quail PH, Lange BM (2006) Integrative analysis of transcript and metabolite profiling data sets to evaluate the regulation of biochemical pathways during photomorphogenesis. Arch Biochem Biophys 448:45–59

Gilmore AM, Bjorkman O (1994) Adenine nucleotides and the xanthophylls cycle in needles. II. Comparison of the effects of CO_2 and temperature limited photosynthesis on photosystem II fluorescence quenching, the adenylate energy charge and violaxanthin de-epoxidation in cotton. Planta 192:537–544

Gilmore AM, Bjorkman O (1995) Temperature-sensitive coupling and uncoupling of ATPase-mediated, nonradiative energy dissipation: similarities between chloroplasts and leaves. Planta 197:646–654

Gilmore AM, Yamamoto HY (1992) Dark induction of zeaxanthin-dependent nonphotochemical fluorescence quenching mediated by ATP. Proc Natl Acad Sci USA 89:1899–1903

Girolomoni L, Bellamoli F, de la Cruz VG, Perozeni F, D'Andrea C, Cerullo G, Cazzaniga S, Ballottari M (2020) Evolutionary divergence of photoprotection in the green algal lineage: a plant-like violaxanthinde-epoxidase enzyme activates the xanthophyll cycle in the green alga *Chlorella vulgaris* modulating photoprotection. New Phytol 228:136–150

Hamberger B, Ohnishi T, Hamberger B, Séguin A, Bohlmann J (2011) Evolution of diterpene metabolism: Sitka spruce CYP720B4 catalyzes multiple oxidations in resin acid biosynthesis of conifer defense against insects. Plant Physiol 157:1677–1695

Hammami A, Farman M, Semmar N (2021) Highlighting aglycone-dependent glycosylation aspects in caryophyllaceae saponins by a simplex simulation approach. Curr Top Med Chem 21:612–627

Holopainen JK, Himanen SJ, Yuan JS, Chen F, Stewart CN (2013) Ecological functions of terpenoids in changing climates. In: Ramawat KG, Mérillon JM (eds) Natural products. Springer, Berlin

Horton P, Ruban AV, Walters RG (1996) Regulation of light harvesting in green plants. Annu Rev Plant Physiol Plant Mol Biol 47:665–684

Jensen SR, Franzyk H, Wallander E (2002) Chemotaxonomy of the Oleaceae: iridoids as taxonomic markers. Phytochemistry 60:213–231

Johnson SR, Bhat WW, Bibik J, Turmo A, Hamberger B, Hamberger B (2019) Genomics EM: a database-driven approach identifies additional diterpene synthase activities in the mint family (Lamiaceae). J Biol Chem 294:1349–1362

Koike K, Joa Z, NIikaido T (1998) Triterpenoid saponins from *Vaccaria segetalis*. Phytochemistry 47(7):1343–1349

Krithika R, Srivastava PL, Rani B, Kolet SP, Chopade M, Soniya M, Thulasiram HV (2015) Characterization of 10-hydroxygeraniol dehydrogenase from *Catharanthus roseus* reveals cascaded enzymatic activity in iridoid biosynthesis. Sci Rep 5:8258

Latowski D, Kuczynska P, Strzalka K (2011) Xanthophyll cycle—a mechanism protecting plants against oxidative stress. Redox Rep 16(2):78–90

Li XP, Björkman O, Shih C, Grossman AR, Rosenquist M, Jansson S, Niyogi KK (2000) A pigment-binding protein essential for regulation of photosynthetic light harvesting. Nature 403:391–395

Li Z, Peers G, Dent RM, Bai Y, Yang SY, Apel W, Leonelli L, Niyogi KK (2016) Evolution of an atypical de-epoxidase for photoprotection in the green lineage. Nat Plants 2:1–6

Lipko A, Paczkowski C, Perez-Fons L, Fraser PD, Kania M, Hoffman-Sommer M, Danikiewicz W, Rohmer M, Poznanski J, Swiezewska E (2023) Divergent contribution of the MVA and MEP pathways to the formation of polyprenols and dolichols in *Arabidopsis*. Biochem J 480:495–520

Maoka T (2009) Recent progress in structural studies of carotenoids in animals and plants. Arch Biochem Biophys 483:191–195

Maoka T (2020) Carotenoids as natural functional pigments. J Nat Med 74:1–16

Maoka T, Etoh H (2016) Biological antioxidation mechanism: quenching of peroxynitrite, in Japanese food. In: Shi J (ed) Functional food ingredients and nutraceuticals processing technologies, 2nd edn. CRC Press, Boca Raton, pp 589–607

Melendez-Martinez AJ, Britton G, Vicario IM, Heredia FJ (2007) Relationship between the coulour and the chemical structure of carotenoid pigments. Food Chemistry 101:1145–1150

Merritt AT, Ley SV (1992) Clerodane diterpenoids. Nat Prod Rep 9:243–287

Miettinen K, Dong L, Navrot N, Schneider T, Burlat V, Pollier J, Woittiez L, van der Krol S, Lugan R, Ilc T, Verpoorte R, Oksman-Caldentey KM, Martinoia E, Bouwmeester H, Goossens A,

Memelink J, Werck-Reichhart D (2014) The seco-iridoid pathway from *Catharanthus roseus*. Nat Commun 5:3606
Miro M (1995) Cucurbitacins and their pharmacological effects. Phytother Res 9:159–168
Monoz O, Delporte C, Backhouse N, Erazo S, Erazo S, Negrete R, Maldonado S, Lopez-Perez JL, San Feliciano A (2000) A new cucurbitacin glycoside from *Kageneckia oblonga* (Rosaceae). Z Naturforsch C Biosci 55:141–145
O'Connor SE, Maresh JJ (2005) Chemistry and biology of monoterpene indole alkaloid biosynthesis. Nat Prod Rep 23:532–547
Ohyama K, Suzuki M, Kikuchi J, Saito K, Muranaka T (2009) Dual biosynthetic pathways to phytosterol via cycloartenol and lanosterol in *Arabidopsis*. Proc Natl Acad Sci USA 106:725–730
Ortega A, Blount JF, Manchand PS (1982) Salvinorin, a new trans-neoclerodane diterpene from *Salvia divinorurn* (Labiatae). J Chem Soc Perkin Trans I:2505–2508
Peltier G, Aro EM, Shikanai T (2016) NDH-1 and NDH-2 plastoquinone reductases in oxygenic photosynthesis. Annu Rev Plant Biol 67:55–80
Piozzi F, Bruno M, Rosselli S, Maggio A (2006) The diterpenoids of the genus *Marrubium* (Lamiaceae). Nat Prod Commun 1(7):585–592
Re EB, Jones D, Learned RM (1995) Coexpression of native and introduced genes reveals cryptic regulation of HMG CoA reductase expression in *Arabidopsis*. Plant J 7:771–784
Rehm S, Enslin PR, Meeuse ADJ, Wessels JM (1957) Bitter principles of the cucurbitaceae. VII. The distribution of bitter principles in this plant family. J Sci Food Agric 8:679–686
Rohdich F, Wungsintaweekul J, Luttgen H, Fischer M, Eisenreich W, Schuhr CA, Fellermeier M, Schramek N, Zenk MH, Bacher A (2000) Biosynthesis of terpenoids: 4-diphosphocytidyl-2-C-methyl-D-erythritol kinase from tomato. Proc Natl Acad Sci USA 97:8251–8256
Sampaio-Santos MI, Kaplan MAC (2001) Biosynthesis significance of iridoids in chemosystematics. J Braz Chem Soc 12(2):144–153
Sarraj-Laabidi A, Messai H, Hammami-Semmar A, Semmar N (2017) Chemometric analysis of inter- and intra-molecular diversification factors by a machine learning simplex approach. A review and research on *Astragalus* saponins. Curr Top Med Chem 17:2820–2848
Sarraj-Laabidi A, Lacaille-Dubois MA, Semmar N (2018) Structural organization of cycloartane-based saponins in the genus *Astragalus* (Fabaceae). Phytochem Rev 17:431–452
Sarraj-Laabidi A, Semmar N (2017) Interspecific chemical differentiation within the genus *Astragalus* (Fabaceae) based on sequential variability of saponin structures. In: Ansari AA, Gill SS (eds) Plant biodiversity: monitoring, assessment and conservation. CABI Publishers
Sawai S, Kazuki Saito K (2011) Triterpenoid biosynthesis and engineering in plants. Front Plant Sci 2(25):1–8
Schun Y, Cordell GA (1985) Studies in the thymelaeaceae. III. Constituents of *Gyrinops walla*. J Nat Prod 48:684–685
Tan Y, Zhang Q, Zhao W, Liu Z, Ma M, Zhong M, Wang M, Xu B (2020) Chlororespiration serves as photoprotection for photo-inactivated oxygen-evolving complex in *Zostera marina*, a marine angiosperm. Plant Cell Physiol 61:1517–1529
Tanaka Y, Sasaki N, Ohmiya A (2008) Biosynthesis of plant pigments: anthocyanins, betalains and carotenoids. Plant J 54:733–749
Thimmappa R, Geisler K, Louveau T, O'Maille P, Osbourn A (2014) Triterpene biosynthesis in plants. Annu Rev Plant Biol 65:225–257
Tholl D (2015) Biosynthesis and biological functions of terpenoids in plants. Adv Biochem Eng Biotechnol 148:63–106
Tsuboi M, Kato K, Etoh H, Nakatugawa H, Kato H, Maejima Y, Matumoto G, Mori H, Hosokawa M, Miyashita K, Tokuda H, Suzuki N, Maoka T (2011) Nitrocapsanthin and nitrofucoxanthin, respective products of capsanthin and fucoxanthin reaction with peroxynitrite. J Agric Food Chem 59:10578–10872
Verhoeven AS, Adams WW, Demmig-Adams B (1999) The xanthophyll cycle and acclimation of *Pinus ponderosa* and *Malva neglecta* to winter stress. Oecologia 118:277–287

Vranova E, Coman D, Gruissem W (2013) Network analysis of the MVA and MEP pathways for isoprenoid synthesis. In: Merchant SS (ed) Ann Rev Plant Biol 64:665–700

Wong RJ, McCormack DK, Russell DM (1982) Plastid 3-hydroxy-3-methylglutaryl coenzyme reductase has distinctive kinetic and regulatory features: properties of the enzyme and positive phytochrome control of activity in pea seedlings. Arch Biochem Biophys 216:631–638

Zerbe P, Chiang A, Dullat H, O'Neil-Johnson M, Starks C, Hamberger B, Bohlmann J (2014) Diterpene synthases of the biosynthetic system of medicinally active diterpenoids in *Marrubium vulgare*. Plant J 79:914–927

Zhou F, Pichersky E (2020) More is better: the diversity of terpene metabolism in plants. Curr Opin Plant Biol 55:1–10

Chapter 6
Nitrogen-Containing Compounds

Nitrogen-containing compounds mainly include three classes of SMs: alkaloids, glucosinolates and cyanogenic glycosides (Fig. 6.1).

Alkaloids are reactive compounds with alkaline (basic) character favored by the occurrence of nitrogenous heterocycle (Fig. 6.1a). They are at the origin of many drugs. Glucosinolates are nitrogenated compounds containing glucose and sulfur (S, SO_3^-) (Fig. 6.1b). They have irritating effect due to their release of isothiocyanate (Fig. 6.1c) (Ishida et al. 2014). Cyanogenic glycosides are nitrogenated and glycosylated compounds containing cyanide group (C≡N) (Fig. 6.1d). They are potentially toxic due to the release of hydrogen cyanide (H–C≡N) after enzymatic hydrolysis (Fig. 6.1e) (Gleadow and Moller 2014).

6.1 Alkaloids

Alkaloids are low molecular weight nitrogenous compounds generally showing alkaline character due to basic nitrogen atom(s). Apart from the three constitutive atoms (carbon, hydrogen, nitrogen), alkaloids can contain oxygen, sulfur and other (less common) atoms such as phosphore, chlorine and bromine.

6.1.1 General Structural Aspects

Alkaloids show more complex structural organization compared with phenolic compounds and terpenes. Combined classification criterion was suggested involving biosynthetic origins with nitrogen location leading to three main families: true alkaloids, proto-alkaloids and pseudo-alkaloids (Fig. 6.2):

N. Semmar, *Secondary Metabolites in Plant Stress Adaptation*, Signaling and Communication in Plants, https://doi.org/10.1007/978-3-031-52595-7_6

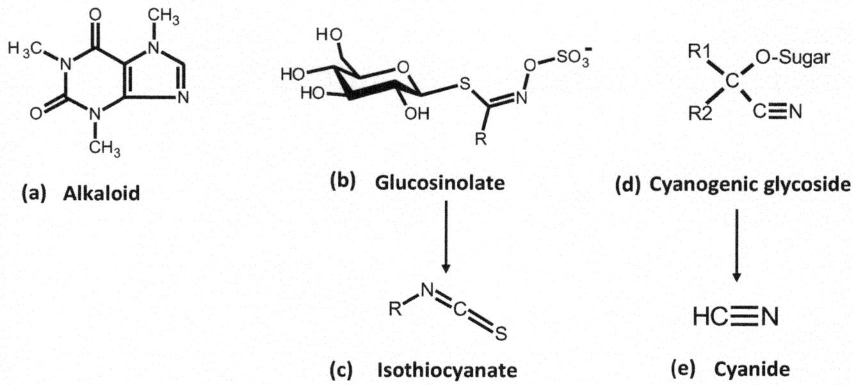

Fig. 6.1 Illustrative structures of alkaloids, glucosinolates and cyanogenic glycosides representing different nitrogen-containing compounds in secondary metabolites

- True alkaloids are characterized by the presence of heterocyclic nitrogen with a scaffold directly issued from amino acids (Fig. 6.2a).
- Proto-alkaloids are issued from amino acid but contain nitrogen in lateral chain rather than in heterocycle (Fig. 6.2b).
- Pseudo-alkaloids have scaffolds that are not derived from amino acids and possess heterocyclic nitrogen (Fig. 6.2c).

This classification of alkaloids combines functional and structural aspects consisting of initial metabolic pathway (precursor type) and final nitrogenous scaffold (derivative compound).

6.1.2 Biosynthesis Processes

Biosynthesis of alkaloids implies four main steps (i–iv) starting with the accumulation of two precursor components _ (i) an amine and (ii) an aldehyde _ at enough levels. Interaction between these precursors results in (iii) the formation of an electrophilic cation. Then, the cation interacts with a nucleophile group (iv) leading to the formation of alkaloid scaffold (Fig. 6.3) (Lichman 2021). Amine group originates from amino acid metabolism (Fig. 6.3a, b); it can be directly incorporated with the amino acid that will constitute the scaffold of alkaloid (in true alkaloids and protoalkaloids) or transmitted from an amino acid by a transamination process (in pseudoalkaloids).

Accumulation of amines results in an efficient metabolic flux that will be directed for alkaloid synthesis (Fig. 6.3b). Amines contributing to iminium formation can be under polyamine chain or aromatic forms: polyamines derive from ornithine, lysine, arginine; aromatic amines are associated with tryptophan and tyrosine (Fig. 6.6c, d).

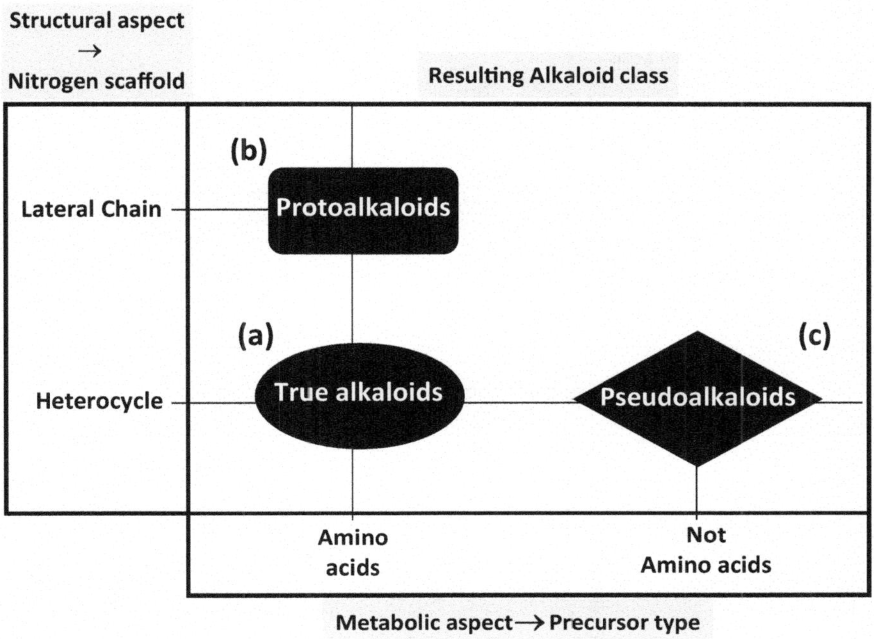

Fig. 6.2 Classification of alkaloids combining precursor type (or scaffold origin) and nitrogen location in the molecule

Independently from the amine precursor, aldehyde precursor is also accumulated at enough level to react with amine for alkaloid biosynthesis (Fig. 6.3c). Aldehyde can be originated from various sources including phenylacetaldehydes, phenylpropanoid derivatives (e.g. benzaldehydes) and segologanin (Fig. 6.4).

Condensation of a primary amine with an aldehyde gives a carbonolamine (Fig. 6.3d) which transforms to iminium cation after hydroxyl loss and water formation (Fig. 6.3e). Iminium cation is a reactive intermediate characterized by rapid formation and highly electrophilic aspect (Fig. 6.3e). Accumulation of iminium is favored in upstream by the p*K*a of the amine precursor (Lichman 2021). Rapid formation of iminium results in its accumulation even if it is not formed at high concentrations. Accumulation of iminium favors its reaction with a nucleophile through an irreversible and energetically favorable process known as Mannich-like reaction (Fig. 6.3f). Mannich-like reaction quenches the reactivity of both electrophile iminium and nucleophile through the formation of a new covalent C–C bond. It provides specific (iminium-nucleophile)-combinations leading to the formation of alkaloid scaffolds (Fig. 6.3g).

A more illustrative example is given in Fig. 6.4 showing the formation of strictosidine (an alkaloid with terpenoid indole scaffold) (Fig. 6.4d) from tryptamine (amine source) (Fig. 6.4b) and secologanin (aldehyde source) (Fig. 6.4c). Tryptamine is

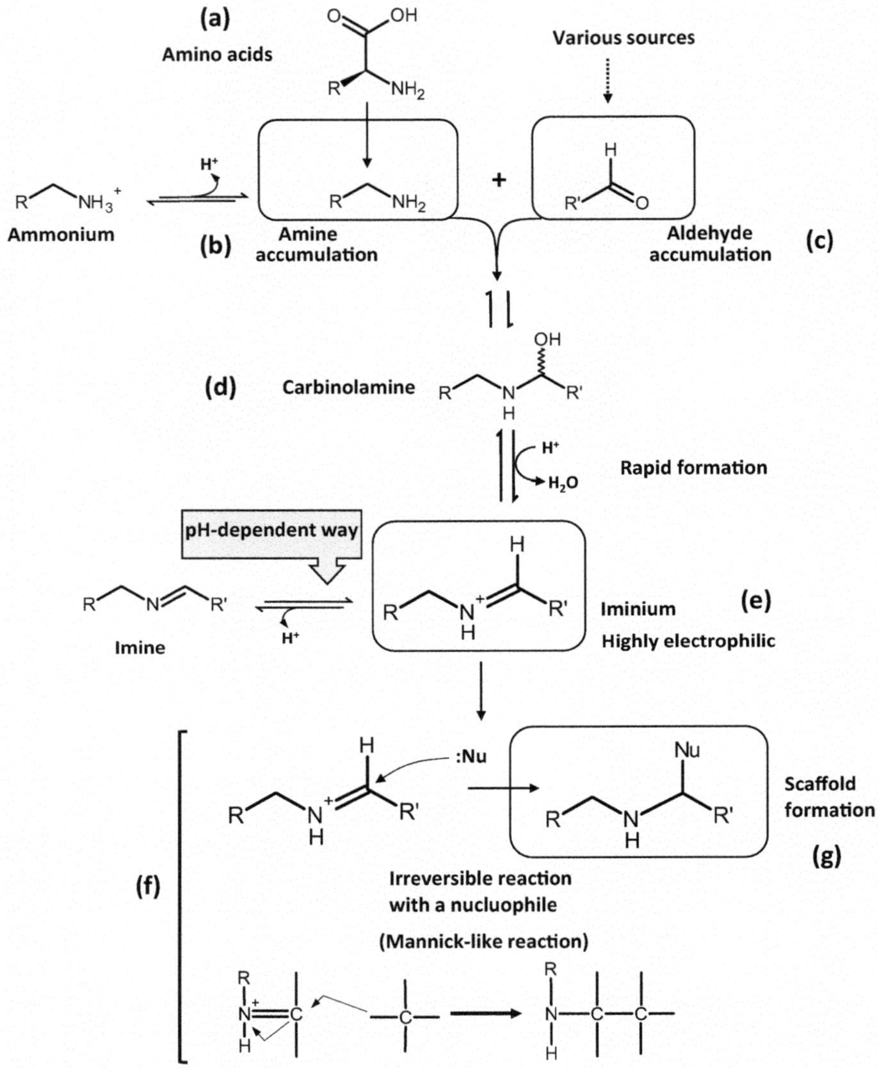

Fig. 6.3 Biosynthesis steps of alkaloid scaffolds by Mannich-like reaction preceded by the formation of electrophile cation by initial condensation of amine with aldehyde precursors

issued from decarboxylation of tryptophan (Fig. 6.4a, b). Secologanin has secoiridoid monoterpene structure (Fig. 5.5k). Tryptamine and secologanin represent amine and aldehyde sources, respectively, and can interact in a metabolic flux leading to their condensation and the formation of strictosidine (Fig. 6.4d). Strictosidine is the first appeared monoterpenoid indole alkaloid and serves as precursor for further products of this metabolic family (Fig. 6.4e).

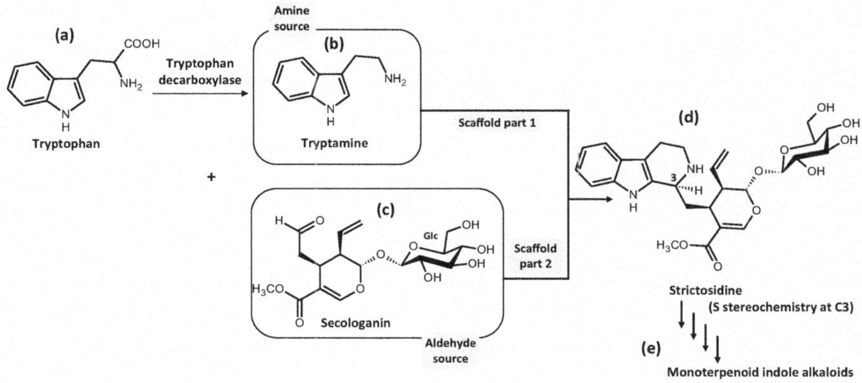

Fig. 6.4 Condensation between tryptamine (amine source) and secologanin (aldehyde source) for the formation of strictosidine (alkaloid with a terpenoid indole scaffold)

6.1.3 Structural Classification

Alkaloids show different classification ways including (combining) metabolic origins, nitrogen location aspect in molecule and molecular scaffold shapes (or nitrogen rings). Metabolic origins include amino-acids and not amino-acids; location aspects include intra-cyclic nitrogen and lateral chain nitrogen. By combining metabolic origin with nitrogen location aspect, alkaloids were classified into three main groups (Fig. 6.2) (Sect. 6.1.1 in Part II—in this chapter): true alkaloids, proto-alkaloids and pseudo-alkaloids.

Scaffolds resulting from condensation of iminium cation with nucleophile results in the formation of scaffolds or aglycones that play governing role in alkaloid identities and downstream reactions. Aglycones are characterized by different shapes associated with different precursors and helpfully serving for identifying alkaloid families.

Heterocyclic scaffolds play key role in classification of true alkaloids (Fig. 6.5). They originate from different amino acids and differ by their numbers of cycles, their cycle sizes and their conjugation levels. Monocyclic heterocycles include (Fig. 6.5): pyrrolidine (numbered *1* in Fig. 6.5), pyridine, piperidine (num. *3*), imidazols (*10*) and pyrimidine (*11*). Moreover, monocyclic scaffolds can differ by their total number of carbons: for instance, pyrrole (*1*) has a pentagon structure with 4 carbons (and 1 nitrogen) whereas piperidine (*3*) has hexagon structure with 5 C and 1 N; however, pyrimidine (*11*) shows also hexagon structure but with 4 C versus 2 N. Bicyclic scaffolds include pyrrolizidine (numbered *2* in Fig. 6.5), tropane (*4*), quinoline (*5*), isoquinoline (*6*), norlupane (*7*), indole (*8*), indolizidine and indolizine (*9*). However, purine (*12*) has also a bicyclic scaffold, but it is closely associated with pseudoalkaloid subtypes deriving from nucleic bases (adenine, guanine) (Fig. 6.15a, b). Scaffolds with similar cyclization levels and shapes can be distinguished by different conjugation levels. This can be illustrated by pyrrole and pyrrolidine (numbered *1* in Fig. 6.5), pyridine and piperidine (*3*), indolizidine and indolizine (*9*). Scaffold with more than two cycles levels exist such as phenenthridine (3 cycles) and aporphine (4 cycles) (Fig. 6.5, numbers *14*, *15*, respectively). They are more common in pseudoalkaloid (e.g. terpenoid alkaloids) than true alkaloids (Fig. 6.14).

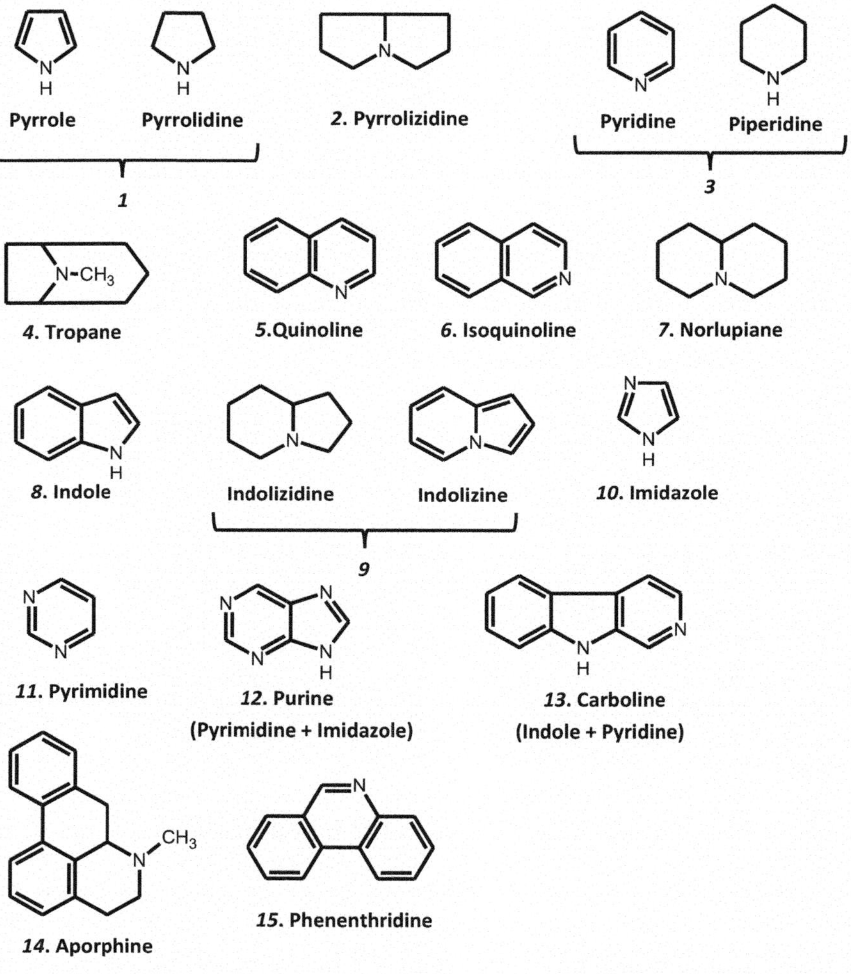

Fig. 6.5 Different nitrogenous heterocycles representing various aglycones (scaffolds) associated with different alkaloid types (mainly observed within the true alkaloids)

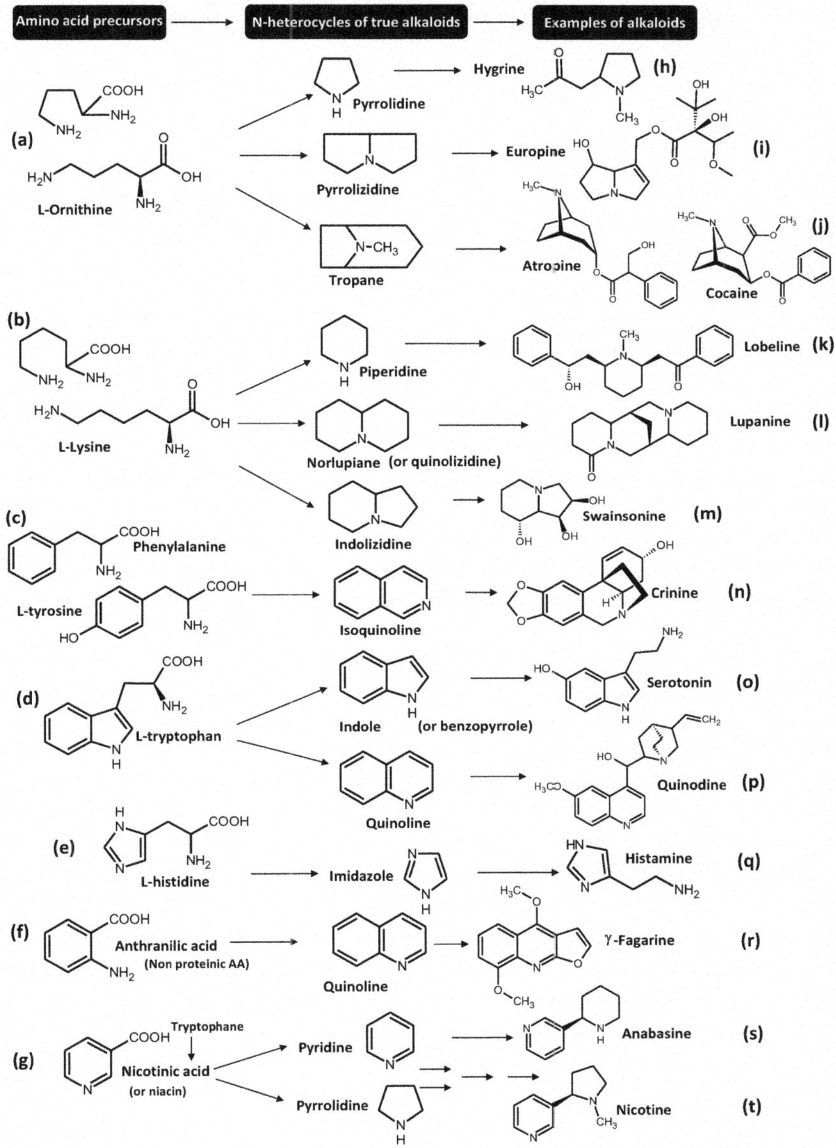

Fig. 6.6 Different scaffold structures (N-heterocycles) of true alkaloids with associated amino acid precursors and illustrative derived alkaloids

6.1.4 Nitrogenated Scaffold Synthesis and Structural Diversification Processes

6.1.4.1 True alkaloids

Different scaffolds of true alkaloids (and protoalkaloids) are originated from different amino acids through several types of metabolic processes (Fig. 6.6). In upstream, amino acids can be subjected to catabolic reactions, duplications, cyclization or chain closure leading to the formation of different true alkaloid scaffolds. Thus, among chain amino acids, L-ornithine gives pyrrolidine, pyrrolizidine and tropane scaffolds by cyclization and condensation (Fig. 6.6a). Also, L-lysine gives piperidine, norlupiane (or quinolizidine) and indolizidine by analogous metabolic processes (Fig. 6.6b). Among aromatic amino acids, L-tyrosine (issued from hydroxylation of phenylalanine) provides isoquinoline, whereas tryptophan is at the origin of indole and quinoline scaffolds (Fig. 6.6c, d).

Nicotinic acid (or niacin), deriving from tryptophan, gives pyridine and pyrrolidine type heterocycles (Fig. 6.6g). Anthranilic acid (an aromatic nonproteinic amino acid) gives quinoline (Fig. 6.6f). L-histidine _characterized by a di-nitrogenated heterocycle_ provides imidazole scaffold (Fig. 6.6e).

In downstream, alkaloid scaffolds can be transformed into many specific alkaloids via different condensation and substitution reactions. These precursor-derivative processes can be illustrated by pyrrolidine-hygrine (Fig. 6.6h), pyrrolizidine-europine (Fig. 6.6i), tropane-atropine and tropane-cocaine (Fig. 6.6j), piperidine-lobeline (Fig. 6.6k), norlupiane-lupanine (Fig. 6.6l), indolizidine-swaisonine (Fig. 6.6m), isoquinoline-crinine (Fig. 6.6n), indole-serotonin (Fig. 6.6o), quinoline-quinodine (Fig. 6.6p), imidazole-histamine (Fig. 6.6q), quinoline-γ-fagarine (Fig. 6.6r), pyridine-anabasine (Fig. 6.6s), (pyridine, pyrrolidine)-nicotine (Fig. 6.6t).

Fingerprints of amino acid precursors in final alkaloid products are further illustrated by transformation cases including arginine, tyrosine and tryptophan (Fig. 6.7). Arginine (chained AA) is degraded to putrescine which is duplicated by condensation to provide homospermidine (Fig. 6.7a). Homospermidine chain cyclizes to give pyrrolizidine scaffold serving for the synthesis of senecionine (Fig. 6.7b). Tyrosine (aromatic AA) transforms into L-DOPA by hydroxylation, then the L-DOPA is condensed and cyclized leading to the formation of betanin (Fig. 6.7c). Betanin (or betacyanin) is a betalain type pigment giving variable color with pH (from red to purple blue). It is responsible for the red color of beet. A third illustrative case is provided by the synthesis of harmaline and harmine from tryptophan via tryptamine (Fig. 6.7d). Tryptophan fingerprint is recognizable in harmaline through the indole scaffold (Fig. 6.7d, bold lines).

An extended list of alkaloids associated with different nitrogenous heterocyclic scaffolds is given in Fig. 6.8.

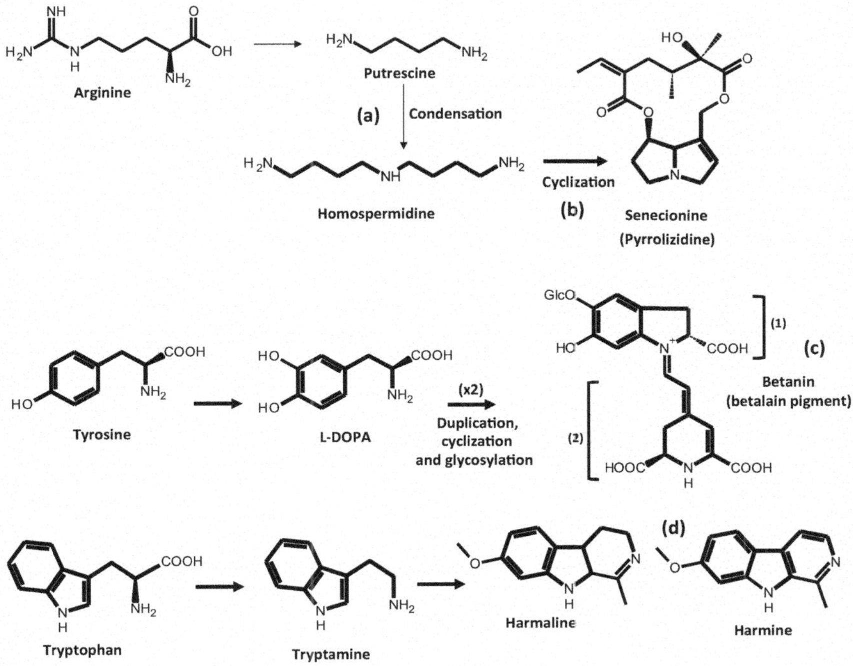

Fig. 6.7 Three illustrative cases of amino acid precursor scaffolds of true alkaloids after different intermediate metabolic processes including condensation, cyclization and chemical substitutions. Fingerprints (portions) of amino acid precursors in alkaloid products are shown in bold lines

6.1.4.2 Protoalkaloids

Protoalkaloids are alkaloids deriving from amino acids but they oppose to true alkaloids by the occurrence of nitrogen in lateral chain rather than in heterocycle (Fig. 6.2). The most common protoalkaloids are issued from aromatic amino acids including phenylalanine and its hydroxylated derivatives , tyrosine and dihydroxy-phenylalanine (or DOPA) (Fig. 6.9a–c) (Robinson 1981). These three amino acids submit decarboxylation leading to phenylethylamine, tyramine and dopamine, respectively (Fig. 6.9d–f). These three amines serve as direct precursors of protoalkaloids (Fig. 6.9g–m). Tyramine (Fig. 6.9e) and its methylated derivatives (Fig. 6.9h–j) are relatively more common in the plant world. Phenylethylamine (Fig. 6.9d) is relatively widespread in plants. Dopamine (Fig. 6.9f), decarboxylation product of DOPA, was found in high concentration in banana peel where it is involved in the browning reaction (Griffiths 1959; Danyen et al. 2009). Several derivative protoalkaloids are synthesized by further hydroxylation and sequential methylation processes: (i) from thyramine, mono-, di- and trimethylation of nitrogen results in the formation of N-methyltyramine, hordenine and candicine, respectively (Fig. 6.9h–j). Due to sequential methylation of nitrogen,

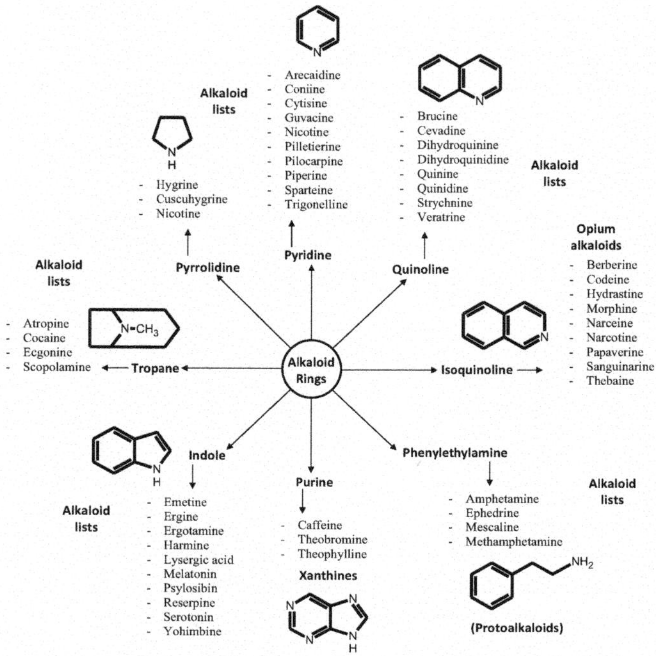

Fig. 6.8 Lists of several alkaloids with different scaffolds based on different nitrogenous heterocycles

these three protoalkaloids have secondary, tertiary and quaternary (cation) amine structures, respectively. (ii) Dopamine is hydroxylated to noradrenaline (Fig. 6.9k) then noradrenaline gives epinephrine (Fig. 6.9l) by methylation of nitrogen. Moreover, dopamine can give mescaline (Fig. 6.9m) by methoxylation of phenolic cycle. (iii) Also, ephedrine (Fig. 6.9g) is issued from phenylethylamine (Fig. 6.9d) by dimethylation of nitrogen and carbon of lateral chain.

6.1.4.3 Pseudoalkaloids

Pseudoalkaloids differ from true alkaloids and protoalkaloids by a downstream introduction of nitrogen in molecules. This late process occurs through enzymatic transamination where N-atom can be provided by an amino acid or another nitrogen source. At upstream stage, biosynthesis of pseudoalkaloids can occur from various sources including (i) precursors or (ii) postcursors (degradation products) of amino acids, and (iii) non-amino acid compounds (Aniszewski 2007).

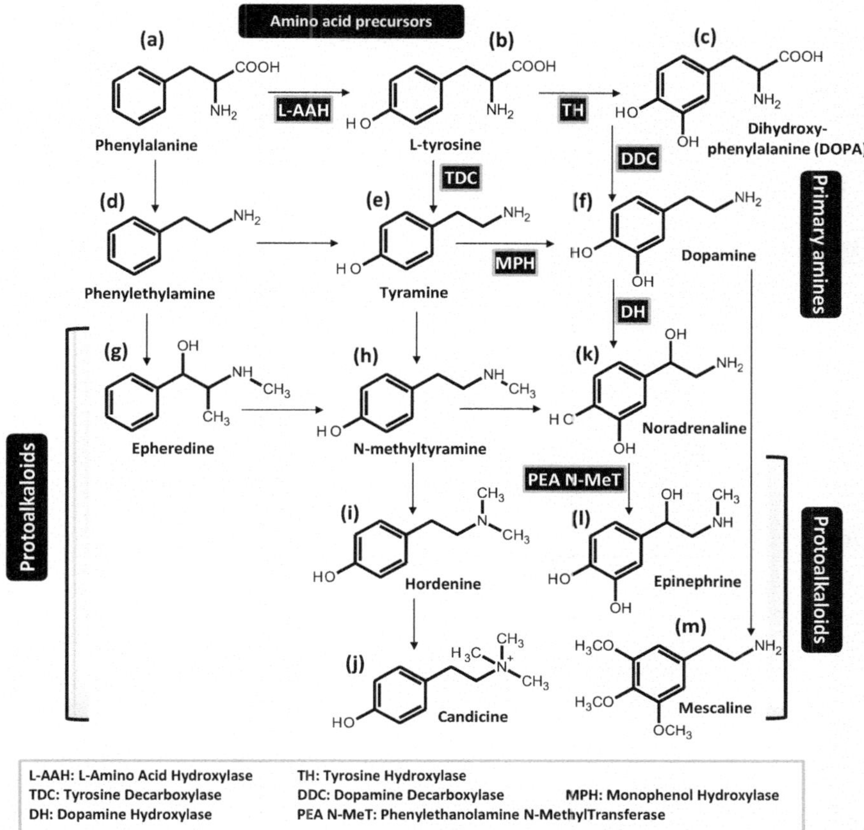

Fig. 6.9 Different metabolic processes leading to sequential synthesis of protoalkaloids from phenylalanine, tyrosine and dihydroxy-phenylalanine (DOPA)

Acetate, aromatic compounds and terpenes can be involved among the precursors of pseudoalkaloids. Lab's experiments on chemical synthesis showed key roles of amine and aldehyde in the synthesis of piperidine ring (Fig. 6.10a). This experience involved four structural components including ammonium acetate (NH_4OAc), two aldehyde forms (formaldehyde, aromatic aldehyde) and malononitrile (commonly used in organic synthesis); NH_4OAc played the role of both catalyst and nitrogen source. These four components subjected a cascade of synthesis reactions in methanol under reflux (for 2h) leading to the formation of 2,6-diaryl-3,3,5,5-tetracyanopiperidine with good to excellent yields (Fig. 6.10b) (Aniszewski 2007). This product has the scaffold of piperidine ring which characterizes several alkaloids including pinidine (found in pines), sedamine (found in *Sedum* genus) and lobeline (in *Lobelia*) (Fig. 6.11).

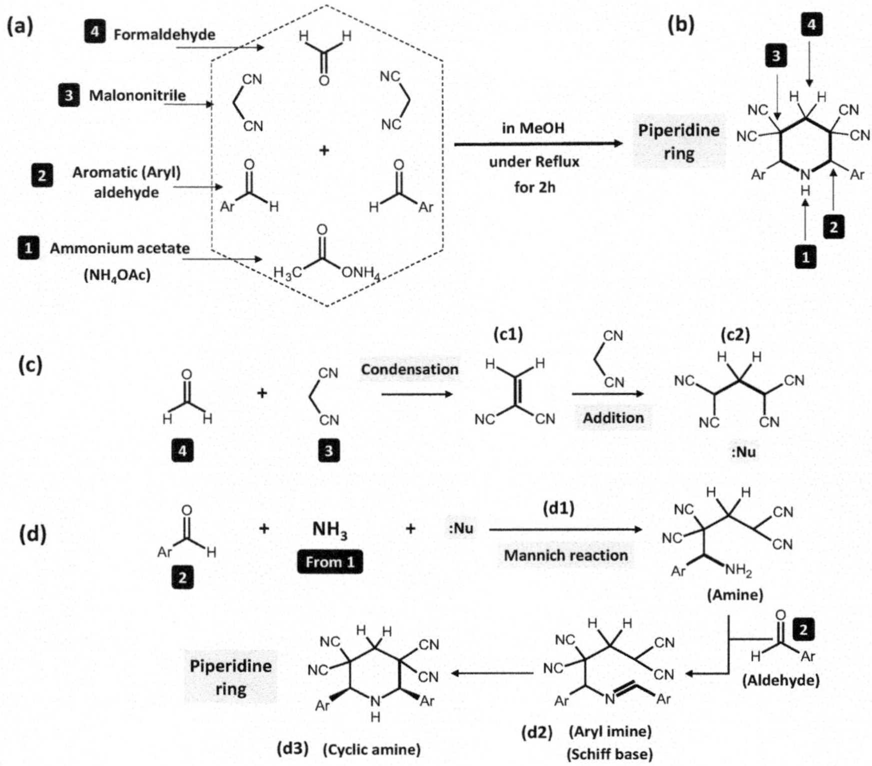

Fig. 6.10 Cascade of chemical synthesis reactions leading to the formation of piperidine (a pseudoalkaloid precursor) from condensation between amine, aldehyde and nucleophile. **a** Structural reactive components; **b** final product; **c**, **d** sequential synthesis processes

Initially malononitrile reacts with formaldehyde to form a condensate (Fig. 6.10c1); this reaction (called Knoevenagel condensation) is favored by catalytic effect of ammonium acetate. The condensate is attacked by a second malononitrile leading to the formation of tetracyanopropane (Fig. 6.10c2) (Vereshchagin et al 2018). Tetracyanopropane (:Nu) represents nucleophile which will interact with iminium in Mannich reaction. This reaction involves the participation of aldehyde, ammonia (issued from NH_4OAc) and nucleophile (Fig. 6.10d) leading to amine formation (Fig. 6.10d1). The process continues by reaction of the formed amine with aldehyde leading to a Schiff base (aryl imine) (Fig. 6.10d2). Finally, aryl imine is cyclized to give the sterically less hindered cyclic amine (Fig. 6.10d3). The scaffold of this final product corresponds to piperidine ring.

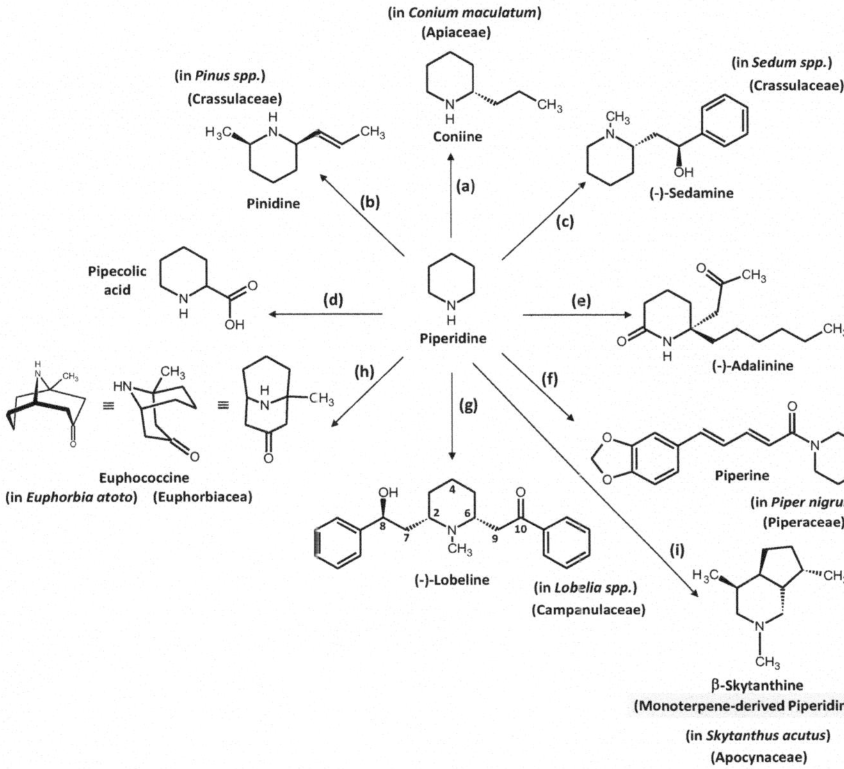

Fig. 6.11 Various pseudoalkaloids based on piperidine scaffold, associated with different plant taxa and showing structural variability governed by different types, levels and positions of chemical substitutions

Pseudoalkaloids are distinct from true alkaloids and protoalkaloids by structural heterogeneity of precursors including piperidine, terpenes, steroids, phenyls, phenols and purine (Figs. 6.11 and 6.12) (Dewick 2009). Piperidine alkaloids are recognizable by the presence of a piperidine nucleus in their structure consisting of a six-membered saturated heterocyclic ring. Piperidine alkaloids have wide botanical distribution with various structures characterized by different chemical substitutions on different carbons or nitrogen of piperidine (Fig. 6.11). The names of piperidine derivatives are often originated from plant species in which they were elucidated. Coniine derives from substitution of a propane chain on piperidine (Fig. 6.11a) whereas pinidine (Fig. 6.11b) is disubstituted by propene and methyl in ortho positions of nitrogen. However, sedamine has methylated nitrogen with a benzo-ethane chain (Fig. 6.11c), and pipecolic acid results from piperidine carboxylation (Fig. 6.11d).

Beyond these simple substitution patterns, more complex molecules are synthesized through more numbers, types and positions of substitutions on piperidine (Fig. 6.11e–h):

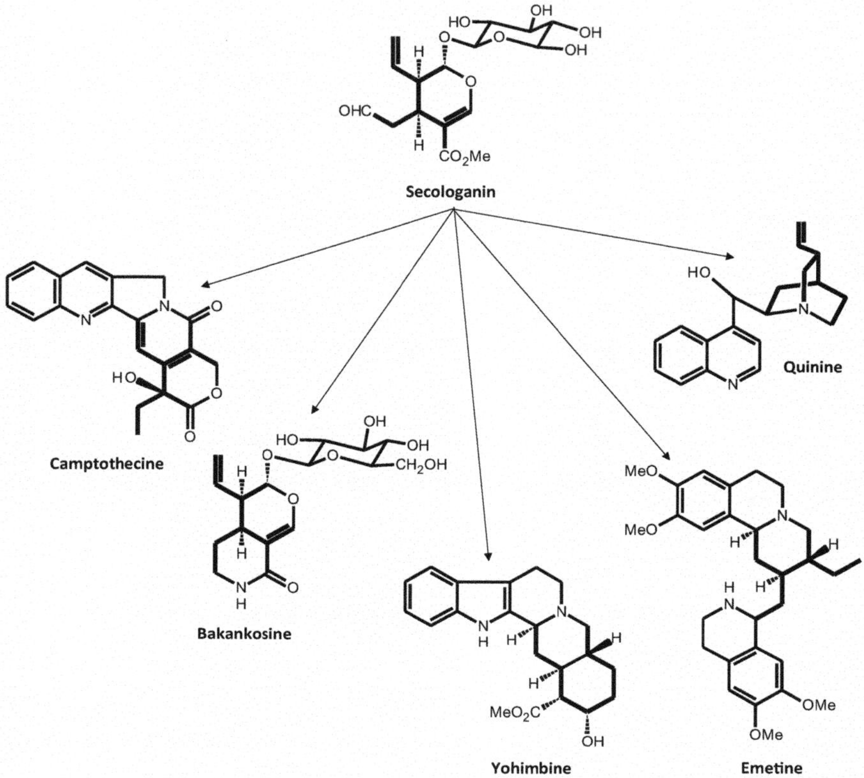

Fig. 6.12 Different monoterpenoid alkaloids derived from secologanin (secoiridoid monoterpene). Monoterpene scaffolds in different derivative alkaloids were shown in bold lines. OMe, methoxyl

- Adalinine (Fig. 6.11e) has quaternary carbon in heterocycle due to its substitution by both hexanal and acetone in addition to CH_2 and NH.
- Lobeline (Fig. 6.11g) contains piperidine ring with central position which is substituted on both sides (at C2 and C6) with two phenyl structures containing lateral chains with hydroxyl and keto functionalities at C8 and C10, respectively. When, hydroxyl and ketone are absent, molecular structure corresponds to lobelane which has symmetrical aspect on both sides of piperidine. Moreover, synthesis chemistry revealed various possible substitutions other than CH_3 occurring on the nitrogen (Zheng et al. 2013). Lobeline is found in *Lobelia* spp. (Campanulaceae) (St-Pierre et al. 2017; Tamboli et al. 2012).
- Piperine (Fig. 6.11f) consists of piperidine connecting an aromatic moiety (3,4-methylenedioxyphenyl) via a conjugated aliphatic chain of 5 carbons. The whole substitution (aliphatic and aromatic parts) corresponds to piperic acid having lost the OH of COOH to fix on piperidin. Piperine occurs naturally in black, green,

and white pepper (*Piper* spp.) (Tripathi et al. 2022). It was revealed to have considerable benefit pharmacological effects (Ul-Haq et al. 2020).

- Euphococcine shows a bicyclic structure where the amine of piperidine is at the intersection of two condensed six-members-rings (Fig. 6.11h). This alkaloid was first elucidated from the Australian coastal plant *Euphorbia atoto* (Hart et al. 1967).
- β-Skytanthine (Fig. 6.11i) shows piperidine nucleus within a monoterpene scaffold providing a fingerprint of monoterpene precursor in upstream of metabolism. Skytanthine was isolated as original monoterpenoid alkaloid from *Skytanthus acutus* Meyen (Apocynaceae) (Djerassi et al. 1962).

Apart from piperidine, several other types of precursors are concerned with biosynthesis of pseudoalkaloids leading to various (heterogenous) scaffolds. Heterogeneous set of pseudoalkaloid precursors include terpenoids, steroids, phenolic compounds and purines; representative structures are presented in the following paragraphs (Figs. 6.12, 6.13, 6.14 and 6.15).

Various monoterpene alkaloids derived from secologanin (secoiridoid monoterpene) are presented in Fig. 6.12 (Yang et al. 2021; Yamazaki et al. 2004; Trenti et al. 2021; Brown et al. 2000; Tietze 1983). The fingerprint of monoterpene scaffold in alkaloids is shown by bold lines.

Sesquiterpene alkaloids (*SAk*) are widely distributed in marine organisms and terrestrial plants (Chen et al. 2020). In marine organisms, they are produced by sponges, soft corrals, nudibranchs and slugs. In terrestrial plants, *SAk* are distributed between different families characterized by specific aglycones (Fig. 6.13) (Chen et al. 2020). Sesquiterpene aglycones of terrestrial plants include eudesmane (Fig. 6.13a), bisabolane (Fig. 6.13b), germacrane (Fig. 6.13c), elemane (Fig. 6.13d), farnesane (Fig. 6.13e), guaiane (Fig. 6.13f). Sesquiterpene alkaloid patterns are based on (i) eudesmane and elemane in Zingiberaceae (Fig. 6.13a, d1), (ii) germacrane, elemane and guaiane in Asteraceae (Fig. 6.13c3, d2, f), (iii) germacrane in Magnoliaceae and Lamiaceae (Fig. 6.13c1, c2), farnesane in Primulaceae and Myoporacrae (Fig. 6.13e), bisabolane in Phyllanthraceae (Fig. 6.13b) (Chen et al. 2020). Finally, plant species of Celastraceae family were specifically characterized by the production of dihydroagarofuran type *SAk* leading to highly distinct chemotaxonomical patterns in the plant world (Fig. 6.13g) (Smith 1977; Chen et al. 2020). Dihydroagarofuran type-*SAk* were found in several genera of Celastraceae including *Maytenus*, *Celastrus*, *Eunymus*, *Tripterygium*, *Monimopetalum*.

Steroid alkaloids (*StA*) represent also a wide category of pseudoalkaloids with high structural variability and chemotaxonomical interest (Zhao et al. 2021; Xiang et al. 2022). Structural variability of this metabolic family is due to many types of steroid scaffolds including conanine, paravallarine, pregnane, cyclopregnane, 9(10 → 19)Abeo-14α-methylpregnane, cevanine, veratamine, spirosolane, solanidine, verazine and samandrine (Fig. 6.14).

Conanine type alkaloids are characterized by a five membered E ring with 18,20-epimino group (Fig. 6.14a). The C3 is generally oxygenated or nitrogenous.

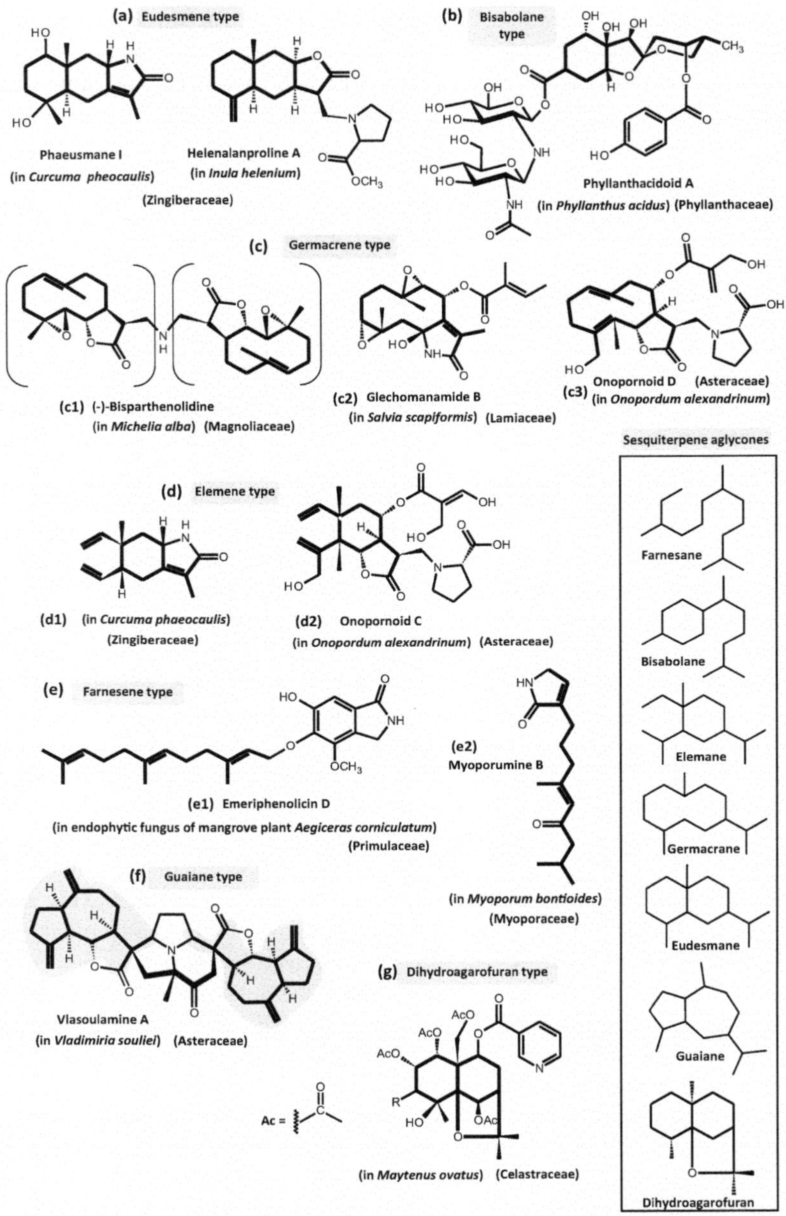

Fig. 6.13 Illustrative cases of sesquiterpene alkaloids based on different sesquiterpene aglycones and issued from different plant species (families)

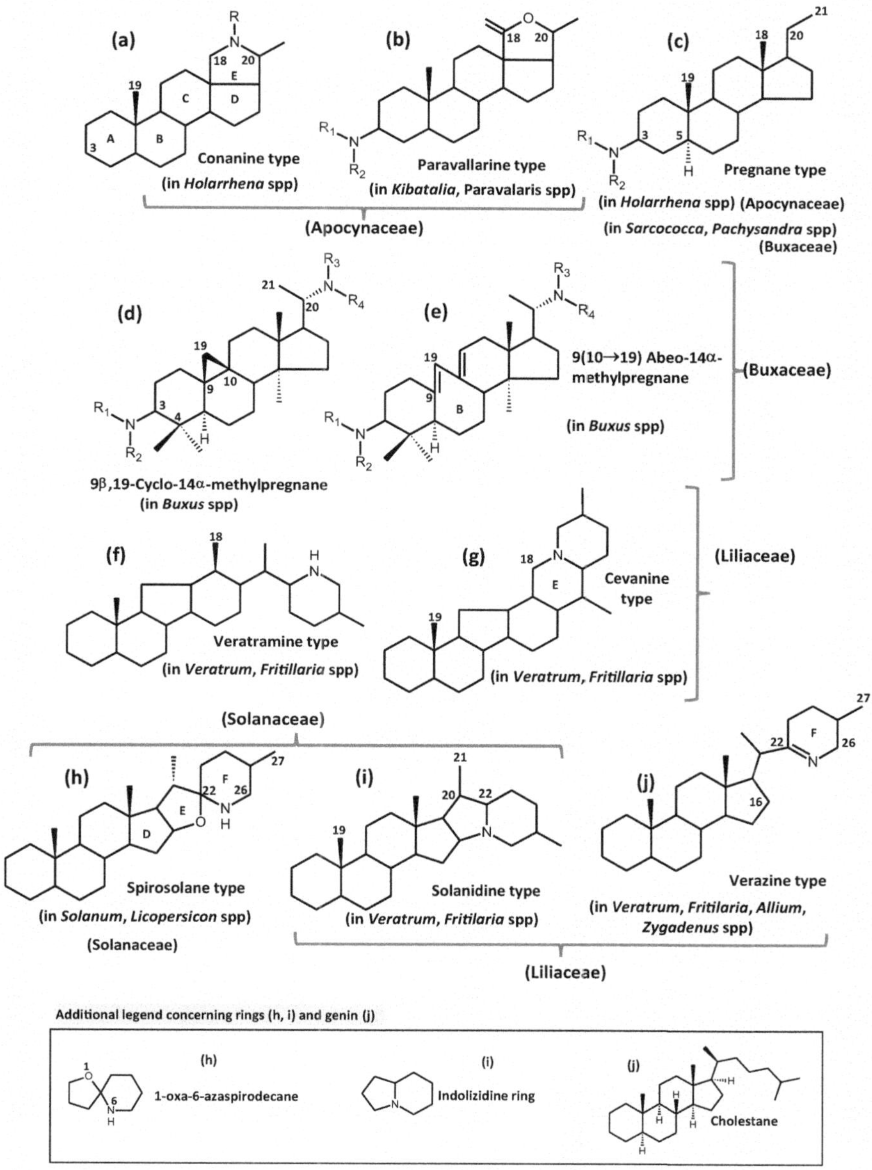

Fig. 6.14 Different steroid alkaloids based on different aglycone types and produced by different plant taxa (genera, families). R, R_1, R_2, R_3, R_4: various chemical substitutions

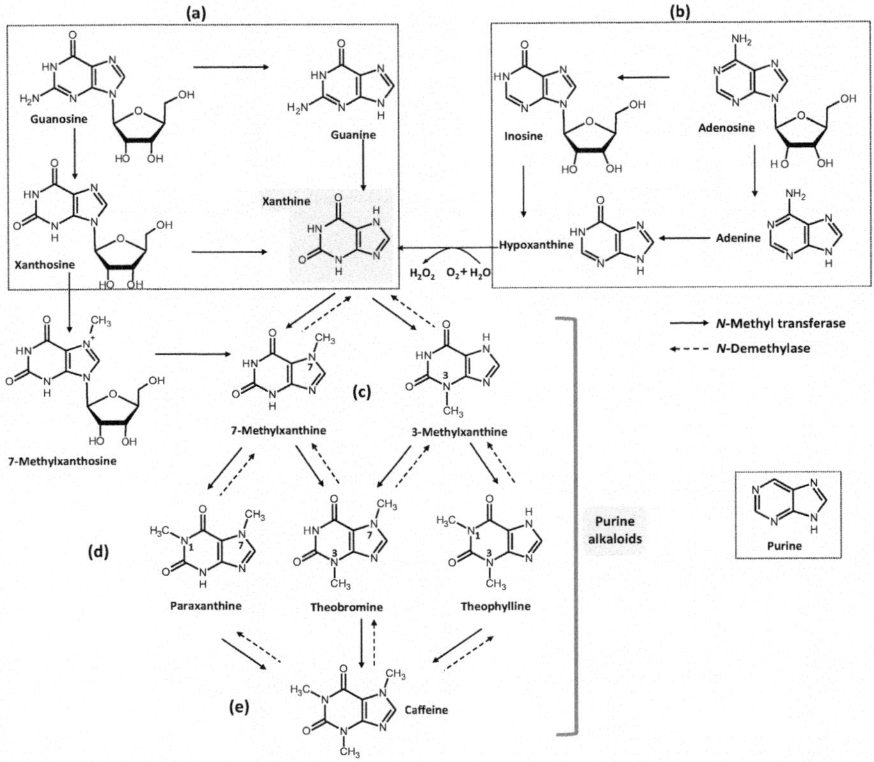

Fig. 6.15 Different purine alkaloids issued from sequential methylations of xanthine synthesized initially from degradation pathways of guanine and adenine

These *StAs* were found in different species associated with the genera *Funtumia, Holarrhena, Malouetia* and *Wrightia* of the Apocynaceae family.

Paravallarine type alkaloids are based on a pregnane-(18 → 20)-lactone skeleton (Fig. 6.14b) (Phi et al. 2011). Paravallarine-based *StAs* were found in plant species of the genera *Kibatalia* and *Paravallaris* of the Apocynaceae family.

Pregnane type alkaloids show a general backbone based on 5α-pregnane skeleton with structural variability due to different chemical substitutions occurring at C3 and C20 including amino function, methyl, benzoyl and aliphatic groups (Fig. 6.14c). These pseudoalkaloids were found in (i) the genus *Holarrhena* of the Apocynaceae family and (ii) the genera *Sarcococca* and *Pachysandra* of Buxaceae.

Cyclopregnane alkaloids are structurally distinguished by a cycloartenol system where C9, C10 and C19 are involved in a cyclopropane (Fig. 6.14d). They are also characterized by a dimethylated C4 and the occurrence of nitrogen at C3 and/or C20. These *StAs* are also known as triterpenoid alkaloids. They are strongly produced by the genus *Buxus* of the Buxaceae family.

9(10 → 19)Abeo-14α-methylpregnane alkaloids are characterized by a tetracyclic system due to a fission of 9,10-bond leading to a seven-membered ring B (Fig. 6.14e). Plant species of the genus *Buxus* (Buxaceae) were characterized by such compounds.

Cevanine type *StAs* have hexacyclic scaffold corresponding to the structure of benzo[7, 8] fluoreno[2,1-β]quinolizine nucleus (Fig. 6.14g) (Li et al. 2006). These compounds were isolated from the genera *Veratrum* and *Fritillaria* within the family Liliaceae.

*V*eratramine type corresponds to an opened cevane in ring E at C18 (Fig. 6.14f). *V*eratramine type pseudoalkaloids were found in two genera of Liliaceae family: *Veratrum* and *Fritillaria*.

Spirosolane-based pseudoalkaloids are characterized by the presence of an oxa-azaspirodecane system associating E and F rings. They have a nitrogen-C-22 bond that can show either α or β configuration resulting in 22αN- and 22βN-spirolane, respectively (Fig. 6.14h). These alkaloids are often glycosylated and they were reported from the genera *Solanum* and *Lycopersicon* associated with the Solanaceae family, and the species *Fritillaria meleagris* and *Lilium longiforum* of the Liliaceae family.

Solanidine type StAs contain an indolizidine ring issued from the cyclization and nitrogenation of the lateral chain of C27 steroid (Fig. 6.14i). These alkaloids were found in the genera *Veratrum* and *Fritillaria* of Liliaceae, and *Solanum* in Solanaceae.

Verazine type StAs are distinguished by a 22,26-epiminocholestane skeleton with absence of ring E and presence of a piperidine ring F containing double bond(s) (Fig. 6.14j) (Xiang et al. 2022); the C16 can be either saturated or substituted. These alkaloids were found in several genera of Liliaceae family including *Veratrum*, *Fritillaria*, *Allium* (*Allium victorialis*) and *Zygadenus* (*Zygadenus sibiricus*).

Beyond terpenoid scaffolds, pseudoalkaloid diversity extends to other aglycone classes including aromatics (phenyls, phenols) and nitrogenous bases (purines, pyrimidines) (Deng and Ashihara 2010; Zhou et al. 2020). Illustrative case is given for purine alkaloids represented by several derivatives of xanthine (Fig. 6.15): xanthine is a purine basis issued from guanine and adenine degradation pathways (Fig. 6.15a, b). Through these pathways, xanthine can be formed from guanine, hypoxanthine and xanthosine by catalysis effect of guanine deaminase, xanthine oxidoreductase and purine nucleoside phosphorylase, respectively (Cicero et al. 2023). Therefore, synthesis of different purine alkaloids occurs through sequential methylations of different nitrogen atoms of xanthine (Fig. 6.15c): xanthine can be initially methylated at position 3 or 7 leading to the formation of 3- and 7-methylxanthine, respectively. Also, xanthosine (glycosylated xanthine) can be subject to 7-methylation followed by a deglycosylation leading to 7-methylxanthine. The N-methyl transferase effect pursues on both 3- and 7-methylxanthine leading to the formation of dimethylated pseudoalkaloids: paraxanthine, theobromine and theophylline (Fig. 6.15d). These three alkaloids are dimethylated at positions (1, 7), (3, 7) and (1, 3), respectively. Finally, each of these alkaloids can be methylated at its nonmethylated nitrogen leading to a trimethylated alkaloid, the caffein (Fig. 6.15e).

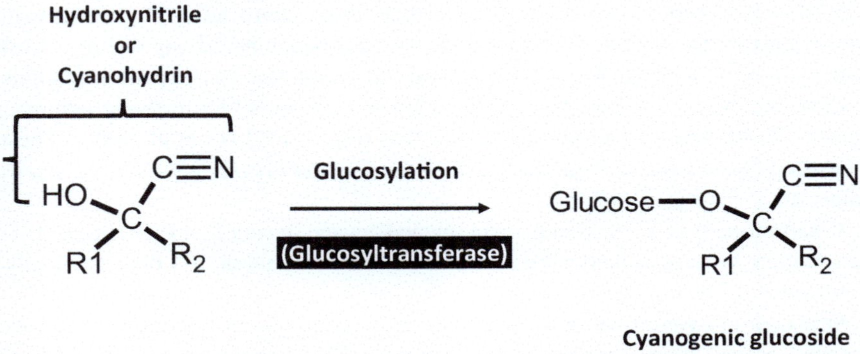

Fig. 6.16 General structural characteristics of cyanogenic glycosides

6.2 Cyanogenic Glycosides

6.2.1 General Structural Characteristics

Cyanogenic glycosides (CGs) correspond to α-hydroxynitriles (or cyanohydrins, $R_2C(OH)(CN)$ that are stabilized by glucosylation catalyzed by a glucosyltransferase (Fig. 6.16).

CG structure derives from the combination between AA and glycosylation involving one or more saccharides (Fig. 6.17a–n). Structural diversification processes of CGs are provided by conjugation, hydroxylation and acylation (Fig. 6.17i, j, o). Many CG structures were characterized by both (R)- and (S)-configurations (Fig. 6.17) (Selmar 2010). These pairs of epimeric GCs are due to the presence of asymmetric (chiral) carbon center bearing the nitrile group (Olafsdottir et al. 1992). Common examples of epimer pairs include (prunasin, sambunigrin), (dhurrin, taxiphyllin), (deidaclin, tetraphyllin A), etc. (Fig. 6.17 e, h, m, q, t).

6.2.2 Biosynthesis and Structural Diversification

Biosynthesis of cyanogenic glycosides occurs from specific amino acids (precursors) and glucose through a series of reactions catalyzed by two cytochrome P450 enzymes (CYP79A1, CYP71E1) and a UDP-glucosyltransferase (Fig. 6.18a) (Bak et al. 1998; Kahn et al. 1997, 1999; Koch et al. 1995; Sibbesen et al. 1995). Amino acids include protein and non-protein precursors (Fig. 6.17) (Clausen et al. 2002; Møller 2010; Selmar 2010):

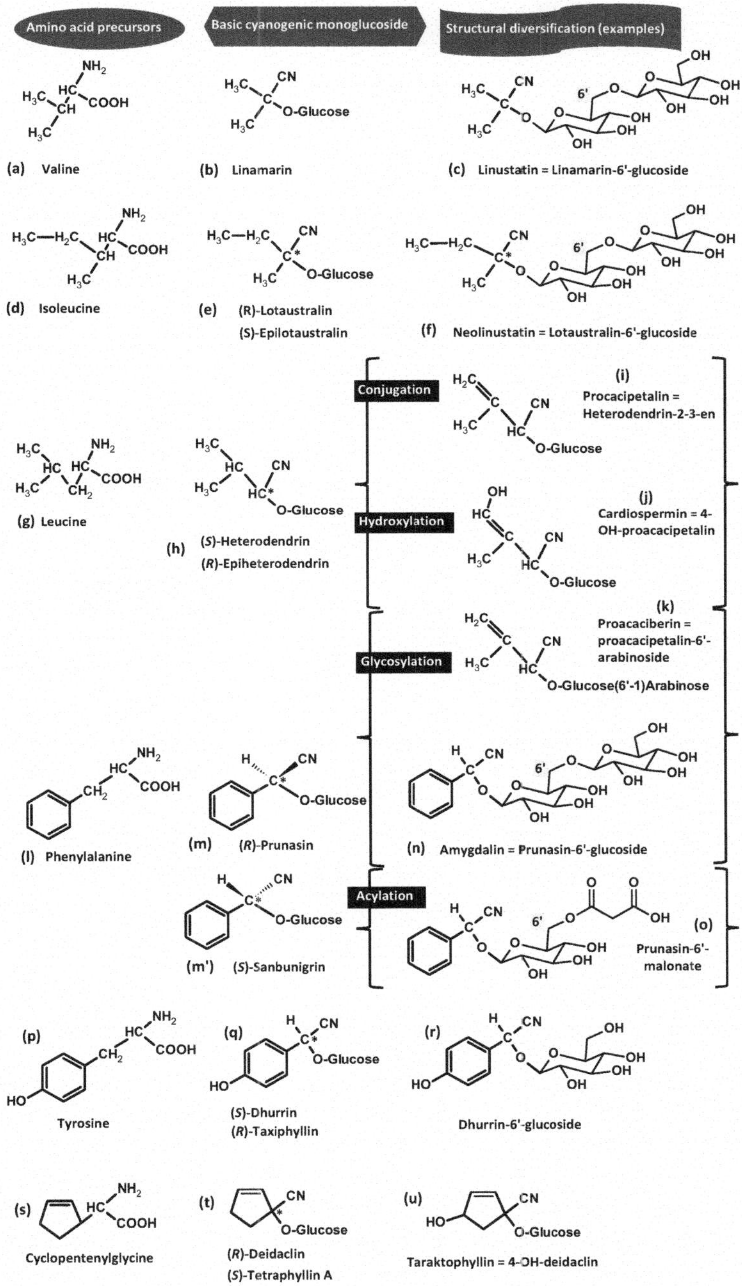

Fig. 6.17 Illustration of different basic monoglucosylated cyanogenic glycosides (CGs) with their amino acid precursors and some derivatives CGs issued from conjugation, hydroxylation, glycosylation and acylation processes

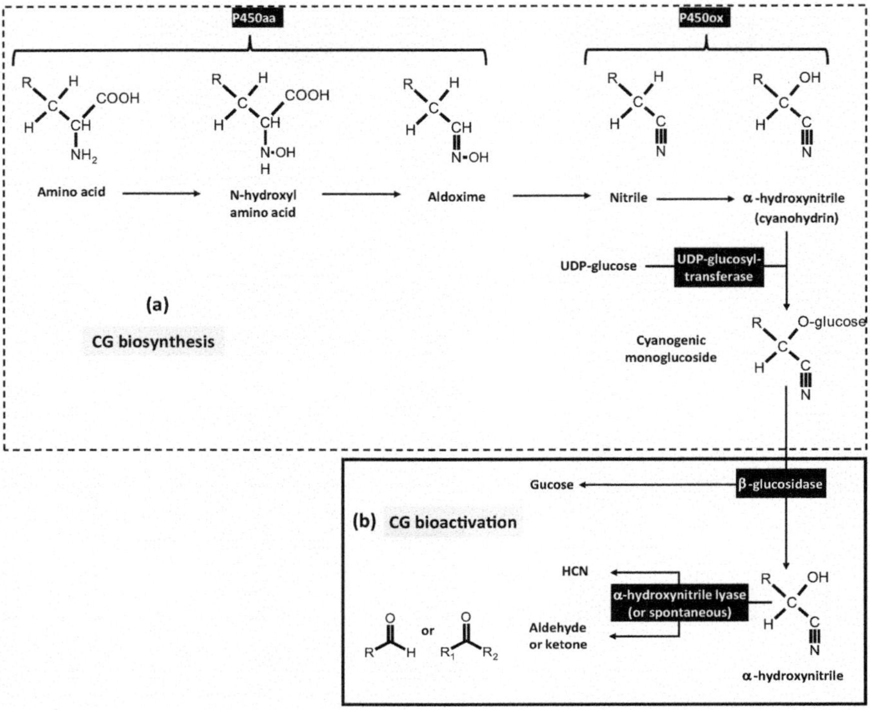

Fig. 6.18 General overview of biosynthesis (**a**) and bioactivation (**b**) of cyanogenic glycosides. Legend of P450-substrates: aa, amino acid; ox, oxime

- Valine, isoleucine, leucine, phenylalanine and tyrosine (protein amino acids) (Fig. 6.17a, d, g, l, p).
- Cyclopentenylglycine (non protein amino acid) (Fig. 6.17s).

Involved cytochrome P450s have multi-functional aspect and are located in endoplasmic reticulum (membrane-bound) (Koch et al. 1995). Intermediate metabolic reactions include, respectively, the formation of (i) aldoxime via N-hydroxyl amino acid and (ii) α-hydroxynitrile (cyanohydrin) via nitrile (Fig. 6.18a). The α-hydroxynitrile provides the basic aglycone structure of CG; after its synthesis by P450aa (CYP79A1) and P450ox (CYP71E1), CG is formed by glucosylation catalyzed by UDP-glucosyltransferase (UGT85B1) (Fig. 6.18a) (Jones et al. 1999).

From a study on *Sorghum bicolor* (Poaceae), a biosynthesis model of CGs was proposed (Fig. 6.19) (McFarlane et al. 1975; Moller and Conn 1979). Starting from tyrosine as precursor amino acid of dhurrin (the CG), the model includes three main steps:

- Conversion of tyrosine to (E)-*p*-hydroxyphenylacetaldoxime through successive hydroxylations and decarboxylation (Fig. 6.19a) (Sibbesen et al. 1995). CYP79A1 catalyzes this step which is rate limiting on dhurrin biosynthesis in young sorghum seedlings.
- Conversion of aldoxime to *p*-hydroxymandelonitrile through a CYP71E1-catalyzed reaction involving dehydration followed by C-hydroxylation reactions (Fig. 6.19b) (Bak et al. 1998).
- Glucosylation of *p*-hydroxymandelonitrile by glucosyltransferase UGT85B1 results in the formation of dhurrin (Fig. 6.19c) (Jones et al. 1999). At neutral or alkaline pH, the *p*-hydroxymandelonitrile shows labile aspect resulting in its dissociation into HCN and *p*-hydroxybenzaldehyde if it is not rapidly glycosylated (Fig. 6.19b′).

Most of CGs are commonly monoglucosylated with structural differentiations essentially due to the aglycone. Further structural variability results from additional

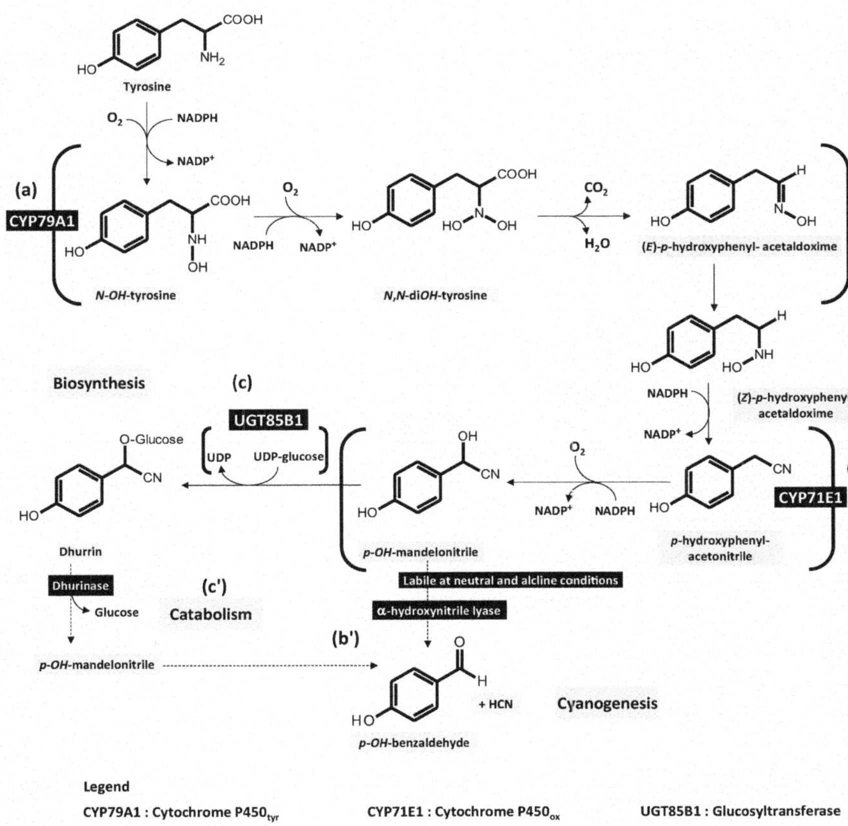

Fig. 6.19 Biosynthesis model of dhurrin (cyanogenic glycoside) in *Sorghum bicolor*

hydroxylation, glycosylation and acylation leading to less common CGs (Fig. 6.17) (Gleadow and Moller 2014).

Glycosylation contributes to structural diversification of CGs. The second glycosylation may involve another glucose moiety which can be attached at C6′ (generally), C2′, C3′ or C4′ of the first glucose residue. Illustrative cases are provided by eucalyptosin A, C, and B, respectively (Fig. 6.20a) (Yulvianti and Zidorn 2021; Neilson et al. 2011). In other CGs, the second sugar residue is not a glucose and can be of pentose type including arabinose, xylose and apiose as in vicianin, lucumin and oxyanthin, respectively (Fig. 6.20b) (Yulvianti and Zidorn 2021; Rockenback et al. 1992; Schwind et al. 1990; Gleadow and Moller 2014).

Further structural diversification of CGs is provided by acylation process which occurs after glycosylation and provides additional way for eventual other glycosylation. Illustrative cases are provided by oxyanthin 5″-benzoate and xeranthin (Fig. 6.21) (Rockenbach et al. 1992; Schwind et al. 1990). Xeranthin is among the most structurally complex CGs consisting of a tetraglycoside CGs with (apiosyl-1 → 4-Xylosyl-1 → 6-glucosyl) chain acylated by a caffeic acid residue occurring at 5-apiosyl and attaching a terminal glucose at position 4 (Fig. 6.21b). It was reported from *Xeranthemum cylindraceum* Sm. (Asteraceae, Cardueae) (Schwind et al. 1990).

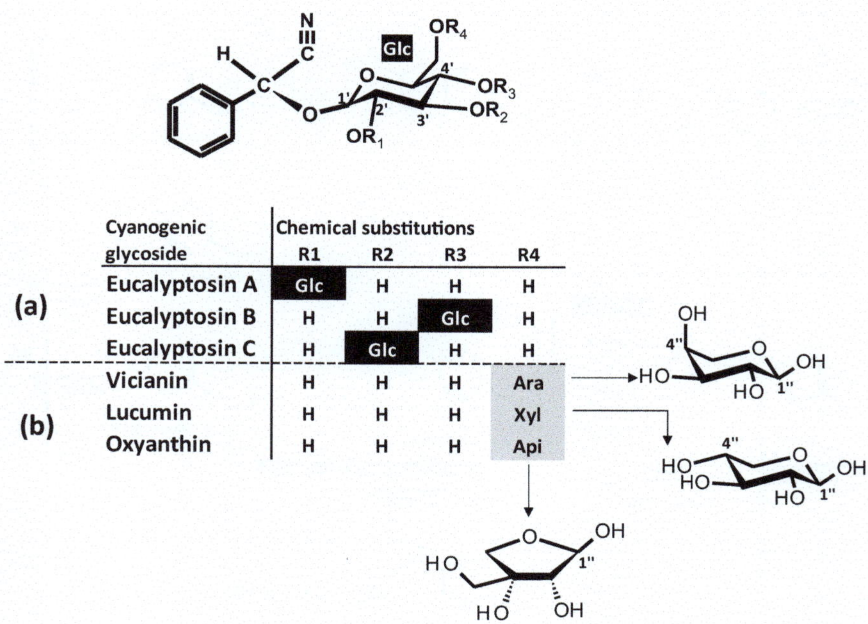

Cyanogenic glycoside	Chemical substitutions R1	R2	R3	R4
Eucalyptosin A	Glc	H	H	H
Eucalyptosin B	H	H	Glc	H
Eucalyptosin C	H	Glc	H	H
Vicianin	H	H	H	Ara
Lucumin	H	H	H	Xyl
Oxyanthin	H	H	H	Api

Fig. 6.20 Chemical structures of cyanogenic glycosides showing different substitution positions of the second glucose (Glc) moiety

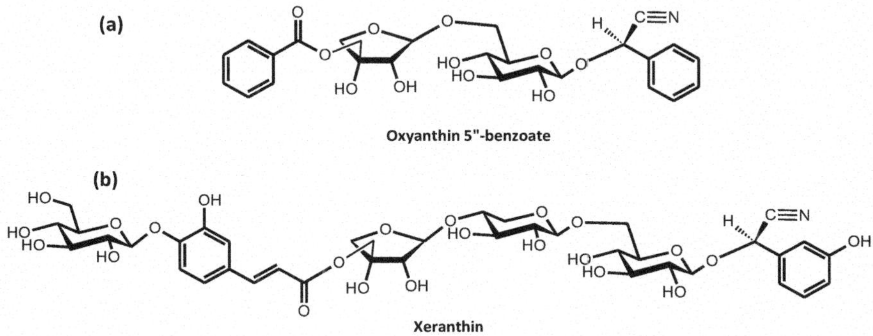

Fig. 6.21 Illustrative chemical structures of acylated glycosylated cyanogenic glycosides

6.2.3 *Cyanogenesis*

Cyanogenesis is the process by which toxic hydrogen cyanide (H–C≡N, HCN) is released from endogenous CGs. This occurs when cell integrity is lost subsequently to damaged plant tissues and disrupted cells. Tissue disruption can be caused by chewing herbivores or by physical processes (freezing, maceration of the plant material) (Gleadow and Woodrow 2002). Consequently, previously stored CGs will be in contact with their hydrolytic enzymes.

Bioactivation of CGs (by cyanogenesis) occurs through reverse processes to those involved in biosynthesis (Fig. 6.18b) (Morant et al. 2008):

- CGs are cleaved by hydrolytic effect of β-glucosidase leading to the loss of glucose and (re)formation of α-hydroxynitrile (Figs. 6.18b and 6.19c′).
- α-hydroxynitrile is either enzymatically or spontaneously dissociated into hydrogen cyanide (HCN) (toxic) and aldehyde or ketone (Figs. 6.18b and 6.19b′). Spontaneous reaction is favored by alkaline pH whereas enzymatic way is catalyzed by an α-hydroxynitrilase (Bjarnholt and Møller 2008).

HCN is toxic because it inhibits metalloenzyme activities (cytochrome c oxidase) leading to damage in the respiratory electron transport chain (Leavesley et al. 2008).

In plants containing diglucosidic CGs, the two saccharide units can be either simultaneously or sequentially lost (Fig. 6.22) (Kuroki et al. 1984).

Sequential hydrolysis can be illustrated by injured seeds of (i) *Prunus serotina* (Rosaceae) and (ii) *Linum usitatissimum* (flax) (Linaceae) containing amygdalin and linustatin, respectively (Fig. 6.23):

(i) In *Prunus serotina*, amygdalin is initially hydrolyzed by amygdalin hydrolase leading to the loss of terminal glucose and the formation of prunasin; then prunasin hydrolase catalyzes hydrolysis of prunasin leading to the release of the other glucose moiety (Fig. 6.23a) (Kuroki and Poulton 1987).

(ii) In *Linum usitatissimum* (flax), linustatin sequentially loses the two glucose units by catalytic actions of linustatinase and linamarase, respectively (Fig. 6.23b).

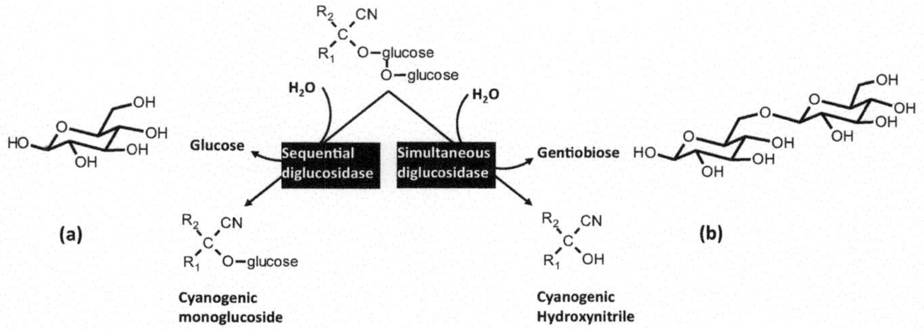

Fig. 6.22 Cleavage of cyanogenic diglycosides by sequential (**a**) or simultaneous (**b**) hydrolytic ways

Simultaneous hydrolysis was highlighted in injured leaves of the fern, *Davallia trichomanoides* (Polypodiaceae, Davaliaceae), producing vicianin (Fig. 6.23c)

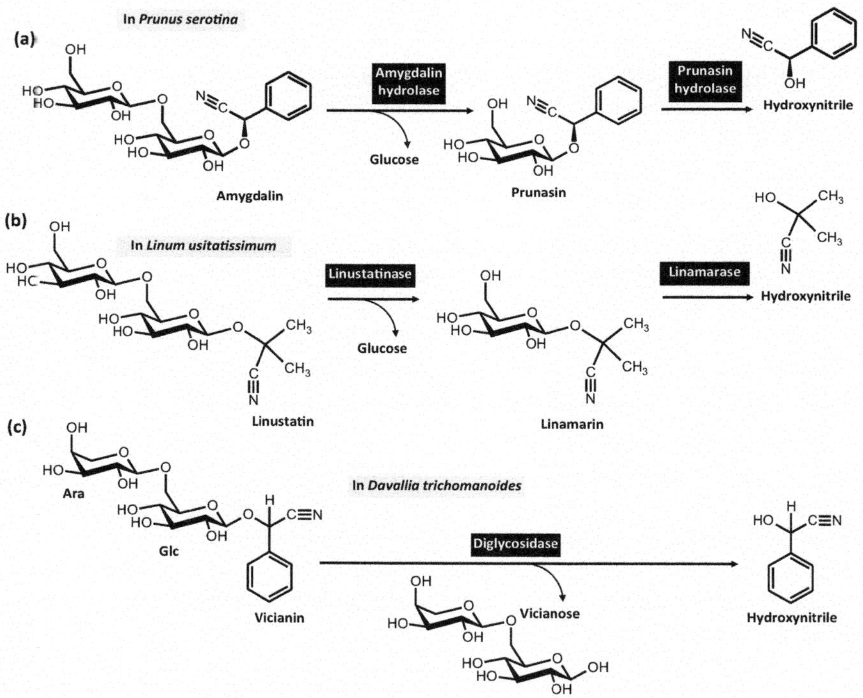

Fig. 6.23 Sequential (**a**, **b**) and simultaneous (**c**) hydrolysis of cyanogenic diglycosides illustrated by three plant species cases: *Prunus serotina* (**a**), *Linum usitatissimum* (**b**) and *Davallia trichomanoides* (**c**)

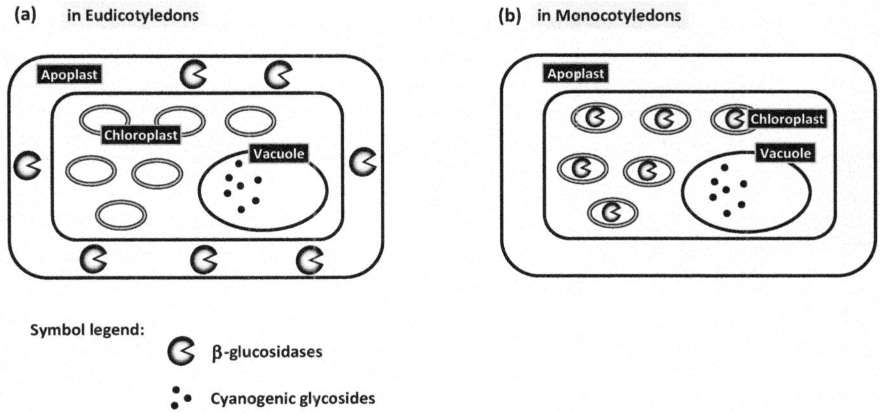

Fig. 6.24 Different subcellular compartments for storage of β-glucosidases in eudicotyledons (**a**) and monocotyledons (**b**); illustrative cases are provided by *Manihot esculenta* and wheat, respectively

(Kuroki et al. 1984). Vicianin is hydrolyzed by diglycosidase leading to the separation of hydroxynitrile from the diglycosyl, vicianose (arabinosyl 1–6 glucoside).

Naturally, in intact (undamaged) plants, CGs and their specific catabolic enzymes are stored into different cell/tissue compartments resulting in their separation to avoid self-poisoning (Morant et al. 2008). Eudicotyledons and monocotyledons were characterized by different subcellular compartmentation (Fig. 6.24). Both plant clades used vacuole for storage of glycosylated defenses (phytoanticipins), whereas they differed in compartmentation ways of β-glucosidases which are localized in:

- Apoplastic space as glycoprotein form in eudicotyledons (Fig. 6.24a).
- Plastid spaces (e.g. chloroplast) as unglycosylated form in monocotyledons (Fig. 6.24b).

6.2.4 Botanical Distribution and Chemotaxonomical Interest

CGs show wide distribution in plant world with nearly 3000 cyanogenic plant species reported until now (Selmar 2010; Møller and Seigler 1998). Different CGs showed restricted or extended botanical distributions associated with family, generic or specific levels (Fig. 6.25) (Yulvianti and Zidorn 2021; Selmar 2010). Some illustrative cases are given:

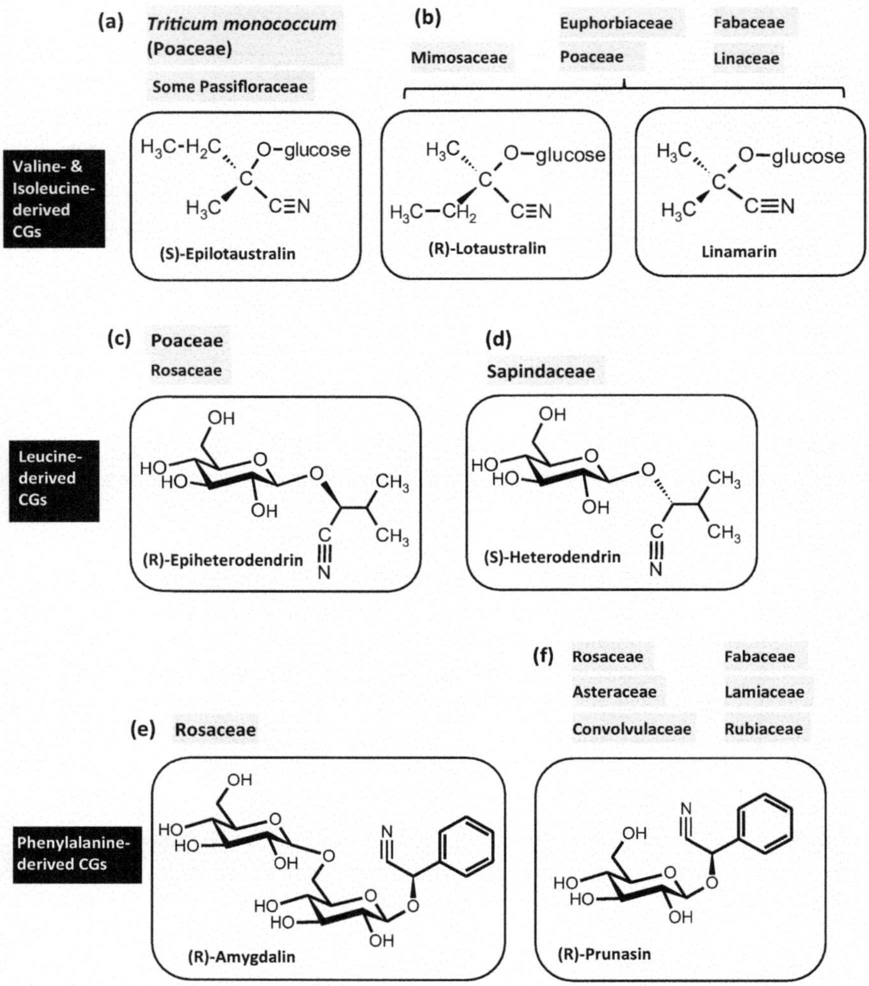

Fig. 6.25 Different cyanogenic glycosides showing specific or extended botanical distributions

In Poaceae family, the species *Triticum monococcum* was specifically characterized by the production of (S)-epilotaustralin (Fig. 6.25a) (Pitsch et al. 1984). (S)-epilotaustralin was also found in some Passifloraceae species (Olafsdottir et al. 1989). However, among valine- and isoleucine-derivative CGs, linamarin and R-lotaustralin showed wider distribution shared by Poaceae, Fabaceae, Linaceae, Mimosaceae and Euphorbiaceae (Fig. 6.25b).

Among leucine-derivative CGs, (R)-epiheterodendrin was commonly found in Poaceae and various species of Rosaceae (Pourmohseni et al. 1993; Lechtenberg et al. 1996) (Fig. 6.25c). However, its (S)-epimer _ (S)-heterodendrin_ was found in Sapindaceae (Fig. 6.25d).

Among phenylalanine-derivative CGs, amygdalin was found in the seeds of the genus *Prunus* (Rosaceae) (Swain et al. 1992; Poulton and Li 1994) (Fig. 6.25e). However, its precursor _ (R)-prunasin_ showed wide botanical distribution including Rosaceae, Fabaceae, Asteraceae, Lamiaceae, Convolvulaceae, Rubiaceae, etc. (Fig. 6.25f).

6.3 Glucosinolates

6.3.1 General Structural Aspect

Like, cyanogenic glucosides, glucosinolates (GS) are glycosylated nitrogen-containing compounds releasing active component (isothiocyanate) by hydrolysis after contact with their hydrolytic enzymes. However, they structurally differ from CGs by the presence of sulfur atoms (Fig. 6.26a).

GS consist of three structural components including a β-thioglucose and a sulfonated oxime moieties with a variable side group (R) responsible for structural diversity (Fig. 6.26a). According to the substitution group R, GS can be categorized into three structural classes including aliphatic, indole and aromatic glucosinolates (Fig. 6.26b–d) (Blažević et al. 2015).

6.3.2 Biosynthesis and Structural Diversification

GS are synthesized by plants into three steps including (Fig. 6.27a–c) (Halkier 1999; Halkier et al. 2002; Grubb and Stephen 2006; Halkier and Gershenzon 2006; Yan and Chen 2007):

(i) Preliminary chain elongation of amino acids leading to the formation of different GS-precursors (Fig. 6.27a).
(ii) Conversion of the obtained precursor into first (parent) GS (Fig. 6.27b).
(iii) Additional modifications leading to structural diversification of GS (Fig. 6.27c).

(i) Preliminary formation of GS-precursors is initiated by transamination leading to conversion of amino acid to keto acid (or oxo acid) (Fig. 6.27a1). Keto acid is subject to condensation by the addition of acetyl group from acetyl coenzyme A (Fig. 6.27a2). This acetylation represents the key step in chain elongation. The formed malate derivative is isomerized (Fig. 6.27a3) then oxidized through

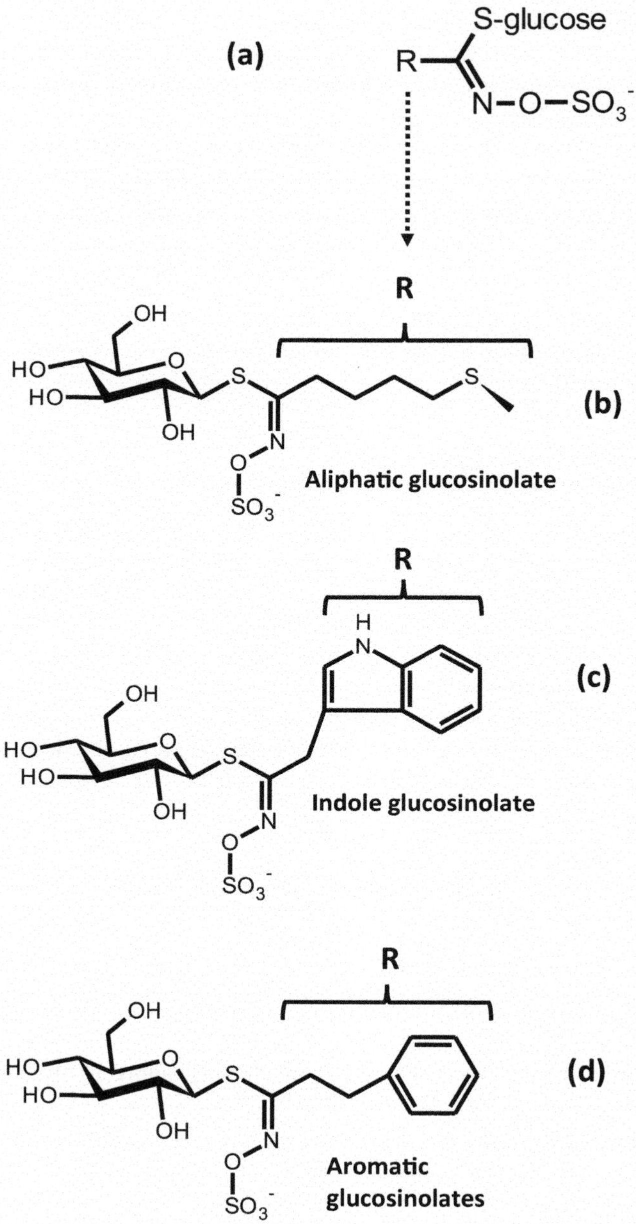

Fig. 6.26 Structural backbone of glucosinolates (GS) (**a**) with three main structural classes depending on the variable substitution side group R (**b**–**d**)

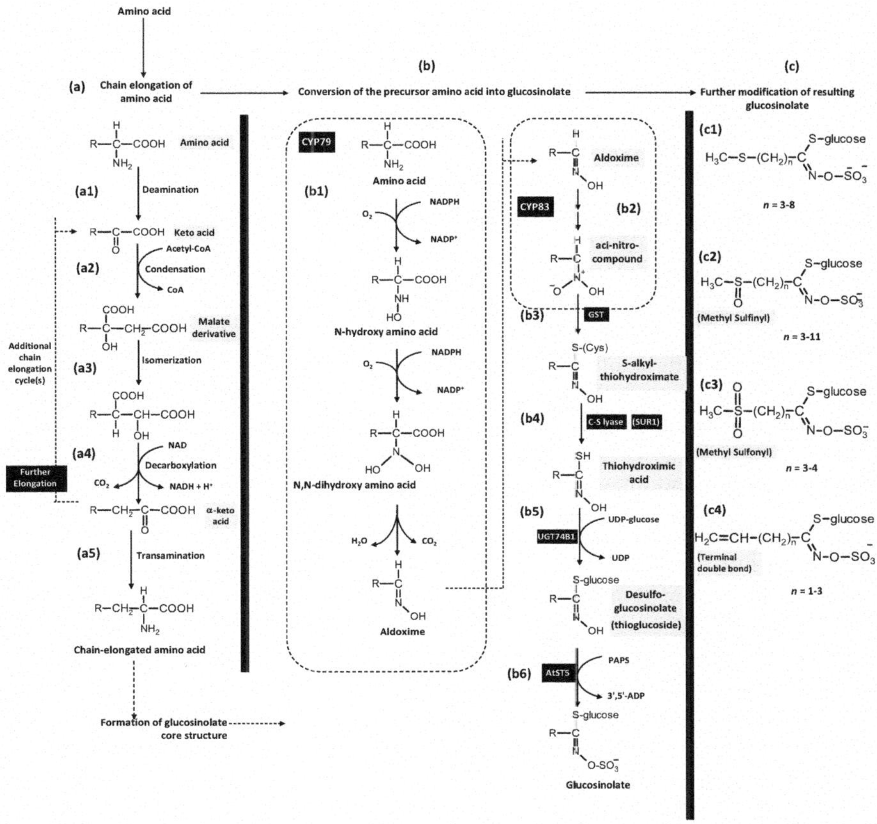

Fig. 6.27 Biosynthesis steps (**a**, **b**) and structural diversification ways (**c**) of glucosinolates

NAD-mediated reaction leading to a decarboxylated intermediate (a new α-keto acid) (Fig. 6.27a4). The new keto-acid can be subject to two bifurcated metabolic ways:

- It can substitute its oxo group by amine leading to the formation of immediate precursor of GS (Fig. 6.27a5).
- It can be subject to additional acetylation-oxidation-amination cycle(s) leading to the formation of more elongated precursor(s) favoring structural diversification of GS (Fig. 6.27a2–a4).

(ii) The formation of GS from (elongated) amino acid precursor involves two sets of reactions catalyzed by multifunctional cytochrome P450s (Fig. 6.27b1 and b2) that are followed by glycosylation and sulphuration reactions leading to GS (Fig. 6.27b3–b6) (Bennett et al. 1993; Du et al. 1995).

Starting from (elongated or not) amino acid precursor, the first set of reactions results in the deprotonation of amine group through two successive hydroxylation processes followed by a loss of one OH and subsequent π-bond formation to generate aldoxime (RCH=NOH) (Fig. 6.27b1). This first set of reactions is catalyzed by CYP79 (Wittstock and Halkier 2002). In *Arabidopsis,* CYP79 family was characterized by five enzyme members responsible for aldoxime production from amino acids including CYP79B2/CYP79B3, CYP79F1/CYP79F2 and CYP79A2 which were associated with tryptophan, chain elongated methionine derivatives and phenylalanine, respectively. Tryptophan is converted to indole-3-acetaldoxime by catalytic effect of both CYP79B2 and CYP79B3 (Mikkelsen et al. 2000; Hull et al. 2000). Variable length chain aliphatic glucosinolates deriving from methionine are metabolized by CYP79F1 and CYP79F2 (Chen et al. 2003; Hansen et al. 2001a): CYP79F1 was concerned with mono- to hexahomomethionine whereas CYP79F2 showed more restricted length range represented by penta- and hexahomomethionines.

The second cytochrome P450s-dependent step leads to conversion of aldoxime to aci-nitro (or nitrile oxide) compound by CYP83 (Fig. 6.27b2). Biochemical characterization revealed that CYP83A1 primarily metabolizes aliphatic aldoximes, whereas both CYP83A1 and CYP83B1 are involved in the metabolism of aromatic aldoximes derived from tryptophan, phenylalanine, and tyrosine (Bak and Feyereisen 2001; Naur et al. 2003). Aci-nitro derivative represents an activated and oxidized form of the aldoxime with instable aspect. It is sulfured by glutathione-S-transferase (GST) leading to the formation of thiohydroximate (Fig. 6.27b3). *In vitro,* it was shown to efficiently react with nucleophilic S-donors leading to the formation of S-alkyl thiohydroximates (Bak et al. 2001; Hansen et al. 2001b). *In vivo* experiments revealed that cysteine was more efficiently incorporated into thiohydroximates than other sulfur compounds making to conclude that cysteine could be the sulfur donor (Wetter and Chisholm 1968). Produced thiohydroximate is separated from its sulfur donor by C-S-lyase (Fig. 6.27b4), followed by its glycosylation by a glucosyl transferase (UGT74B1) (Fig. 6.27b5). Finally, glucosinolate is formed after sulfurylation of the desulfoglucosinolate by 3′ phosphoadenosine-5′-phosphosulfate (PAPS) under the catalytic effect of sulfotransferase (AtST5) (Fig. 6.27b6). AtST5 showed three homologs (AtST5a, AtST5b, AtST5c) characterized by their affinities toward different ligands: AtST5a preferentially interacts with desulfoglucosinolates deriving from tryptophan and phenylalanine, whereas AtST5b and AtST5c are concerned with long chain aliphatic desulfoglucosinolates (Piotrowski et al. 2004).

(iii) ***Structural diversification***. Apart from the preliminary chain elongation of their core structures, GS can be further diversified by later oxidations of sulfur of thiomethyl group (Fig. 6.27c1) to sulfinyl or sulfonyl groups (Figs. 6.27c2 and 6.27c3) (Dalgaard et al. 1977). Other transformations can result from the loss of the thiomethyl group followed by the formation of terminal double bond (Fig. 6.27c4). Moreover, GS side chains can be subject to hydroxylation, desaturation or methoxylation representing other structural diversification ways. Further diversifications can be achieved at the hydroxyl groups of the side chain through esterification or acylation processes (Hogge et al. 1988; Ettlinger and Kjxr 1968).

6.3.3 Degradation Products of Glucosinolates

Glucosinolates (GS) are decomposed by β-glucosidases generally called myrosinases. Myrosinases catalyze the hydrolysis of the thioglucoside linkage in GS making the glucose to be released and resulting in the formation of an unstable aglycone (Rask et al. 2000; Bones and Rossiter 1996). According to its side chain structure, the aglycone rearranges in presence of additional proteins and cofactors leading to the formation of different hydrolysis products responsible for different biological activities. GS-degradation products include isothiocyanates, oxazolidine-2-thiones, nitriles, epithionitriles, and thiocyanates (Fig. 6.28). These different GS-hydrolysis products influence the composition of liberated mustard oils and initially depend on reaction conditions combined with the influence of involved protein factors:

- Isothiocyanates are the most common products of GS hydrolysis during which the side chain (R) is shifted from the oxime carbon to the neighbor nitrogen by Lossen rearrangement (Fig. 6.28a).
- Less common rearrangement of GS hydrolysis results in the formation of thiocyanates in which side substituent R is migrated on the sulfur (Fig. 6.28b). Thiocyanates were found to be formed from some benzyl-, allyl-, and 4-methylsulfinylbutyl-glucosinolates. These formations involve the participation of thiocyanate forming protein (TFP), a special enzyme catalyzing the rearrangement of R position (Burow et al. 2006).
- Nitriles are analogously formed by the effect of a special enzyme called epithiospecifier protein (ESP) (Fig. 6.28c) (Foo et al. 2000; Bernadi et al. 2000; MacLeod and Rossiter 1985). *In vitro*, nitrile formation is favored at $pH < 3$ or in the presence of Fe^{2+} ions (Galletti et al. 2001; Gil and MacLeod 1980).
- ESP is also involved in the formation of epithionitriles from GS having a terminal double bond in their side chains (R) (Fig. 6.28d) (Tookey 1973; Foo et al. 2000).
- When side chains are hydroxylated, intermediate isothiocyanates can be spontaneously cyclized leading to the formation of oxazolidine-2-thiones (Fig. 6.28e).

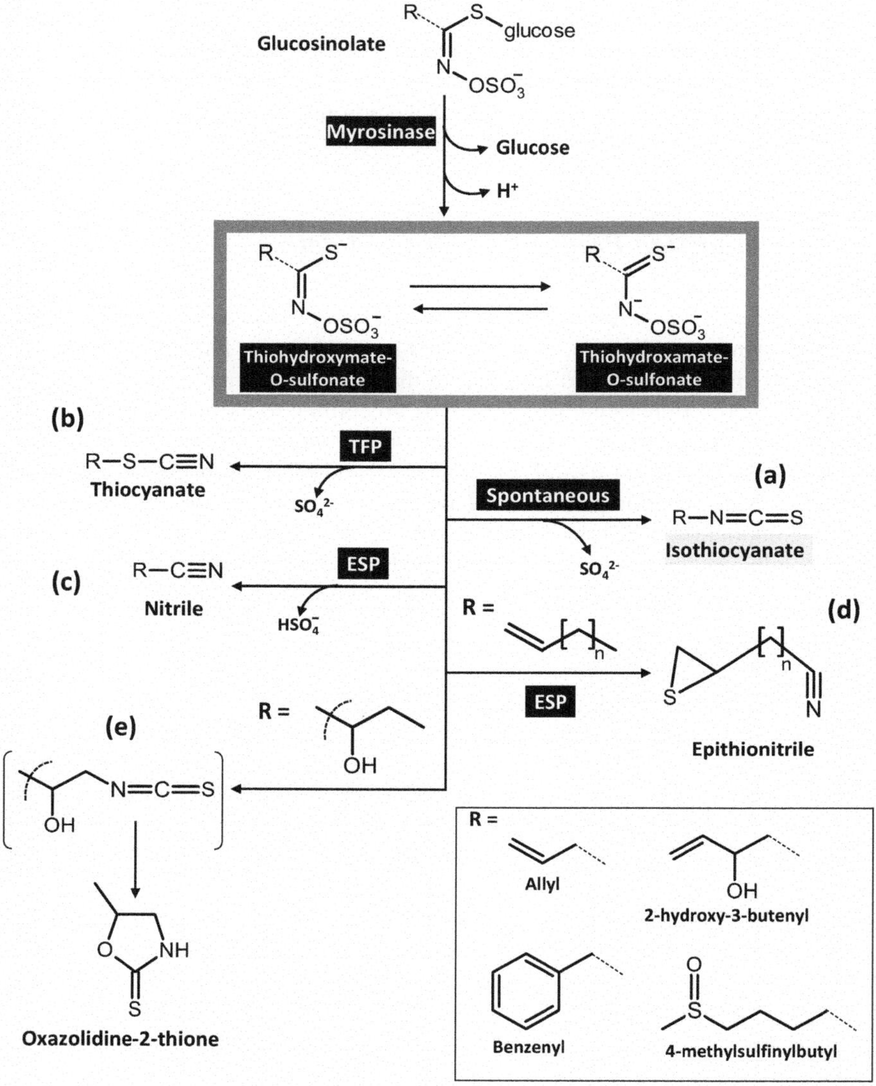

Fig. 6.28 Different products issued from the hydrolysis of glucosinolates by spontaneous (**a**) or enzymatic (**b-e**) ways. ESP: EpithioSpecifier Protein; TFP: Thiocyanate Forming Protein

6.3.4 *Botanical Distribution and Chemotaxonomical Interest*

Glucosinolates (GS) show restricted botanical distribution compared with cyanogenic glycosides although the two metabolic families have analogous structural aspects due to their nitrogen- and sugar-based compositions. Most of GS are found in the plant order of Capparales including four families, Brassicaceae, Capparaceae, Resedaceae and Tovariaceae (Rodman 1991). Sporadic occurrences of GS were revealed in some members of other plant families including Caricaceae, Euphorbiaceae and Sterculiaceae (Rodman 1991; Fahey et al. 2001). An extensive review on GS-compositions of producing plants with botanical and geographical distributions can be consulted in Blažević et al. (2015).

Brassicaceae were particularly well studied for their GS compositions because of their high nutritional interests. Twelve Brassicaceae vegetables from Singapore nutrition market were analyzed for their global GS compositions in terms of relative compositions of aliphatic, indole and aromatic glucosinolates (Fig. 6.29) (Liang et al. 2018). GS phytochemical patterns revealed differentiations at low plant taxonomic levels including species, subspecies and varieties. GS compositions of the twelve vegetables revealed metabolic regulation trends governed by gradual opposite variations between aliphatic and indole structures in presence of generally low levels of aromatic type:

Indole GS were strongly regulated in three varieties of *Brassica oleraceae* (var. *botrytis, capitata, italica*) (Fig. 6.29a) and in three subspecies of *Brassica rapa* (subsp. *chinensis, parachinensis, pekinensis*) (Fig. 6.29b). The two species were

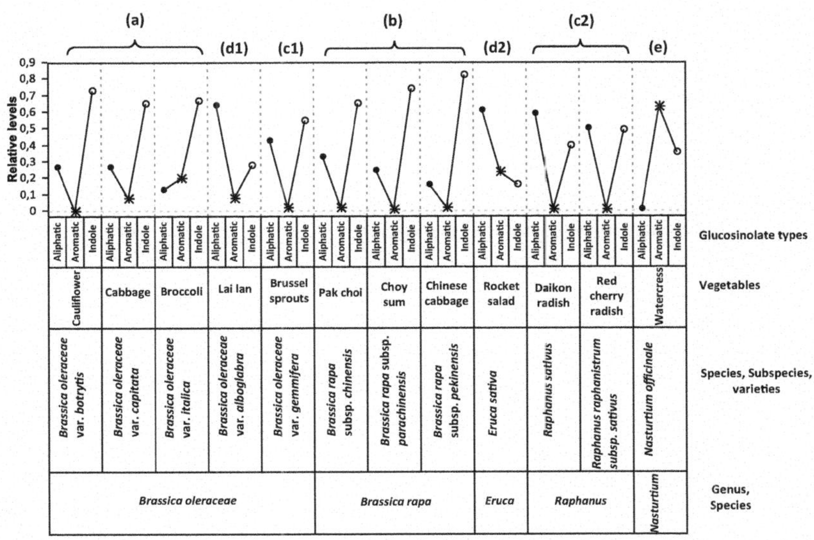

Fig. 6.29 Different glucosinolate patterns showing different relative levels of aliphatic, indole and aromatic side groups in different Brassicaceae vegetables

differentiated by relatively higher regulation ratios of aromatic GS in *B. oleraceae* despite the low relative levels of this GS-type at the whole population scale. At intraspecific level of *B. oleraceae*, the variety *gemmifera* was distinguished by balanced regulation ratios between indole and aliphatic GS (Fig. 6.29c1). Such a metabolic pattern was found in the two species of *Raphanus* genus: *R. sativus* and *R. raphanistrum* (Fig. 6.29c2). However, the variety *alboglabra* of *B. oleraceae* showed opposite metabolic trend to the other varieties of *B. o.* through significantly higher regulation of aliphatic than indole GS (Fig. 6.29d1). Such a pattern was found with more extreme trend in *Eruca sativa* where aliphatic and indole GS showed maximal and minimal regulations, respectively (Fig. 6.29d2). These results highlighted high variability of GS regulation trends within the species *Brassica oleraceae* resulting in characteristic phytochemical patterns at both intraspecific and interspecific levels. Finally, the species *Nasturtium officinale* was strongly differentiated by original metabolic peak of aromatic GS (Fig. 6.29e).

Beyond the variations at GS family level, phytochemical differentiations between Bracicaceae plants were highlighted by different metabolic regulations of individual molecules. Illustrative study case is given by Vioque et al. (1994) who analyzed glucosinolate composition in the seeds of different species of the genus *Coincya* Rouy (Brassicaceae) living in the Iberian Peninsula. Four species with eleven intraspecific taxa were studied including (Fig. 6.30):

(**1**) *Coincya transtagana* (Coutinho), (**2**) *C. longirostra* (Boiss.), (**3**) *C. rupestris* subsp. *rupestris*, (**4**) *C. rupestris* subsp. *leptocarpa* (Gonzalez-Albo), (**5**) *C. monensis* subsp. *cheiranthos* var. *recurvata* (All.), (**6**) *C. monensis* subsp. *cheiranthos* var. *johnstonii* (Samp.), (**7**) *C. monensis* subsp. *cheiranthos* var. *setigera* (J. Gay ex Lange), (**8**) *C. monensis* subsp. *cheiranthos* var. *granatensis* (O. E. Schulz), (**9**) *C. monensis* subsp. *orophila* (Franco), (**10**) *C. monensis* subsp. *puberula* (Pau), (**11**) *C. monensis* subsp. *nevadensis* (Willk.).

The eleven *Coincya* taxa were characterized by four GS (progoitrin, epiprogoitrin, gluconapin and glucosinalbin) showing different metabolic regulation ratios (Fig. 6.30). Progoitrin and epiprogoitrin showed opposite regulations resulting in differential variations where one of them is dominant or both have balanced ratios: high regulation trends of progoitrin at the expense of epiprogoitrin were observed in *Coincya momensis* (*C.m*) subspecies *cheiranthos* varieties *granatensis* and *johnstonii* (Fig. 6.30a1). The variety *johnstonii* was distinguished from *granatensis* by frankly lower regulation ratio of epiprogoitrin. Moreover, glucosinalbin was expressed in the variety *johnstonii* but it was absent in *granatensis*. Dominant progoitrin trend was also observed in *C.m.* subsp. *nevadensis* and *C. transtagana* (Fig. 6.30a2, a3). This metabolic trend favoring progoitrin on epiprogoitrin was also detected in *C. longirostra* but with less extreme ratios because of improvement of epiprogoitrin regulation (Fig. 6.30a4). Intermediate metabolic trends were found in *C.m.* subsp. *cheiranthos* var. *recurvata* and *setigera* which showed comparable regulation ratios between progoitrin and epiprogoitrin (Fig. 6.30b1). Also, *C.m.* subsp. *orophila* showed comparable GS pattern but with slightly higher regulation of epiprogoitrin than progoitrin (Fig. 6.30b2). Other extreme metabolic regulation pattern is provided by *C.m.* subsp. *puberula* showing dominant epiprogoitrin at the expense of progoitrin (Fig. 6.30c). Moreover, this plant variety showed some low expression of gluconapin by opposition to the previously described taxa where this GS is absent (Fig. 6.30c vs 6.30b2). Finally, the species *C. rupestris* was particularly characterized by a GS pattern showing comparable levels of progoitrin, epiprogoitrin and glucosinalbin (Fig. 6.30d); gluconapin had a smaller (minor) regulation peak (Fig. 6.30d).

Further analysis was caried out at intra-specific level and geographical aspect (Vioque et al. 1994). The GS-based chemotaxonomic study of *Coincya momensis* subsp. *orophila* was completed by analysis of geographical variation of (progoitrin : epiprogoitrin) metabolic ratio (Fig. 6.31). This taxon was further analyzed because its global GS pattern showed balanced ratios between progoitrin and epiprogoitrin (Fig. 6.30b2). Therefore, several subsp. *orophila* taxa were individually analyzed by considering their geographical origins. This fine phytochemical analysis revealed that progoitrin and epiprogoitrin were regulated by opposite

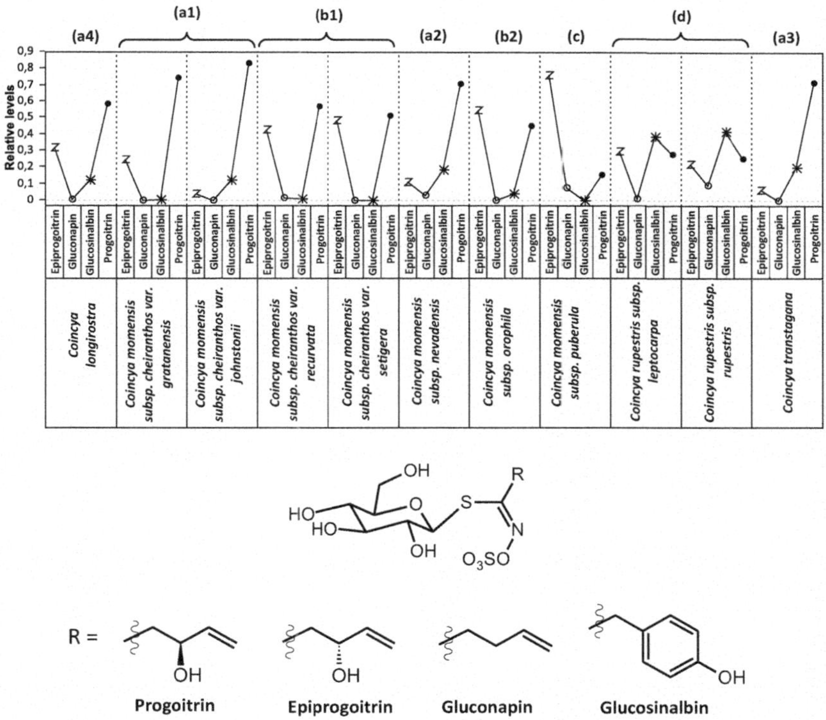

Fig. 6.30 Metabolic regulation patterns of four glucosinolates in different species and subspecies of the genus *Coincya* (Brassicaceae) living in Iberian Peninsula

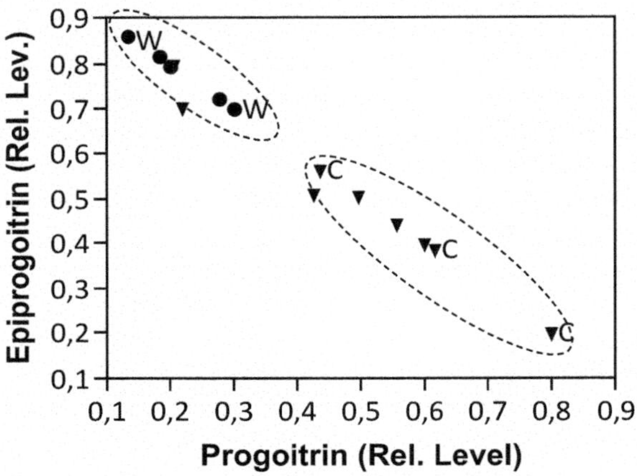

Fig. 6.31 Effect of geographical area (global environmental conditions) on metabolic regulation levels of progoitrin and epiprogoitrin (aliphatic glucosinolates) in *Coincya momensis* subsp. *orophila* (Brassicacea) living in Center and West regions of Iberian Peninsula. C, Centre region; W, West region

ways depending on the geographical area (Fig. 6.31): West regions of Iberian Peninsula favored higher regulations of epiprogoitrin whereas Center regions favored relatively more progoitrin. These results highlighted significant effect of environment conditions on GS metabolism at intra-specific level.

References

Aniszewski T (2007) Alkaloids—secrets of life; alkaloid chemistry, biological significance, applications and ecological role. Elsevier, Amsterdam, p 335

Bak S, Feyereisen R (2001) The involvement of two P450 enzymes, CYP83B1 and CYP83A1, in auxin homeostasis and glucosinolate biosynthesis. Plant Physiol 127:108–118

Bak S, Kahn RA, Nielsen HN, Møller BL, Halkier BA (1998) Cloning of three A-type cytochromes P450, CYP71E1, CYP98, and CYP99 from *Sorghum bicolor* (L.) Moench by a PCR approach and identification by expression in *Escherichia coli* of CYP71E1 as a multifunctional cytochrome P450 in the biosynthesis of the cyanogenic glucoside dhurrin. Plant Mol Biol 36:393–405.

Bak S, Tax FE, Feldmann KA, Galbraith DW, Feyereisen R (2001) CYP83B1, a cytochrome P450 at the metabolic branch paint in auxin and indole glucosinolate biosynthesis in *Arabidopsis*. Plant Cell 13:101–111

Bennett RN, Donald AM, Dawson GW, Hick AJ, Wallsgrove RM (1993) Aldoxime-forming microsomal enzyme systems involved in the biosynthesis of glucosinolates in oil seed rape leaves. Plant Physiol 102:1307–1312

Bernardi R, Negri A, Ronchi S, Palmieri S (2000) Isolation of the epithiospecifier protein from oil-rape (*Brassica napus* ssp. *oleifera*) seed and its characterization. FEBS Lett 467:296–298

Bjarnholt N, Møller BL (2008) Hydroxynitrile glucosides. Phytochemistry 69:1947–1961

Blažević I, Montaut S, Burčul F, Rollin P (2015) Glucosinolates: novel sources and biological potential. In: Mérillon J-M, Ramawat KG (eds) Glucosinolates, reference series in phytochemistry. Springer International Publishing, Switzerland, pp 1–58. https://doi.org/10.1007/978-3-319-26479-0_1-1

Bones AM, Rossiter JT (1996) The myrosinase-glucosinolate system, its organisation and biochemistry. Physiol Plant 97:194–208

Brown RT, Pratt SB, Richards P (2000) Enantiospecific synthesis of (–)-3-iso-19,20-dehydro-byohimbine from secologanin: a route to normal and pseudo stereoisomers of yohimbine. Tetrahedron Lett 41:5627–5630

Burow M, Muller R, Gershenzon J, Wittstock U (2006) Altered glucosinolate hydrolysis in genetically engineered *Arabidopsis thaliana* and its influence on the larval development of *Spodoptera littoralis*. J Chem Ecol 32:2333–2349

Chen S, Glaswischnig E, Jorgensen K, Naur P, Jorgensen B et al (2003) CYP79F1 and CYP79F2 have distinct functions in the biosynthesis of aliphatic glucosinolates in *Arabidopsis*. Plant J 33:923–937

Chen DL, Wang BW, Sun ZC, Yang JS, Xu XD, Ma GX (2020) Natural nitrogenous sesquiterpenoids and their bioactivity: a review. Molecules 25(2485):1–32

Cicero AFG, Fogacci F, Di Micoli V, Angeloni C, Giovannini M, Borghi C (2023) Purine metabolism dysfunctions: experimental methods of detection and diagnostic potential. Int J Mol Sci 24(7027):1–14

Clausen V, Frydenvang K, Koopmann R, Jørgensen LB, Abbiw DK et al (2002) Plant analysis by butterflies: occurrence of cyclopentenylglycines in passifloraceae, flacourtiaceae, and turneraceae and discovery of the novel nonproteinogenic amino acid 2-(3′-cyclopentenyl)glycine in Rinorea. J Nat Prod 65:542–547

Dalgaard L, Nawaz R, Sørensen H (1977) 3-methylthiopropylamine and (R)-3-methylsulphinylpropylamine in *Iberis amara*. Phytochemistry 16:931–932

Danyen SB, Boodia N, Ruggoo A (2009) Interaction effects between ascorbic acid and calcium chloride in minimizing browning of fresh-cut green banana slices. J Food Process Preserv 33:12–26

Deng WW, Ashihara H (2010) Profiles of purine metabolism in leaves and roots of *Camellia sinensis* seedlings. Plant Cell Physiol 51(12):2105–2118

Dewick PM (2009) Alkaloids. In: Medicinal natural products: a biosynthetic approach. Wiley, pp 311–420

Djerassi C, Kutney JP, Shamma M (1962) Alkaloid studies-XXXII: studies on *Skytanthus acutus* meyen. The structure of the monoterpenoid alkaloid skytanthine. Tetrahedron 18(1):183–188

Du L, Lykkesfeldt J, Olsen CE, Halkier BA (1995) Involvement of cytochrome P450 in oxime production in glucosinolate biosynthesis as demonstrated by an in vitro microsomal enzyme system isolated from jasmonic acid-induced seedlings of *Sinapis alba* (L.). Proc Natl Acad Sci North Am 92:12505–12509

Ettlinger MG, Kjxr A (1968) Sulfur compounds in plants. Recent Adv Phytochem 1:59–144

Fahey JW, Zalcmann AT, Talalay P (2001) The chemical diversity and distribution of glucosinolates and isothiocyanates among plants. Phytochemistry 56:5–51

Foo HL, Gronning LM, Goodenough L, Bones AM, Danielsen BE et al (2000) Purification and characterisation of epithiospecifier protein from *Brassica napus*: enzymic intramolecular sulphur addition within alkenyl thiohydroximates derived from alkenyl glucosinolate hydrolysis. FEBS Lett 468:243–246

Galletti S, Bernardi R, Leoni O, Rollin P, Palmieri S (2001) Preparation and biological activity of four epiprogoitrin myrosinase-derived products. J Agric Food Chem 49:471–476

Gil V, MacLeod AJ (1980) The effects of pH on glucosinolate degradation by a thioglucoside glucohydrolase preparation. Phytochemistry 19:2547–2551

Gleadow RM, Møller BL (2014) Cyanogenic glycosides: synthesis, physiology, and phenotypic plasticity. Annu Rev Plant Biol 65:155–185

Gleadow RM, Woodrow IE (2002) Constraints on effectiveness of cyanogenic glycosides in herbivore defense. J Chem Ecol 28:1301–1313

Griffiths LA (1959) Detection and identification of the polyphenoloxidase substrate of the banana. Nature 184:58–59

Grubb CD, Steffen A (2006) Glucosinolate metabolism and its control. Trends Plant Sci 11:89–100

Halkier BA (1999) Glucosinolates. In: Ikan R (ed) Naturally occurring glycosides. Wiley, Chichester, pp 193–223

Halkier BA, Gershenzon J (2006) Biology and biochemistry of glucosinolates. Annu Rev Plant Biol 57:303–333

Halkier BA, Hansen CH, Naur P, Mikkelsen MD, Wittstock U (2002) Biosynthesis and evolution of glucosinolates—the role of cytochromes P450. In: Romeo JT, Dixon RA (eds) Recent advances in phytochemistry—phytochemistry in the genomics and post-genomics eras. Pergamon, Oxford

Hansen CH, Wittstock U, Olsen CE, Hick AJ, Pickett JA, Halkier BA (2001a) Cytochrome P450 CYP79F1 from *Arabidopsis* catalyzes the conversion of dihomomethionine and trihomomethionine to the corresponding aldoximes in the biosynthesis of aliphatic glucosinolates. J Biol Chem 276:11078–11085

Hansen CH, Du LC, Naur P, Olsen CE, Axelsen KB et al (2001b) CYP83B1 is the oxime-metabolizing enzyme in the glucosinolate pathway in *Arabidopsis*. J Biol Chem 276:24790–24796

Hart NK, Johns SR, Lamberton JA (1967) (+)-9-Aza-l-methylbicyclo[3,3,1]nonan-3-one, a new alkaloid from *Euphorbia atoto* forst. Aust J Chem 20:561–563

Hogge LR, Reed DW, Underhill EW, Haughn GW (1988) HPLC separation of glucosinolates from leaves and seeds of *Arabidopsis thaliana* and their identification using thermospray liquid chromatography/mass spectrometry. J Chromatogr Sci 26:551–556

Hull AK, Vij R, Celenza JL (2000) *Arabidopsis* cytochrome P450s that catalyze the first step of tryptophan-dependent indole-3-acetic acid biosynthesis. Proc Natl Acad Sci USA 97:2379–2384

Ishida M, Hara M, Fukino N, Kakizaki T, Morimitsu Y (2014) Glucosinolate metabolism, functionality and breeding for the improvement of Brassicaceae vegetables. Breed Sci 64:48–59

Jones PR, Møller BL, Hoj PB (1999) The UDP-glucose: *p*-hydroxymandelonitrile-O-glucosyltransferase that catalyzes the last step in synthesis of the cyanogenic glucoside dhurrin in *Sorghum bicolor*: isolation, cloning, heterologous expression, and substrate specificity. J Biol Chem 274:35483–35491

Kahn RA, Bak S, Svendsen I, Halkier BA, Møller BL (1997) Isolation and reconstitution of cytochrome P450ox and in vitro reconstitution of the entire biosynthetic pathway of the cyanogenic glucoside dhurrin from sorghum. Plant Physiol 115:1661–1670

Kahn RA, Fahrendorf T, Halkier BA, Møller BL (1999) Substrate specificity of the cytochrome P450 enzymes CYP79A1 and CYP71E1 involved in the biosynthesis of the cyanogenic glucoside dhurrin in *Sorghum bicolor* (L.) Moench. Arch Biochem Biophys 363:9–18

Koch BM, Sibbesenn O, Halkier BA, Svendsen I, Møller BL (1995) The primary sequence of cytochrome P450tyr, the multifunctional N-hydroxylase catalyzing the conversion of L-tyrosine to *p*-hydroxyphenylacetaldehyde oxime in the biosynthesis of the cyanogenic glucoside dhurrin in *Sorghum bicolor* (L.) Moench. Arch Biochem Biophys 323:177–186

Kuroki GW, Poulton JE (1987) Isolation and characterization ofmultiple forms of prunasin hydrolase from black cherry (*Prunus serotina* Ehrh.) seeds. Arch Biochem Biophys 255:19–26

Kuroki G, Lizotte PA, Poulton JE (1984) Catabolism of (R)-amygdalin and (R)-vicianin by partially purified β-glucosidases from *Prunus serotina* Ehrh. and *Davallia trichomanoides*. Z Naturforsch 39:232–239

Leavesley HB, Li L, Prabhakaran K, Borowitz JL, Ison GE (2008) Interaction of cyanide and nitric oxide with cytochrome c oxidase: implications for acute cyanide toxicity. Toxicol Sci 101:101–111

Lechtenberg M, Nahrstedt A, Fronczek FR (1996) Leucine-derived nitrile glucosides in the Rosaceae and their systematic significance. Phytochemistry 41:779–785

Li HJ, Jiang Y, Li P (2006) Chemistry, bioactivity and geographical diversity of steroidal alkaloids from the Liliaceae family. Nat Prod Rep 23:735–752

Liang X, LeeHW LZ, Lu Y, Zou L, Ong CN (2018) Simultaneous quantification of 22 glucosinolates in 12 brassicaceae vegetables by hydrophilic interaction chromatography−tandem mass spectrometry. ACS Omega 3:15546–15553

Lichman BR (2021) The scaffold-forming steps of plant alkaloid biosynthesis. Nat Prod Rep 38:103–129

MacLeod AJ, Rossiter JT (1985) The occurence and activity of epithiospecifier protein in some cruciferae seeds. Phytochemistry 24:1895–1898

McFarlane IJ, Lees EM, Conn EE (1975) The in vitro biosynthesis of dhurrin, the cyanogenic glycoside of *Sorghum bicolor*. J Biol Chem 250:4708–4713

Mikkelsen MD, Hansen CH, Wittstock U, Halkier BA (2000) Cytochrome P450 CYP79B2 from *Arabidopsis* catalyzes the conversion of tryptophan to indole-3-acetaldoxime, a precursor of indole glucosinolates and indole-3-acetic acid. J Biol Chem 275:33712–43317

Møller BL (2010) Functional diversifications of cyanogenic glucosides. Curr Opin Plant Biol 13:337–346

Møller BL, Conn EE (1979) The biosynthesis of cyanogenic glucosides in higher plants: nhydroxytyrosine as an intermediate in the biosynthesis of dhurrin by *Sorghum bicolor* (L.) Moench. J Biol Chem 254:8575–8583

Møller BL, Seigler DS (1998) Biosynthesis of cyanogenic glucosides, cyanolipids and related compounds. In: Singh B (ed) Plant amino acids: biochemistry and biotechnology. Decker, New York, pp 563–609

Morant AV, Jørgensen K, Jørgensen C, Paquette SM, Sanchez-Pérez R et al (2008) β-glucosidases as detonators of plant chemical defense. Phytochemistry 69:1795–1813

Naur P, Petersen BL, Mikkelsen MD, Bak S, Rasmussen H et al (2003) CYP83A1 and CYP83B1, two nonredundant cytochrome P450 enzymes metabolizing oximes in the biosynthesis of glucosinolates in *Arabidopsis*. Plant Physiol 133:67–72

Neilson EH, Goodger JQD, Motawia MS, Bjarnholt N, Frisch T et al (2011) Phenylalanine derived cyanogenic diglucosides from *Eucalyptus camphora* and their abundances in relation to ontogeny and tissue type. Phytochemistry 72:2325–2334

Olafsdottir S, Andersen JV, Jaroszewski JW (1989) Cyanohydrin glycosides of passifloraceae. Phytochemistry 28:127–132

Olafsdottir ES, Jorgensen LB, Jaroszewski JW (1992) Substrate specificity in the biosynthesis of cyclopentanoid cyanohydrin glucosides. Phytochemistry 31:4129–4134

Phi TD, Pham VC, Mai HDT, Litaudon M, Guéritte F, Nguyen VH, Chau VM (2011) Cytotoxic steroidal alkaloids from *Kibatalia laurifolia*. J Nat Prod 74:1236–1240

Piotrowski M, Schemenewitz A, Lopukhinat A, Mueller A, Janowitz T et al (2004) Desulfoglucosinolate sulfotransferases from *Arabidopsis thaliana* catalyze the final step in the biosynthesis of the glucosinolate core structure. J Biol Chem 279:50717–50725

Pitsch C, Keller M, Zinsmeister HD, Nahrstedt A (1984) Cyanogenic glycosides from *Triticum monococcum*. Planta Med 50:388–390

Poulton JE, Li CP (1994) Tissue level compartmentation of (*R*)-amygdalin and amygdalin hydrolase prevents large-scale cyanogenesis in undamaged *Prunus* seeds. Plant Physiol 104:29–35

Pourmohseni H, Ibenthal WD, Machinek R, Remberg G, Wray V (1993) Cyanoglucosides in the epidermis of *Hordeum vulgare*. Phytochemistry 33:295–297

Rask L, Andreasson E, Ekbom B, Eriksson S, Pontoppidan B, Meijer J (2000) Myrosinase: gene family evolution and herbivore defense in Brassicaceae. Plant Mol Biol 42:93–113

Robinson T (1981) The biochemistry of alkaloids. Springer, New York

Rockenbach J, Nahrstedt A, Wray V (1992) Cyanogenic glycosides from *Psydrax* and *Oxyanthus* species. Phytochemistry 31:567–570

Rodman JE (1991) A taxonomic analysis of glucosinolate-producing plants, part 1: phenetics. Syst Bot 16(4):598–618

Schwind P, Wray V, Nahrstedt A (1990) Structure elucidation of an acylated cyanogenic triglycoside, and further cyanogenic constituents from *Xeranthemum cylindraceum*. Phytochemistry 29:1903–1911

Selmar D (2010) Biosynthesis of cyanogenic glycosides, glucosinolates and non-protein amino acids. Annu Plant Rev 40:92–181

Sibbesen O, Koch B, Halkier BA, Møller BL (1995) Cytochrome P-450TYR is a multifunctional hemethiolate enzyme catalyzing the conversion of L-tyrosine to p-hydroxyphenylacealdehyde oxime in the biosynthesis of the cyanogenic glucoside dhurrin in *Sorghum bicolor* (L.) Moench. J Biol Chem 270:3506–3511

Smith RM (1977) The Celastraceae Alkaloids. In: Manske RHF (ed) The alkaloids: chemistry and physiology, vol 16. Academic Press, New York, pp 215–248

St-Pierre A, Lajeunesse A, Desgagné-Penix I (2017) Determination of piperidine alkaloids from Indian Tobacco (*Lobelia inflata*) plants and plant-derived products. Austin Biochem 2(2) id1014:1–8.

Swain E, Li CP, Poulton JE (1992) Tissue and subcellular localization of enzymes catabolizing (R)-amygdalin in mature *Prunus serotina* seeds. Plant Physiol 100:291–300

Tamboli AM, Rub RA, Ghosh P, Bodhankar SL (2012) Antiepileptic activity of lobeline isolated from the leaf of *Lobelia nicotianaefolia* and its effect on brain GABA level in mice. Asian Pac J Trop Biomed 2(7):537–542

Tietze LF (1983) Secologanin, a biogenetic key compound-synthesis and biogenesis of the iridoid and secoiridoid glycosides. Angew Chem Int Ed Engl 22:828–841

Tookey HL (1973) Crambe thioglucoside glucohydrolase (EC 3.2.3.1): separation of a protein required for epithiobutane formation. Can J Biochem 51:1654–1660

Trenti F, Yamamoto K, Hong B, Paetz C, Nakamura Y, O'Connor SE (2021) Early and late steps of quinine biosynthesis. Org Lett 23:1793–1797

Tripathi AK, Ray AK, Mishra SK (2022) Molecular and pharmacological aspects of piperine as a potential molecule for disease prevention and management: evidence from clinical trials. J Basic Appl Sci 11–16

Ul Haq I, Imran M, Nadeem M, Tufail T, Gondal TA, Mubarak MS (2020) Piperine: a review of its biological effects. Phytother Res 2020:1–21

Vereshchagin AN, Karpenko KA, Elinson MN, Goloveshkin AS, Ushakov IE, Egorov MP (2018) Four-component stereoselective synthesis of tetracyano-substituted piperidines. Res Chem Intermed 44: 5623–5634. https://doi.org/10.1007/s11164-018-3444-7

Vioque J, Pastor JE, Alaiz M, Vioque E (1994) Chernotaxonornic study of seed glucosinolate composition in *Coincya* Rouy (Brassicaceae). Bot J Linn Soc 116:343–350

Xiang ML, Hu BY, Qi ZH, Wang XN, Xie TZ, Wang ZJ, Ma DY, Zeng Q, Luo XD (2022) Chemistry and bioactivities of natural steroidal alkaloids. Natural Products and Bioprospecting 12(23). https://doi.org/10.1007/s13659-022-00345-0

Wetter LR, Chisholm MD (1968) Sources of sulfur in the thioglucosides of various higher plants. Can J Biochem 46:931–935

Wittstock U, Halkier BA (2002) Glucosinolate research in the *Arabidopsis* era. Trends Plant Sci 7:263–270

Yamazaki Y, Kitajima M, Arita M, Takayama H, Sudo H, Yamazaki M, Aimi N, Saito K (2004) Biosynthesis of camptothecin. In silico and in vivo tracer study from [1-^{13}C] glucose. Plant Physiol 134:161–170

Yan X, Chen S (2007) Regulation of plant glucosinolate metabolism. Planta 226:1343–1352

Yang M, Wang Q, Liu Y, Hao X, Wang C, Liang Y, Chen J, Xiao Y, Kai G (2021) Divergent camptothecin biosynthetic pathway in *Ophiorrhiza pumila*. BMC Biol 19(122):1–16

Yulvianti M, Zidorn C (2021) Chemical diversity of plant cyanogenic glycosides: an overview of reported natural products. Molecules 26:719

Zhao DK, Zhao Y, Chen SY, Kennelly EJ (2021) Solanum steroidal glycoalkaloids: structural diversity, biological activities, and biosynthesis. Nat Prod Rep 38:1423–1444

Zheng G, Horton DB, Penthala NR, Nickell JR, Culver JP, Deaciuc AG, Dwoskin LP, Crooks PA (2013) Exploring the effect of N-substitution in nor-lobelane on the interaction with VMAT2: discovery of a potential clinical candidate for treatment of methamphetamine abuse. Med Chem Commun 4:564–568

Zhou MZ, Yan CY, Zeng Z, Luo L, Zeng W, Huang YH (2020) N-methyltransferases of caffeine biosynthetic pathway in plants. J Agric Food Chem 68(52):15359–15372

Part III
Secondary Metabolites as Plant Responses to Abiotic Stresses

Chapter 7
Temperature-Linked Constraints and Plant Protection Responses

Temperature stresses include excessive heat and excessive cold. These thermic stresses can be classified according to their intensities, duration and variation rhythms. One distinguishes sustained high temperatures (gradually established), heat chock (suddenly appeared), chilling (cold stress above 0 °C) and freezing (cold stress bellow 0 °C) (Sung et al. 2003). Harmful effects are of different types (biochemical, cellular, physiological, mechanical) including triggering of oxidative processes, molecular degradations, cell membrane damages, reduction of growth and reproduction (Fig. 7.1a) (Pradhan et al. 2017).

Oxidative reactions are generated by both heat and cold stress and result in production of reactive oxygenated species (ROS). ROS show unstable peripheral electronic states leading to electron removal risk from plant growth molecules followed by metabolic destabilization and subsequent structural degradation (Fig. 7.1a). Antioxidants are produced by plants to cope with threats associated with ROS (Sung et al. 2003; Camejo et al. 2006; Havaux and Kloppstech 2001).

At membrane cell level, high and low temperatures show opposite acclimation responses from plants consisting of increase of membrane rigidity and fluidity, respectively (Sung et al. 2003). High and low temperature signals tend to be perceived by different membrane components. This seems to be linked to functional associations between saturated membrane lipids (fatty acids) and heat on one hand, and between unsaturated fatty acids and cold on other hand. This hypothesis was confirmed by several experimental studies on plant mutants. Thus, a strong tolerance to high temperature was acquired in:

- Soybean mutant deficient in fatty acid unsaturation (Alfonso et al. 2001).
- *Arabidopsis* mutants showing deficiency of unsaturated fatty acids in thylakoid membranes (Hugly et al. 1989).
- Tobacco in which lipid saturation was increased by silencing a $\omega - 3$ desaturase gene (Murakami et al. 2000).

N. Semmar, *Secondary Metabolites in Plant Stress Adaptation*, Signaling and Communication in Plants, https://doi.org/10.1007/978-3-031-52595-7_7

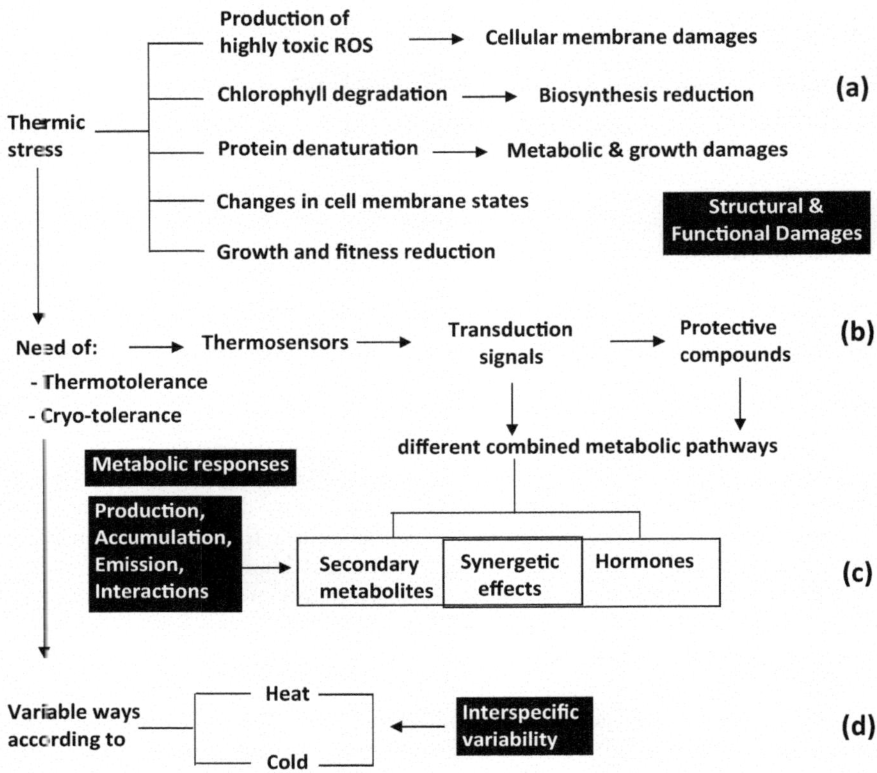

Fig. 7.1 Different harmful risks associated with temperature stresses and subsequent thermotolerance responses in plants

Also, membrane lipid unsaturation appears as key factor for optimum membrane function at low temperature. Cold stress tolerance was reduced in:

- *Arabidopsis faB.1* (fatty acid biosynthesis) mutant having more saturated membranes. It showed reduced quantum efficiency of photosystem II (PSII), chlorophyll content and chloroplast glycerolipids after prolonged exposure to low temperature (Wu et al. 1997).
- *Arabidopsis faB.5* and *fad 6* mutants containing more saturated membranes and showing chlorotic states and growth retardation under low temperature conditions (Hugly and Somerville 1992).

Among abiotic factors, temperature is particularly distinguished by sensitive responses in all the cell components implying thermosensors followed by activation of downstream biochemical processes (Fig. 7.1b) (Bahuguna and Jagadish 2015). Among thermosensors, variation in cell membrane fluidity can play key role in temperature stress perception in plants (Fig. 7.2): decrease of fluidity (rigidity

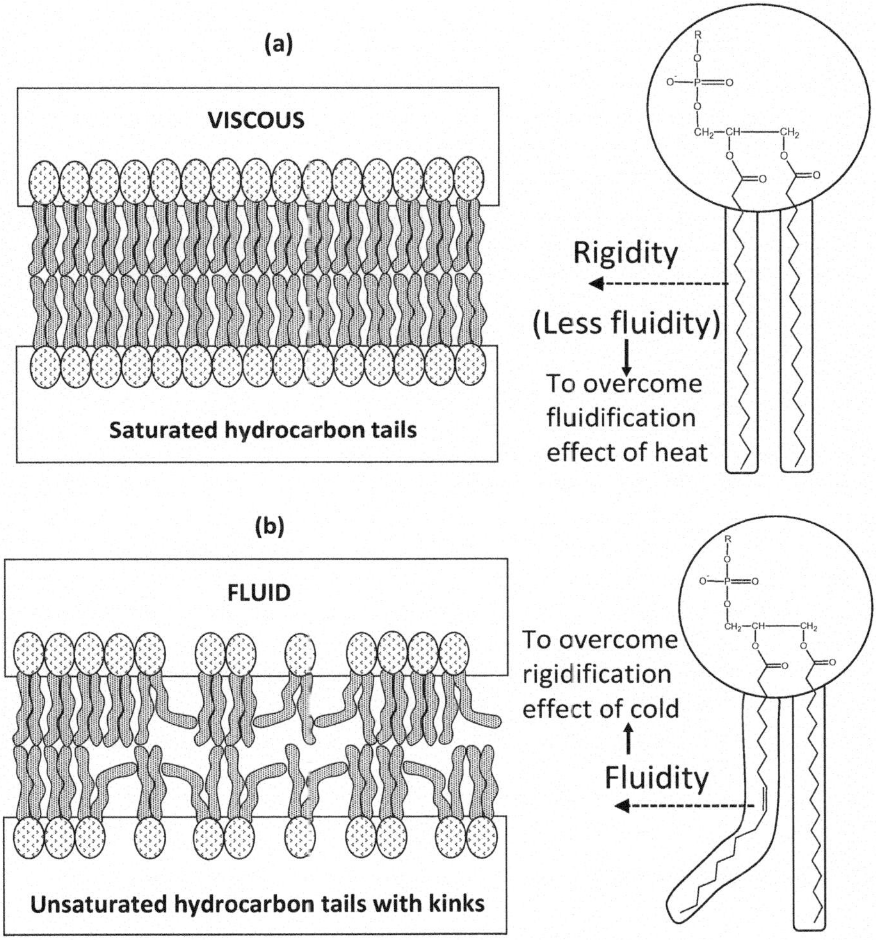

Fig. 7.2 Effect of saturated (**a**) and unsaturated (**b**) fatty acid composition on fluidity aspect of cellular membrane

trend) contribute to cold stress perception, whereas excess of fluidity represents a sensitive way toward temperature elevation (Sung et al. 2003). Therefore, modulation of membrane fluidity helps for the development of thermo- and cryo-tolerance, respectively.

Saturated FAs are expected to rigidify cell membrane by decreasing membrane fluidity due to their greater hydrophobic effect than unsaturated FA. Thus, saturated FAs are used to overcome membrane fluidification effect of heat. However, unsaturated FAs contribute for maintaining membrane fluidity. Thus, rigidification trend of cell membranes by cold stress could be handled by plants through increasing ratio of unsaturated to saturated fatty acids, leading to adjusting membrane fluidity.

Other cold perception ways include Ca^{2+} channels that release Ca^{2+} into the cytosol (Raza et al. 2023). Therefore, cold signaling is transmitted by Ca^{2+} and ROS decoders to nucleus followed by transcription regulation.

Following temperature stress and subsequent transduction signaling pathways, downregulations of energy, protein activation and metabolic reactions are triggered in plants as preventive ways. This leads to production, accumulation (compartmentation) and high regulations of protective compounds including mainly SMs (Fig. 7.1c) (Hancock et al. 2014; Schonhof et al. 2007). Accumulation of secondary (and primary) metabolites represents a strong way for reaching tolerance to environmental stresses (Havaux 1998; Rivero et al. 2001; Sharkey 2005). On this basis, carbon tends to be relatively reallocated from growth to defense resulting in characteristic phytochemical patterns associated with temperature stress (Zobayed et al. 2005). The degree of such reallocation can be associated with the stress level.

Apart from preventive (anticipated) effects, SMs play key curative roles against ROS-linked toxic effects such as lipid peroxidation (Shohael et al. 2006b). Lipid peroxidation is commonly used as biochemical marker of damage level induced by oxidative stress. Protective metabolic responses to temperature stresses tend to be rapid by occurring within the first 30 min of exposure (Kaplan et al. 2004).

Acquisition of tolerance to relatively high or low temperature was revealed to have a multigenic aspect (Sung et al. 2003). Biochemical bases of thermotolerance imply synergetic interactions between different protective metabolic pathways including SMs and hormones (e.g. abscisic acid) (Fig. 7.1c). Heat and cold stresses induce multiple metabolic pathways without necessarily opposite patterns (Fig. 7.1d) (Kaplan et al. 2004). Moreover, thermotolerance -SMs patterns show high interspecific variability.

7.1 Heat Stress

7.1.1 Effects of Heat Stress

Heat stress results in different types of damages to plants occurring at tissue, physiological, cell and molecular scales (Fig. 7.3). Growth and yield of crop plants can be more or less limited by continuous or transient heat stress. Temperature elevation leads to spatiotemporal disturbances in crops by alterating both their geographical distribution and growing season. This affects threshold of growth temperatures making plant maturity to be reached sooner (Porter 2005).

At tissue level, high temperatures cause leaf and stem scorch, leaf abscission and tissue senescence (Fig. 7.3a) (Wahid 2007). At physiologic level, heat represents a constraint opposing to specific temperatures required for different life events (e.g. germination, flowering, etc.). It decreases the net photosynthesis but increase stomatal conductance and transpiration (Fig. 7.3b). Excessive evapotranspiration leads to disturbance in osmotic balance. These different harmful effects lead to

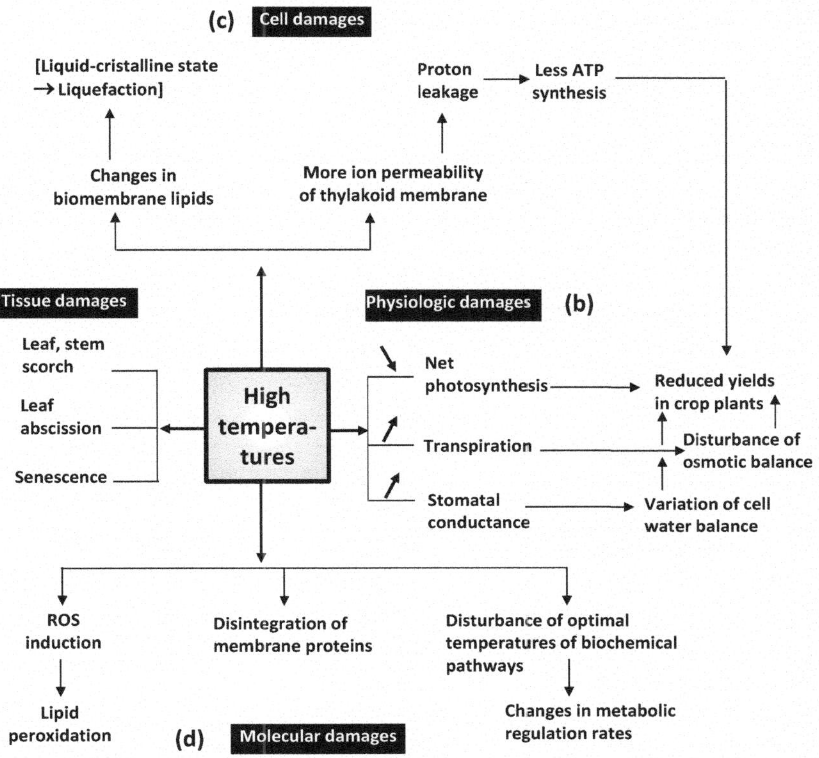

Fig. 7.3 Different types of damages caused by heat stress in plants

reductions of yields in crop plants. At cell scale, high temperatures have negative effects on cell membranes (Fig. 7.3c) (Boncan et al. 2020): they affect thylakoid membranes resulting in proton leakage and consequent less ATP synthesis. Also, heat causes changes in biomembranes from liquid-cristalline state to liquefaction.

At molecular scale, metabolism tends to increase under rising temperatures leading to more needs of energy for growth (Rawson 1988; Karim et al. 2000). Metabolic pathways are affected by high temperatures through oxidative reactions causing cell damages (Fig. 7.3d): elevated temperatures induce the formation of reactive oxygenated species (ROS) leading to lipid peroxidations. Moreover, heat conditions disturb optimal temperatures of biochemical pathways leading to changes in metabolic regulation rates and subsequently unbalanced states between growth metabolism and ROS. Also, high temperatures disintegrate membrane protein complexes like photosystem II leading to severe reduction of photosynthetic rate. At genetic level, heat shock was shown to cause plant sterility through irreversible biochemical alterations (e.g. in *Hordeum vulgare* L., *Arabidopsis thaliana* L.) (Bahuguna and Jagadish 2015).

7.1.2 General Plant Responses to Heat Stress

Oxidative stress and cell membrane perturbation resulting from heat chock make plants to synthesize antioxidant and membrane reparation compounds (Fig. 7.4). Such compounds essentially include phenolics and terpenes. Phenolic compounds play key antioxidant roles, whereas terpenes have diversified roles due to several properties including antioxidative ability, volatility (for leaves' protection) and liposolubility (for membrane interactions and stabilization) (Loreto and Velikova 2001).

Antioxidative effects lead to ROS scavenging (Fig. 7.4a). This property is found in phenolic compounds due to their conjugated double bonds and occurrence of

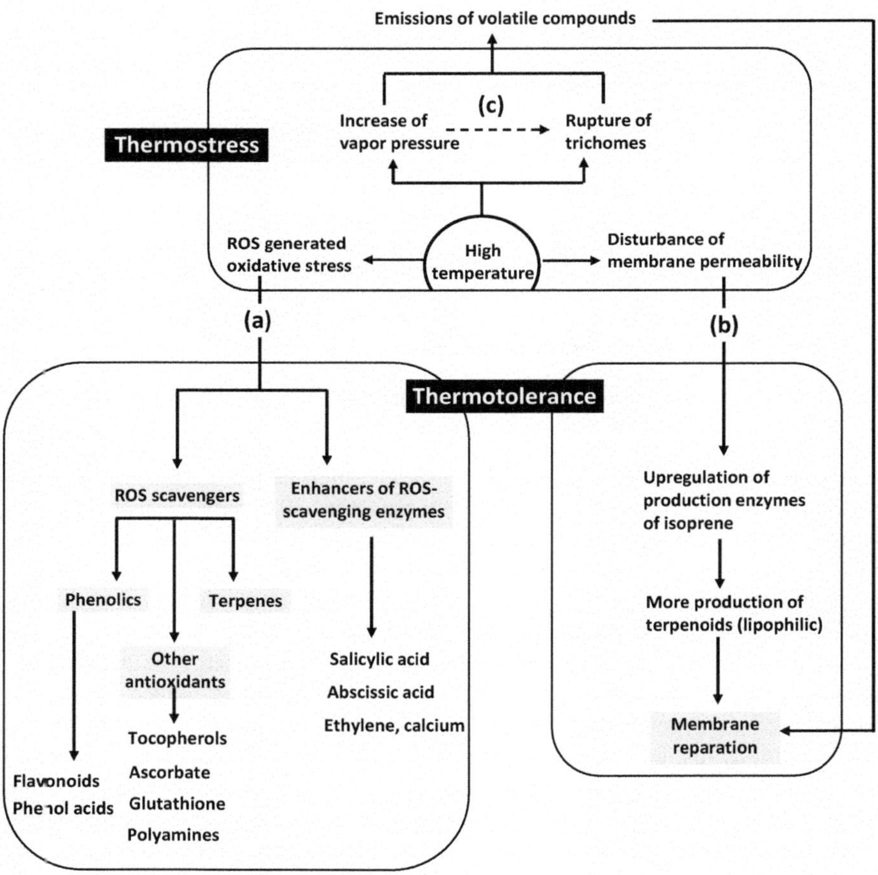

Fig. 7.4 Different protective metabolic ways associated with different molecular (**a**), cell (**b**) and tissue (**c**) disturbances generated by heat stress in plant world

hydroxyl groups. The antioxidant ability is increased by the presence of ortho-di-OH structures favoring metal chelation (Fig. 3.12d, f, Part I). Terpenes have also antioxidant property due to conjugated double bonds originated from their common precursor, isoprene (or hemiterpene) (Velikova et al. 2005). Antioxidant activity can be increased by enhancers of ROS-scavenging enzymes (Fig. 7.4a). Enhancers include salicylic acid, abscisic acid, ethylene, calcium.

Moreover, due to its slight molecular weight, isoprene has volatility property which is implied in protection of leaves living under heat stress. This property concerns all the slight terpenes including monoterpenes (essentially) and sesquiterpenes. When plants living under heat stress were fumigated by volatile terpenes, they showed an increase of their photosynthetic rates. This reparation of photosynthesis was associated with ability of terpenes to interact with cell membrane (Fig. 7.4b) (Sharkey and Singsaas 1995; Tardy and Havaux 1997): due to their lipophilic property favoring their partition in membrane, terpenes are able to act as solutes in the membrane. This favors the stabilization of membrane structure at high temperature leading to reduce proton leakage. This effect was highlighted as well with volatile terpenes (isoprene, monoterpenes) as higher weight terpenes (e.g. xanthophylls) (Singsaas et al. 1997). Zeaxanthin (xanthophyll) in thylakoid membrane was revealed to reduce proton leakage occurring at high temperature (Tardy and Harvaux 1997).

Also, thermic stress results in increase of vapor pressure favoring rupture of trichomes and emission of volatile SMs which contribute to protect plant from high temperatures (Fig. 7.4c). Thermotolerance effect of isoprene was highlighted by the following experience: isoprene synthesis in leaves was initially inhibited by fosmidomycin (Singsaas et al. 1997; Kuzuyama et al. 1998; Zeidler et al. 1998; Sharkey et al. 2001); results showed (i) significant heat damage in inhibited leaves compared with leaves not fed fosmidomycin; (ii) moreover, when blocked leaves were supplied with isoprene (fumigation with exogenous isoprene), they showed restoration of thermotolerance. The thermotolerance varied by a (3–10 °C)-range depending on the applied isoprene concentration. Moreover, thermotolerance occurred in leaves that do not normally emit isoprene after their fumigation with isoprene (Sharkey et al. 2001).

Emitted volatile compounds can be captured by neighbor plants helping them in thermotolerance (Holopainen et al. 2013): monoterpenes, showing lower volatility than isoprene, can be taken up and temporarily stored by neighboring non-emitting plants. In non-storing plants, monoterpenes are generally emitted in a light-dependent way, whereas storing plants can emit monoterpenes in darkness under sufficiently high temperature.

Thermo-protection effect was also shown for other volatile terpenes under heat conditions: oat (*Quercus ilex*) showed increasing photosynthesis after fumigation of its leaves by a monoterpenes' mixture (Loreto et al. 1998). Moreover, fumigated leaves with some terpenes show increased concentrations with some ability of additional metabolism. More generally, thermotolerance effects were revealed to be favored by compounds with double bonds (alkenes) (Sharkey et al. 2008).

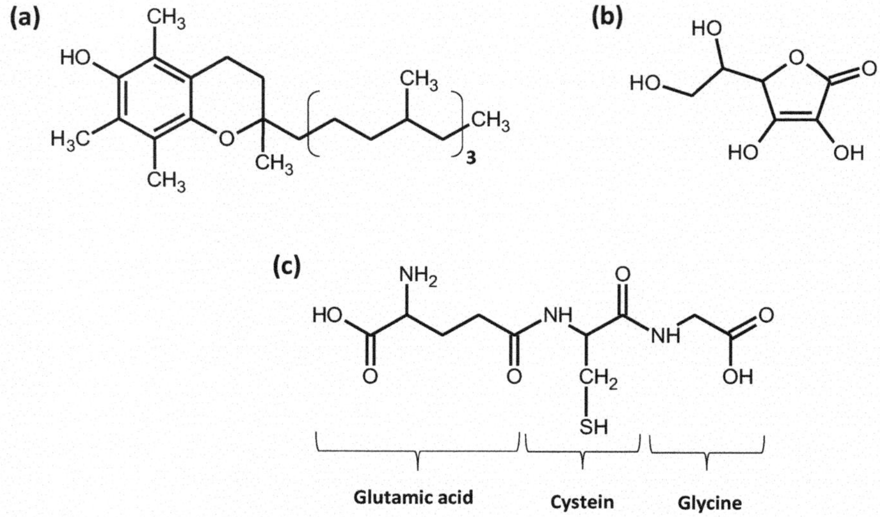

Fig. 7.5 Different ubiquitous antioxidants in plant world. **a** α-tocopherol; **b** ascorbic acid; **c** glutathione

Apart from phenolics and terpenes, plants synthesize some ubiquitous antioxidants including glutathione, polyamines and tocopherol (Fig. 7.5). Tocopherol (vitamin E), ascorbate (vitamin C) and glutathione are ubiquitously synthesized in plant world (Fig. 7.5a–c). Tocopherol merges phenolic structure with diterpene one (Fig. 7.5a).

Ascorbate (Fig. 7.5b) participates to many growth and protective reactions including photosynthesis, photoprotection, cell wall growth, cell expansion and synthesis of ethylene, gibberellins (phytohormones), anthocyanins and hydroxyproline. It plays strong antioxidant activity by giving protons and electrons leading to both radical neutralization and its self-transformation into dehydroascorbic acid (DHA) (Fig. 7.6). DHA can be hydrolyzed to inactive 2,3-diketogulonic acid which is further degraded to oxalic acid.

The redox potential of ascorbic acid offers interactions with many reactive or radical compounds including oxygen singlet ($^1O_2$), oxygen peroxide ($O_2^{\cdot}$), hydroxyl radical ($OH^{\cdot}$), glutathione radicals and tocopherol radicals (Noctor and Foyer 1998). This contributes to strongly limit lipid peroxidation (Bilska et al. 2019; Munné-Bosch 2005).

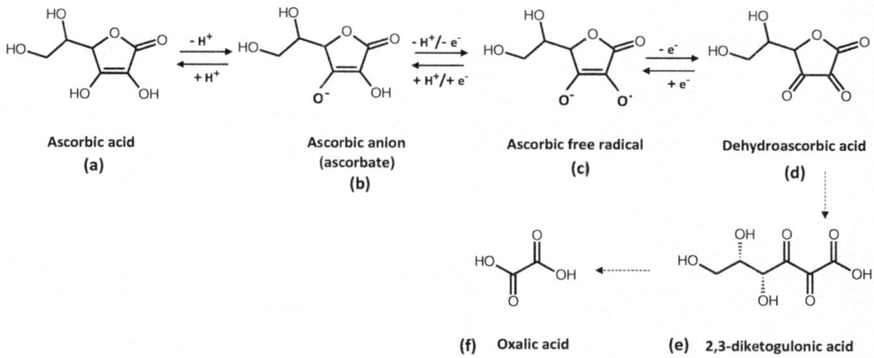

Fig. 7.6 Oxidation reactions of ascorbic acid leading to neutralization of radicals with formation of chemically stable compounds

7.1.3 Specific Phytochemical Patterns of Plant Species Under Heat Stress

Plants react to heat conditions by qualitative and quantitative variations of different SMs. In some metabolic families, increasing or decreasing trends can form dominant patterns, but they fundamentally depend on the plant species with possible intra-specific variations that can be associated with strain, sex and tissue (Figs. 7.7, 7.8, 7.9, 7.10 and 7.11).

7.1.3.1 Phenolic Patterns

Variations of phenolic products in plants with temperature showed either positive or negative trends according to the biological systems and phytochemical compounds (Figs. 7.7 and 7.9). Illustrative cases of positive trends associated with high temperature are presented in Fig. 7.7:

The species *Astragalus compactus* (Fabaceae) and *Hypericum perforatum* (Clusiaceae) responded by increasing their total phenolic amounts (Fig. 7.7a) (Naghiloo et al. 2012; Radusiene et al. 2012). Also, flavonoid content increased in *Ocimum basilicum* L. (Lamiaceae) (Fig. 7.7b) (Al-Huqail et al. 2020). Other studies revealed increase of specific phenolic families or compounds: flavonols increased in *Polygonum minus* (Polygonaceae) (Fig. 7.7c) (Goh et al. 2015); among flavonol glycosides, kaempferol glycosides were favored by high temperatures in *Brassica campestris* (Brassicaceae) (Chinese cabbage) (Fig. 7.7d) (Neugart et al. 2012; Harbaum-Piayda et al. 2010). These examples show multiple analysis levels and interpretability aspects of phenolic patterns going from whole contents to specific molecules.

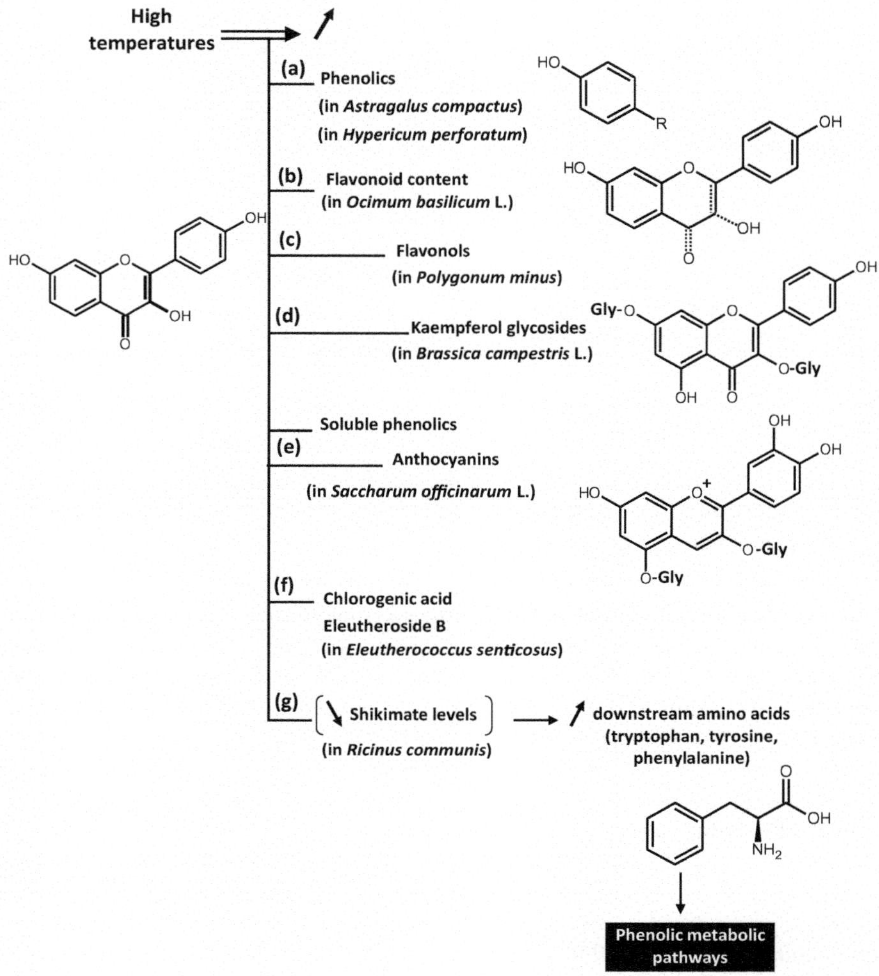

Fig. 7.7 Illustrative study cases of increasing effects of heat conditions or elevated temperature on phenolic metabolism in different plant species. Gly, glycosyl

A study on 1-month-old *Saccharum officinarum* (sugarcane) (Poaceae) revealed significant increase of soluble phenolics and anthocyanins in leaves when plants were exposed to 40 °C for 72 h compared with control conditions (28 °C) (Fig. 7.7e) (Wahid 2007). The two types of metabolites showed rapid increase within the first 24 h of applied heat conditions. However, the positive trend of anthocyanins was more sharped than that of soluble phenolics leading to an increase of (anthocyanins/soluble phenolics)-regulation ratio (Fig. 7.8a). This indicated strong physiological implications in the accumulation of both soluble phenolics and anthocyanins under heat conditions, with more sensitive trend for anthocyanins.

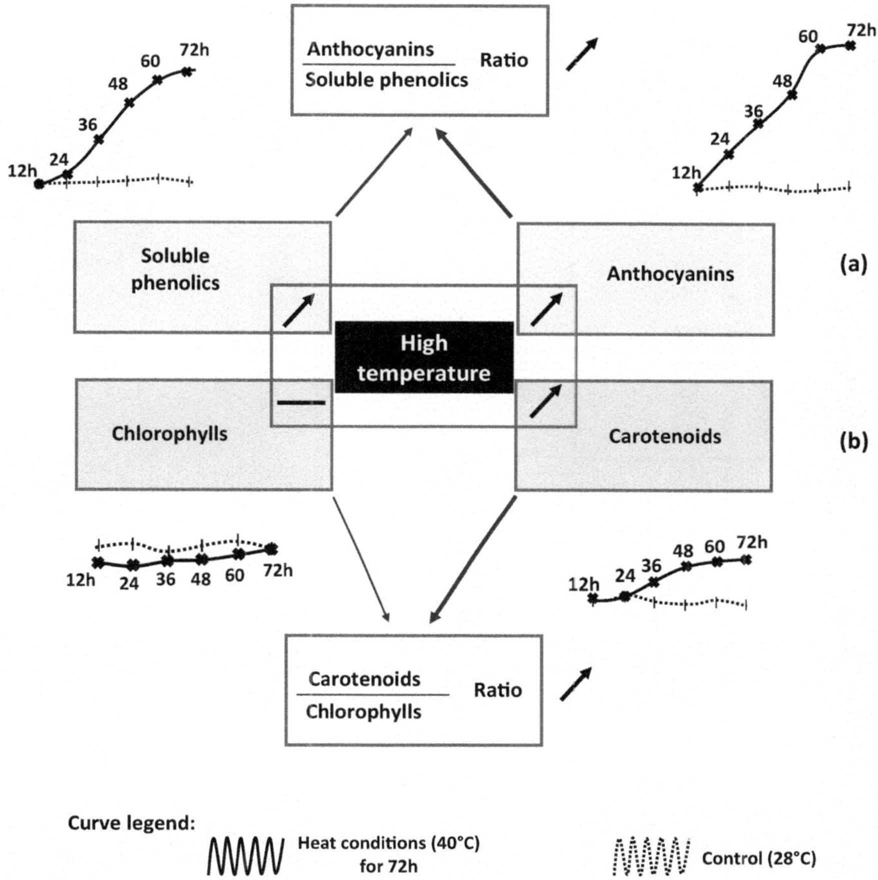

Fig. 7.8 Increasing effects of heat conditions on different secondary metabolites (phenolics, anthocyanins, carotenoids) of *Saccharum officinarum* (Poaceae) exposed at 40 °C for 72 h. Continuous and drop lines correspond to experimental (40 °C) and control (28 °C) conditions

Positive trend of anthocyanins under heat conditions is less common than their increase under low temperatures in the plant world (Oren-Shamir and Nissim-Levi 1999; Sachray et al. 2002). In this sense, the high positive trend of anthocyanins in sugarcane under heat stress has remarkable aspect. The fast and sharp increase in anthocyanins leading to gradual increase of their regulation level relatively to that of soluble phenolics, could indicate that anthocyanins tend to be preferentially biosynthesized among the water-soluble phenolics in *Saccharum officinarum*. Positive trend between anthocyanin levels and growth variables could be attributed with high hydrophily and osmotic activity of anthocyanins (Chalker-Scott 1999).

Moreover, positive trends were revealed between growth and metabolic parameters in *Saccharum officinarum* (Poaceae) cultivated under heat stress versus no

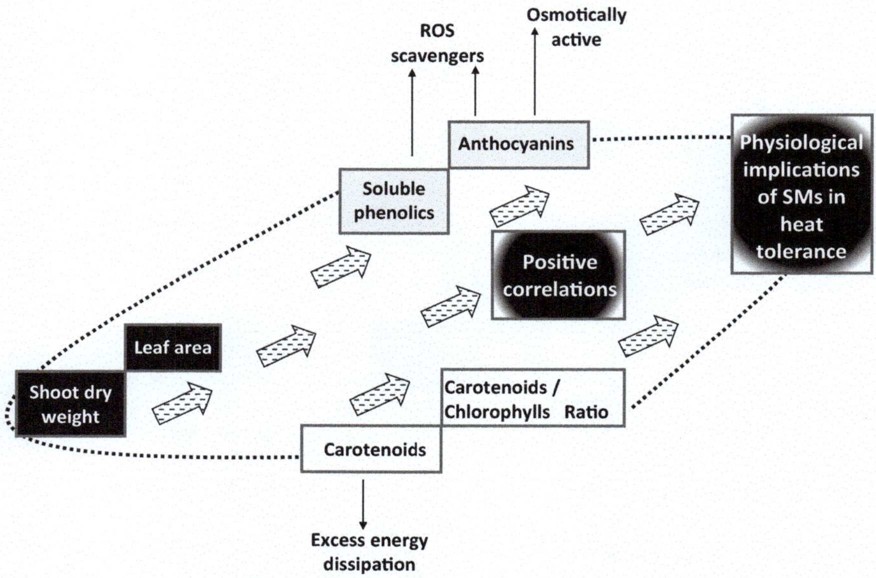

Fig. 7.9 Positive trends between growth variables and secondary metabolites in sugarcane cultivated under heat stress (40 °C) for 72 h

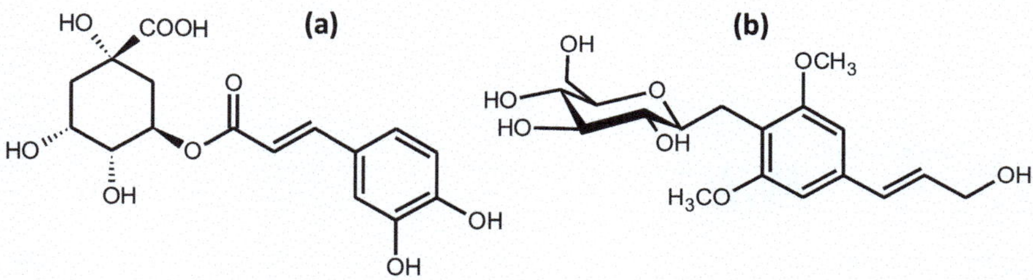

Fig. 7.10 Structures of chlorogenic acid (**a**) and eleutheroside B (**b**) showing increased amounts in *Eleutherococcus senticosus* (Araliaceae, Rosidae) after temperature elevation from 12 to 24 °C

significant trends under control conditions (Fig. 7.9) (Wahid 2007): Levels of soluble phenolics, anthocyanins and carotenoids, as well as carotenoids/chlorophylls regulation ratio, showed positive trends with leaf area and shoot dry weight (Fig. 7.8b). All these parameters shared temporal increasing levels under heat stress providing further proofs on the physiological implications of SMs in heat tolerance of sugarcane sprouts. Since phenolics are metabolized in the cytosol, they would be favorite to play the scavenging role of ROS (Moyer et al. 2002).

In *Eleutherococcus senticosus* (Siberian ginseng) (Araliaceae), chlorogenic acid and eleutheroside B (phenylpropanoid derivatives) increased under heat conditions (Figs. 7.7f and 7.10) (Shohael et al. 2006a).

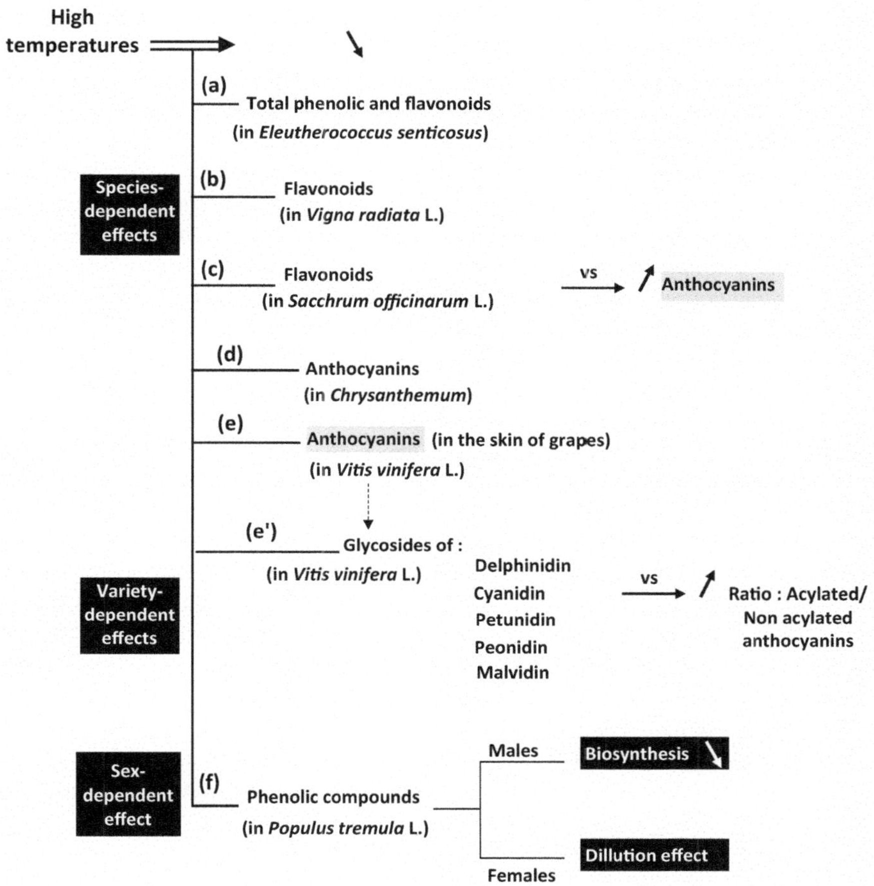

Fig. 7.11 Illustrative study cases on decreasing effects (↘) of heat conditions or elevated temperature on phenolic metabolism in different plant species

In *Ricinus communis* (castor bean) (Euphorbiaceae), temperature elevation had positive effect on phenolic metabolism and other antioxidants through an increase of their precursor amino acids (tryptophan, tyrosine and phenylalanine) (Fig. 7.7g) (Ribeiro et al. 2014). In upstream, this increase was associated with a decrease of shikimate levels due to its transformation into aromatic amino acids (favored by temperature elevation).

In other plants, phenolic compounds showed negative trends with temperature elevation (Fig. 7.11). Such trends may differ between plant species (or varieties), tissues and sex:

In *Eleutherococcus senticoccus* (Siberian ginseng), total phenolic content and flavonoids decreased with temperature elevation (Fig. 7.11a) (Shohael et al. 2006a). Such variations were opposite to those of some phenylpropanoid derivatives in the same species (Fig. 7.11 vs. Fig. 7.7, 7.10). These two opposite patterns provide more variability for analysis of temperaure elevation effects on plant responses. This is advantageous for further characterization and monitoring of gradual or differential changes in environment. Heat conditions resulted also in decrease of flavonoid levels in *Vigna radiata* (mung bean) (Fabaceae) and *Saccharum officinarum* (sugarcane) (Poaceae) (Fig. 7.7e, 7.11b, c) (Wahid 2007; Reardon and Qaderi 2017). However, anthocyanin levels in *Saccahrum officinarum*, increased highlighting specific responses of some metabolic pathways in a same plant species (Fig. 7.11c). An opposite (negative) variation trend of anthocyanins was revealed in *Chrysanthemum* sp. (Asteraceae) which responded by a decrease of anthocyanins under temperature elevation (Fig. 7.11d) (Shibata et al. 1988). These study cases show the interest of multi-species analysis for assessment of environmental states and changes by independent, complementary and confirmative phytochemical aspects.

Also, *Vitis vinifera* (grapevine) (Vitaceae) showed sensitive decrease of anthocyanins in berry skin under high temperature (Figs. 7.11e and 7.12) (Mori et al. 2005; Movahed et al. 2016). Moreover, the decrease pattern of anthocyanins in grape skin varied significantly between different grapevine varieties. Among anthocyanins, non-acylated structures were the most susceptible to degradation under high temperature (Fig. 7.11e′, 7.12). Thus, acylated structures tend to be more thermostable than non-acylated ones. Temperature-dependent degradation of anthocyanins in grapevine is mediated by the action of peroxidase enzymes (Movahed et al. 2016). Also, it has been suggested that biosynthesis rate of anthocyanins in red grape tends to be slowdowned by high temperature.

Phenolic content in *Populus tremula* L. (aspen) (Salicaceae) was reduced by high temperature by sex-dependent ways (Fig. 7.11f) (Sobuj et al. 2018): In male plants, phenolic biosynthesis frankly decreased, whereas in females, the decrease was originated from a dilution effect.

7.1.3.2 Terpene Patterns

High temperatures essentially result in increase of terpene productions most of which are volatile and emitted in air as thermotolerance reactions (Fig. 7.13). Thermotolerance is associated with ability to protect or repair cell membranes against disturbances linked to state changes or lipidic peroxidation. Apart from emission of volatile compounds, heat stress provokes accumulation (storage) of higher weight terpenes (e.g. tetraterpenes, triterpenes) in some plant species (Fig. 7.13). In other cases, some terpenes show negative trends with temperature elevation (Fig. 7.14).

Emitted terpenes can be initially stored in specialized structures from which emission occur under the control of temperature (Singsaas 2000). This is the case of conifers and Lamiaceae (aromatic plants). In other plants, emission occurs spontaneously without previous storage because of the absence of storage structures. Concerned plants are those emitting isoprene (non-stored terpene) and some Mediterranean oaks (e.g. *Quercus ilex*) (Loreto et al. 1996). In this case, non-stored terpenes are emitted under the effect of external temperature and internal control of their

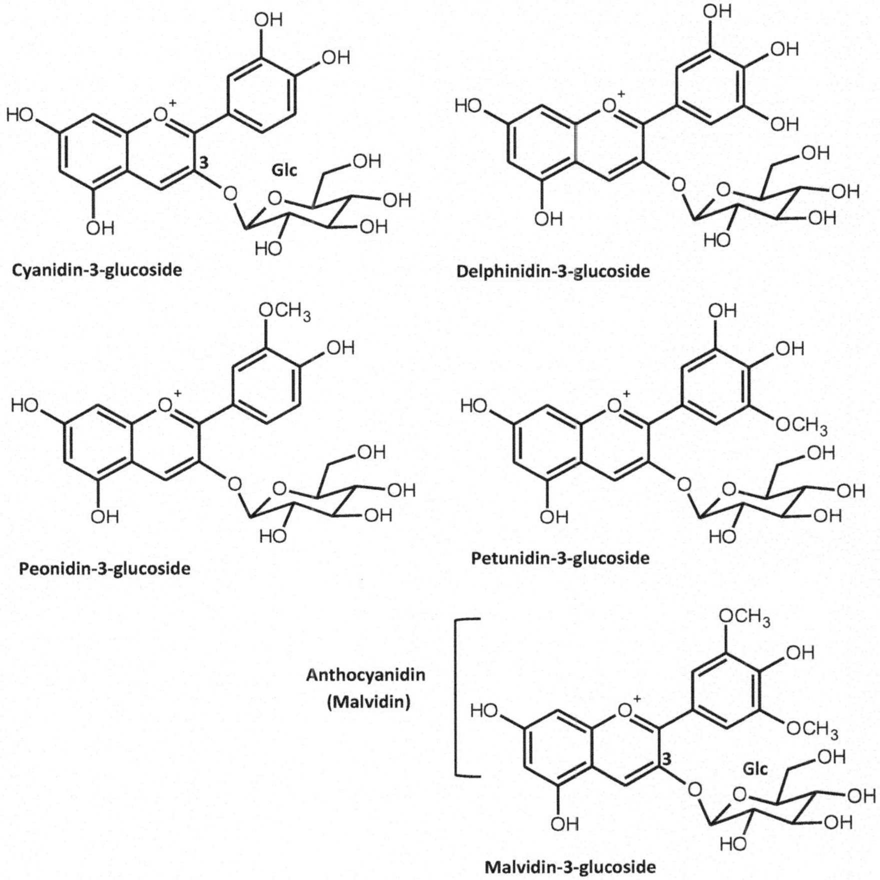

Fig. 7.12 Different non-acylated anthocyanins (3-glucosylated anthocyanidins) found in skin of grape (*Vitis vinifera*) and varying with temperature stress and according to different plant varieties

synthesis rates. For isoprene, the emission capacity as leave expand tends to be controled by the expression of isoprene synthase gene (Sharkey et al. 2008).

Emissions of monoterpenes are more frequent in plant world, but exchanges between biosphere and atmosphere are quantitatively dominated by the amounts of emitted isoprene (smaller molecule) (Sharkey et al. 2008). Isoprene-emitting plants tend to be more tolerant toward rapid heating of leaves induced by sunlight (heat flecks) compared with non-emitting ones. Isoprene emission-linked tolerance (protection) manifests also under oxidative effects caused by ozone and ROS.

Isoprene emission was highlighted in *Salix pylicifolia* (sage) (Salicaceae) living at 30 °C (Fig. 7.13a) (Rinnan et al. 2011). Emissions of volatile terpenes were reported

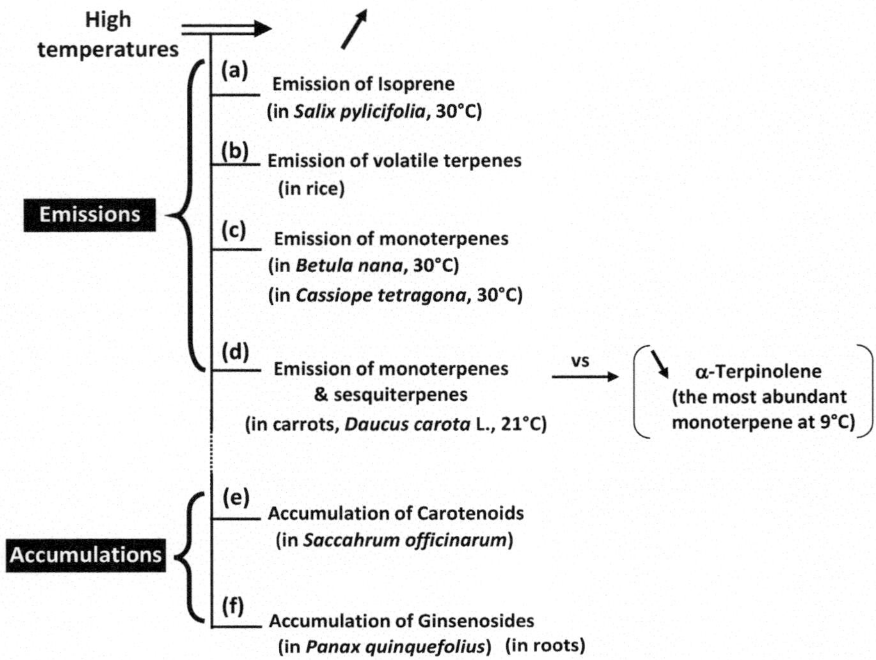

Fig. 7.13 Illustrative study cases of increasing (↗) terpene levels in different plant species under heat conditions

in *Onza sativa* (rice) (Poaceae) (Fig. 7.13b) (Goh et al. 2015). *Betula nana* (Betulaceae) and *Cassiope tetragona* (Ericaceae) living at 30 °C showed emissions of monoterpenes (Fig. 7.13c) (Rinnan et al. 2011).

Temperature elevation resulted in higher levels of both monoterpenes and sesquiterpenes in *Daucus carota* (carrot) (Figs. 7.13d and 7.14) (Rosenfeld et al. 2002). The most significant increase of monoterpenes concerned β-pinene, myrcene, α-ocimene and β-ocimene (Fig. 7.14). Among the sesquiterpenes, β-farnesene, α-humulene (or α-caryophyllene) and trans-caryophyllene (or β-caryophyllene) showed the highest positive trends (Fig. 7.14). Terpene pattern of carrot under temperature elevation was also characterized by a significant decrease of α-terpinolene at 21 °C whereas this cyclic monoterpene represents the most abundant terpene of carrot at 9 °C (Figs. 7.13d and 7.14).

Other volatile terpenes of carrot did not show significant variations with temperature including α-pinene, L-limonene and α-farnesene (Fig. 7.14). Under other aspect, this study showed that the increase of some terpenes in carrot roots has direct repercussion on its bitterness leading to lesser perception of sweet taste. This resulted in less preferred carrots by North European consumers. This calls for warning against undesirable effects of climate changes on organoleptic quality of fruits and legumes.

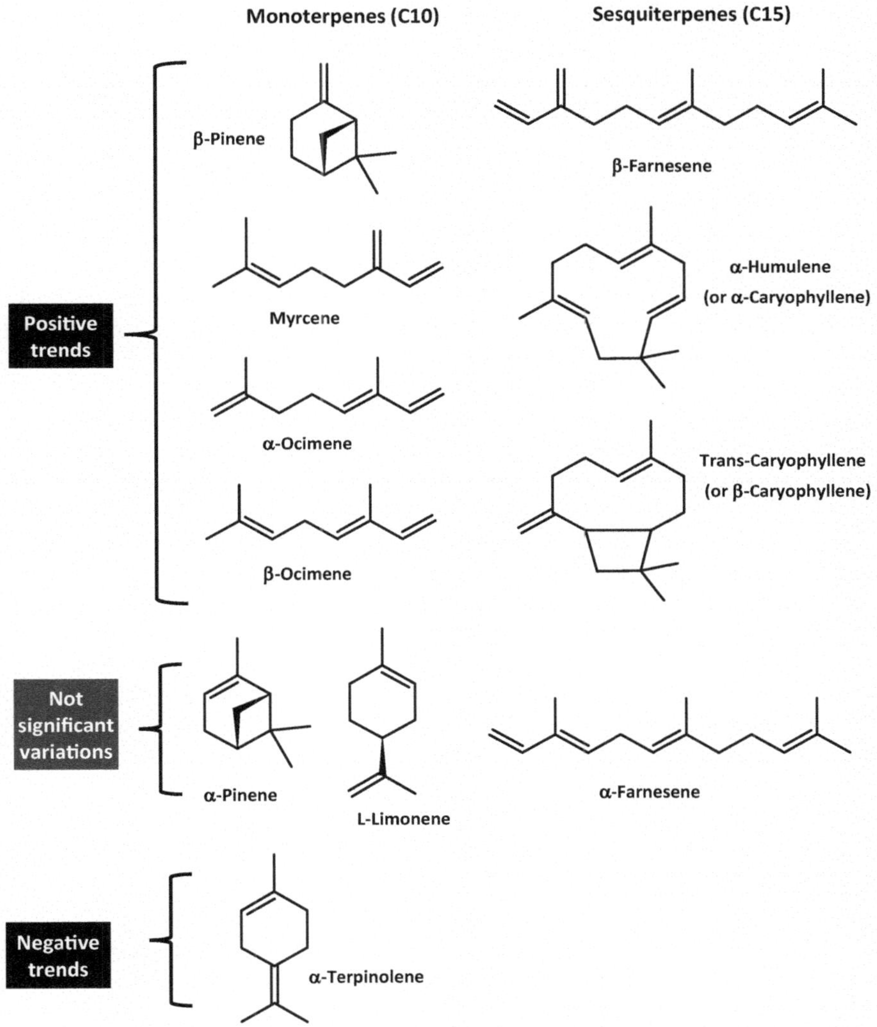

Fig. 7.14 Different monoterpenes and sesquiterpenes found in *Daucus carota* (carrot) showing positive, negative or not significant trends with temperature elevation (from 9 to 21 °C) (Rosenfeld et al. 2002)

Apart from volatile terpenes, heat conditions cause accumulation of higher weight metabolites (Fig. 7.13e, f).

Saccahrum officinarum (Poaceae) showed accumulation of carotenoids under heat conditions (Fig. 7.13e) (Wahid 2007). Moreover, significant increase of regulation ratio of carotenoids on chlorophylls was observed (Fig. 7.8). During 72 h of heat conditions, carotenoid levels in leaves increased within the first 36 h and reached a steady state at 48 h (Fig. 7.8). In addition to harvesting light, carotenoids could protect against heat-induced oxidative damage (Wahid 2007): as they are biosynthesized in chloroplast, carotenoids could protect the thylakoid lamellae from heat stress linked-lipid peroxidation.

At season scale, *Panax quinquefolius* (Araliaceae) was subject to several months duration of higher (26.5 °C, night vs. 31.2 °C, day) or lower temperatures (21.2 °C, night vs. 26.8 °C, day). Higher temperatures resulted in higher concentrations of ginsenosides in roots compared with lower temperatures (Fig. 7.13f, 7.15) (Jochum et al. 2007). Such a concentration process was directly linked to volume reduction of storage in plant roots at high temperature; inversely, at low temperature, preparation process to overwinter results in greater root biomass leading to a dilution effect of ginsenosides. Thus, ginsenosides tend to be well maintained despite the root biomass variation indicating key roles of these phytocompounds in thermotolerance. Ginsenosides of *Panax quinquefolius* are saponins with a dammarane-type aglycone hydroxylated at C3 and C12 (called protopanaxatriol) and substituted at C6 and C20 by two-four sugars including glucose (Glc) and rhamnose (Rha). (Fig. 7.15) (Zhu et al. 2011). These seasonal variations of ginsenosides were concluded as good monitoring parameters of climate changes (Jochum et al. 2007).

Short time-studies were carried out on rapid phytochemical responses of *Solanum lycopersicum* L. (tomato) (Solanaceae) following some minutes of sudden heat conditions (Copolovici et al. 2012): the leaves of tomato were gently immersed for 5 min in hot distilled water of six different temperatures (30, 37, 41, 46, 49, 51 °C). Emissions of several monoterpenes and sesquiterpenes by the leaves were revealed to be rapid plant responses to heat stress with different kinetic profiles (Figs. 7.16 and 7.17).

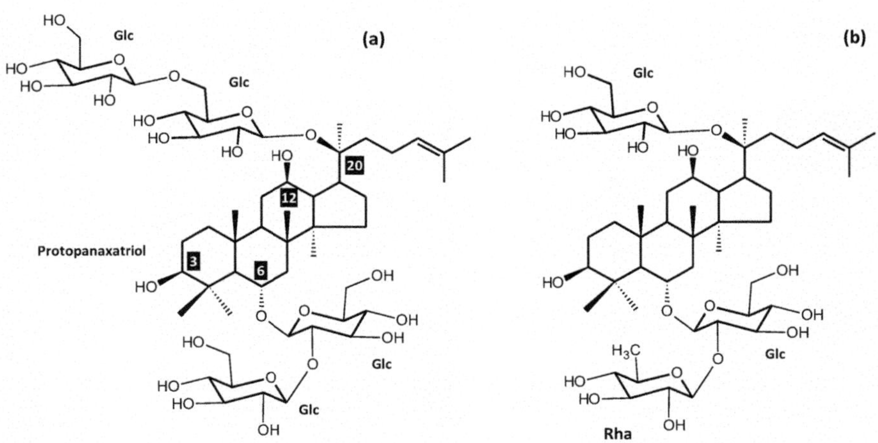

Fig. 7.15 Ginsenoside saponins _RB.1 (**a**) and Re (**b**)_ in the root of *Panax quinquefolius* (Araliaceae) showing positive regulation trends with high temperatures occurring during growth season

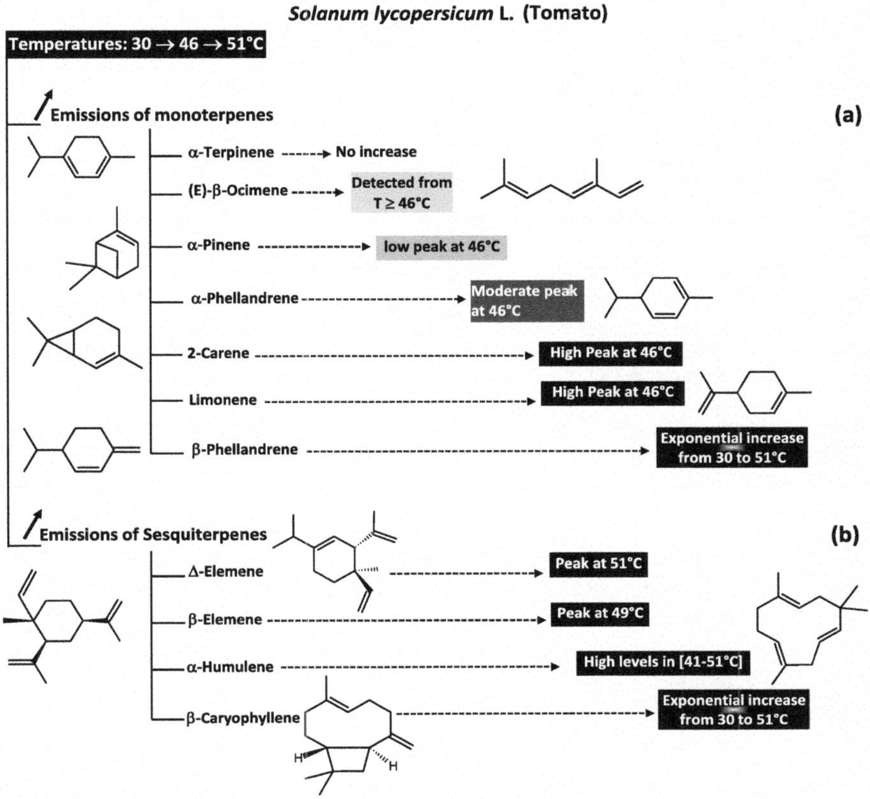

Fig. 7.16 Different monoterpenes (**a**) and sesquiterpenes (**b**) showing peaks or monotonous emissions under increasing temperatures in the leaves of *Solanum lycospersicum* (tomato) (Solanaceae)

Monoterpenes concerned with temperature-dependent emission levels included (in increasing order) (Fig. 7.16a):

- α-terpinene without significant increase.
- (E)-β-ocimene detected under heat conditions ≥ 46 °C.
- α-pinene showing a low emission peak at 46 °C.
- α-phellandrene with a moderate peak at 46 °C.
- 2-carene and limonene showing higher peaks at 46 °C.
- β-phellandrene showing monotonous increasing with temperature from 30 to 51 °C.

β-phellandrene dominated the monoterpene emissions because of its exponential increase with temperature (Fig. 7.17a). Its increase significantly influenced the whole emission levels of monoterpenes with the temperature. The different monoterpenes that were concerned with emission peaks under temperature elevation were also

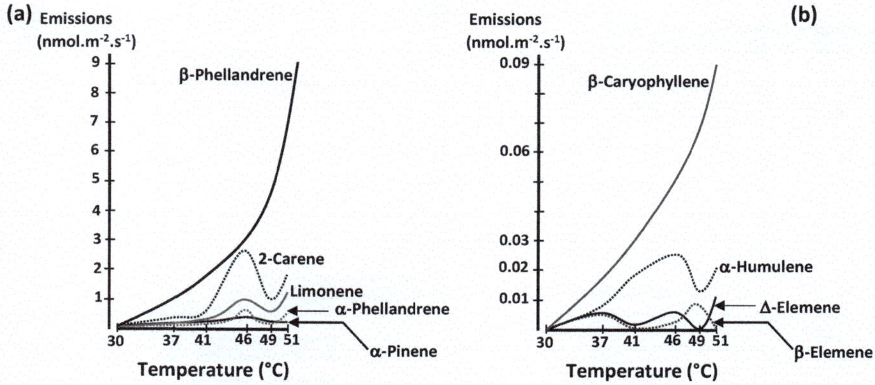

Fig. 7.17 Kinetic curves of monoterpenes (**a**) and sesquiterpenes (**b**) emitted by the leaves of *Solanum lycospersicum* (tomato) submitted at increasing temperature (30–51 °C) for 5 min (Copolovici et al. 2012)

constitutively emitted (in non-stressed plants) from glandular trichomes (their storage tissues) (Schilmiller et al. 2009, 2010; Jansen et al. 2009; Maes and Debergh 2003).

Apart from monoterpenes, sesquiterpenes (SQT) of *Solanum lycospersicum* showed also increase of emission levels with temperature but with more dispersed peaks (Fig. 7.18):

- β-elemene showed emission peak at 49 °C.
- Δ-elemene showed maximal emission at 51 °C.
- α-humulene (or α-caryophyllene) alternated between decreasing trend and stationarity in the temperature range 41–51 °C.
- β-caryophyllene emission, however, showed gradual increase from 30 to 51 °C.

SQT emission was found to occur from glandular trichomes in different genotypes of tomato (Schilmiller et al. 2010).

Beside information provided by emission peaks of different SQT, the volatile pool pattern can be further characterized by the relative levels (ratios) of different compounds by reference to the total emission (Fig. 7.18). By comparing the increasing rates of β-caryophyllene and α-humulene (also called α-caryophyllene) under different temperatures, the former SQT showed a positive regulation trend with temperature by opposition to the latter. Moreover, β-elemene and humulene tended to be regulated the one at the expense of the other under different temperatures. These inverse regulation trends provide key information on the role of relative levels for better characterization of phytochemical patterns helping to highlight induced metabolic pathways under heat (or other environement) constraint.

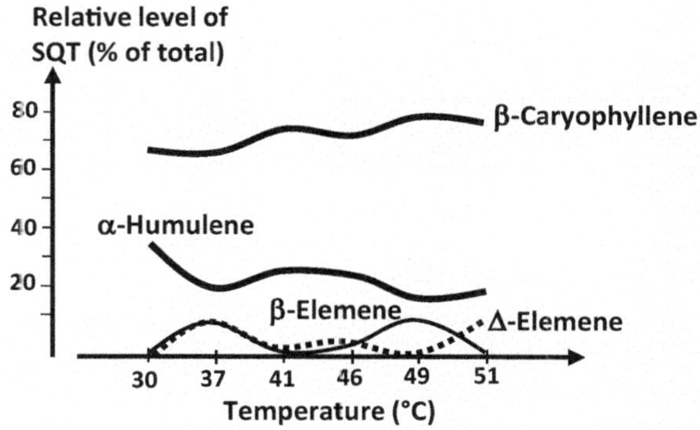

Fig. 7.18 Variation of relative levels of different volatile sesquiterpenes (SQT) with increasing temperature in *Solanum lycospersicum* (tomato)

Enhanced emissions of volatile terpenes (monoterpenes and sesquiterpenes) under heat conditions revealed the temperature effects on volatility of already stored products. Similar emission trends of monoterpenes and sesquiterpenes were associated with shared response system starting by mechanical damages and channeled to isoprenoid biosynthesis pathway. Temperature has direct effect on biosynthesis enzyme leading to de novo formation of volatiles (Niinemets et al. 2010). Moreover, temperature elevation can result in higher cuticular permeability through partial melting of cuticular lipids (Gibbs 2002; Matas et al. 2004; Riederer 2006). Also, rupture of glandular trichomes can occur by the effect of increase of vapor pressure with temperature (Copolovici and Niinemets 2005).

7.1.3.3 Nitrogen-Containing Compounds' Patterns

Nitrogen-containing compounds include alkaloids, glucosinolates and cyanogenic glycosides (Fig. 3.1, Part I).

Effects of elevated temperature on glucosinolate levels were studied in sprouts of *Brassica oleracea* (broccoli) (Brassicaceae) during 6 successive days of plant development (post-emergence) stage. Seedlings were cultivated under three (day/night)-temperature regimes including 30/15, 22/15 and 18/12 °C. Analysis of glucosinolates (GS) in sprouts revealed a main negative variation trend with age (days) in which higher temperature conditions resulted in higher GS levels (Fig. 7.19) (Pereira et al. 2002). GS naturally decreased with age but at each sprout development state, the highest levels of total GS corresponded to the highest temperature conditions (30/15 °C, day/night) (Fig. 7.20a). This trend was also observed for aliphatic and indole GS (Fig. 7.20b, c) with more pronounced positive temperature effect in the initial days (in younger sprouts). However. aromatic GS were rare and minor without clear variation in relation to age or temperature (Fig. 7.20d). Positive effect of high temperature on GS was also observed for broccoli sprouts which were grown under a constant temperature of 29° or 33.1 °C (Pereira et al. 2002). These results highlighted positive effect of supraoptimal growth temperature on induction of high GS-levels in broccoli (Pereira et al. 2002; Rosa et al. 1996; Rosa and Rodrigues 1998).

Aliphatic (MethylThioAlkyl) GS showed higher levels than indole or aromatic (minor) ones. Among aliphatic GS, glucoraphanin (GR) was dominant (Figs. 7.21a and 7.22a); it was followed by glucoiberin (GI) (Figs. 7.21a and 7.22b). These two metabolites represented 61.3 and 21.6% of total GS content, respectively. Although, glucoerucin (GE) was minor among the aliphatic GS, its level significantly increased with temperature elevation (Figs. 7.21a, c and 7.22c). Indole structures represented 8.4% of total GS among which 4-OH-glucobrassicin (4-OH-GB) showed high sensitivity to temperature elevation (Figs. 7.21b and 7.22e). Finally, glucobracissin (GB, indole GS) and gluconasturtiin (GN, aromatic GS) dis not manifest increase under the applied temperature regime (Figs. 7.21a and 7.22d, f).

Effect of temperature on N-containing compounds concerned also ricinine (alkaloid) in cotyledon and roots of *Ricinus communis* (castor bean, Euphorbiaceae) seedling cultivated at 20 and 35 °C for four days (after germination at 25 °C) (Ribeiro et al. 2014). Ricinine is an alkaloid with α-pyridine ring (Fig. 7.23a). Seeds of *Ricinus communis* were germinated at 25 °C in the dark then transferred to moist vermiculite for growing at 20 °C in continuous light for 10 days. Half of the 10-day-old seedling were transferred for growing at 35 °C versus 20 °C for the other moiety. After 4 days of development at 20 or 35 °C, roots and cotyledons were collected for phytochemical analysis. The experiments were applied to different genotypes of *R. communis*.

In cotyledons, ricinine showed higher levels under growing temperature of 35 °C compared with 20 °C (Fig. 7.23b). Moreover, ricinine levels were higher in cotyledons than roots where the alkaloid tended to decrease with temperature. This highlighted tissue-dependent variations of a same SM under same temperature conditions. However, genotype effect on ricinine was less pronounced than those of tissue and

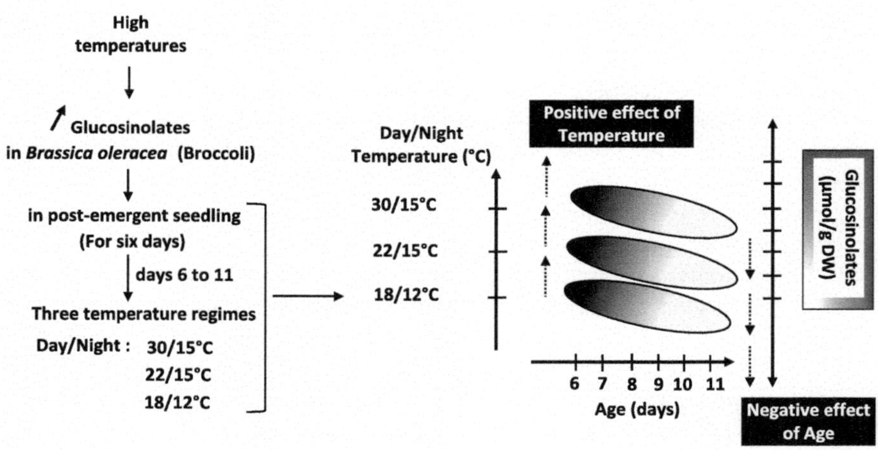

Fig. 7.19 Positive effect of treatment temperature and negative effect of development state (age) on glucosinolate content of *Brassica oleraceae* (broccoli) (Brassicaceae) (Pereira et al. 2002)

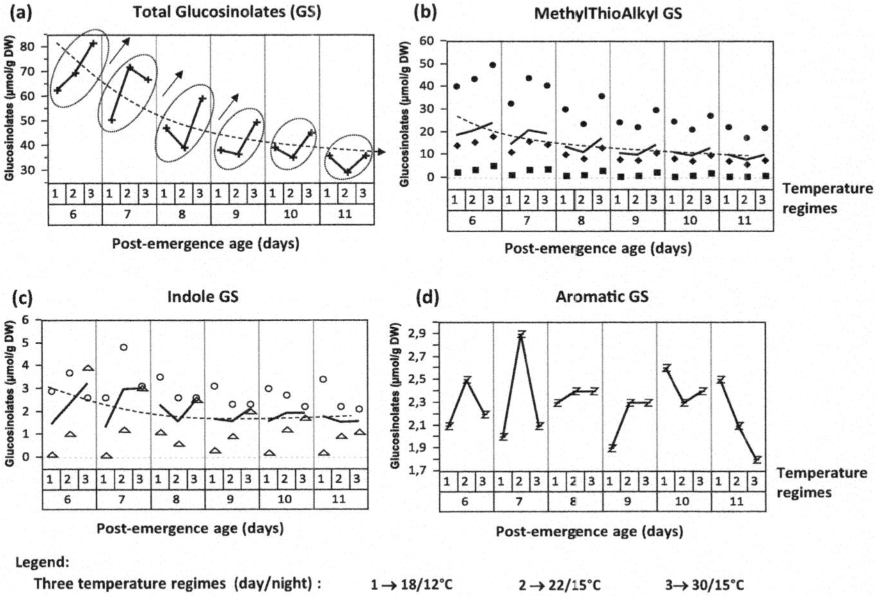

Fig. 7.20 Variations of different types of glucosinnolates (GS) with age and temperature conditions in broccoli (*Brassica oleraceae*) sprouts. Total (**a**), aliphatic (**b**) and indole (**c**) glucosinolates showed negative global trends with age versus positive local effect of temperature. No significant trend was observed for aromatic glucosinolates (**d**)

temperature. These results showed that ricinine could be used as good metabolic marker of temperature conditions in *Ricinus communis*. The survey role of ricinine was extended to other environmental conditions including defensive response against phloem feeders as aphids (Olaifa et al. 1991).

Positive effects of temperature on nitrogen-containing compounds concerned also the amino acids which showed increasing trends by opposition to carbohydrate (Fig. 7.24) (Ribeiro et al. 2014). Amino acids issued from different metabolic pathways were strongly favored by temperature elevation (35 °C) compared with their relative levels at 20 °C. They include aromatic amino acids (phenylalanine, tryptophan, tyrosine) issued from shikimate pathway, aliphatic amino acids (alanine, isoleucine, leucine, valine), those issued from downstream TCA cycle (asparagine, lysine, threonine) and glutamate-derivatives (proline, glutamine). However, glutamate did not show difference between both temperatures (20 °C vs. 35 °C).

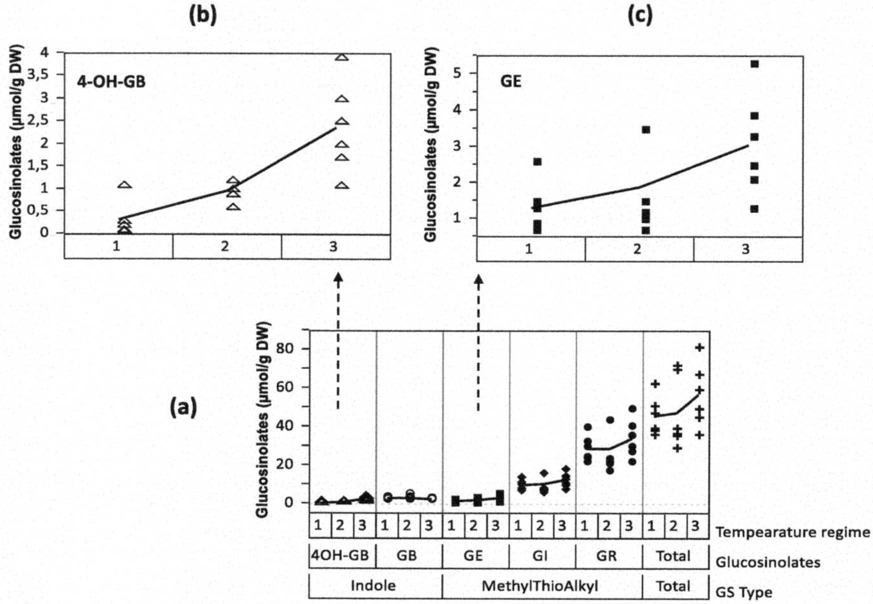

Fig. 7.21 Amount variations of different types of glucosinolates in the sprouts of *Brassica oleracea* grown under three applied temperature regimes. Temperature regimes (day/night): 1 → 5/12 °C; 2 → 22/15 °C; 3 → 30/15 °C. GB, glucobrassicin; GE, glucoerucin; GI, glucoiberin; GR, glucoraphanin

7.2 Cold Stress

7.2.1 General Effects of Cold Stress

Cold stress is linked to low temperatures which influence the plants by different aspects including temperature level, duration and variation rate. Low temperature levels can be associated with chilling (< 20 °C) or freezing (< 0 °C) (Thakur and Nayyar 2013). Plants species are well differentiated by their various optimal temperature levels (ranges) favoring their physiological, metabolic and biomolecular processes. Moreover, tolerance limits to low temperature levels vary between plant species which also require different times for acclimatization. Tropical and subtropical plant species are more sensitive to cold stress than plant living in temperate areas. For a given plant species, different reactions can occur according to cold degree and its apparition in environment (Fig. 7.25).

Decrease of temperature has direct effect on photosynthetic reactions in chloroplast leading to disturbance in electron transport chain. This results in increase of oxygenation reaction in chloroplast leading to more oxidation rate and more ROS production causing oxidative damages (Fig. 7.25a). Such damages are

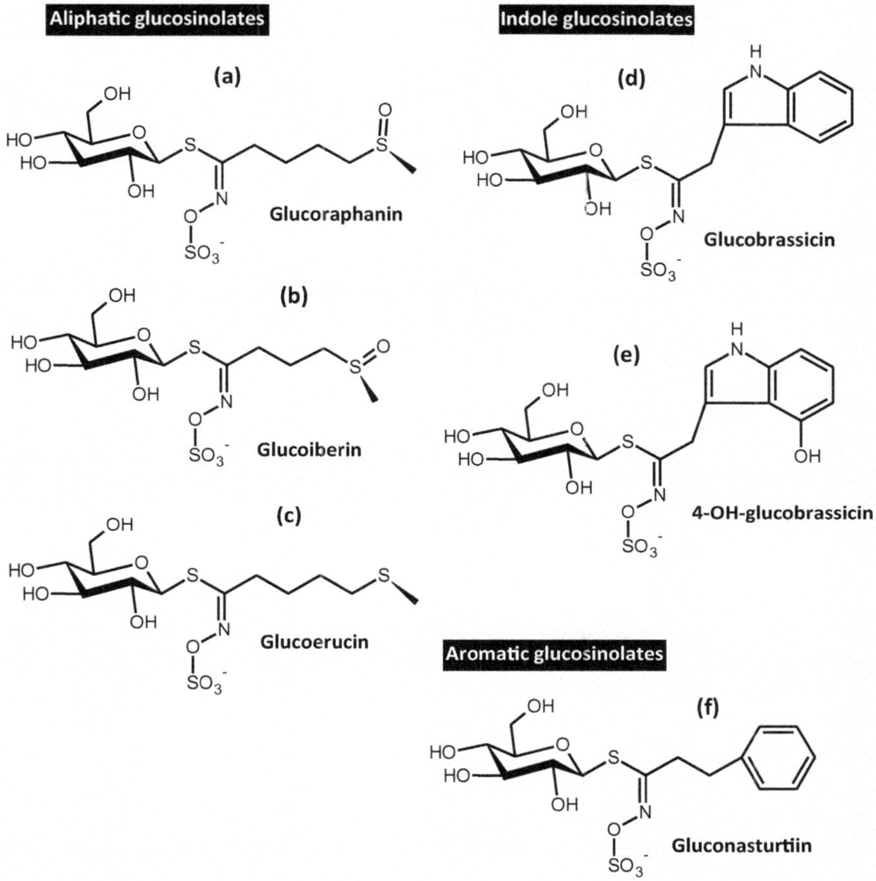

Fig. 7.22 Chemical structures of aliphatic (**a-c**), indole (**d**, **e**) and aromatic (**f**) glucosinolates found in *Brassica oleraceae* (broccoli)

reduced in plants by means of antioxidant metabolites (in addition to antioxidant enzymes) which act by scavenging and neutralizing ROS to maintain well-balanced redox potential (Fig. 7.25e) (Fig. 5.26, Part II). Antioxidant metabolites mainly include phenolic compounds because of their conjugated double bonds and hydroxylated states, in addition to terpenes showing conjugated double bonds (Fig. 3.12, 3.13, Part I).

In plant tissues, rapid decrease of temperature results in intracellular freezing initiated by the formation of ice crystals in the protoplasm (Fig. 7.25b). The development of ice crystals results in smaller hydrated cytoplasmic space and less available water form (solid form) for photosynthesis. This process leads to the disruption and rupture of plasma membrane that can be indicated by yellow/brow necrose on the plant tissues. This mechanical damage (or cell lyse) issued from rapid decrease of

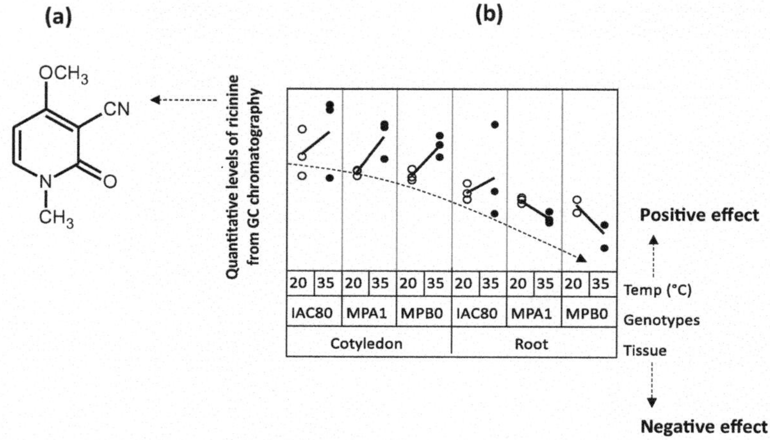

Fig. 7.23 (**a**) Ricinine structure (alkaloid) and (**b**) its quantitative variations in cotyledons and roots of three genotypes of *Ricinus communis* (Euphorbiaceae) initially growing at 20°C for ten days (after germination at 25°C) then transferred to temperature conditions of 20°C (white circles) or 35 °C (black circles) for 4 days

temperature is less frequent than that issued from gradual thermic decrease. Gradual decrease of temperature leads to water freezing in extracellular space (Fig. 7.25c). The subsequent liquid water deficit in this space triggers water flux from the most diluted sap to extracellular space. Such process leads to intracellular dehydration, ion leakage, and can be extended to plasmolysis. Finally, cold stress indirectly results in water stress (Fig. 7.25d). In these cases, cryoprotection in plants is provided by osmoregulation and membrane protection ways (Fig. 7.25f, g).

Osmoregulatory protection against cell dehydration calls for hydrophilic solutes to concentrate intracellular space and reduce water outlet (Fig. 7.25f). Sugars (carbohydrates) and anthocyanins which are glycosylated metabolites play key role in osmotic potential regulations helping the plants to acclimate in cold environments.

Moreover, temperature bellow optimum reduces the membrane movements to slow down and results in membrane transition from liquid crystalline to gel phase (Fig. 7.25g). This process acts as alerting signal and is followed by initialization of temperature-signaling pathways.

Membrane lipid desaturation in bacteria (by desaturase enzyme) was revealed to help microorganisms to acquire cold-tolerance (Wada et al. 1990). Inversely, saturation of membrane lipid is expected to rigidify the membrane leading to less membrane fluidity and less cold tolerance.

Moreover, in *Arabidopsis thaliana* suspension cell cultures, low temperature was immediately followed by activation of membrane phospholipase enzyme leading to rapid increase of phosphatidic acid (PtdOH) (Ruelland et al. 2002). PtdOH acts as preliminary cold-induced signal by playing a mediating role in nitric oxide (NO) signaling cascade (Testerink and Munnik 2005). This shows that change in membrane fluidity tends to be initiating event of downstream low temperature induced signaling

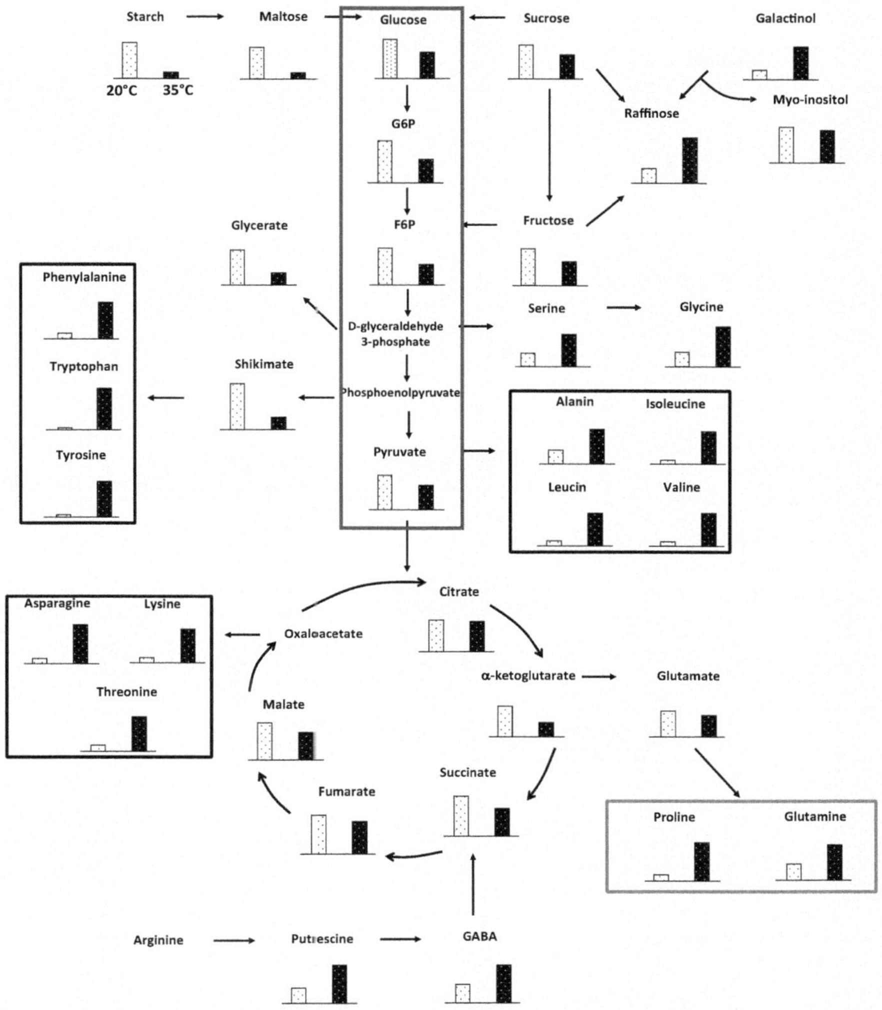

Fig. 7.24 Effect of temperature elevation on amino acids and carbohydrates of *Ricinus communis* (Euphorbiaceae). White bars, 20 °C; black bars, 35 °C

pathways. Such process was also shown to cause changes in membrane proteins followed by signaling cascade.

Membrane protection calls for lipophilic compounds that can be integrated inside to reinforce its stability and oppose to its denaturation. Terpenes and steroids have lipophilic aspect favoring their interactions with cell membranes (Fig. 7.25g). Cell membrane play key role as communication interface between intra- and inter-cellular spaces. It represents the primary reception and perception site of cell toward environment conditions.

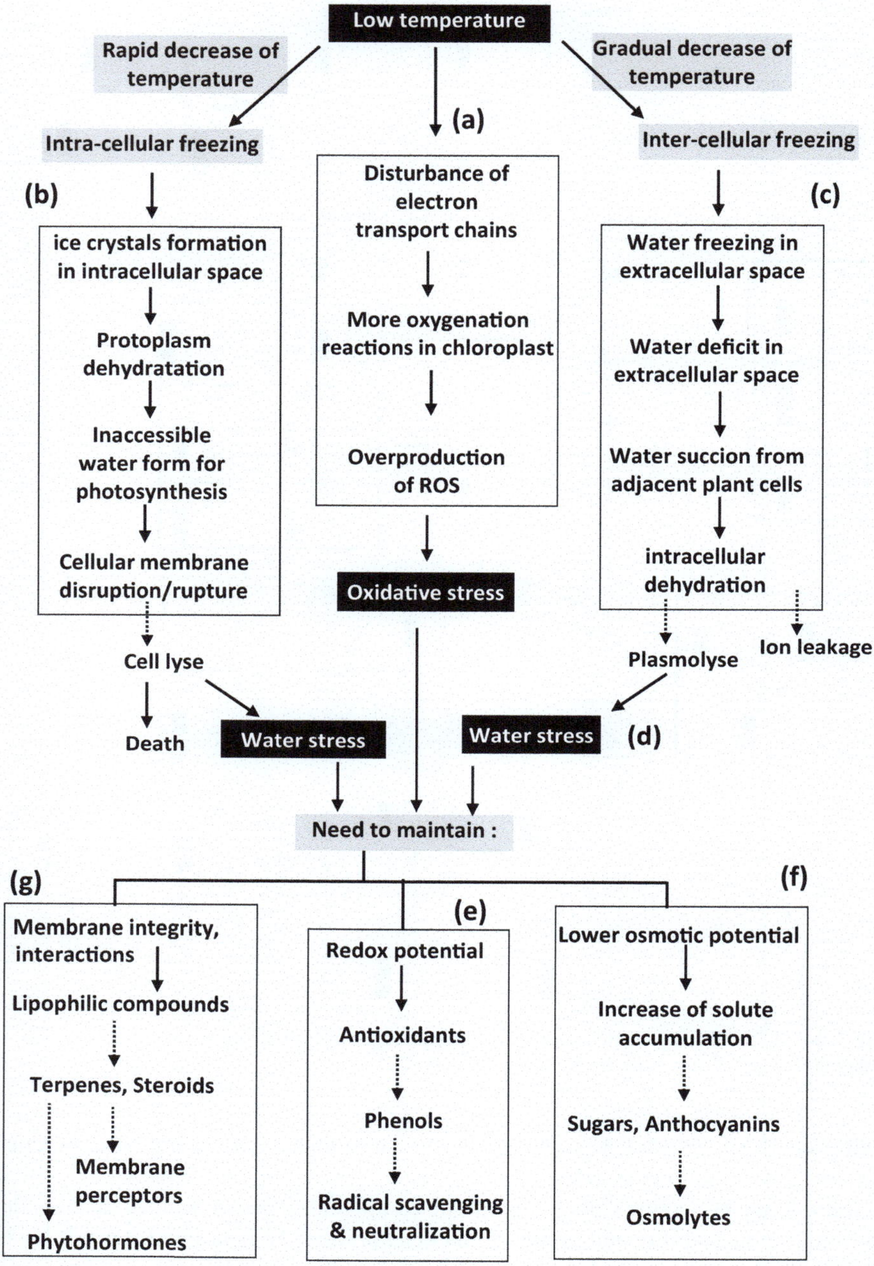

Fig. 7.25 Main risks of cold conditions on plant tissues resulting in oxidative and water stresses leading to cryoprotection need via antioxidant and osmoregulatory metabolites

7.2.2 General Plant Responses to Cold Stress

Variations of membrane fluidity, osmotic potential and redox potential provide primary perception centers of cold stress in plants; their control represents key protection processes against this disturbance factor. Variations of membrane fluidity, osmotic and redox potentials are followed by biochemical variations which trigger protective and reparative processes of plant growth against cold stress. These homeostasis processes imply phytochemical compounds which play phytohormone roles by triggering signaling cascades leading to expression of cold genes including: expression of C-repeat-binding factors (CBFs) which activate cold responsive genes (COR) (Fig. 7.26) (Shi et al. 2018; Barrero-Gil and Salinas 2017; Eremina et al. 2015). In absence of cold stress, normal plant growth is associated with low expression levels of CBFs. Under cold stress, however, CBFs are rapidly induced by transcription factors.

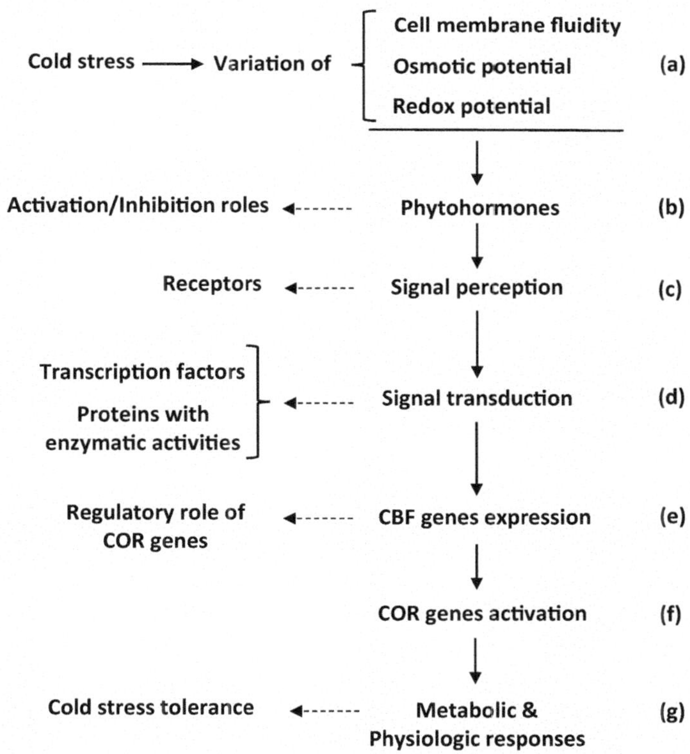

Fig. 7.26 Cold stress effects (**a**) on cascade system initialized by phytohormones (**b**) followed by signal perception (**c**), signal transduction (**d**), CBF and COR genes activation (**e**, **f**) and leading to metabolic and physiologic protective responses and freezing tolerance in plants (**g**)

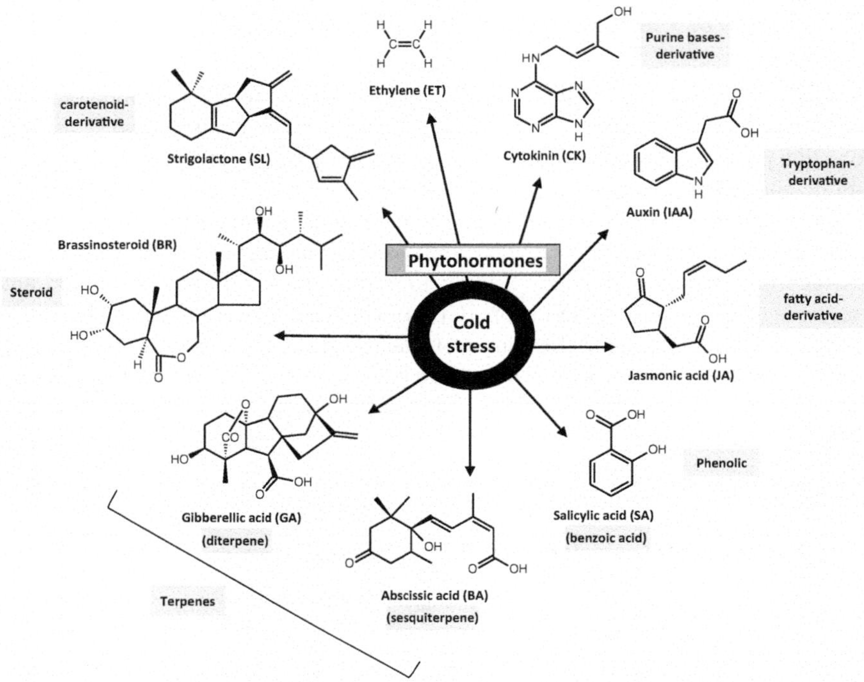

Fig. 7.27 Different phytohormones implied in transcription activation and metabolic responses of plants to cold stress

Different phytohormones are associated with different signal perceptions and transductions (Fig. 7.26b–d) (Raza et al. 2023). Multiway expression of the complex cold gene network leads to production and regulation of many secondary metabolites depending on cold level, duration, variation rate on one hand, and plant species, geographical area (temperate, tropical), growth phase (vegetative, reproductive), plant tissues/organs (leaves, shoots, roots) (Raza et al. 2023). Implied phytohormones in cold stress include abscisic acid (ABA), brassinosteroids (BR), gibberellic acid (GA) (and derivatives, gibberellins), salicyclic acid (SA), jasmonic acid (JA), strigolactones (SL), auxin (indole-3-acetic acid, IAA), cytokinin (CK) and ethylene (ET) (Fig. 7.27). Several of these hormones are derived from SM pathways: ABA (sesquiterpene), GA (diterpene), SL (tetraterpene derivative), SA (phenol). Thus, secondary metabolites reveal to play roles in both upstream and downstream responses to cold stress through signaling phytohormones and protective-communicative phytocompounds, respectively.

Control of redox potential is provided by antioxidant enzymes and metabolites (Fig. 7.28a). Antioxidant enzymes include superoxide dismutase (SOD) and peroxidase (POD), whereas antioxidant metabolites include phenolic compounds essentially. Cold stress was revealed to be followed by activation of phenylalanine ammonia-lyase (PAL) pathway giving phenylalanine which represents precursor amino acid of phenolic compounds (Fig. 4.2, Part II). PAL bridges primary and secondary metabolism. It is initially favored by funneling shikimic acid into the shikimate pathway (Fig. 7.24) (Fig. 4.2, Part II). Moreover, the quantum yield of PSII tends to be boosted through an increase of (Fv/Fm)-ratio. Decrease of oxidative stress is generally revealed by lower H_2O_2, ROS and malondialdehyde (MDA) (oxidative stress marker) levels as well as decrease of oxidase enzymes (e.g. polyphenol oxidase).

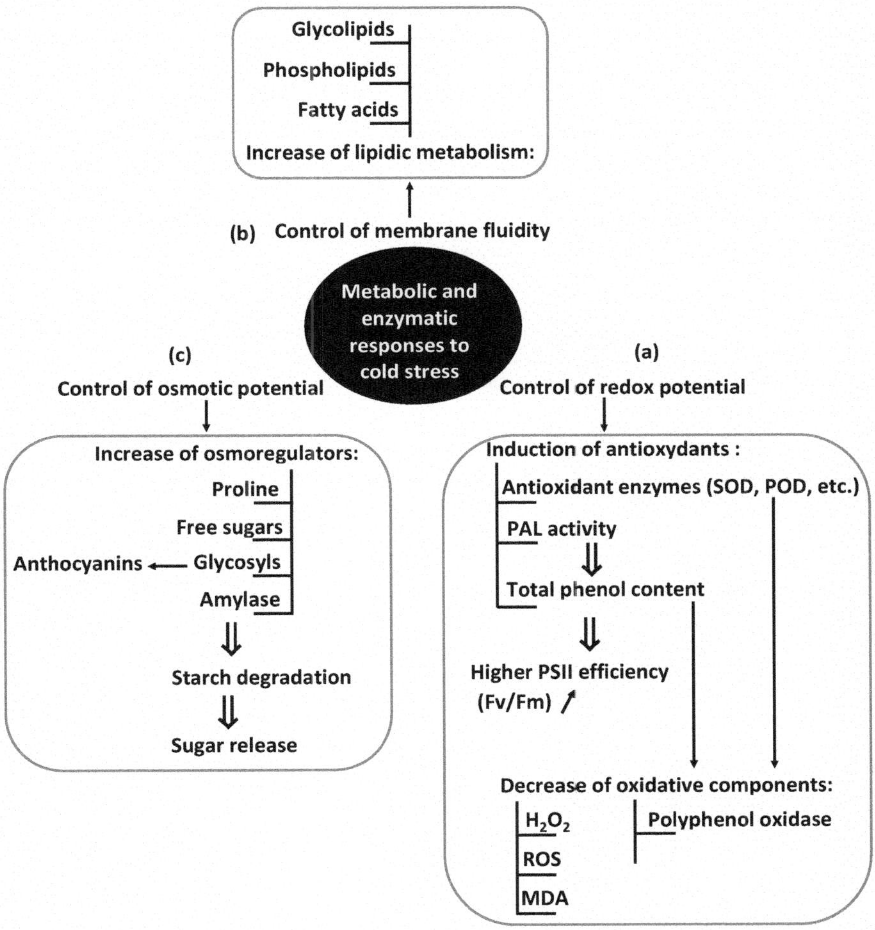

Fig. 7.28 Downstream enzymatic and metabolic responses of plants to cold stress induced by phytohormones at upstream level. PAL, phenylalanine ammonia lyase

Control of cellular membrane fluidity is performed by increasing the contents of hydrophobic metabolites having similar structural properties than membrane (Fig. 7.28b). This result in relatively higher production of phospholipids (including phosphatidylcholine and phosphatidylethalonamine) and glycolipids which are important cell membrane components (Fig. 7.29). Moreover, metabolism of unsaturated fatty acids (e.g. α-linoleic acid) tends to be increased. Also, free sugars (glucose, fructose) play key role in the maintain of membrane phospholipids in a liquid-crystalline state contributing to membrane fluidity (Duan et al. 2022).

Hormonal inductions concern multiples enzymatic and metabolic processes which play key roles in the control of cellular membrane fluidity and osmotic and redox potentials (Fig. 7.28). Control of transmembrane osmotic potential occurs by osmoregulatory metabolites which are hydrosoluble including typically free sugars (e.g. glucose, fructose), proline and glycosides (Fig. 7.28c). Among the glycosides, anthocyanins highly participate to osmoregulation because of their high glycosylation levels (increasing their hydrophily), in addition to the ionic state of their heterocyclic oxygen (Fig. 4.4, Part II). Also, sugar production is favored by activation of amylase which catalyzes the hydrolysis of starch.

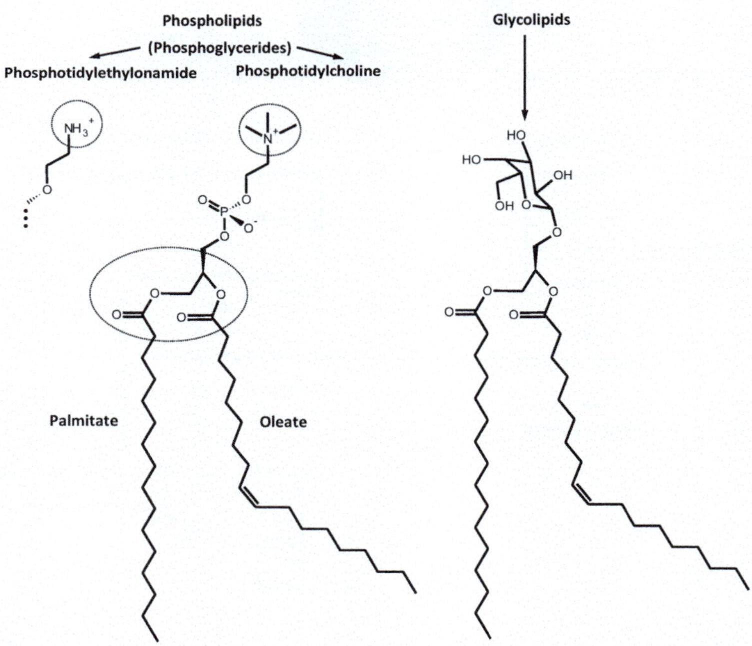

Fig. 7.29 Chemical structures of phospholipids and glycolipids representing important structural components of cellular membranes

Also, flavonoids (including anthocyanins) were revealed to be strongly stimulated by free sugars via induction of different biosynthesis genes. This was highlighted by an integrative study combining genomic, transcriptomic and metabolomic aspects of phenolic compounds in *Arabidopsis thaliana* cultivated under different sugars' concentrations (Solfanelli et al. 2006). Culture middles were enriched with three sugars (sucrose, glucose, fructose) separately with increasing concentrations (0, 7.5, 15, 30, 60, 90 mM). Sucrose showed higher induction levels of genes coding for biosynthesis enzymes of flavonoids (Fig. 7.30). Enzymatic pathways favoring the biosynthesis of anthocyanidins and anthocyanins were revealed to be upregulated in upstream by sucrose induction. This was highlighted by high expression levels of dihydroflavonol reductase (DFR), leucoanthocyanidin dioxygenase (LDOX) and UDP-Glc:flavonoid 3-O-glucosyltransferase (UF3GT) (Fig. 7.30e–g). These enzymes are responsible for the biosynthesis of leucoanthocyanidins, anthocyanidins and anthocyanins, respectively (Fig. 7.31).

More generally, upregulation of glucosytransferase (UF3GT) can also benefit for glycosylation of other flavonoids including flavonols (e.g.; kaempferol-3-glucoside) (Fig. 7.31). Effect of sucrose on upregulation of flavonoid metabolic pathway was also revealed by resulting increase of accumulation level of mRNA associated with genes coding for biosynthesis enzymes of flavonoids (Fig. 7.32). Cultures subjects to 90 mM (sucrose) showed higher accumulation levels of mRNA associated with different flavonoid biosynthesis genes compared with control conditions (cultures without sucrose) (Fig. 7.32a vs. Fig. 7.32b). The highest accumulation levels were measured for chalcone synthase (CHS) which is responsible for the formation of chalcone which represents a common precursor of flavonoids (Fig. 7.31).

7.2.3 *Specific Patterns of Plant Species Under Cold Stress*

Many analyses of metabolic variations were carried out in different plant species under different conditions of cold stresses. Cold-induced phytochemical patterns were associated with different extrinsic and intrinsic factors including temperature level, plant species, tissues, development state, etc.

7.2.3.1 Phenolic Patterns

Responses of cork oak (*Quercus suber*) (Fagaceae) leaves to contrasting temperatures were studied by analyzing transcript accumulation of four key enzymes involved in phenolic pathways: chorismite synthase (CS), phenylalanine ammonia-lyase (PAL), NADPH-dependent cinnamyl alcohol dehydrogenase (CAD) and chalcone synthase (CHS) (Chaves et al. 2011). Oak forests play key role in soil, water and biodiversity conservation. In addition to its great ecological importance, cork oak has a high economic impact in Mediterranean area due to the cork's value.

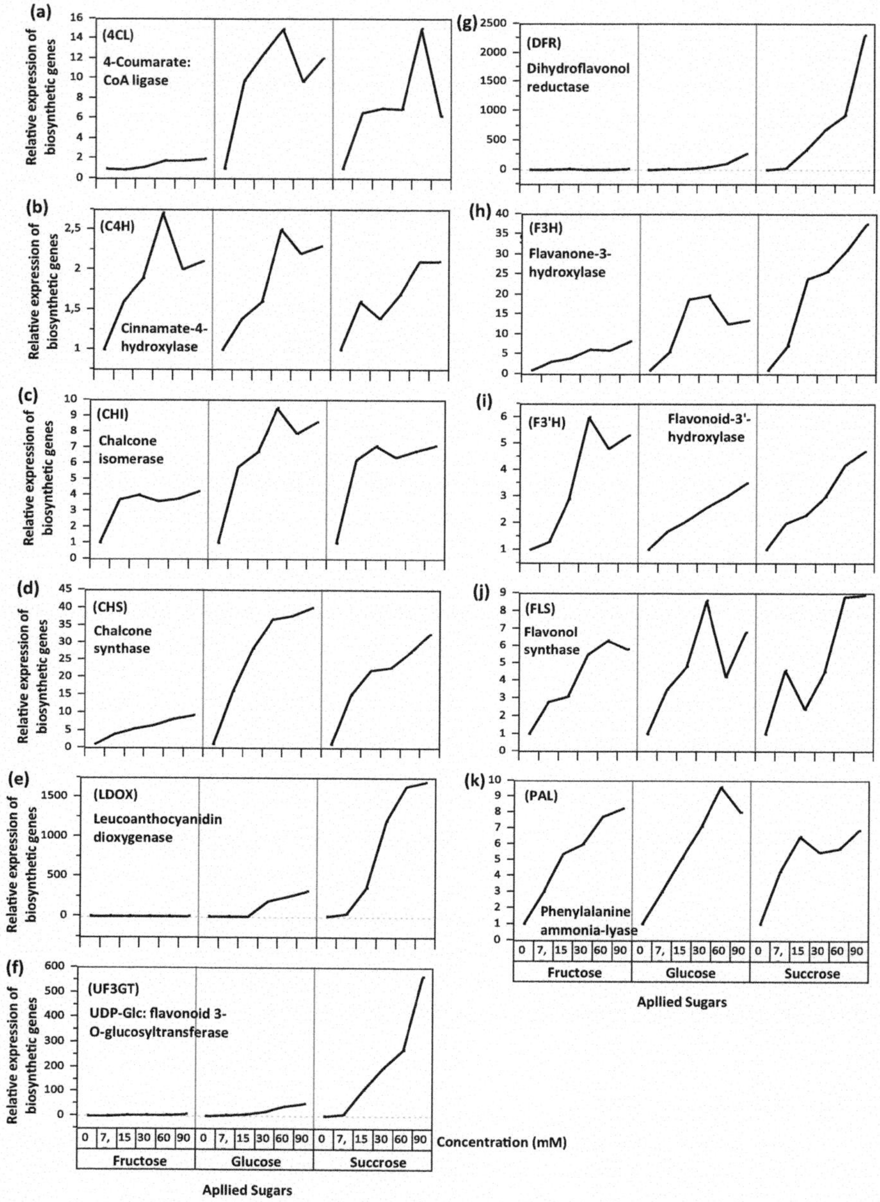

Fig. 7.30 Effect of glucose, fructose and sucrose at different concentrations on the expression of genes involved in biosynthesis pathways of flavonoids in cultivated *Arabidopsis thaliana*

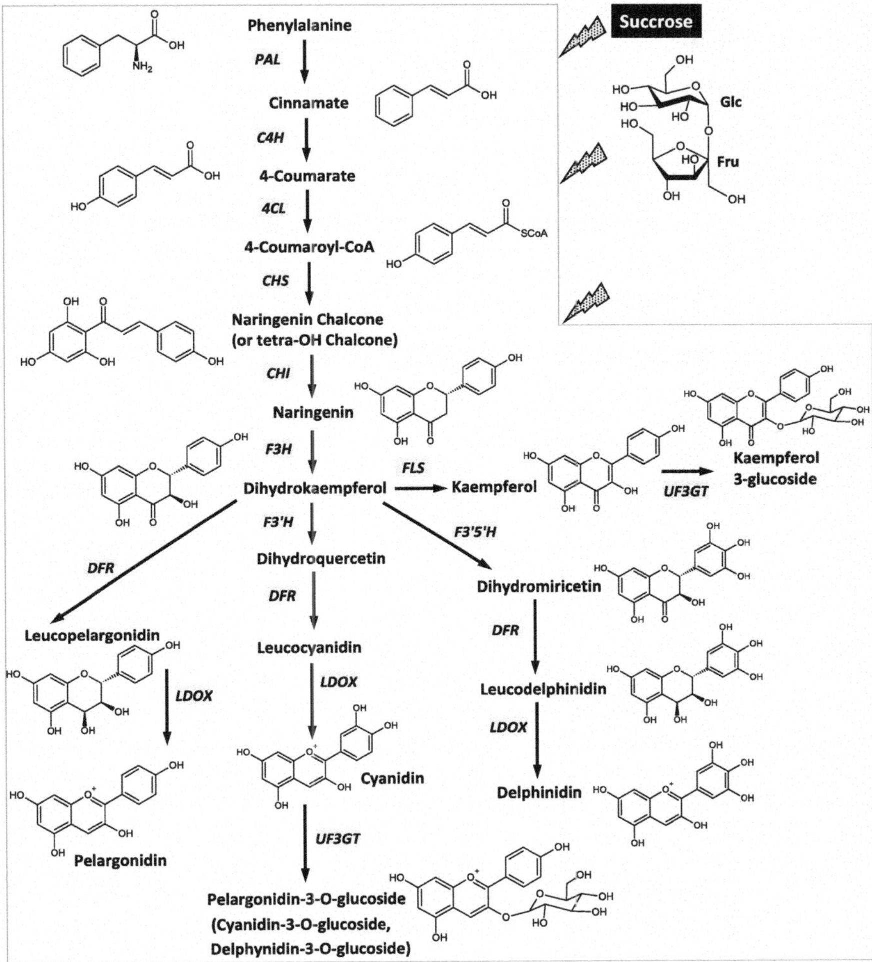

Fig. 7.31 Increase of anthocyanin regulation pathway among other flavonoids under inductive effect of sucrose on genes involved in flavonoid synthesis in *Arabidopsis*. PAL, phenylalanine ammonia-lyase; C.4H, cinnamate-4-hydroxylase (C4H); 4CL, 4-Coumarate-CoA-ligase; CHS, chalcone synthase; CHI, chalcone isomerase; F3H, flavanone-3-hydroxylase; FLS, flavonol synthase; F3′H, flavonoid-3′-hydroxylase; F3′5′H, flavonoid-3′5′-hydroxylase; DFR, dihydroflavonol reductase; LDOX, leucoanthocyanidin dioxygenase; UF3GT, UDP-Glc:flavonoid 3-O-glucosyltransferase

Three months-old plants growing at 15 °C were then submitted for 53 days to constant low (10 °C) and high (28 °C) temperatures representing winter and summer conditions, respectively (with 50% humidity, 12 h photoperiod, 500 $\mu E\ m^{-2}\ s^{-1}$) (Fig. 7.33a). Therefore, preexisting (PE) and newly formed (NF) leaves were

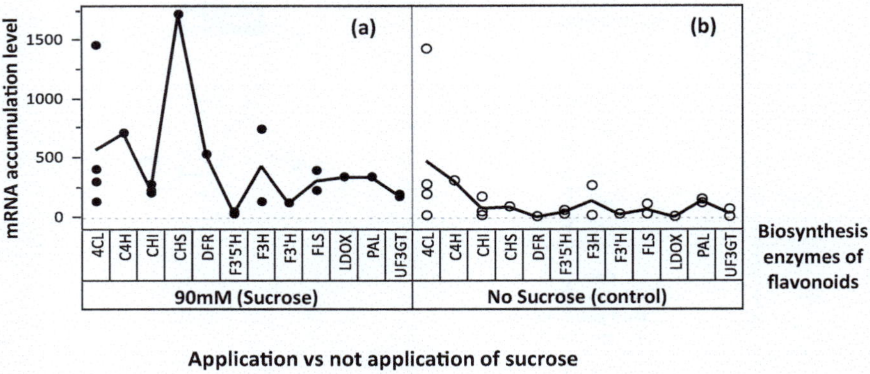

Fig. 7.32 Accumulation levels of mRNA for genes coding for biosynthesis enzymes of flavonoids in Arabidopsis seedlings. *Arabidopsis* seedling was treated without (control) or with 90 mM of sucrose subsequently to germination (for 3 days). 4CL, 4-Coumarate-CoA-ligase; C4H, cinnamate-4-hydroxylase (C4H); CHI, chalcone isomerase; CHS, chalcone synthase; DFR, dihydroflavonol reductase; F3′5′H, flavonoid-3′, 5′-hydroxylase; F3H, flavanone-3-hydroxylase; F3'H, flavonoid-3'-hydroxylase; FLS, flavonol synthase; LDOX, leucoanthocyanidin dioxygenase; PAL, phenylalanine ammonia-lyase; UF3GT, UDP-Glc:flavonoid 3-O-glucosyltransferase

harvested. Four main water-soluble metabolites were revealed by NMR including glucose, sucrose, quinic acid and quercitol (Fig. 7.33b):

Sucrose accumulated at higher concentrations both in old (PE) and young (NF) leaves when plants were grown at 10 °C (Fig. 7.34b). This revealed significant role of sucrose under cold conditions in *Quercus suber*. This metabolic trend was also found for quercitol which showed relatively high accumulation in NF leaves at 10 °C. These results highlighted development state-dependent response to cold stress in *Q. suber*. An annual variation study showed that field-grown plants accumulated higher sucrose in April where leaves develop than June where leaves are old (Passarinho et al. 2006). Quercitol and other cyclitols were revealed to play cryoprotectant role of thylakoid membrane (Orthen and Pop 2000). Beside cyclitols, flavonoid pathway was sensitive to low temperature with relative increase of catechins at 10 °C (Fig. 7.33c and 7.34b) (Chaves et al. 2011). At enzymatic level, higher catechin regulations were consistent with higher expression levels of genes coding for chalcone synthase (CHS) and phenylalanine ammonia-lyase (PAL) at 10 °C (Figs. 7.33c, 7.34a and 7.35) (Chaves et al. 2011).

However, quinic acid showed opposite trend to quercitol with more accumulation at 28 °C than 10 °C (Fig. 7.34b) (Chaves et al. 2011; Passarinho et al. 2006). Quinic acid and its derivatives (e.g. chlorogenic acid) were suggested to accumulate under energetically favorable conditions (Herrmann 1995). At enzymatic level, chorismate synthase (CS) (Fig. 7.33) showed higher expression at 28 °C (not showed).

Another integrative study highlighted the effect of cold duration on phenylpropanoid and primary metabolisms in *Fagopyrum tataricum* (tatary buckwheat) (Polygonaceae) (Jeon et al. 2018). Sterilized seeds were grown on agar plates with

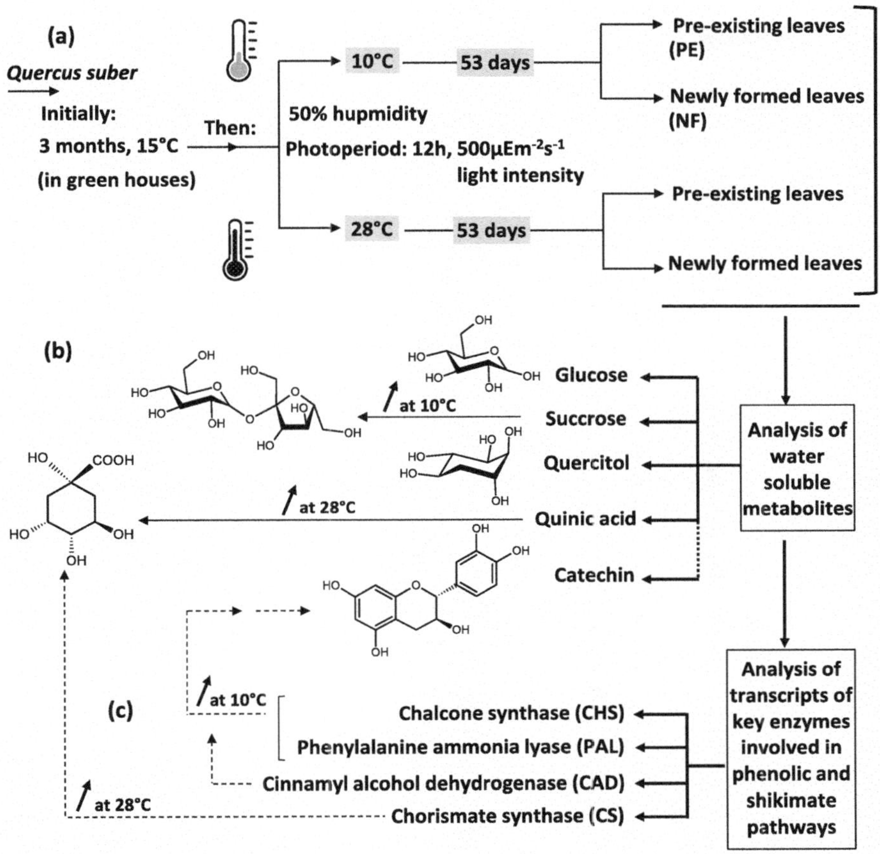

Fig. 7.33 Experimental design aiming for analysis contrasting temperature-responses in young *Quercus suber* (cork oak) subject to cold (10 °C) or heat (28 °C) conditions for 53 days (**a**) through temperature-induced hydrophilic metabolites (**b**) and transcript levels of involved enzymes (**c**) in both old and young leaves

pH 5.7 at 25 °C for 16 h of photoperiod (light intensity of 150 μmol m^{-2} s^{-1}) (Fig. 7.36a). After six growth days, plants were incubated at 4 °C for different durations varying from 2 to 8 days (Fig. 7.36b). Expression of genes coding for phenylpropanoid pathway was regularly performed by measuring their relative fold changes at 0, 1, 2, 4, 8, 24, 48 and 96 h of applied cold (Fig. 7.36c). Extensive metabolomic analysis covered both primary and secondary metabolites including (Fig. 7.36d, e):

- Sugars, polyols, organic acids and amino acids.
- Phenolic metabolites including anthocyanins, proanthocyanidins, cinnamic and benzoic acid derivatives.

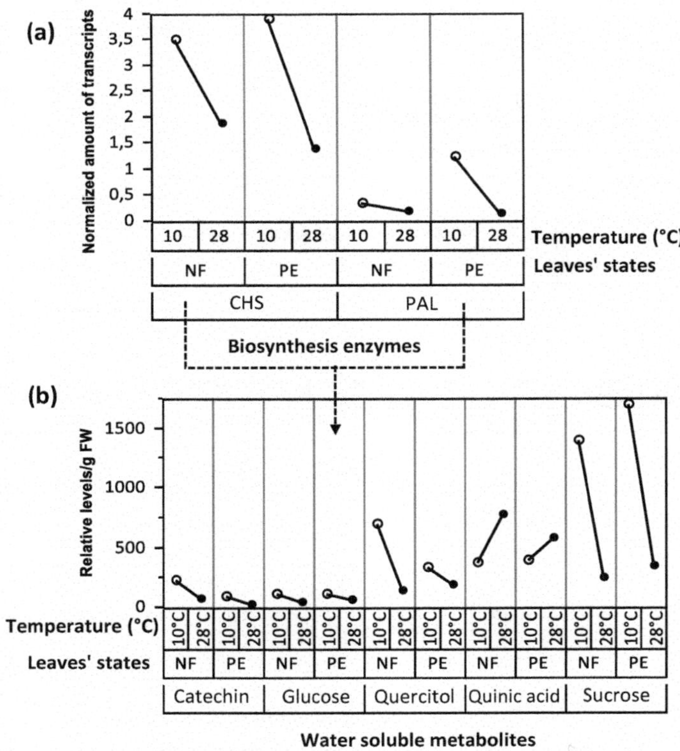

Fig. 7.34 Effect of cold (10 °C) and heat (28 °C) on phenylpropanoid and shikimate pathways in young *Quercus suber* grown for 53 days in thermic stress. **a** Accumulation levels of transcripts of genes encoding key enzymes of the shikimate pathways. **b** Accumulation levels of different water-soluble metabolites. CHS, chalcone synthase; PAL, phenylalanine ammonia lyase; NF, new formed leaves; PE, pre-existing leaves

Moreover, antioxidant activity was followed for 8 days by measuring and comparing scavenging capacity of DPPH in plant grown at 4 and 25 °C (control) (Fig. 7.36f).

Genes coding for biosynthesis enzymes of phenylpropanoid pathway showed expression peaks at 24 or 48 h (Fig. 7.37). The peaks provided fold change of gene expressions relative to their reference states at time = 0 (just before cold application). Enzymes located in upstream of phenylpropanoid pathway showed peaks at 24 h; they concerned phenylalanine ammonia-lyase (PAL), cinnamate-4-hydroxylase (C.4H), 4-Coumarate-CoA-ligase (4CL), chalcone synthase (CHS) (Fig. 7.37a–d). Enzymes located in downstream metabolism showed gene expression peaks at 48 h; they include chalcone isomerase (CHI), flavanone-3-hydroxylase (F3H), flavonol

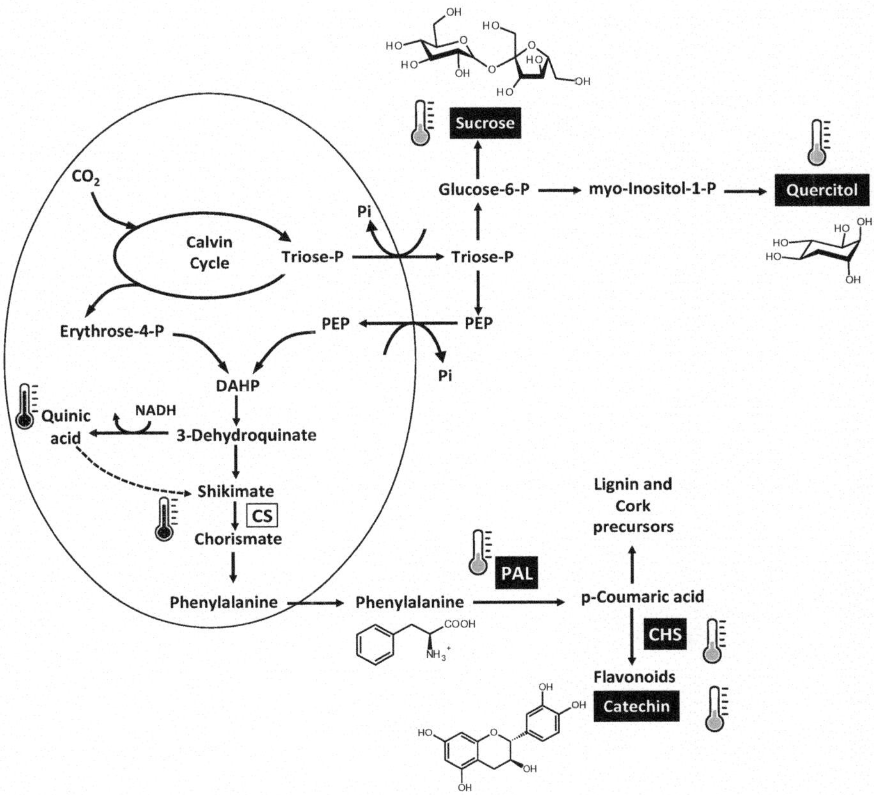

Fig. 7.35 Saccharide, cyclitol, shikimate and phenylpropanoid pathways involved in the response of *Quercus suber* leaves to cold (10 °C) and heat (28 °C) growing conditions for long exposure duration (53 days). Black and grey thermometers indicate favourable heat and cold conditions, respectively. PEP, phosphoenolpyruvate

synthase 1 (FLS1), flavonoid-3′-hydroxylase 1 and 2 (F3′H 1, F3′H 2), anthocyanin synthase (ANS) (Fig. 7.37e–i, l). However, dihydroflavonol reductase (DHR) (governing leucoanthocyanidins) showed low expression with slight peak at 24 h (Fig. 7.37j). This could be due to stronger biosynthesis trends of anthocyanidins and anthocyanins located in downstream of phenylpropanoid pathway (Fig. 7.31). The gene coding for ANS enzyme responsible for the biosynthesis of anthocyanidins was upregulated by an induction factor more than 15.2-fold (Fig. 7.37l). Apart from the anthocyanidins, flavonol pathway was revealed to be highly induced in response to cold through a strong upregulation of the gene coding for the flavonol synthase 2 (FLS 2) (key enzyme of flavonol synthesis) (Fig. 7.37k). This upregulation exceeded 24-fold after 96 h of cold stress application.

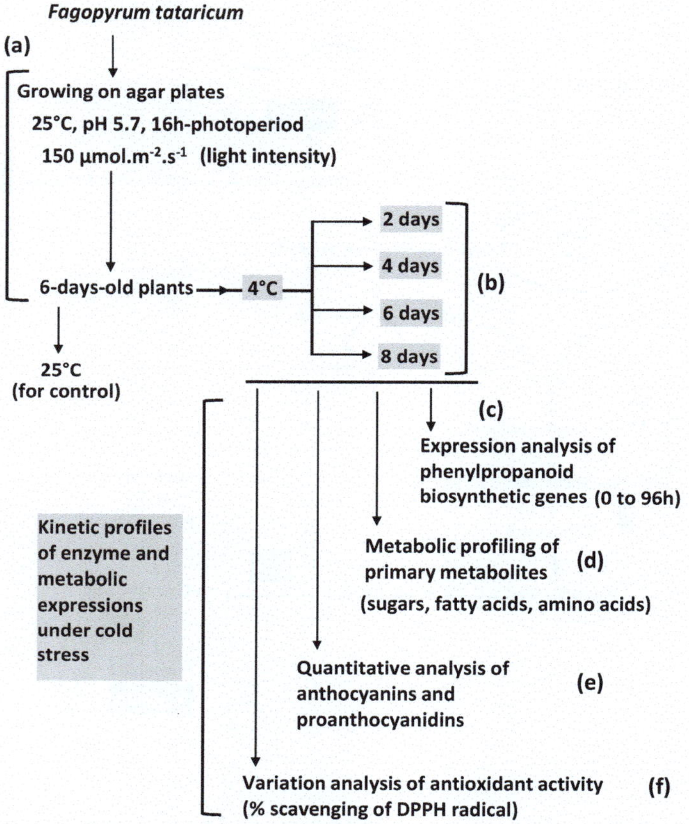

Fig. 7.36 Experimental design for analysis of primary and secondary metabolic responses in young seedling of *Fagopyrum tataricum* exposed to cold stress (4 °C) for variable durations (2–8 days) (Jeon et al. 2018)

Quantitative analysis of flavonoids revealed strong increase of the anthocyanins, cyanidin-3-O-glucoside and cyanidin-3-O-rutinoside, at 4 °C compared with 25 °C (Fig. 7.38c, d). These two metabolites are the only anthocyanins found in *Fagopyrum tataricum* (Thwe et al. 2014). They continuously accumulated with cold duration from 0 to 8 days. Moreover, epicatechin showed similar kinetic profile with continuous increase at 4 °C versus negligible variation at 25 °C (Fig. 7.38b). The highest productions concerned cyanidin-3-O-glucoside which accumulated 8–10 folds more than the other flavonoids (Fig. 7.38d). Moreover, cyanidin-3-O-glucoside showed the strongest kinetic variation by increasing 33.8-fold through 8 days-cold application ($33.8 = 2^{5.08}$; 2 is the basis of logarithmic basis (Fig. 7.39); 5.08 is the ratio between accumulated amounts at time = 8 days and time = 0) (Fig. 7.39b). However, catechin showed positive trend with time at 4 °C but with very lower levels compared with the three other flavonoids (0.025, 0.5, 4 mg/g DW) (Fig. 7.38a).

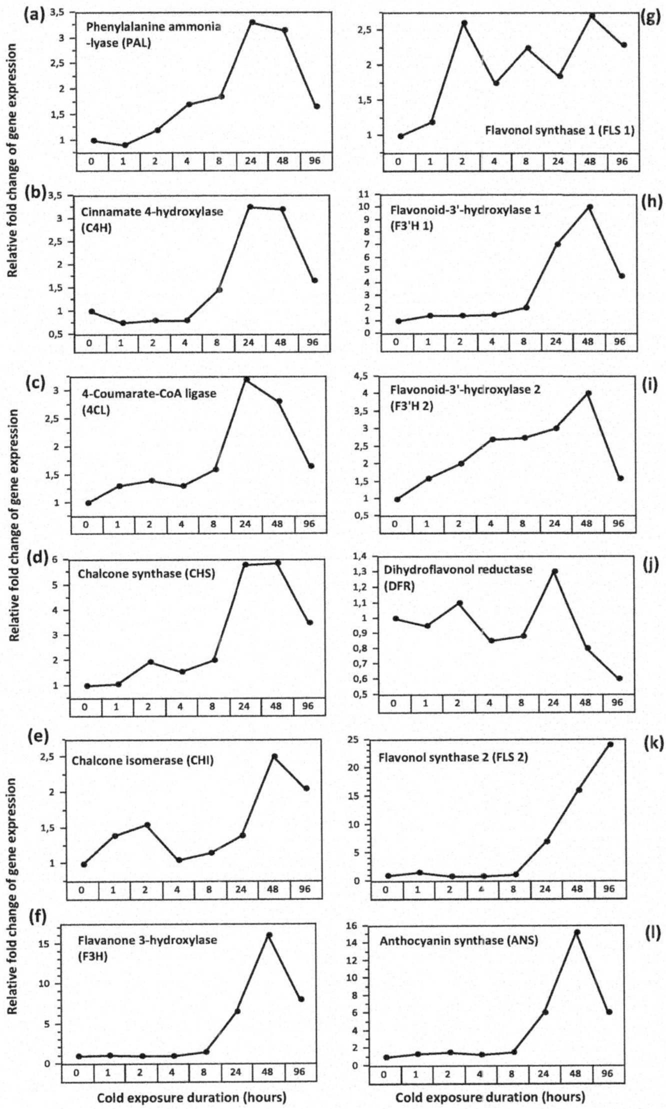

Fig. 7.37 Expression of most phenylpropanoid biosynthesis genes in young seedling of *Fagopyrum tataricum* subjected to cold stress of 4 °C for variable duration (2–8 days). The measures of gene expressions were regularly followed from 0 to 96 h (Jeon et al. 2018)

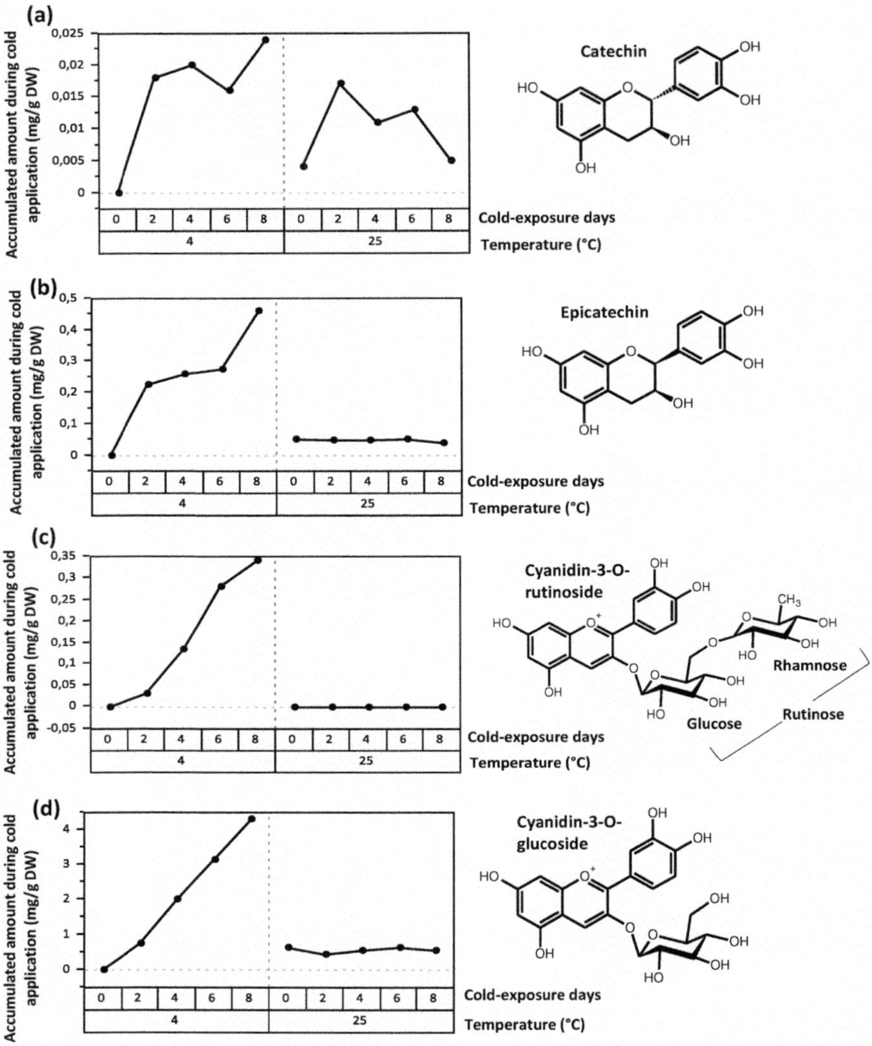

Fig. 7.38 Variation of accumulation levels of catechin (**a**), epicatechin (**b**) and anthocyanins (**c**, **d**) in young seedling of *Fagopyrum tataracum* subjected to cold stress of 4 °C and control conditions (25°C) for variable duration (2–8 days). The four flavonoids showed positive variation trend with cold exposure duration at 4 °C

Analysis of primary metabolites revealed cold-induced positive trends for sugars, polyols and organic acids versus negative trends for amino acid and nitrogen-containing compounds (Figs. 7.39 and 7.40): production levels of metabolites were measured at 2, 5 and 8 days (2, 4, 6, 8 days for phenolics), then they were relativized by reference to the initially observed levels (at time = 0). Levels at initial and subsequent

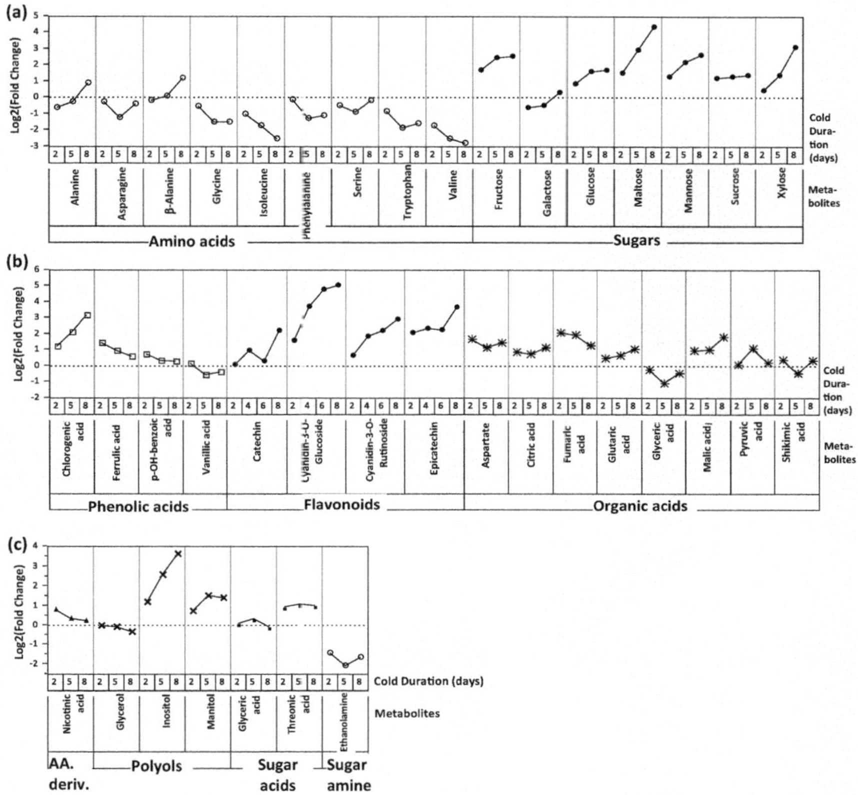

Fig. 7.39 Positive or negative accumulation rates of different metabolites belonging to different biochemical classes in young seedlings of *Fagopyrum tataracum* grown under cold stress (4 °C) for 8 days compared with initial (normal) growth conditions at 25 °C. Accumulation rates were expressed by the log2 of ratios between accumulation levels at 4 °C (for variable durations) and 25 °C (initial time)

times were used to calculate ratios which were standardized by their log2-values. Positive and negative Log-values indicated positive and negative variation trends by reference to the fold 2; the log2-value quantifies how much the accumulation level of considered metabolite was doubled at time t by reference to the initial level (at time = 0).

Sugars, polyols and organic acids showed positive trends of log-ratios with cold duration indicating their strong induction by cold stress (Fig. 7.39a–c). Maltose had the highest induction level increasing 20.6-fold (i.e. $2^{4.37}$) within eight successive cold days (at 4 °C) compared with pre-cold conditions (at 25 °C) (Fig. 7.39a). Also, inositol (Cyclitol) showed the highest increase among the polyols (Fig. 7.39c). However, the log2-values were negative for most of amino acids (Fig. 7.39a). Also,

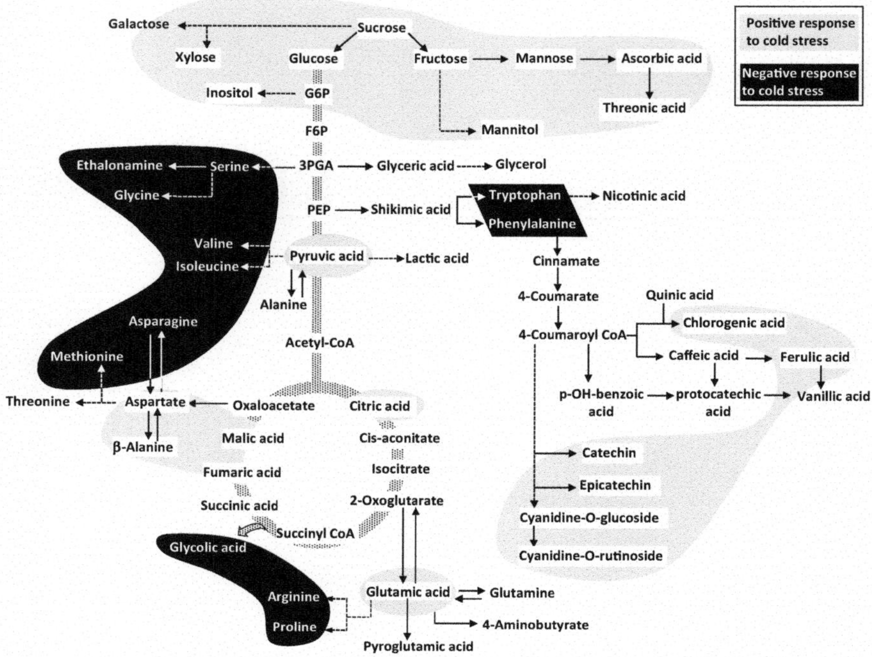

Fig. 7.40 Chart showing positive and negative trends of different metabolic pathways in *Fagopyrum tataracum* in response to cold stress (4 °C for 2–8 days)

ethanolamine (N-containing) compound showed strong negative trend under cold stress (Fig. 7.39c).

These results highlighted acclimation strategy based on (i) production of energetic (sugars) and antioxidant (phenolics) metabolites versus (ii) reduction of growth governed by decrease of amino acids and their derivatives (Fig. 7.40). A field study on *Fagopyrum tataracum* revealed that plants cultivated on higher altitude (7500–7540 ft → 2286–2298 m) showed less nitrogen-free extracts (NFE) than plants of lower altitude (6700–6950 ft → 2042–2118 m). This negative trend between NFE and altitude (cold factor) showed the role of modulating nutrient balance for (cold) stress tolerance.

At higher time scale (inter-monthly, seasonal or inter-annual scales), an extensive phytochemical study was carried out on *Brassica oleracea* (kale) (Brassicaceae) to analyze variation trends of its flavonols under the effects of low temperatures and radiation (Neugart et al. 2012). *Brassica oleracea* shows high diversity of glycosylated and acylated flavonols. This phytochemical diversity provides a variability basis for sensitivity analysis of plants toward climate change trends.

Brassica oleracea was cultivated in experimental fields then fully developed plants were harvested in four successive months (October to January) showing differentially decreasing weather temperature (Fig. 7.41). The experiments were applied for two successive years (2007, 2008). Field temperatures and PAR (photosynthetically active radiation) were measured every two and five minutes, respectively. Cumulated measured data in the four weeks prior each monthly harvest served to calculate daily mean temperature and daily mean PAR associated with each sampling month (Fig. 7.41a, b). Temperature and radiation decreased from October to January in both sampling years (2007, 2008).

Dried harvested plants were analyzed for their flavonoid contents by HPLC–DAD-MS which revealed highly diversified glycosylated acylated flavonols (Figs. 7.42 and 7.43); flavonol aglycones included kaempferol, quercetin and isorhamnetin (3′-O-CH_3-quercetin) (Fig. 7.42). Glycosylations of kaempferol and quercetin were due to glucose or sophorose (glucose 1–2 glucose) occurring at carbons 3 and/or 7 (Table 7.1; Fig. 7.43). Acylation was due to different coumaric acid-derivatives including sinapic, hydroxyferulic and ferulic acids (Fig. 7.43). Acylation was further diversified by substitution position and levels (mono- and di-acyls).

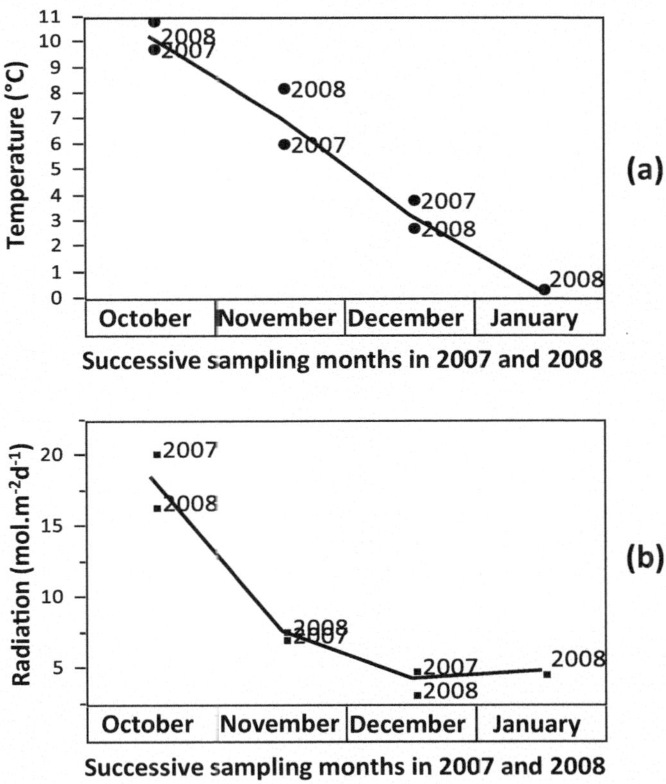

Fig. 7.41 Variations of mean daily temperature (**a**) and photosynthetically active radiation (PAR) (**b**) in the experimental filed of growth of *Brassica oleracea* during four sampling months of the years 2007, 2008

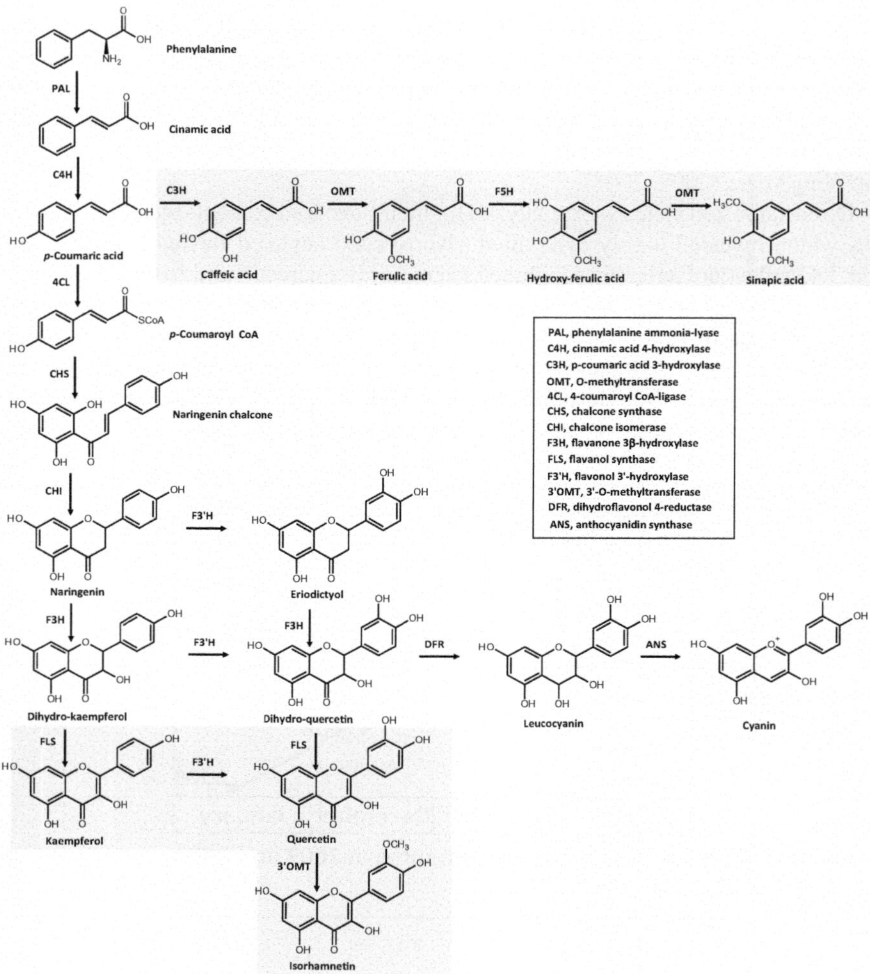

Fig. 7.42 Phenylpropanoid metabolic network showing the main invested biosynthesis pathways (in grey) occurring in *Brassica oleracea* and leading to acylations of flavonol glycosides (kaempferol, quercetin glycosides) by sinapic, hydroxyferulic and ferulic acids

Analysis of quantitative variations of flavonoids was carried out on standardized levels calculated by dividing amounts (mg g^{-1} DM) by the mean of the whole sampling period (in 2007 and 2008). Standardization makes metabolic variations to be analyzed taking into account the different biosynthesis magnitudes of the flavonoids. For instance, flavonoid A with amount of 1.5 mg g^{-1} in a given month and a mean of 2 mg g^{-1} calculated on all the sampling month corresponds to a relative level of 1.5/2 = 0.75 fold the mean. However, a flavonoid B with higher local

Table 7.1 Names of glycosylated and acylated glycosylated kaempferol and quercetin found in the leaves of *Brassica oleracea* (Brassicaceae)

Metabolites	Names
1	Kaempferol-3-O-sophoroside-7-O-D-glucoside
2	Quercetin-3-O-sophoroside-7-O-D-glucoside
3	Kaempferol-3-O-(2-sinapoyl-sophoroside)-7-O-D-glucoside
4	Quercetin-3-O-(2-sinapoyl-sophoroside)-7-O-D-glucoside
5	Kaempferol-3-O-(4-sinapoyl-sophoroside)-7-O-diglucoside
6	Quercetin-3-O-(4-sinapoyl-sophoroside)-7-O-diglucoside
7	Kaempferol-3-O-(2-ferruloyl-sophoroside)-7-O-D-glucoside
8	Kaempferol-3-O-(2-hydroxyferruloyl-sophoroside)-7-O-D-glucoside
9	Kaempferol-3-O-disinapoyl-triglucoside-7-O-D-glucoside
10	Quercetin-3-O-disinapoyl-triglucoside-7-O-D-glucoside

amount = 5 mg g^{-1} and mean = 10 mg g^{-1} correspond to lesser relative variation = 5/10 = 0.5.

Multivariate statistical analysis (by principal component analysis, PCA) of relative levels of different flavonol-glycosides in the leaves of *B. oleracea* highlighted different metabolic responses associated with different variation trends of temperature (Figs. 7.44 and 7.45).

Quercetin pathway was revealed to be more favored by lower temperatures than kaempferol pathway. Quercetin and its derivatives (projected at the left in PCA plot) showed opposite variation way to kaempferol and derivative (project at the right in PCA). Monoacylated quercetin triglucoside **4** and diacylated quercetin tetraglucoside **10** were favored by continuous decrease of temperature from 11 °C to sub-zero (Fig. 7.44). In general, quercetin glycosides showed frank decrease when temperature varies from 0 to 11 °C (Fig. 7.45a). Sub-zero temperatures were relatively more inductive of quercetin glycosides.

However, within the range of low temperatures (0–12 °C) of cultivation field, kaempferol glycosides were relatively more manifested in response to higher temperatures (Fig. 7.44). The most concerned kaempferol glycosides were **1**, **7**, **8**, (Fig. 7.44b), i.e. kaempferol-3-O-sophoroside-7-O-D-glucoside, kaempferol-3-O-(2-ferruloyl-sophoroside)-7-O-glucoside and kaempferol-3-O-(2-hydroxyferruloyl-sophoroside)-7-O-glucoside, respectively (Table 7.1). These flavonol glycosides were associated (in PCA) with temperatures varying in the range (8–11 °C) (Fig. 7.44a). These results showed that kaempferol glycosides were induced by lesser cold conditions than quercetin glycosides (Fig. 7.44b). In PCA (Fig. 7.44b), quercetin glycosides were closer to low temperatures than corresponding kaempferol glycosides: quercetin aglycone (Q) versus kaempferol aglycone (K); glycoside **2** versus glycoside **1**; acylated glycoside **4** versus acylated glycoside **3** (Table 7.1). This could

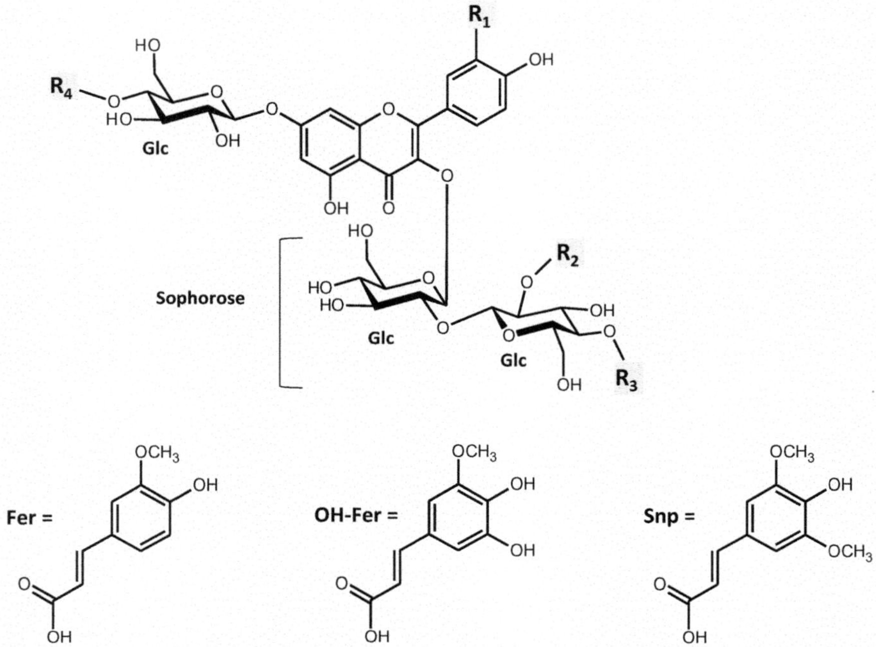

Metabolites	R1	R2	R3	R4
1	H	H	H	H
2	OH	H	H	H
3	H	Snp	H	H
4	OH	Snp	H	H
5	H	H	β-D-Snp-Glc	H
6	OH	H	β-D-Snp-Glc	H
7	H	Fer	H	H
8	H	OH-Fer	H	H
9	H	Snp	β-D-Snp-Glc	H
10	OH	Snp	β-D-Snp-Glc	H

Fig. 7.43 Glycosides and acyl-glycosides of kaempferol and quercetin (flavonols) found in the leaves of *Brassica oleracea* (Brassicaceae). Fer, ferrulic acid; Snp, sinapic acid; Glc, glucose

be linked to differential protective responses against oxidative (ROS) stress occurring under cold stress: quercetin having additional OH with ortho-di-OH system has higher antioxidative activity than kaempferol.

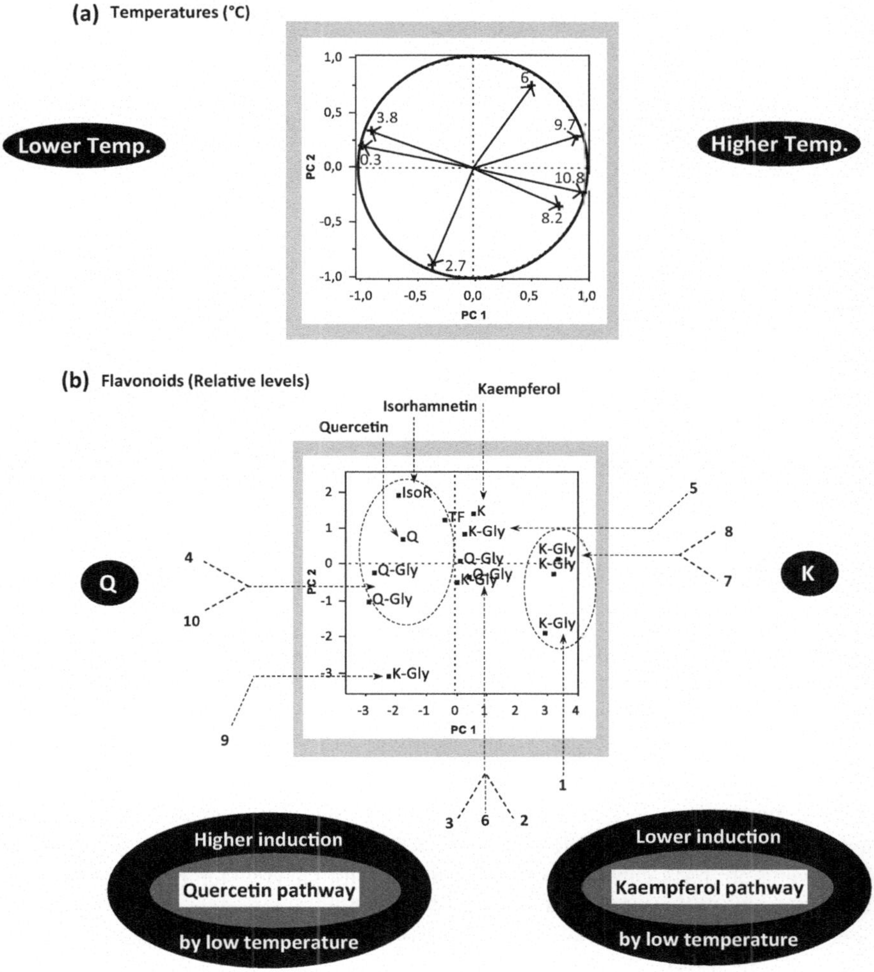

Fig. 7.44 Principal component analysis highlighting differential sensitivity of quercetin (Q) and kaempferol (K) pathways (**b**) to relatively low temperature trends (**a**) in *Brassica oleracea*

Plotting relative levels of kaempferol glycosides versus temperature showed different peaks that could indicate various phytochemical responses to cold conditions associated with different optimal temperatures (Fig. 7.45b, c). Kaempferol glycosides acylated by sinapic acid (**3**, **9**) showed peaks at lower temperatures (at 2.7 and 3.8 °C) compared with those acylated by ferulic and hydroxyferulic acids (**7**, **8**) (peaks at 9.7 °C). These flexible phytochemical responses in *Brassica oleracea* provide strong bases of sensitive model plant helping to monitor gradual variations of temperature associated with cold conditions.

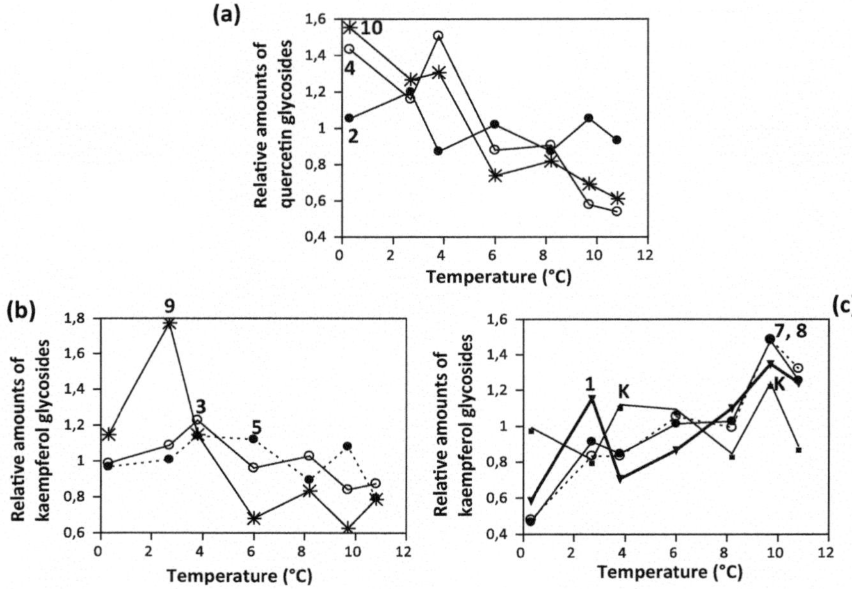

Fig. 7.45 Variation of different quercetin (**a**) and kaempferol (**b**, **c**) glycosides in the leaves of *Brassica oleracea* with field variable low temperature. (**a**) continuous decreasing trend of quercetin glycosides; (**b**, **c**) low temperature-linked optimum peaks in kaempferol glycosides. The bold numbers associated with the curves correspond to different flavonol glycosides (see Fig. 7.43 for structures)

These flavonol glycosides patterns highlighted multiple adaptation responses of plants to living in high altitudes and latitudes. Other field studies revealed similar defensive trend: back cabbage cultivated at high altitude (low temperature and high radiation space) showed higher biosynthesis levels of quercetin-3-O-(2-caffeoyl-sophoroside)-7-O-glucoside and quercetin feruloylglycoside than at medium altitude (Romani et al. 2003).

Variability (flexibility) of phytochemical profiles in *Brassica oleracea* under low temperatures is influenced by interaction with PAR which represents another inductive factor of ROS and antioxidant metabolites (Neugart et al. 2012). Relative variations of different flavonol glycosides were revealed to vary under multiple effects combining temperature with radiation levels on one hand and phytochemical structures on other hand (Figs. 7.46 and 7.47).

Compounds **7** (kaempferol-3-O-(2-ferruloyl-sophoroside)-7-O-D-glucoside) showed high relative level (> 1.3) under relatively high temperature and radiation (Fig. 7.46a). Compound **8** (kaempferol-3-O-(2-hydroxyferruloyl-sophoroside)-7-O-D-glucoside) issued from hydroxylation of **7** was also favored by this positive trend between temperature and radiation (Fig. 7.46b). However, **3** (Kaempferol-3-O-(2-sinapoyl-sophoroside)-7-O-D-glucoside) (resulting from methylation of **8**) showed opposite trend to **7** and **8**, and was induced by both low temperature

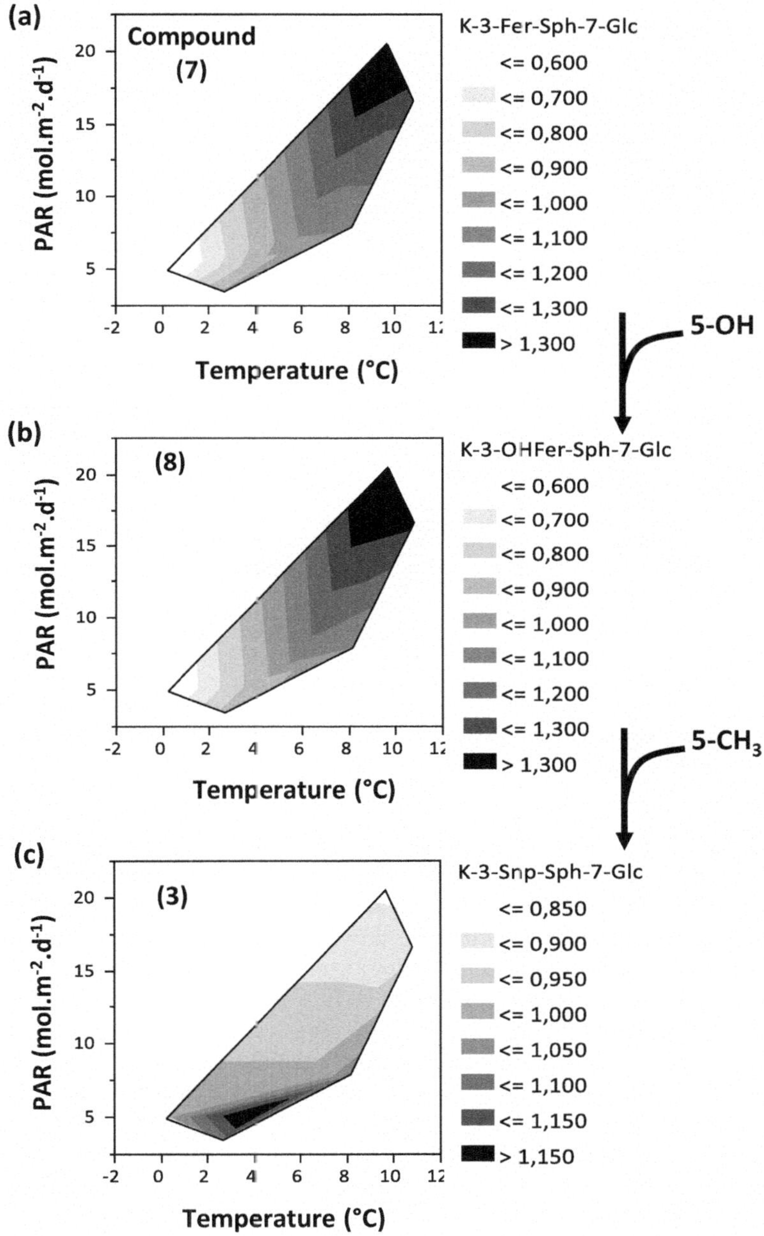

Fig. 7.46 Variations of relative amounts of kaempferol glucosides acylated by ferulic (**a**), hydroxyferulic (**b**) and sinapic (**c**) acids in *Brassica oleracea* in response to both temperature and photosynthetic active radiation (PAR). Structural transitions from (**a**) to (**b**) and (**c**) occur by hydroxylation and methoxylation, respectively, cf carbon C5 of ferulic acid. Structures and complete names of compounds are given in Fig. 7.43 and Table 7.1

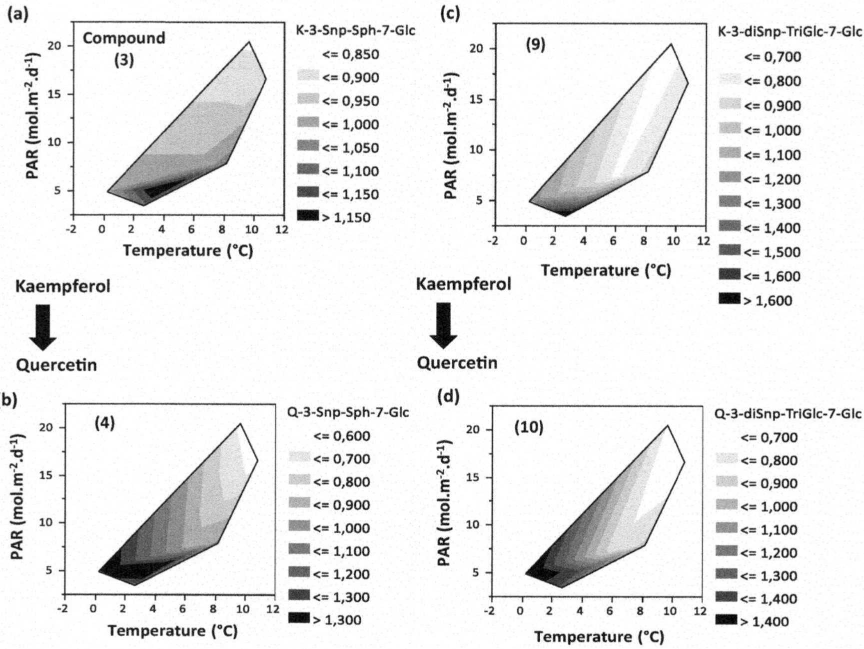

Fig. 7.47 Variations of relative amounts of acylated glycosides of kaempferol and quercetin in *Brassica oleracea* as reponses to low temperature and radiation. (**a**, **b**) Mono-sinapoil glycosides of kaempferol (**3**) and quercetin (**4**); (**c**, **d**) di-sinapoil tetraglycosides of kaempferol (**9**) and quercetin (**10**) by sinapic acid

and radiation (Fig. 7.46c). These results seem to show that kaempferol glycosides acylated by sinapic acid would be more induced by (low temperature-low PAR)-conditions than kaempferol glycosides acylated by ferulic and hydroxyferulic acids. Additional argument to this hypothesis is provided by compound **9** (Kaempferol-3-O-disinapoyl-triglucoside-7-O-D-glucoside) which relatively increased (> 1.6 folds the mean) under both low temperature and PAR (Fig. 7.47c).

Beside the acylation effect, the aglycone type seems to play role in response strength (stability) to both low temperature and radiation: (kaempferol → quercetin)-transition by C3′-hydroxylation resulted in extended peak areas where the maximum levels of flavonoids were maintained (well-invested) (**3** → **4** and **9** → **10**) (black area in plot; Fig. 7.47). This variational space expansion of maximum flavonoid levels could be indicative of more stable defensive phytochemical responses favoring higher tolerance to cold stress. This could be due to higher anti-oxidant activity of quercetin than kaempferol because of ortho-dihydroxyl structure in lateral phenyl (Figs. 7.42 and 7.43) (see also Fig. 3.12 and 3.13, Part I; Fig. 4.4, Part II).

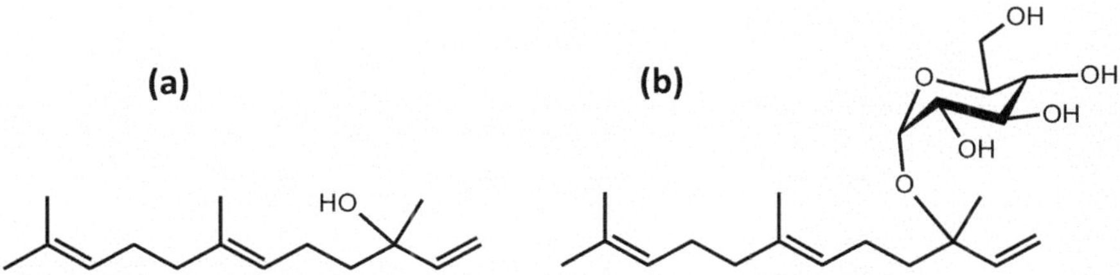

Fig. 7.48 Chemical structure of nerolidol (**a**) and its glucoside (**b**). Nerolidol is a widespread sesquiterpene occurring in floral essential oils of various plants

7.2.3.2 Terpene Patterns

In *Camelia sinensis* (tea) (Teaceae), cold stress induces glucosylation of nerolidol (sesquiterpene) which frequently occurs under free volatile form in absence of such a stress (Fig. 7.48) (Zhao et al. 2021). Nerolidol is a widespread sesquiterpene occurring in floral essential oils of various plants. By its odor property, it plays key role in plant-plant and plant-animal communicative signals (Jing et al. 2019). Moreover, nerolidol has many biological and pharmacological activities including antioxidant, antimicrobial and insecticidal and anti-inflammatory properties (Chan et al. 2016). Sesquiterpene glycosylation is relatively rare compared with glycosylation of monoterpenes and diterpenes in plant world. On this basis, sesquiterpene glycosylation provides a highly original phytochemical pattern of productive plants.

One-year-old tea plants (growing under optimal temperature, 25°C) were subjected to gradually decreasing temperature for a serial of seven successive days by applying different conditions (Fig. 7.49a) (Zhao et al. 2021):

(1) Cold stress (CS) at 4 °C for 6 h or
(2) Chilling acclimation (CA) at 4 °C for 7 days, followed by
(3) Freezing acclimation (FA) at 0 °C for additional 7 day before
(4) De-acclimation (DA) by moving to optimal conditions (25 °C) for 7 days.

Cold stress-induced accumulation of nerolidol glucoside which was positively correlated with strong expression of specific UDP-glucosyltransferase, UGT91Q2, in tea (Fig. 7.50). Induced transcription of UGT91Q2 reached 60-fold under CS and 240-fold under CA conditions. However, mRNA expression level reached a peak under CA condition then it decreased under FA and was further reduced under DA. These results suggested that variation of UGT91Q2 manifests as highly sensitive response to cold stress variation.

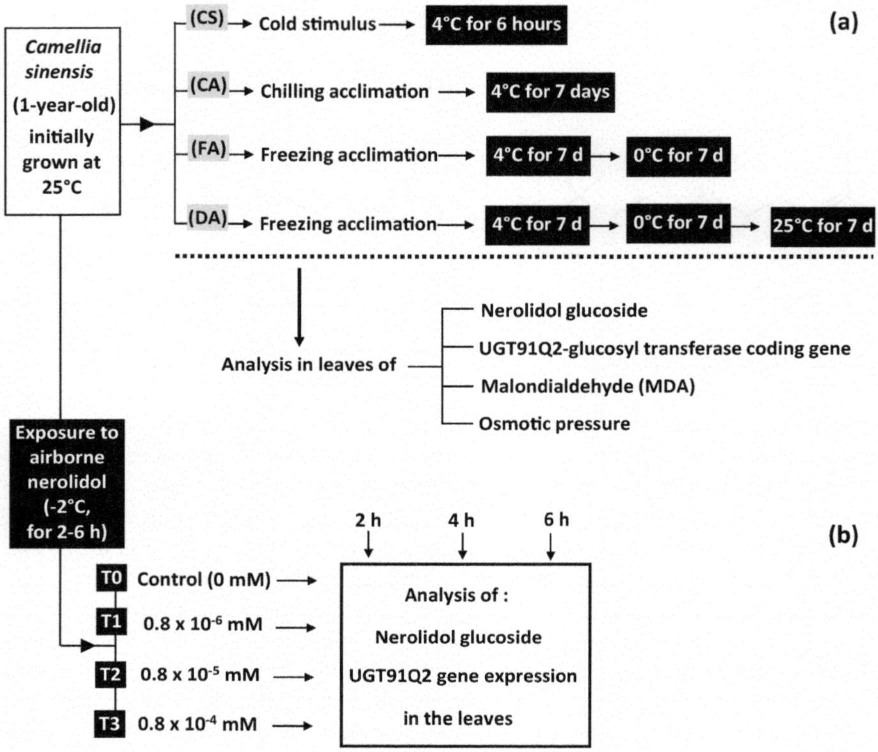

Fig. 7.49 Experimental design applied to 1-year old *Camelia sinensis* (tea) (Theaceae) to analyze the variation of nerolidol, nerolidol glucoside and glycosyl transferase UGT91Q2 in response to cold stress. (**a**) Application of different cold regimes; (**b**) application of different concentrations and exposure times of airborn nerolidol under cold stress

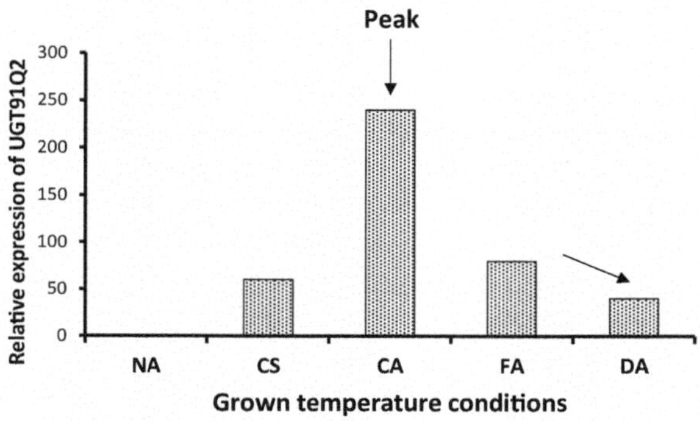

Fig. 7.50 Variation of relative transcription level of glucosyl transferase UGT91Q2 in 1-year-old tea plants exposed to cold stress. Cold stimulus (CS); chilling acclimation (CA); freezing acclimation (FA); De-acclimation (DA)

Coding gene for UGT91Q2 was identified as induced by low temperatures and contributes for the regulation of cold stress tolerance (Zhao et al. 2021). Suppression of expression of UGT91Q2 gene (by AsODN solution) (Fig. 7.51a) resulted in significantly reduced accumulation level of nerolidol and nerolidol glucoside (Fig. 7.51b, c). In active gene (wild) variety, nerolidol accumulation mainly occurred as water-soluble form (glycoside) (Fig. 7.51b, c); the aglycone form represented negligible fraction of glycosylated one (Fig. 51b vs 51c). Glycosylation of metabolites enhance their stability, water solubility and transport resulting in better control/regulation of responses to various biotic and abiotic stresses (Bowles et al. 2005).

Nerolidol incubated with UGT91Q2 resulted in higher ROS-scavenging ability of nerolidol glucoside compared with that of free nerolidol (Fig. 7.52a). This suggests that glycosylation of nerolidol tends to increase its antioxidant activity. Moreover, tea plants with down-regulation of UGT91Q2 showed lower ROS-scavenging capacity leading to lesser cold stress tolerance (Fig. 7.52b). This highlighted upstream factor

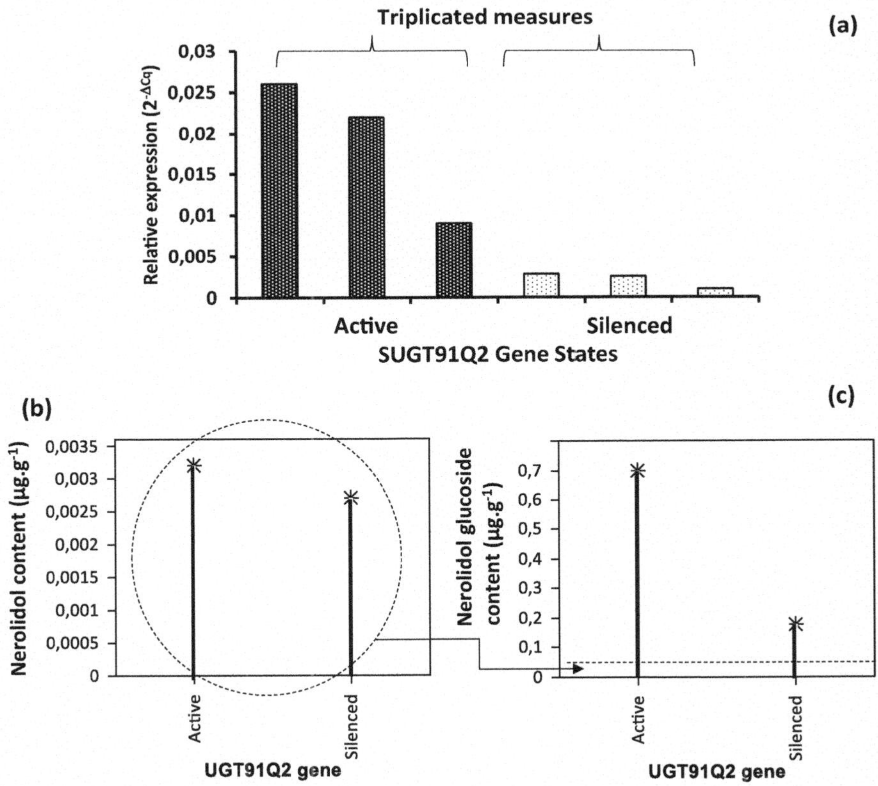

Fig. 7.51 Variations of relative transcript level of UGT91Q2 (**a**) and biosynthesis levels of nerolidol (**b**) and its glucoside (**c**) in wild and silent UGT91Q2 varieties of tea under cold conditions

controlling relationship between low nerolidol glucoside in UGT91Q2-silenced plants and weak ROS-scavenging leading to lower cold tolerance in tea.

Moreover, under cold stress, tea with silenced gene showed significant decrease of F_v/F_m ratio (variable fluorescence/maximal fluorescence) compared with control tea plants. This indicated reduction of quantum efficiency of photosystem II and suggested key role of UGT91Q2 in (i) response to cold stress and (ii) PSII protection.

Cold stress tolerance in *Camellia sinensis* was also revealed to be favored by ability of the plant to capture volatile nerolidol and to convert it to glycosylated form: 1-year-old plants were put for 2–6 h into enclosed glass vessel (5000 cm^3) containing variable nerolidol concentrations (T1 = 0.8×10^{-6} mM; T2 = 0.8×10^{-5} mM; T3 = 0.8×10^{-4} mM) (Fig. 7.49b) (Zhao et al. 2021). Plant exposed to

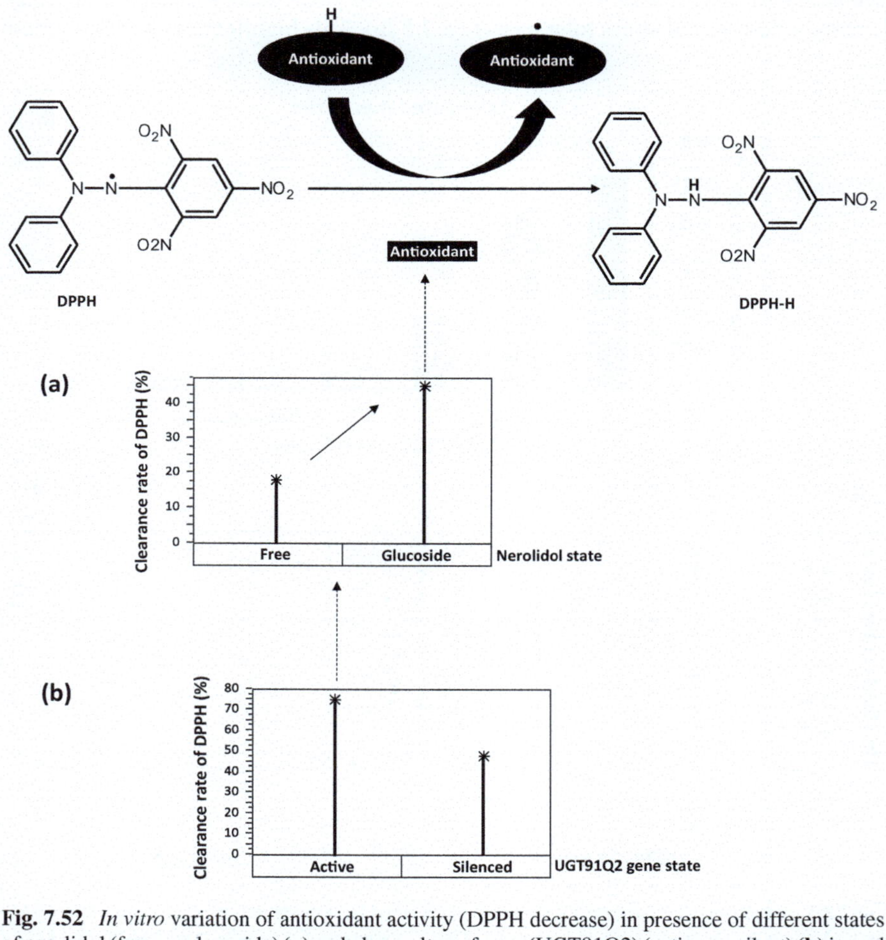

Fig. 7.52 *In vitro* variation of antioxidant activity (DPPH decrease) in presence of different states of nerolidol (free vs. glycoside) (**a**) and glycosyltransferase (UGT91Q2) (active vs. silent) (**b**) issued from *Camelia sinensis*

airborne nerolidol showed higher accumulation levels of both nerolidol and nerolidol glucoside than control (nonexposed) plants (Fig. 7.53a). However, induced nerolidol glucoside showed threefold level compared with that of aglycone, suggesting that absorbed fee nerolidol is converted into glycosylated form. Moreover, UGT91Q2 gene showed higher expression levels with increasing airborne nerolidol (Fig. 7.53b). This highlighted implication of this UGT in the formation of nerolidol glucoside from absorbed airborne nerolidol. Also, the increasing expression of UGT91Q2 gene positively correlated with that of related transcription factors (CBF1) in tea leaves (Fig. 7.53c).

Inversely, malondialdehyde content (MDA) (oxidative stress indicator) decreased with the increase of airborne nerolidol (Fig. 7.54a). Decrease of MDA revealed less oxidative stress subsequent to exposure of the tea leaves to nerolidol indicating better cold stress tolerance. However, osmotic pressure in tea leaves was maintained constant (compared with control conditions T_0); this indicated strong endogenous protective process in plant against cold stress in addition to the benefic role provided by exogenous nerolidol (Fig. 7.54b).

A synthesis of different protective role of nerolidol glucoside in tea under cold stress is given in Fig. 7.55. Under cold stress, nerolidol is emitted as volatile compound to transmit alerting cold signal between plants (Zhou et al. 2017). Receiving plants capture nerolidol and transform it into glycosylated form helping them to improve their cold stress tolerance. Absorption of airborne nerolidol followed by its glucosylation were also revealed in tobacco (*Nicotiana benthamiana*) for cold stress tolerance (Zhao et al. 2021).

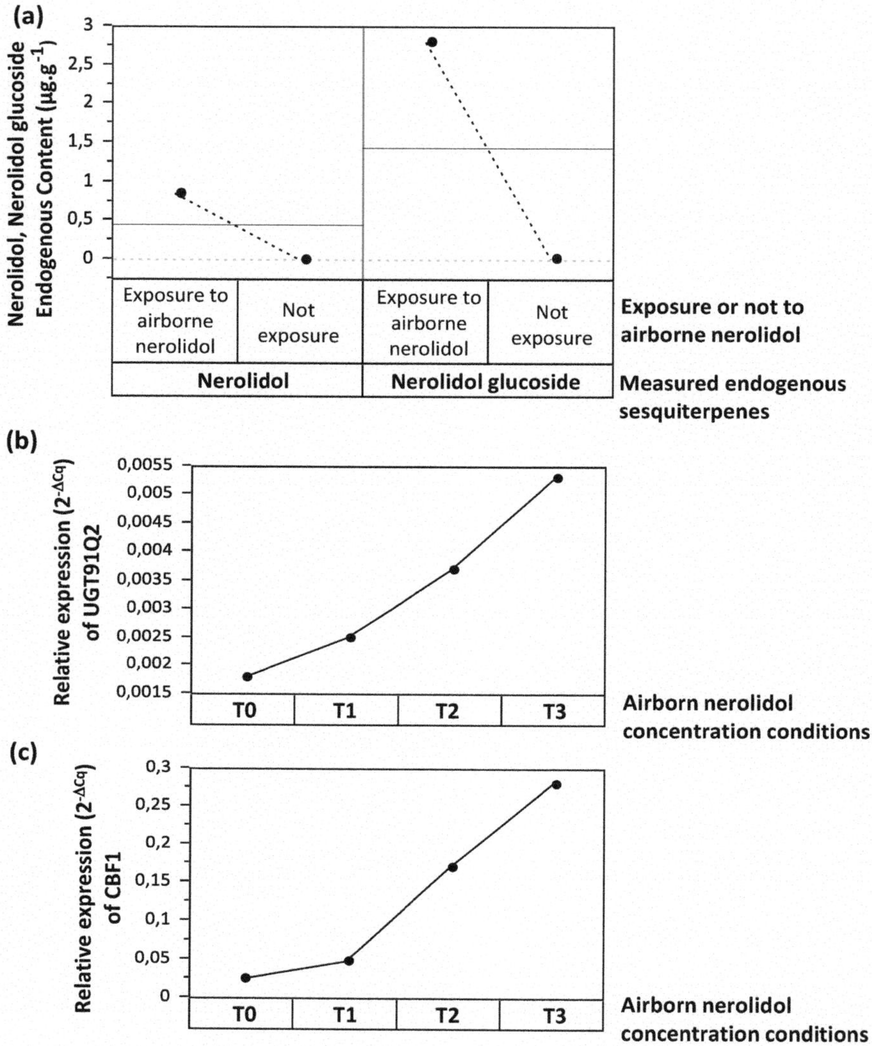

Fig. 7.53 Effects of exogenous nerolidol on the variation of endogenous levels of nerolidol and nerolidol glucoside (**a**), glycosyl transferase UGT91Q2 (**b**) and CBF1 factors (cold-sensitive genes) (**c**) in 1-year-old plants *Camelia sinensis*. Tea plants were exposed to − 5 °C for 2, 4 and 6 h in presence of variable concentrations of nerolidol: T1 = 0.8×10^{-6} mM; T2 = 0.8×10^{-5} mM; T3 = 0.8×10^{-4} mM; T0, control (not applied exogenous nerolidol). Average values issued from the cold experiments applied for 2, 4 and 6h are represented

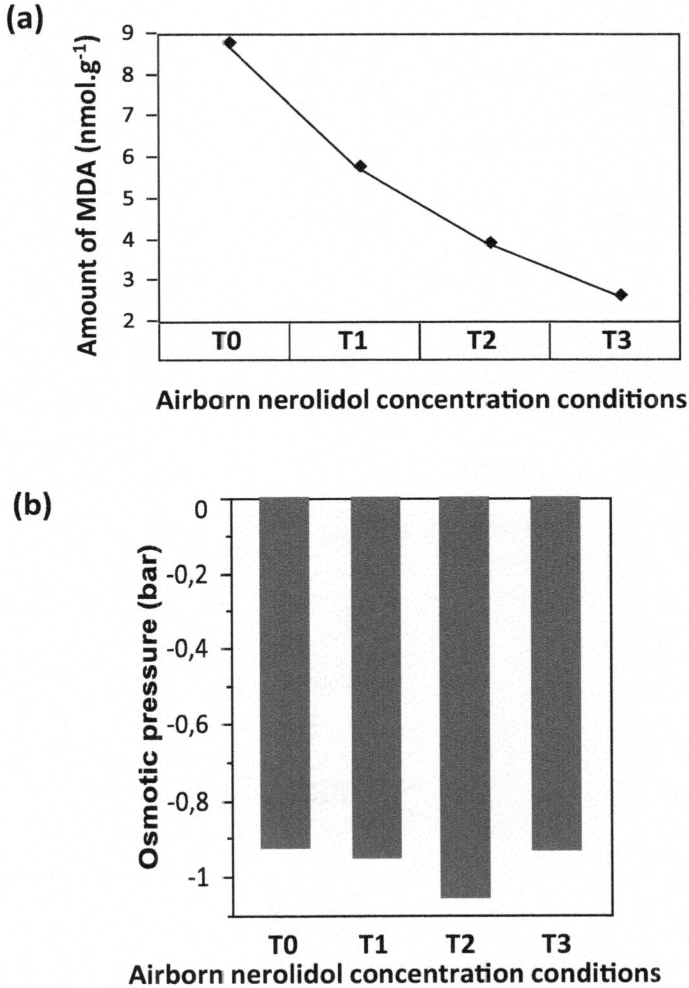

Fig. 7.54 Decreasing malondialdehyde (MDA) (**a**) and osmotic pressure stabilization (**b**) observed after application of exogenous nerolidol at different concentrations in 1-year-old *Camelia sinensis*. Tea plants were exposed to − 5 °C for 2, 4 and 6 h in presence of variable concentration of nerolidol: T1 = 0.8×10^{-6} mM; T2 = 0.8×10^{-5} mM; T3 = 0.8×10^{-4} mM; T0, control (not applied exogenous nerolidol). Average values are represented in the two plots (**a, b**)

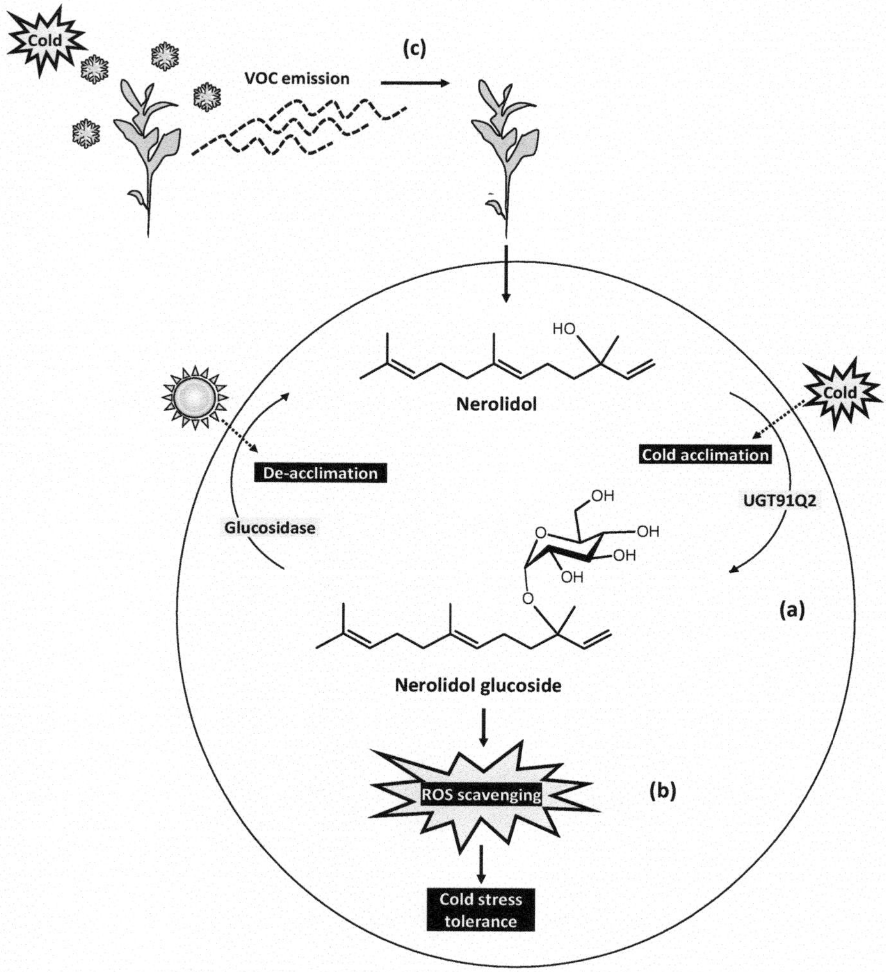

Fig. 7.55 Various processes through which nerolidol (sesquiterpene) is implied in cold tolerance in tea plants, including UGT91Q2 induction and nerolidol glucosylation (**a**) which favors higher ROS-scavenging (**b**) and which is improved by inter-plant emission and absorption of nerolidol (in VOC) (**c**)

References

Alfonso M et al (2001) Unusual tolerance to high temperatures in a new herbicide-resistant D1 mutant from *Glycine max* (L.) Merr. cell cultures deficient in fatty acid desaturation. Planta 212:573–582

Al-Huqail A, El-Dakak RM, Sanad MN, Badr RH, Ibrahim MM, Soliman D, Khan F (2020) Effects of climate temperature and water stress on plant growth and accumulation of antioxidant compounds in Sweet Basil (*Ocimum basilicum L.*) leafy vegetable. Scientifica 2020:1–12

Bahuguna RN, Jagadish KSV (2015) Temperature regulation of plant phenological development. Environ Exp Bot 111:83–90

Barrero-Gil J, Salinas J (2017) CBFs at the crossroads of plant hormone signaling in cold stress response. Mol Plant 10:542–544

Bilska K, Wojciechowska N, Alipour S, Marzena Kalemba E (2019) Ascorbic acid—the little-known antioxidant in woody plants. Antioxidants 8:645

Boncan DAT, Tsang SSK, Li C, Lee IHT, Lam HM, Chan TF, Hui JHL (2020) Terpenes and terpenoids in plants: interactions with environment and insects. Int J Mol Sci 21:7382

Bowles D, Isayenkova J, Lim EK, Poppenberger B (2005) Glycosyltransferases: managers of small molecules. Curr Opin Plant Biol 8:254–263

Camejo D, Jimenez A, Alarcon JJ, Torres W, Gomez JM, Sevilla F (2006) Changes in photosynthetic parameters and antioxidant activities following heat-shock treatment in tomato plants. Funct Plant Biol 33:177–187

Chalker-Scott L (1999) Environmental significance of anthocyanins in plant stress responses. Photochem Photobiol 70:1–9

Chan WK, Tan LTH, Chan KG, Lee LH, Goh BH (2016) Nerolidol: a sesquiterpene alcohol with multi-faceted pharmacological and biological activities. Molecules 21:529

Chaves I, Passarinho JAP, Capitão C, Chaves MM, Fevereiro P, Ricardo CPP (2011) Temperature stress effects in *Quercus suber* leaf metabolism. J Plant Physiol 168:1729–1734

Copolovici LO, Niinemets Ü (2005) Temperature dependencies of Henry's law constants and octanol/water partition coefficients for key plant volatile monoterpenoids. Chemosphere 61:1390–1400

Copolovici L, Kännaste A, Pazouki L, Niinemets A (2012) Emissions of green leaf volatiles and terpenoids from *Solanum lycopersicum* are quantitatively related to the severity of cold and heat shock treatments. J Plant Physiol 169:664–672

Duan X, Zhu Z, Yang Y, Duan J, Jia Z, Chen F, Sang Z, Ma L (2022) Salicylic acid regulates sugar metabolism that confers freezing tolerance in *Magnolia wufengensis* during natural cold acclimation. J Plant Growth Regul 41:227–235

Eremina I, Aloyan A, Arutyunyan V, Larin I, Chubarova N, Yermakov A (2015) Acidity and mineral composition of precipitation in Moscow: influence of deicing salts. Izv Atmos Ocean Phys 51(6):624–632

Gibbs AG (2002) Lipid melting and cuticular permeability: new insights into an old problem. J Insect Physiol 48:391–400

Goh HH, Khairudin K, Sukiran NA, Normah MN, Baharum SN (2015) Metabolite profiling reveals temperature effects on the VOCs and flavonoids of different plant populations. Plant Biol 18:130–139

Hancock RD, Morris WL, Ducreux LJM, Morris JA, Usman M, Verrall SR, Fuller J, Simpson CG, Zhang R, Hedley PE et al (2014) Physiological, biochemical and molecular responses of the potato (*Solanum tuberosum* L.) plant to moderately elevated temperature. Plant Cell Environ 37:439–450

Harbaum-Piayda B, Walter B, Bengtsson GB, Hubbermann EM, Bilger W, Schwarz K (2010) Influence of pre-harvest UV-B irradiation and normal or controlled atmosphere storage on flavonoid and hydroxycinnamic acid contents of pak choi (*Brassica campestris* L. ssp. chinensis var. communis). Postharvest Biol and Technol 155:290–297

Havaux M (1998) Carotenoids as membrane stabilizers in chloroplasts. Trend Plant Sci 3:147–151

Havaux M, Kloppstech K (2001) The protective functions of carotenoid and flavonoid pigments against excess visible radiation at chilling temperature investigated in *Arabidopsis* NPQ and TT mutants. Planta 213:953–966

Herrmann KM (1995) The shikimate pathway—early steps in the biosynthesis of aromatic compounds. Plant Cell 7(7):907–919

Holopainen JK, Himanen SJ, Yuan JS, Chen F, Stewart CN (2013) Ecological functions of terpenoids in changing climates. In: Ramawat KG, Mérillon JM (eds) Natural products. Springer, Berlin

Hugly S, Somerville C (1992) A role for membrane lipid polyunsaturation in chloroplast biogenesis at low temperature. Plant Physiol 99:197–202

Hugly S et al (1989) Enhanced thermal tolerance of photosynthesis and altered chloroplast ultrastructure in a mutant of *Arabidopsis* deficient in lipid desaturation. Plant Physiol 90:1134–1142

Jansen RMC, Miebach M, Kleist E, van Henten EJ, Wildt J (2009) Release of lipoxygenase products and monoterpenes by tomato plants as an indicator of *Botrytis cinerea* induced stress. Plant Biol 11:859–868

Jeon J, Kim JK, Wu Q, Park SU (2018) Effects of cold stress on transcripts and metabolites in tartary buckwheat (*Fagopyrum tataricum*). Environ Exp Bot 155:488–496

Jing T, Zhang N, Gao T, Zhao M, Jin J, Chen Y, Xu M, Wan X, Schwab W, Song C (2019) Glucosylation of (Z)-3-hexenol informs intraspecies interactions in plants: a case study in *Camellia sinensis*. Plant Cell Environ 42:1352–1367

Jochum GM, Mudge KW, Thomas RB (2007) Elevated temperatures increase leaf senescence and root secondary metabolite concentration in the understory herb *Panax quinquefolius (Araliaceae)*. Am J Bot 94:819–826

Kaplan F, Kopka J, Haskell DW, ZhaoW SKC, Gatzke N, Sung DY, Guy CL (2004) Exploring the temperature-stress metabolome of Arabidopsis. Plant Physiol 136:4159–4168

Karim MA, Fracheboud Y, Stamp P (2000) Effect of high temperature on seedling growth and photosynthesis of tropical maize genotypes. J Agron Crop Sci 184:217–223

Kuzuyama T, Shimizu T, Takahashi S, Seto H (1998) Fosmidomycin, a specific inhibitor of 1-deoxy-D-xylulose 5-phosphate reductoisomerase in the nonmevalonate pathway for terpenoid biosynthesis. Tetrahedron Lett 39:7913–7916

Loreto F, Velikova V (2001) Isoprene produced by leaves protects the photosynthetic apparatus against ozone damage, quenches ozone products, and reduces lipid peroxidation of cellular membranes. Plant Physiol 127:1781–1787

Loreto F, Ciccioli P, Cecinato A, Brancaleoni E, Frattoni M, Fabozzi C, Tricoli D (1996) Evidence of the photosynthetic origin of monoterpenes emitted by *Quercus ilex* L. leaves by 13C labeling. Plant Physiol 110:1317–1322

Loreto F, Forster A, Durr M, Csiky O, Seufert G (1998) On the monoterpene emission under heat stress and on the increased thermotolerance of leaves of *Quencus ilex* L. fumigated with selected monoterpenes. Plant Cell Environ 21:101–107

Maes K, Debergh PC (2003) Volatiles emitted from in vitro grown tomato shoots during abiotic and biotic stress. Plant Cell Tiss Org 75:73–78

Matas AJ, Cuartero J, Heredia A (2004) Phase transitions in the biopolyester cutin isolated from tomato fruit cuticles. Thermochim Acta 409:165–168

Mori K, Sugaya S, Gemma H (2005) Decreased anthocyanin biosynthesis in grape berries grown under elevated night temperature condition. Sci Hortic 105:319–330

Movahed N, Pastore C, Cellini A, Allegro G, Valentini G, Zenoni S, Cavallini E, D'Incà E, Tornielli GB, Filippetti I (2016) The grapevine VviPrx31 peroxidase as a candidate gene involved in anthocyanin degradation in ripening berries under high temperature. J Plant Res 129:513–526

Moyer RA, Hummer KE, Finn CE, Frei B, Wrolstad RE (2002) Anthocyanins, phenolics, and antioxidant capacity in diverse small fruits: *vaccinium*, *rubus*, and *ribes*. J Agric Food Chem 50:519–525

Munné-Bosch S (2005) The role of a-tocopherol in plant stress tolerance. J Plant Physiol 162:743–748

Murakami Y et al (2000) Trienoic fatty acids and plant tolerance of high temperature. Science 287:476–479

Naghiloo S, Movafeghi A, Delazar A, Nazemiyeh H, Asnaashari S, Dadpour MR (2012) Ontogenetic variation of total phenolics and antioxidant activity in roots: leaves and flowers of *Astragalus compactus* Lam. (Fabaceae). Bioimpacts 2(2):105–109

Neugart S, Kläring HP, Zietz M, Schreiner M, Rohn S, Kroh LW, Krumbein A (2012) The effect of temperature and radiation on flavonol aglycones and flavonol glycosides of kale (*Brassica oleracea* var. *sabellica*). Food Chem 133:1456–1465

Niinemets Ü, Monson RK, Arneth A, Ciccioli P, Kesselmeier J, Kuhn U et al (2010) The leaf-level emission factor of volatile isoprenoids: caveats, model algorithms, response shapes and scaling. Biogeosciences 7:1809–1832

Noctor G, Foyer CH (1998) Ascorbate and glutathione: keeping active oxygen under control. Ann Rev Plant Physiol Plant Mol Biol 49:249–279

Olaifa JI, Matsumura F, Zeevaart JAD, Mullin CA, Charalambous P (1991) Lethal amounts of ricinine in green peach aphids (*Myzus persicae*) (suzler) fed on Castor bean plants. Plant Sci 73(2):253–256

Oren-Shamir M, Nissim-Levi A (1999) Temperature and gibberellin effect on growth and anthocyanin pigmentation in photinia leaves. J Hortic Sci Biotechnol 74:355–360

Orthen B, Popp M (2000) Cyclitols as cryoprotectants for spinach and chickpea thylakoids. Environ Exp Bot 44:125–132

Passarinho JAP, Lamosa P, Baeta JP, Santos H, Ricardo CPP (2006) Annual changes in the concentration of minerals and organic compounds of *Quercus suber* leaves. Physiol Plant 127:100–110

Pereira FM, Rosa E, Fahey JW, Stephenson KK, Carvalho R, Aires A (2002) Influence of temperature and ontogeny on the levels of glucosinolates in broccoli (*Brassica oleracea* Var. *italica*) sprouts and their effect on the induction of mammalian phase 2 enzymes. J Agric Food Chem 50(21):6239–6244

Porter JR (2005) Rising temperatures are likely to reduce crop yields. Nature 436:174

Pradhan J, Sahoo SK, Laloltra S, Sarma RS (2017) Positive impact of abiotic stress on medicinal and aromatic plants. Int J Plant Sci 12(2):309–313

Radusiene J, Karpaviciene B, Stanius Z (2012) Effect of external and internal factors on secondary metabolites accumulation in St. John's Worth. Bot Lith 18(2):101–108

Rawson HM (1988) Effect of high temperatures on the development and yield of wheat and practices to reduce deleterious effects. In: Klatt AR (ed) Wheat production constraints in tropical environments. CIMMYT, Mexico City, pp 44–62

Raza A, Charagh S, Najafi-Kakavand S, Abbas S, Shoaib Y, Anwar S, Sharifi S, Lu G, Siddique KHM (2023) Role of phytohormones in regulating cold stress tolerance: physiological and molecular approaches for developing cold-smart crop plants. Plant Stress 8:100152

Reardon ME, Qaderi MM (2017) Individual and interactive effects of temperature, carbon dioxide and abscisic acid on mung bean (*Vigna radiata*) plants. J Plant Interact 12:295–303

Ribeiro PR, Gonzaga Fernandez L, Delmondez de Castro R, Ligterink W, Hilhorst HWM (2014) Physiological and biochemical responses of *Ricinus communis* seedlings to different temperatures: a metabolomics approach. BMC Plant Biol 14:223

Riederer M (2006) Thermodynamics of the water permeability of plant cuticles: characterization of the polar pathway. J Exp Bot 57:2937–2942

Rinnan R, Rinnan Å, Faubert P, Tiiva P, Holopainen JK, Michelsen A (2011) Few long-term effects of simulated climate change on volatile organic compound emissions and leaf chemistry of three subarctic dwarf shrubs. Environ Exp Bot 72:377–386

Rivero RM, Ruiz JM, Garcya PC, Lopez-Lefebre LR, Sanchez E, Romero L (2001) Resistance to cold and heat stress: accumulation of phenolic compounds in tomato and watermelon plants. Plant Sci 160:315–321

Romani A, Pinelli P, Galardi C, Corti G, Agnelli A, Vincieri FF, Heimler D (2003) Flavonoids in leaves of black cabbage (*Brassica oleracea* var. *acephala* DC. subvar. *viridis* cv. Serotina) grown on different soils and at different elevations. Ital J Food Sci 15:197–205

Rosa E, Rodrigues PMF (1998) The effect of light and temperature on glucosinolate concentration in the leaves and roots of cabbage seedlings. J Sci Food Agric 78:208–212

Rosa EAS, Heaney RK, Portas CAM, Fenwick GR (1996) Changes in glucosinolate concentrations in Brassica crops (*B. oleracea* and *B. napus*) throughout growing seasons. J Sci Food Agric 71:237–244

Rosenfeld HJ, Aaby K, Lea P (2002) Influence of temperature and plant density on sensory quality and volatile terpenoids of carrot (*Daucus carota* L.) root. J Sci Food Agric 82:1384–1390

Ruelland E, Cantrel C, Gawer M, Kader JC, Zachowski A (2002) Activation of phospholipases C and D is an early response to a cold exposure in Arabidopsis suspension cells. Plant Physiol 130:999–1007

Sachray L, Weiss D, Reuveni M, Nissim-Levi A, Shamir MO (2002) Increased anthocyanin accumulation in aster flowers at elevated temperatures due to magnesium treatment. Physiol Plant 114:559–565

Schilmiller AL, Schauvinhold I, Larson M, Xu R, Charbonneau AL, Schmidt A et al (2009) Monoterpenes in the glandular trichomes of tomato are synthesized from a neryl diphosphate precursor rather than geranyl diphosphate. Proc Natl Acad Sci USA 106:10865–10870

Schilmiller A, Shi F, Kim J, Charbonneau AL, Holmes D, Jones AD et al (2010) Mass spectrometry screening reveals widespread diversity in trichome specialized metabolites of tomato chromosomal substitution lines. Plant J 62:391–403

Schonhof I, Kläring H-P, Krumbein A, Claußen W, Schreiner M (2007) Effect of temperature increase under low radiation conditions on phytochemicals and ascorbic acid in greenhouse grown broccoli. Agr Ecosyst Environ 119(1–2):103–111

Sharkey TD (2005) Effect of moderate heat stress on photosynthesis: importance of thylakoid reactions, rubisco activation, reactive oxygen species and thermotolerance provided by isoprene. Plant Cell Environ 28:269–277

Sharkey TD, Singsaas EL (1995) Why plants emit isoprene? Nature 374:769

Sharkey TD, Chen XY, Yeh S (2001) Isoprene increases thermotolerance of fosmidomycin-fed leaves. Plant Physiol 125:2001–2006

Sharkey TD, Wiberley AE, Donohue AR (2008) Isoprene emission from plants: why and how. Ann Bot 101:5–18

Shi Q, Zhang H, Song X, Jiang YE, Li G (2018) Functional characterization of the maize phytochrome-interacting factors PIF4 and PIF5. Front Plant Sci 8:2273–2285

Shibata M, Amano M, Kawata J, Uda M (1988) Breeding process and characteristics of 'summer queen', a spray-type chrysanthemum. Bull Nat Res Inst Veg Ornament Plants Tea Ser A 2:245–255

Shohael ALM, Ali MB, Yu K, Hahn E, Paek K (2006a) Effect of temperature on secondary metabolites production and antioxidant enzyme activities in *Eleutherococcus senticosus* somatic embryos. Plant Cell Tissue Organ Cult 85:219–228

Shohael AM, Ali MB, Yu KW, Hahn EJ, Islam R, Paek KY (2006b) Effect of light on oxidative stress, secondary metabolites and induction of antioxidant enzymes in *Eleutherococcus senticosus* somatic embryos in bioreactor. Process Biochem 41:1179–1185

Singsaas EL (2000) Terpenes and the thermotolerance of photosynthesis. New Phytol 146:1–4

Singsaas EL, Lerdau M, Winter K, Sharkey TD (1997) Isoprene increases thermotolerance of isoprene-emitting leaves. Plant Physiol 115:1413–1420

Sobuj N, Virjamo V, Zhang Y, Nybakken L, Julkunen-Tiitto R (2018) Impacts of elevated temperature and CO_2 concentration on growth and phenolics in the sexually dimorphic *Populus tremula* (L.). Environ Exp Bot 146:34–44

Solfanelli C, Poggi A, Loreti E, Alpi A, Perata P (2006) Sucrose-specific induction of the anthocyanin biosynthetic pathway in Arabidopsis. Plant Physiol 140:637–646

Sung DY, Kaplan F, Lee KJ, Guy CL (2003) Acquired tolerance to temperature extremes. Trends Plant Sci 8:179–187

Tardy F, Havaux M (1997) Thylakoid-membrane fluidity and thermostability during the operation of the xanthophyll cycle in higher-plant chloroplasts. Biochim Biophys Acta Biomembr 1330:179–193

Testerink C, Munnik T (2005) Phosphatidic acid: a multifunctional stress signaling lipid in plants. Trends Plant Sci 10:368–375

Thakur P, Nayyar H (2013) Facing the cold stress by plants in the changing environment: sensing, signaling and defensing mechanisms. In: Tuteja N, Singh GS (eds) Plant acclimation to environment stress. Springer, New York, pp 29–69

Thwe AA, Kim YB, Li X, Seo JM, Kim SJ, Suzuki T, Chung SO, Park SU (2014) Effects of light-emitting diodes on expression of phenylpropanoid biosynthetic genes and accumulation of phenylpropanoids in *Fagopyrum tataricum* sprouts. J Agric Food Chem 62:4839–4845

Velikova V, Pinelli P, Pasqualini S, Reale L, Ferranti F, Loreto F (2005) Isoprene decreases the concentration of nitric oxide in leaves exposed to elevated ozone. New Phytol 166(2):419–426

Wada H, Gombos Z, Murata N (1990) Enhancement of chilling tolerance of a cyanobacterium by genetic manipulation of fatty acid desaturation. Nature 347:200–203

Wahid A (2007) Physiological implications of metabolite biosynthesis for net assimilation and heat-stress tolerance of sugarcane (*Saccharum officinarum*) sprouts. J Plant Res 120:219–228

Wu J et al (1997) Low-temperature damage and subsequent recovery of fab1 mutant *Arabidopsis* exposed to 28C. Plant Physiol 113:347–356

Zeidler J, Schwender J, Müller C, Wiesner J, Weidemeyer C, Beck E et al (1998) Inhibition of the non-mevalonate 1-deoxy-D-xylulose-5-phosphate pathway of plant isoprenoid biosynthesis by fosmidomycin. Z Naturforsch C 53:980–986

Zhao DK, Zhao Y, Chen SY, Kennelly EJ (2021) Solanum steroidal glycoalkaloids: structural diversity, biological activities, and biosynthesis. Nat Prod Rep 38:1423-1444

Zhou Y, Zeng L, Liu X, Gui J, Mei X, Fu X, Dong F, Tang J, Zhang L, Yang Z (2017) Formation of (E)-nerolidol in tea (*Camellia sinensis*) leaves exposed to multiple stresses during tea manufacturing. Food Chem 231:78–86

Zhu GY, Li YW, Hau DKP, Jiang ZH, Yu ZL, Fong WF (2011) Protopanaxatriol-type ginsenosides from the root of *Panax ginseng*. J Agric Food Chem 59:200–205

Zobayed SMA, Afreen F, Kozai T (2005) Temperature stress can alter the photosynthetic efficiency and secondary metabolite concentrations in St. John's Wort. Plant Physiol Biochem 43:977–984

Chapter 8
Water-Linked Constraints and Plant Protection Needs

8.1 General Effects of Drought Stress on Plants

Drought stress causes water deficit to plants leading to constraining effects that can be classified in four types (Fig. 8.1): morphological, physiological, biophysical and biochemical effects. These effects show some general aspects at plant world level with interspecific and intraspecific variations governed by stress intensity and duration (external factors) on one hand, and plant growth stage and taxon (intrinsic factors) on other hand (Miranda et al. 2011).

8.1.1 Morphological Effects

Morphological effects are linked to growth variations affecting volume, size and mass of plant tissues. These variations can be due to several (combined) functional factors including primarily loss of turgor, ABA hormone activation, impaired enzyme activities, and decreased energy supply (Fig. 8.2) (Farooq et al. 2009a, 2012).

Loss of turgor results in lower leaf water potential, affecting cell division and cell enlargement leading to decrease of plant height (growth) and productivity (Fig. 8.2a) (Anjum et al. 2011; Skirycz and Inzé 2010). Reduced plant height under drought effect was reported in different studies including lily, maize, cane and rice (Li et al. 2020; Anjum et al. 2017; Misra et al. 2020; Patmi et al. 2020).

Apart from the decrease of height due to impaired mitosis and decreased cell expansion, morphological variations of leaves provide strong indications on water shortage effect. These variations include smaller leaf areas, larger leaf thickness, and denser leaf tissue (Werner et al. 1999).

N. Semmar, *Secondary Metabolites in Plant Stress Adaptation*, Signaling and Communication in Plants, https://doi.org/10.1007/978-3-031-52595-7_8

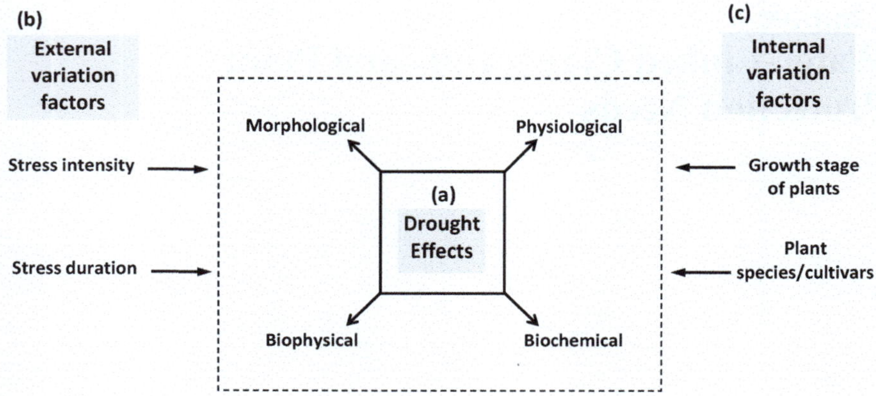

Fig. 8.1 Different types of drought effects (**a**) on plants with variability linked to external and internal factors (**b**)

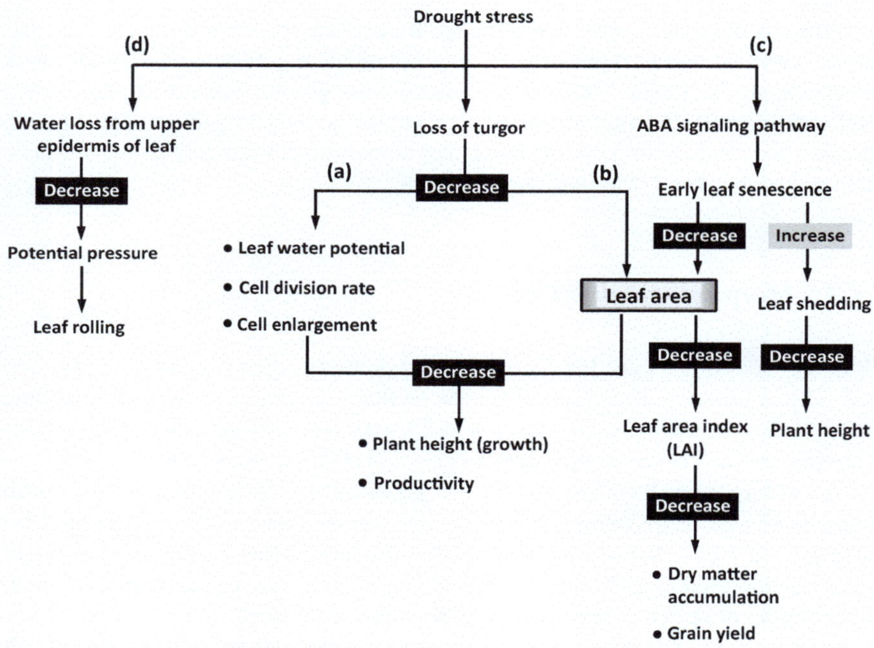

Fig. 8.2 Effects of drought stress on morphological-linked aspects of plant parts. ABA, abscissic acid

Loss of leaf turgor pressure results in decrease of leaf area and subsequently leaf area index (LAI) leading to lower photosynthesis rate and less accumulated dry matter and lower grain yield (Fig. 8.2b). Turgor pressure represents the pressure that protoplasmic fluid exerts on the cell wall; it influences the mechanisms governing the growth and division of plant cell (Brett and Waldron 1996). LAI is defined as the ratio of leaf area to ground area; it evaluates the assimilatory power extent of crops under field conditions. Apart from the loss of turgor factor, leaf area is also reduced as a consequence of drought-induced early senescence via the activation of ABA signaling pathway (Fig. 8.2c) (Nooden 1988). This results in increase of leaf shedding which negatively influences the plant height.

Other morphological variation of leaves associated with drought stress consists of leaf enrolling which occurs as consequence of water loss from the upper epidermis of leaf leading to a decrease of potential pressure (Fig. 8.2d) (Yang et al. 2021). Wheat tends to undergo severe rolling of its flag leaf under drought stress (Willick et al. 2018).

Moreover, drought tends to result in reduction of fresh and dry weights of shoot, flower and root according to the plant species (Asrar and Elhindi 2011).

8.1.2 Physiological Effects

Under physiological aspect, drought stress has significant effects on phenology which concerns transitions between different developmental phases with variable (modulable) durations (Farooq et al. 2012). Drought stress tends to shorten growth cycles of crops leading to a decrease in total growth duration (Fig. 8.3a). These effects were revealed in bread wheat (*Triticum aestivum* L.) and barley (*Hordeum vulgare* L.) (McMaster and Wilhelm 2003).

Also, water deficit induces early switching from vegetative phase to reproductive phase (Fig. 8.3b). Moreover, drought effects are various depending on both developmental phase and plant species (Fig. 8.3c): for example, quinoa (*Chenopodium quinoa* Wild.) and bread wheat (*Triticum aestivum* L.) exposed to drought stress during pre-anthesis phase showed delayed flowering (Geerts et al. 2008). Similar effect occurs in drought-exposed rice (*Oryza sativa* L.) during anthesis phase (Fukai 1999). However, drought stress during anthesis phase resulted in increase of pod abortion in soybean (*Glycine max* L.) (Liu et al. 2003). However, if this plant species is drought-exposed during grain filling phase, it shows hastened maturity with small grains and subsequent lower yield (Desclaux and Roumet 1996).

8.1.3 Biophysical Effects

Drought stress results in lack of water in rhizosphere (around the roots) leading to (i) a decrease in water potential (ψ_w) of environment and (ii) increase in osmotic potential (ψ_π) of plant cells (Fig. 8.4a) (Pamungkas et al. 2022). Water potential measures

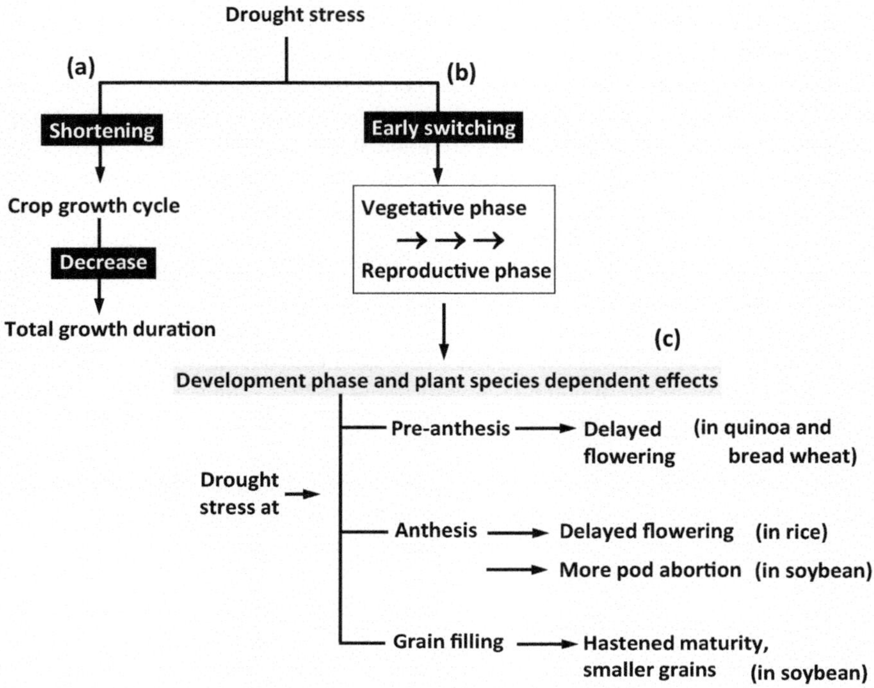

Fig. 8.3 Different effects of drought stress on physiological states of exposed plants at different development periods

the concentration of free water molecules whereas osmotic potential ψ_π measures the trend of a solution to withdraw water from pure water through a semi-permeable membrane (by osmosis) ($\psi_\pi = -\pi$, i.e. water potential = − osmotic pressure). ψ_π decreases and increases by concentration and dilution of solutes in pure water, respectively. Drought causes concentration of environmental (extra-cellular) space leading to relative dilution of intra-cell compartment. This results in increase of ψ_π within plant cells leading to a loss of water from plant to environment (Fig. 8.4b). Beside the solute effect (Ψ_π), water flux goes from higher to lower potential (Ψ_w); therefore, low Ψ_w in environment results in water flux from plant to environment leading to water loss by plants.

The water loss from plant results in decrease of both its water potential and turgor pressure (ψ_p) (Fig. 8.4c). Turgor pressure (or hydrostatic pressure) concerns the water which is exerted against the cell wall; it is associated with adhesion process of water molecules to surfaces or non-dissolved structures. Turgor pressure should be positive and this requires balance between ψ_w and ψ_π according to the equation (Fig. 8.4d):

$$\psi_w = \psi_p + \psi_\pi \rightarrow \text{Water potential} = \text{Turgor pressure} + \text{Osmotic potential}$$

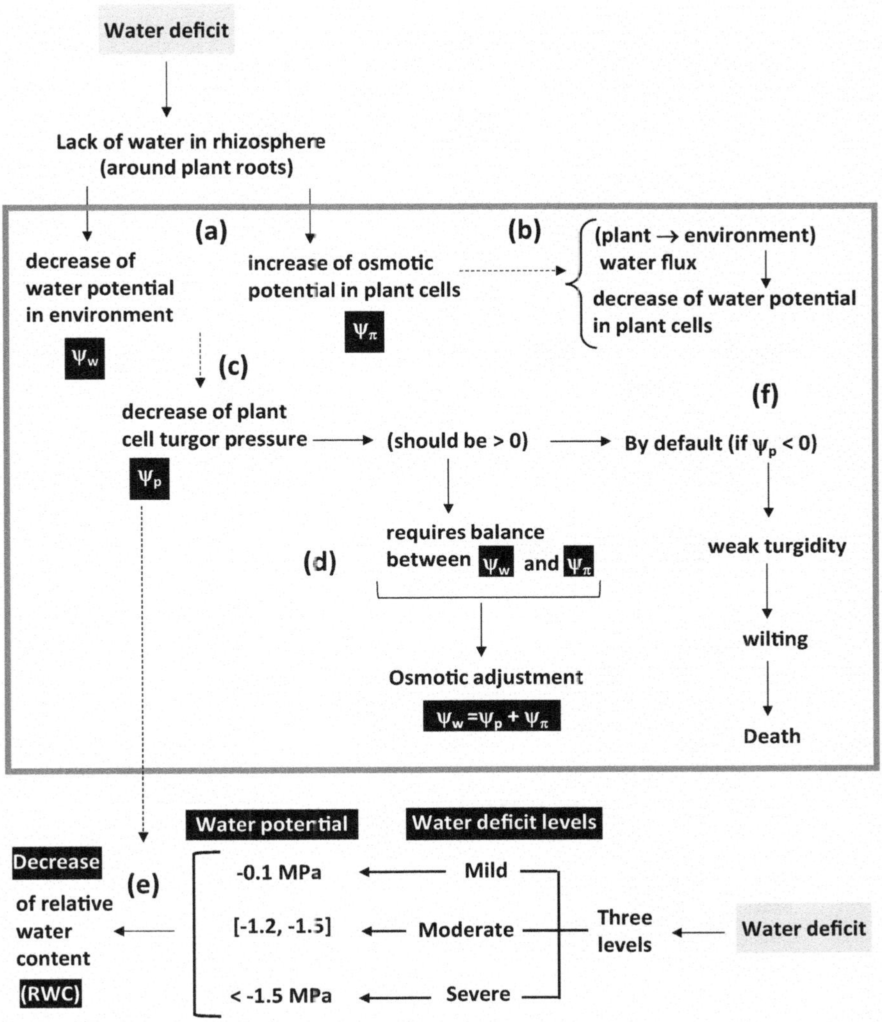

Fig. 8.4 Effects of water deficit on biophysical state of plants due to the variation of different osmotic parameters

Cell wall is sensitive to variation of turgor pressure through stretching process that can be elastic (reversible stretching) or plastic (irreversible/viscous stretching) (Brett and Waldron 1996). Under normal water conditions, the polysaccharide polymers of cell walls are well separated within the aqueous solution (apoplastic fluid) (Fig. 8.5a). Under water deficit conditions, polymers tend to be in close proximity to each other leading to adhesion states (Fig. 8.5b). Pronounced evolution of adhesion process under further water loss results in enhanced H-bounding between polymers causing altered biophysical properties of cell wall followed by impaired growth.

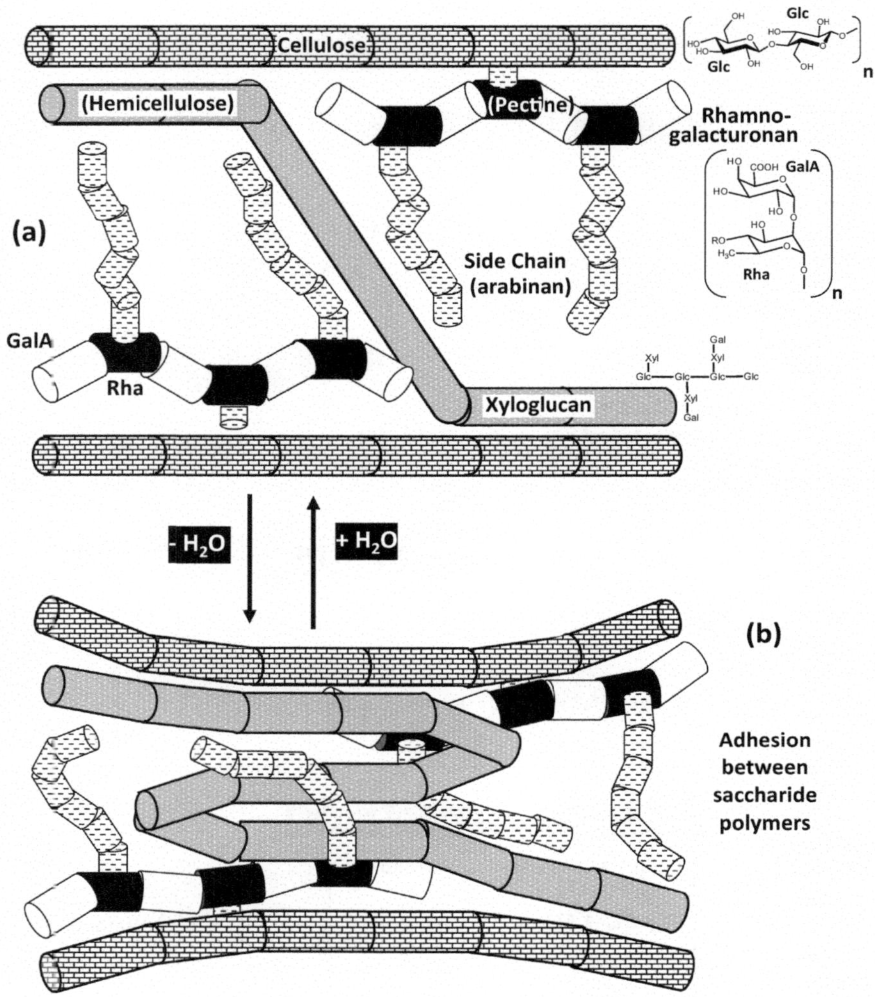

Fig. 8.5 Disturbance of cell wall polymeric organization in (**a**) normal conditions and (**b**) under water deficit effect

Decrease of ψ_p results in that of relative water content (RWC) leading to lower water potential (Ψ_w) in plant cells (Fig. 8.4e). RWC provides an evaluation of turgidity level and a measure of the plant water status (Anjum et al. 2011). Decrease of RWC results in relative plant dehydration. This decline depends on the drought severity which can be classified into three levels: mild ($\psi_w = -0.1$ MPA), moderate ($\psi_w = [-1.2, -1.5]$), severe ($\psi_w < -1.5$ MPA). However, negative turgor pressure ($\psi_p < 0$) results in a weak turgidity leading to leaf wilting and death (Fig. 8.4f).

Under drought stress, plants need to reduce their osmotic potentials to avoid water loss and plasmolysis. Also, turgor pressure can be declined to prevent water loss independently from solute accumulation strategy for osmotic potential reduction (Verslues et al. 2006).

8.1.4 Biochemical Effects

Drought stress has many effects on plant biochemical system (Fig. 8.6). Multiple effects consist of reduction of nutrient transport, impairments of energy production and enzyme activities, and oxidative damages:

Drought stress results in less available water for nutrient dissolution and translocation leading to limited nutrient uptake (Fig. 8.6a) (Farooq et al. 2009b). Also, drought causes stomate closure as response to acute water deficit by reduction/preventing transpiration-linked water loss (Mansfield and Atkinson 1990). This process results in lower CO_2 assimilation and photosynthesis rates (Fig. 8.6b) (Bota et al. 2004; Loreto et al. 1995). This process has a direct enzymatic effect on Rubisco which tends to act as oxygenase under high O_2 relatively to CO_2 (Fig. 8.6c). This results in increase of photorespiration at the expense of carbon fixation (Cornic and Massacci 1996). Moreover, at enzymatic level, drought cause impaired activities of Rubisco and enzymes implied in nutrient assimilation such as glycine synthase or nitrate reductase (Fig. 8.6d). Water loss by plant cell results in higher concentration of solutes and smaller cellular volume. This process favors the crowding of cytoplasmic components and increases the cytoplasmic viscosity (Hoekstra et al. 2001). Consequently, molecular interactions may significantly increase causing protein denaturations and membrane fusion. These intra-cell rheological changes may impair the functioning of enzymes (including those of photosynthetic system).

Concerning the effects on energy, drought cause impaired ATP synthesis leading to lower assimilation rate of CO_2 which results in lower oxidation of NADPH through Calvin cycle (Fig. 8.6e). Consequently, $NADP^+$ (primary electron acceptor) decreases favoring the increase of ROS production and leading to oxidative damage risks to photosynthesis and constitutive molecules (proteins, lipids, carbohydrate, DNA) (Fig. 8.6f).

8.2 General Responses of Plants to Drought Stress

Plant responses to drought stress show high diversity with highly organized cascade processes that can be classified according to four main ways (Fig. 8.7): osmotic variations, ABA effects, antioxidative protection and water control channels. Osmotic variations include drought-induced water loss from cell (decreased water potential and RWC) which plays the role of initial perception signal of water deficit state

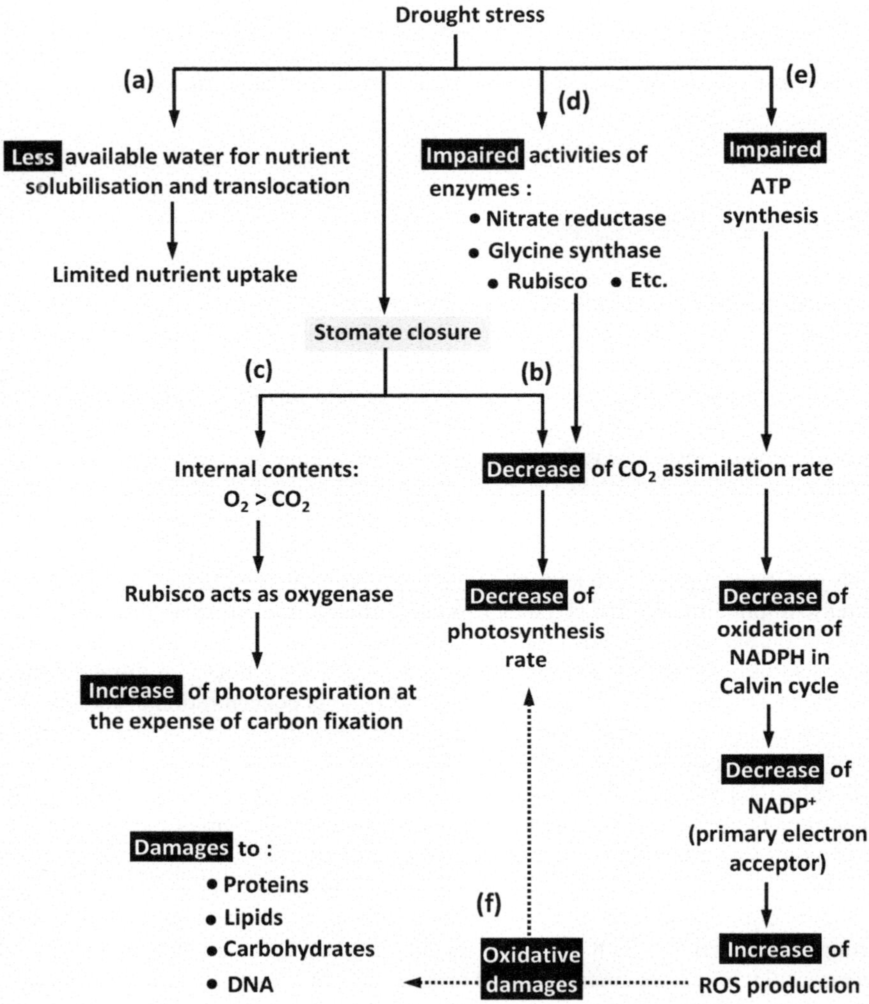

Fig. 8.6 Various effects of drought stress on plants consisting of impairments of nutrient and CO_2 assimilations (**a**, **b**), assimilation and photosynthesis enzymes (**c**, **d**), energy production (**e**) and oxidative damage (**f**)

(Figs. 8.4 and 8.7a). These initial osmotic variations are followed by osmotic adjustments through several processes aiming for reduction of water loss by plant cells (Fig. 8.7a). These processes imply the accumulation of compatible solutes (hydrosoluble compounds) in cytoplasm helping for maintaining stable water state in plant. Moreover, water deficit perception through osmotic variations is followed by cascade biochemical signals induced by ABA and leading to many protective processes against drought stress (Fig. 8.7b). ABA, ROS and Ca^{2+} (among other initially induced

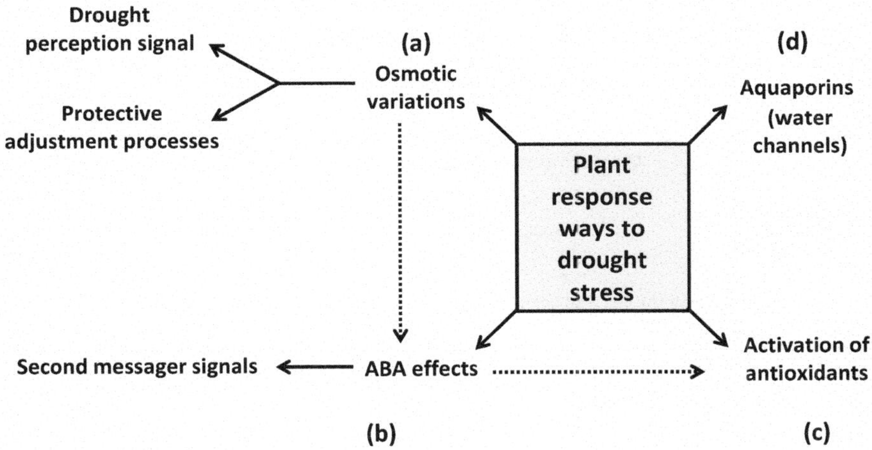

Fig. 8.7 Global ways of plant responses to drought stress

compounds) serve as second messenger signals leading to the expression of protective enzymes and metabolites including antioxidative structures (Fig. 8.7c). Water control in plant can be also made by aquaporins (AQP) which are transmembrane proteins playing channel roles for further flux control of water (essentially) and other compounds (Fig. 8.7d). AQP provide more accelerate and less resistant water transport way compared with the transmembrane transport; it accelerates migration rate of water between cells governed by water potential gradient (Yang et al. 2021).

8.2.1 *Drought Perception Signal and Downstream Processes*

Reduction of water availability in soil and humidity in air are the sources of two initial signals of drought perception in roots and leaves, respectively. Drought stress causes different intra-cellular water variations which bring perception signals of water deficit. These signals include (i) decreasing trends of water and osmotic potentials in plant cell, (ii) needs and difficulties of roots in absorbing ground water, and (iii) alteration trend of biophysical properties of cell wall (Fig. 8.8a) (Brett and Waldron 1996). These perception signals result in further variations in the cell consisting, respectively, of (Fig. 8.8b):

(i) Hydraulic signal stimulation resulting in increase of hydro-soluble metabolites helping for more water retention in the cell (Fig. 8.8 b1). These metabolites play the roles of osmoprotectant by increasing solute concentration in the cell leading to better control of pressure turgor, limitation of water loss and preservation of membrane cell integrity (Fig. 8.9).

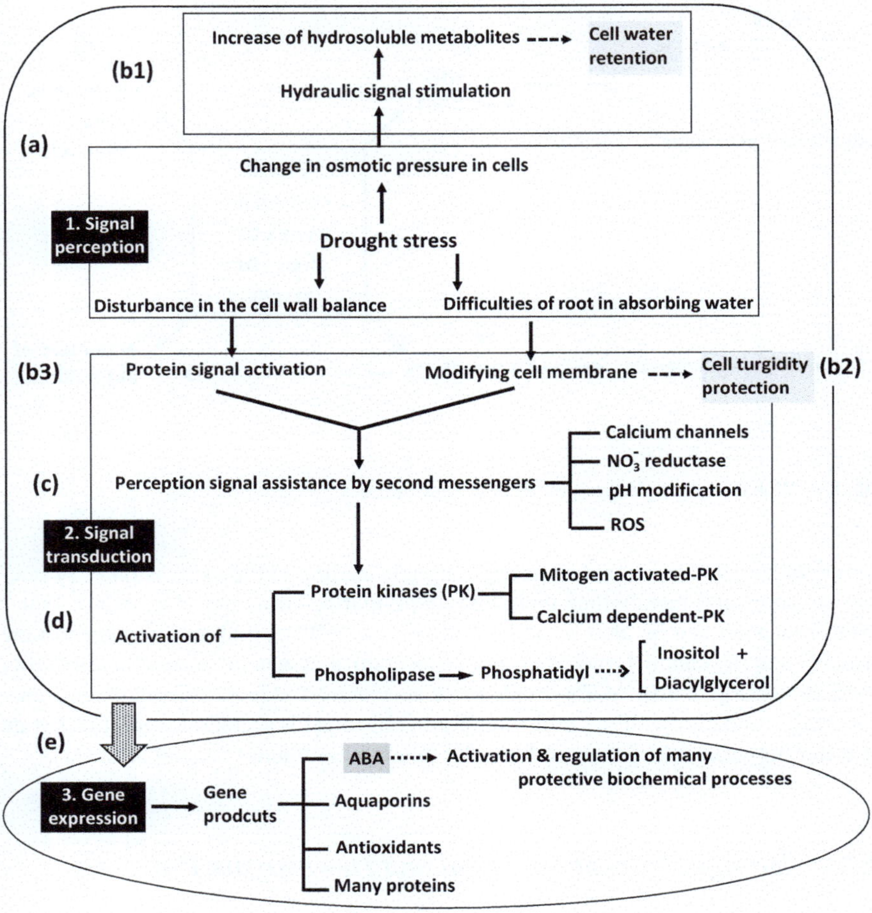

Fig. 8.8 Sequential steps of plant responses to drought stress implying signal perception, signal transduction (stimulation/amplification) and signal transcription (gene expression) leading to later morphological, physiological and metabolic adaptive processes

(ii) Modification of cell membrane through variation of lipidic and protein compositions helping for not lose cell turgidity (Saladin 2014) (Fig. 8.8 b2). Initially, membrane integrity is affected by the loss of water which play a key role in H-bonds with phosphate groups of phospholipids (Fig. 8.10). Consequently, phospholipids tend to become fluidized affecting the membrane integrity. This process serves as initial perception signal of water deficit.

(iii) Sensor protein activation (Fig. 8.8 b3). Disturbance in cell wall balance triggers sensory ways of drought stress which induce cascade processes of transduction,

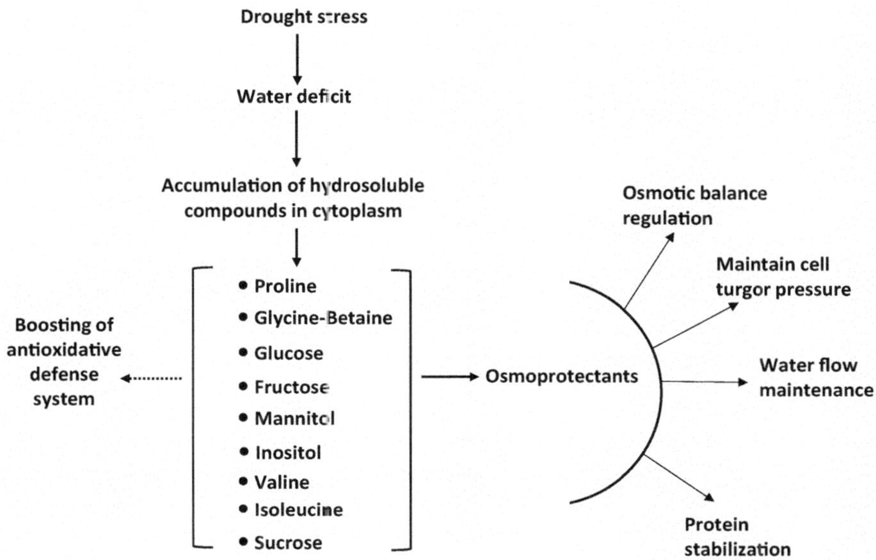

Fig. 8.9 Key roles of osmoprotectant metabolites in the water balance of plant cell

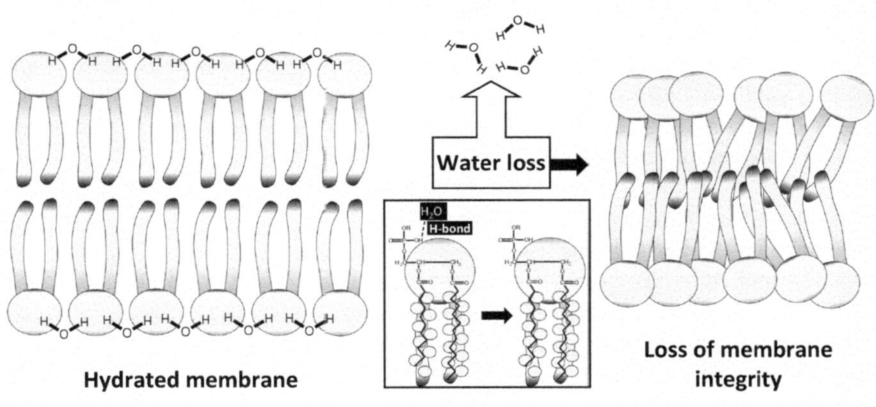

Fig. 8.10 Disorganization (fluidization) of phospholipid constituents under water loss effect affecting the cellular membrane integrity

activation and regulation of different plant responses (morphological, physiological, metabolic) to overcome drought stress (Fig. 8.8c). Modification or alteration of membrane organization by phospholipid fluidization results in (Fig. 8.8c).

- Increase of Ca^{2+} flux through calcium channels leading to accumulation of calcium ions in cytoplasm. This accumulation plays the role of a second messenger in drought signaling. Calcium accumulation occurs also by the release of stored calcium from organelles (vacuole, endoplasmic reticulum) (Boudsocq and Sheen 2010).
- Increase of cytosolic and xylem pH resulting from decrease of nitrate reductase activity (less NO_3^- uptake). More generally, the mineral absorption flux in roots is reduced under water deficit. This results in pH variation which contributes to the amplification of drought stress signal (Schachtman and Goodger 2008).
- ROS production playing the role of other second messenger (like Ca^{2+} flux). ROS are essentially produced from electron transport chains in mitochondria (respiration) and chloroplast (photosystems for photosynthesis) (Wahid 2007).

Accumulation of calcium ions leads to binding of Ca^{2+} to sensor proteins the activation of which leads to amplification of drought stress signal. Sensor proteins include calmodulin (CaM), CaM-like protein, calcium-dependent protein kinases (CDPKs), mitogen-activated protein kinases (MAPKs), calreticulin (Fig. 8.8 d) (Boudsocq and Sheen 2010; Tuteja 2009). These kinases are activated by Ca^{2+}, calmodulin (calciprotein) or both (Fig. 8.8d). Moreover, Ca^{2+} activates phospholipase C, a membrane protein which converts phosphatidylinositol 4,5-biphosphate (Fig. 8.11) into inositol 1,4,5-triphosphate (ITP) and 1,2-diacylglycerol (DAG) (Fig. 8.8d). ITP and DAG play key roles in drought signal amplification by activating channels of endoplasmic reticulum (RE) and protein kinase C, respectively. RE-channels make further release of Ca^{2+} in cytosol. Protein kinases such as MAPKs can be activated by ROS which also play inhibiting effect on phosphatases. After their activation by Ca^{2+} or ROS, kinases phosphorylate other kinases leading to signal amplification. These sequential phosphorylation processes lead to the activation of several transcription factors (TFs) (Fig. 8.8e). These TFs are of two categories: abscisic acid (ABA)-independent and ABA-dependent pathways (more frequent). ABA-dependent pathway stimulates various drought defense mechanisms.

Genes implied in ABA synthesis are early stimulated because ABA is needed for amplification and acceleration of several drought stress response pathways. Formed ABA brings signal which is perceived by ABA receptor then transmitted through sequential regulatory factors in the cell. This forms a complex intracellular signal transduction network leading to downstream physiological processes (Hauser et al. 2011). *Arabidopsis thaliana* showed several hundred TFs regulated by ABA in relation to drought level (Harb et al. 2010).

ABA is a sesquiterpene type (C15) compound derived from β-carotene (C40) through a metabolic pathway starting in chloroplast and ending in the cytosol (Fig. 8.12) (Nambara and Marion-Poll 2005). The β-carotene is initially hydroxylated to zeaxanthin (a xanthophyll, i.e. oxygenated carotenoid) (Figs. 8.12a and 8.13a). Zeaxanthin is converted to violaxanthin (Figs. 8.12b and 8.13b) which is transformed to neoxanthin by neoxanthin-synthase (Fig. 8.13c). Then, isomerization occurs transforming neoxanthin from 9-trans to 9-cis configuration (Figs. 8.12c and 8.13d); also, violaxanthin form is kept by isomerization. Chloroplast metabolic

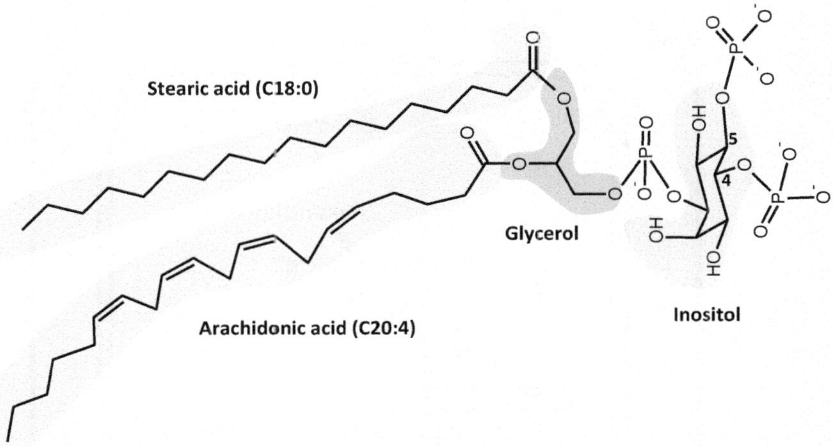

Fig. 8.11 Structure of phosphatidylinositol 4,5-biphosphate, a membrane phospholipid playing key role in signal amplification of response to stress

pathway is ended by oxidative cleavage of 9-cis-neoxanthin and/or 9-cis-violaxanthin to xanthoxin through catalytic effect of 9-cis-epoxycarotenoid dioxygenase (NCED) (Fig. 8.12c and 8.13e). Therefore, neoxanthin is moved from chloroplast to cytosol (Fig. 8.12d) to be sequentially converted to abscisic aldehyde (Figs. 8.12e and 8.13f) and ABA (Figs. 8.12f and 8.13g) via xanthine dehydrogenase (ABA2) and abscisic aldehyde oxidase (AAO3), respectively.

Apart from the ABA biosynthesis pathway, ABA can be also accumulated from the hydrolysis of its glycosyl ester (Fig. 8.12g) (Dietz et al. 2000; Seiler et al. 2011). ABA-glycosyl ester represents inactive form of ABA and can be used for maintaining high levels of ABA during drought stress (Lee et al. 2006; Han et al. 2020).

Under drought stress, ABA plays key role in the induction, stimulation, activation and regulation of many protective processes at morphological, physiological, metabolic and molecular aspects (Fig. 8.14).

Under morphological aspect, increased levels of ABA inhibit shoot growth and lateral root formation versus root elongation favoring more efficient uptake and optimal use of deficient water (Fig. 8.14a) (Yao and Finlayson 2015). Under this aspect, ABA play key communicative role between root and shoot during drought or salt stress (Davies and Zhang 1991).

ABA accumulation results in water flux reduction through stomata closure via its effects on cellular biochemical fluxes and tissue conductance: for the flux variation way, ABA accumulation stimulates Ca^{2+} accumulation and pH modification which are followed by inhibition of H^+-ATPase activity leading to potential membrane reduction (Fig. 8.14b). This results in closure of cell influx vs opening of efflux of K^+ (and other ions, e.g. Cl^-) (Swamy 1999; Raghavendra et al. 2010). The increase of K^+ efflux leads to a decrease of turgor in guard cell favoring stomata closure and subsequent evapotranspiration limitation.

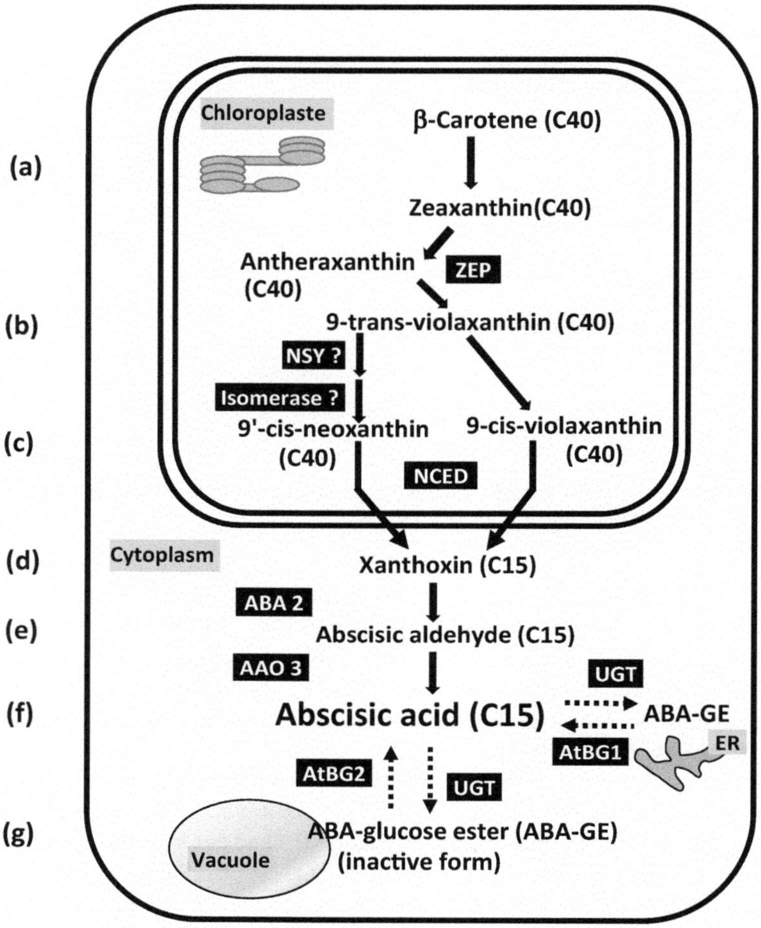

Fig. 8.12 Chloroplast-and cytosol-occurring biosynthesis pathway of abscisic acid (ABA)

ABA-promoted stomata closure can also occur by induced decrease of water permeability in the leaf vascular tissues (Fig. 8.14c) (Pantin et al. 2013). Under moderate and severe drought stress, ABA was revealed to be involved in the reduction of stomatal conductance helping for better regulated net photosynthesis and intrinsic water use efficiency (WUE) (Li et al. 2021). This effect was revealed to be plant species-dependent. Comparison between wild type and ABA-deficient mutant of *Nicotiana plumbaginifolia* revealed crucial effect of leaf ABA accumulation on the decrease of mesophyll conductance under drought conditions (Fig. 8.14 c) (Mizokami et al. 2015). Water potential and ABA tend to be interactive factors for better plant response to drought stress: as water potential decreases in the guard cells and the mesophyll cells, the needed ABA levels in these cells tends to be small for stomatal

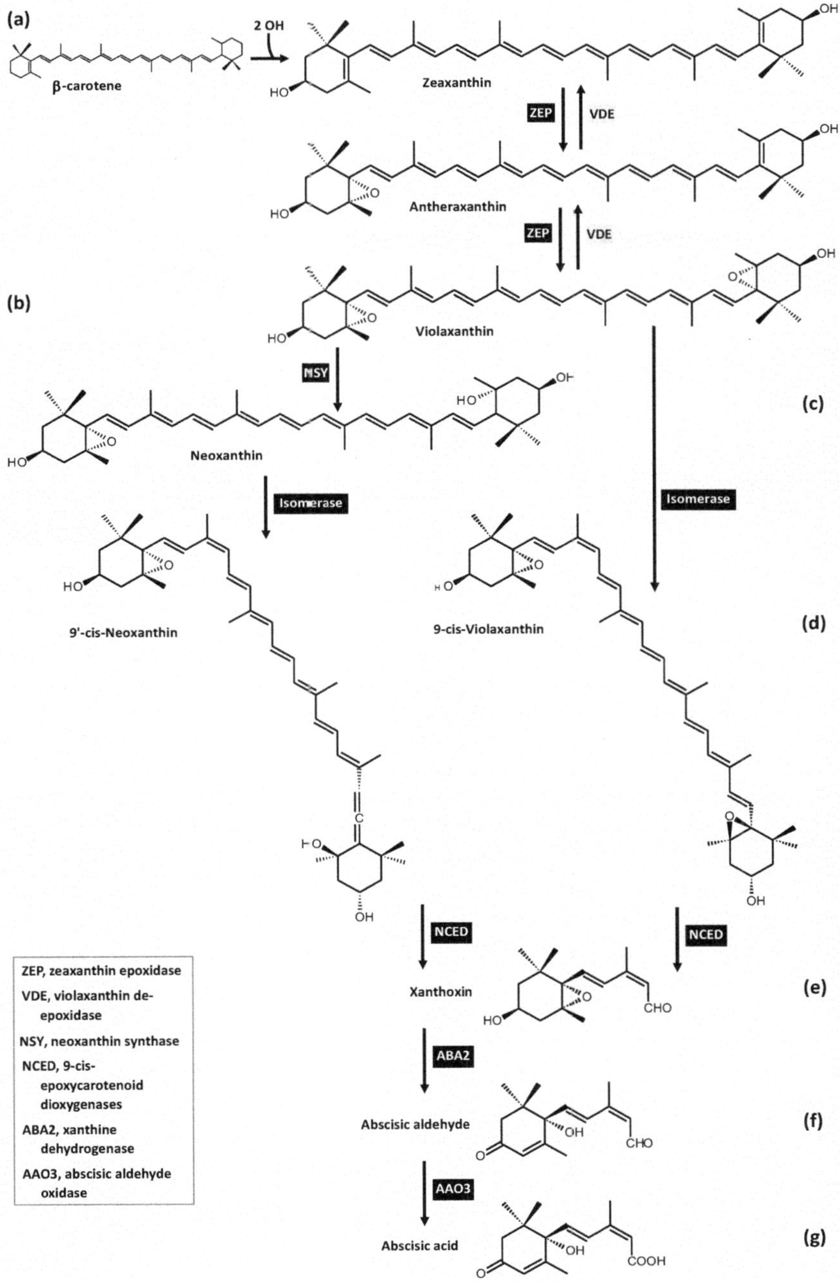

Fig. 8.13 Chemical structures of different β-carotene-derived compounds involved in abscisic acid biosynthesis

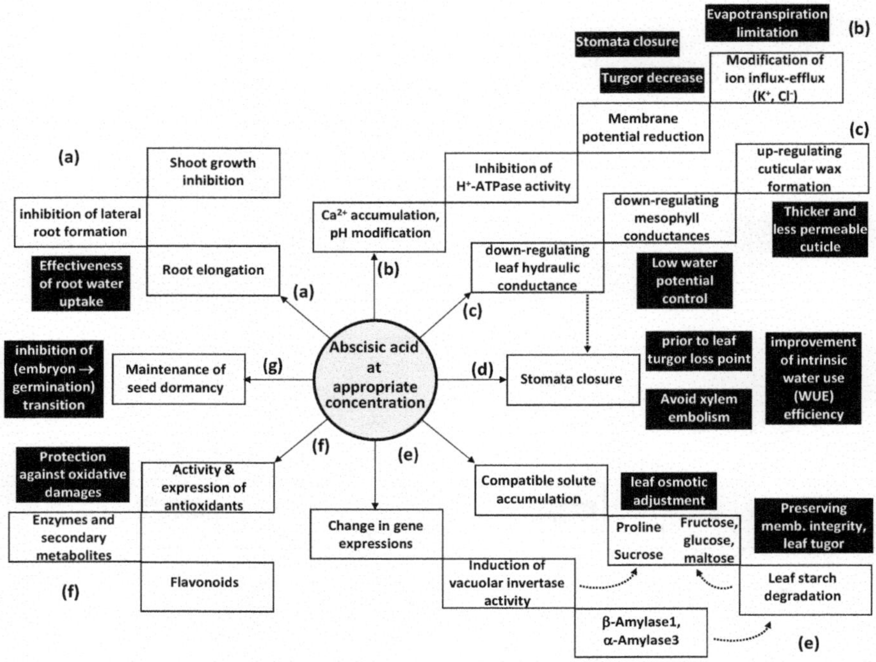

Fig. 8.14 Different effects of ABA helping plants for efficient responses to drought stress under morphological, physiological, metabolic and molecular aspects

closure even prior to leaf turgor loss point (Fig. 8.14d) (Cardoso et al. 2020a). This makes the plants to avoid xylem embolism and to improve intrinsic WUE under moderate to severe drought conditions (Cardoso et al. 2020b).

ABA contributes to osmoregulation balance through its involvement in transcriptional and post-transcriptional regulations of genes associated with enzymes of carbohydrate metabolism (Fig. 8.14e) (Urano et al. 2009; Kempa et al. 2008). Carbohydrates play important role in osmotic adjustment, including sucrose, fructose, glucose and maltose. Metabolic regulation of sucrose occurs via induction and up-regulation of gene expression of vacuolar invertase by ABA. Invertase is a glycoside hydrolase catalyzing the hydrolysis of sucrose into glucose and fructose. Maize mature leave exposed to water deficit stress showed increase of vacuolar acid invertase activity preceding the decrease of gas exchanges due to stomatal closure (Trouverie et al. 2003). The increase of this glycoside hydrolase was positively correlated with (i) xylem sap ABA concentration and (ii) rapid accumulation of glucose and fructose in leaf (eightfold the control leaf value). Also, ABA controls the activity of β-amylase 1 (BAM1) and α-amylase 3 (AMY3) which are synergically involved in the degradation of leaf starch leading to release glucose under drought stress (Thalmann et al. 2016). These mechanisms result in release of hexoses which are compatible solute accumulating for osmotic adjustment (Fig. 8.14e).

Also, proline (amino acid) is an osmoregulatory which showed correlated accumulation with variation levels of ABA (Fig. 8.14 e) (Yang et al. 2000; Ashraf et al. 2018; Yaish 2015). Proline metabolism is controlled by P5CR (Δ1-pyrroline-5-carboxylate reductase) and P5CS (Δ1-pyrroline-5-carboxylate synthetase) enzymes which synthesize proline from glutamate. However, proline is catabolized into glutamate by ProDH (proline dehydrogenase) and P5CDH (Δ1-pyrroline-5-carboxylate dehydrogenase) (Szabados and Savouré 2010). Drought, dehydration and salinity stresses induce expression of P5CS1 in plants and suppress that of ProDH (Verslues and Sharma 2010). Moreover, supplemented ABA was shown to induce the expression of P5CS1 in vegetative tissues (Ábrahám et al. 2003). In *Arabidopsis*, supplemented abscisic acid (ABA) was followed by a striking activation of P5CS1 expression leading to stimulation of proline synthesis. However, ProDH showed inhibition at different extent of supplied ABA in shoots and roots of light-grown plants.

At epigenetic level, proline accumulation was linked to DNA methylation changes (demethylation) which induces δ-OAT and P5CS (proline metabolism-related genes) (Zhang et al. 2013). It was also shown that proline accumulation tends to be a signal of stress memory by manifesting at next plant generation (Kavi Kishor and Sreenivasulu 2014).

Proline is distinguished from the other amino acids by different structural and functional points: (i) it contains a α-amino group as secondary amine; (ii) it exists as zwitterion due to positive and negative charges at carboxylic and nitrogen groups, respectively. The zwitterion property is shared by other compatible solutes playing key role in osmotic adjustment (Farahani et al. 2013). Proline shows greater water solubility compared with other amino acids (Kavi Kishor and Sreenivasulu 2014). Proline biosynthesis and consumption tend to increase in plants subject to low or high water potentials (Raymond and Smirnoff 2002). Proline accumulation was shown to increase with drought intensity in *Medicago saliva* L.; this positive trend was suggested to respond to homeostasis needs including osmotic adjustment, protein stabilization and cytosolic acidity (pH buffering) (Irigoyen et al. 1992).

Apart from osmotic adjustment role, proline accumulation and turnover can act as cellular redox buffer factor during drought (de Carvalho et al. 2013; Sekmen et al. 2014). Its ability of ROS scavenging was shown under different stresses including drought and saline conditions (Fig. 8.9) (Szabados and Savouré 2010; Natarajan et al. 2012; Bandurska et al. 2017).

ABA is also involved in regulation of redox balance by influencing the activities of enzymatic and non-enzymatic anti-oxidants (Fig. 8.14f).Wheat subject to drought stress showed increased activity of superoxide dismutase (SOD) and peroxidase (POD) after ABA application (Bano et al. 2012). In the leaves of *Zea mays* (maize), water stress was associated with a monotonous increase of ABA content, a peak of ROS followed by up-regulation of antioxidant enzymes (SOD, APX, catalase, glutathion reductase) (Jiang and Zhang 2002). In *Arabidopsis* seedling, ABA treatment induced an up-regulation of genes encoding enzymes involved in the ROS-scavenging ascorbate–glutathione cycle (Fig. 8.15): ascorbate peroxidase, APX2 (gene At3g09640); monodehydroascorbate reductase, MDAR (At3g27820); glutathione reductase, GR (At3g24170); dehydroascorbate reductase,

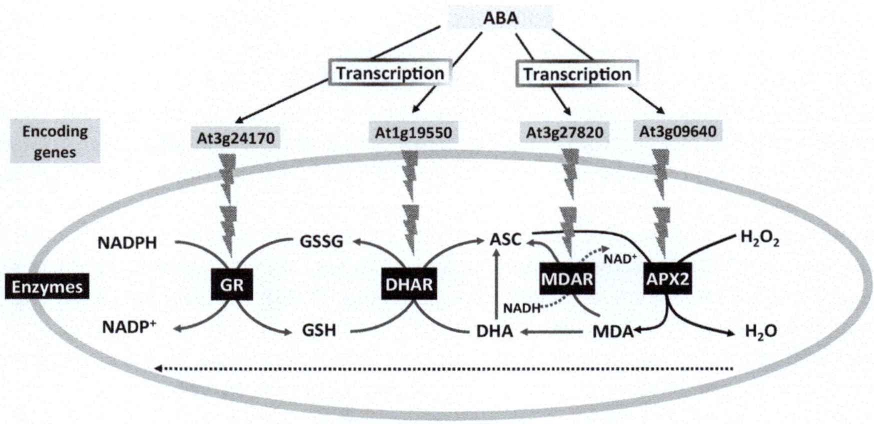

Fig. 8.15 Implication of ABA in transformation (for detoxication) of H_2O_2 through transcription of genes involved in ascorbate–glutathione cycle. ASC, ascorbate; MDA, monodehydroascorbate; DHA, dehydroascorbate; GSSG, oxidized glutathione; APX, ascorbate peroxidase; MDAR, monodehydroascorbate reductase; DHAR, dehydroascorbate reductase; GR, glutathione reductase

DHAR (At1g19550) (Ghassemian et al. 2008). Application of exogenous ABA on tea leaves under drought conditions resulted in induction of biosynthesis of flavonoids which are characterized by their anti-oxidative activities (Fig. 8.14f) (Gai et al. 2020).

Finally, ABA plays a key role in dessication tolerance and more particularly in the maintenance of seed dormancy and inhibition of germination in drought conditions (Fig. 8.14g). This role is antagonistic with that of gibberellins (GAs) which play key role in activation of germination.

8.2.2 *Plant Management Strategies to overcome Drought Stress*

Plants manage water deficit conditions by different strategies which depend on drought severity level and plant development stage (Cooper et al. 2006). According to the water deficit level, plants adopt three strategies consisting of drought escape, drought avoidance and drought tolerance which are associated with local, moderate and severe drought stress, respectively (Fig. 8.16).

(1) When escape drought strategy is adopted, plants tend to complete their vegetative cycle in a short time (shortened life cycle) (Fig. 8.16a). This results in quick generative stage followed by early flowering and seed formation (Turyagyenda et al. 2013). (2) Through drought avoidance strategy, plants show different morphological variations including roots elongation to reach water, stomata closing, leaf rolling (Fig. 8.16b). These changes help for efficient use of available limited water by plants (Nezhadahmadi et al. 2013). (3) When drought stress becomes severe, plants develop

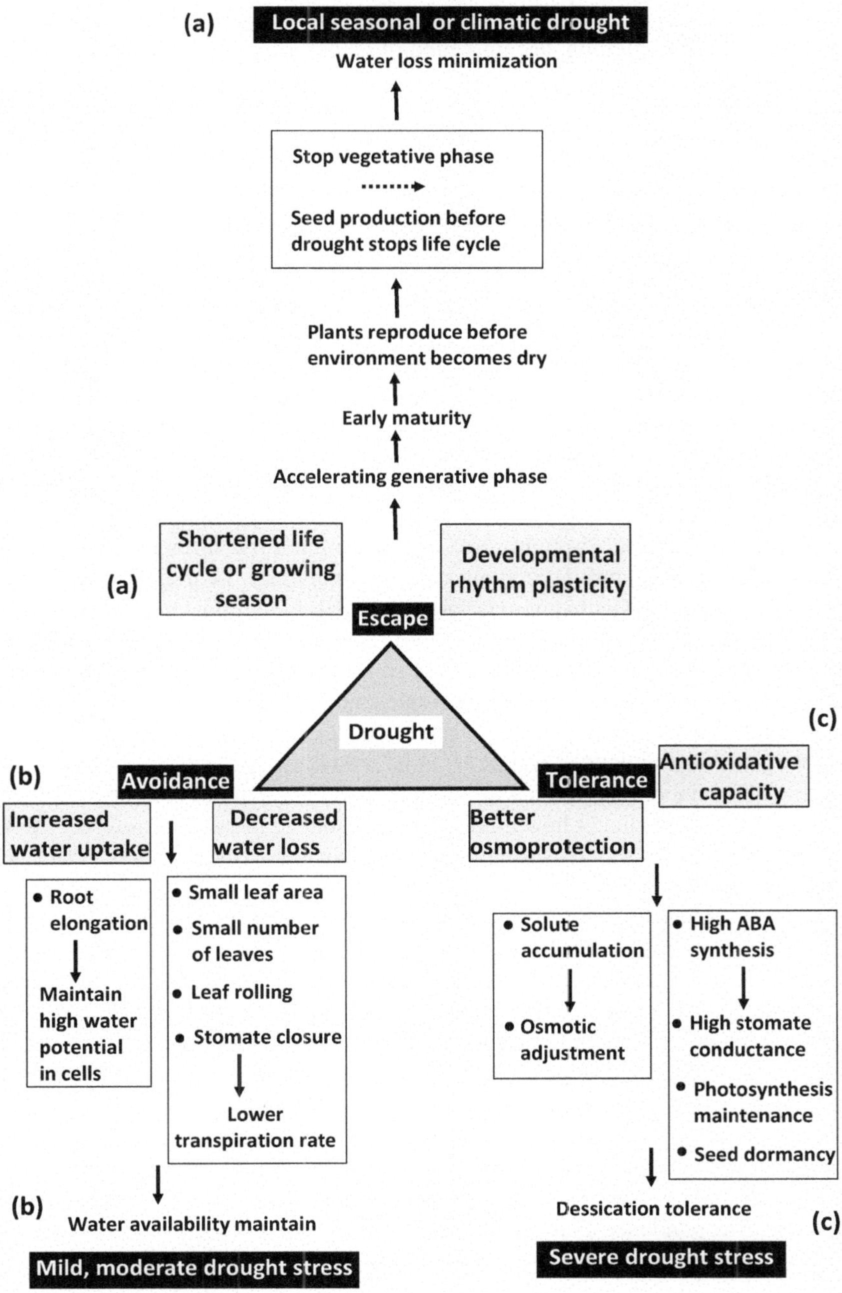

Fig. 8.16 Different strategies adopted by plants to confront drought stress under different increasing levels of water deficit

tolerance strategy by combining molecular, biochemical, and physiological processes including osmotic regulation, accumulation of osmoprotectants, ABA biosynthesis, and stomatal activities (Fig. 8.16c).

8.3 Secondary Metabolites Profiles Associated with Drought Stress: Illustrative Study Cases

Variable drought conditions were applied for 55 days to several trees of *Pinus sylvestris* L. in which water potential (ψ_w) was regularly measured in shoot revealing four increasing ranges (Sancho-Knapik et al. 2017): [-0.39, 0] MPa (control conditions, C), [-0.6, -0.39] MPa (level *1*, ψ_1), [-1.8, -0.6] MPa (level *2*, ψ_2), [-2.4, -1.8] MPa (level *3*, ψ_3). Higher absolute value of ψ_w indicated more severe drought stress. Levels *1*, *2* and *3* were associated with mild, moderate and severe water deficit conditions, respectively. Level *2* resulted in significant reduction of stomatal conductance whereas level *3* showed completely suppressed photosynthesis.

Trees under different drought stresses were monitored through their terpene contents. The chemical profiles contained 12 monoterpenes, 14 sesquiterpenes and 7 diterpenes (acid resins) with amounts varying with the water potential. The amounts ($\mu g\ g^{-1}$) of the 33 SMs were converted into regulation levels by relativizing them to both their (i) row and (ii) column sums: row sum represents the total metabolic amount in plant profile *k*; column sum represents the total amount of considered metabolite M_j in the whole plant population. (i) Relative value of an amount on its row sum gives the regulation level of M_j in its productive tree by considering the 33 co-occurring metabolites. (ii) However, relativization of amount of M_j by reference to the column sum evaluates how much M_j was regulated in tree *k* by reference to its whole synthesis in all the trees. Regulation trend analysis is statistically performed by correspondence analysis (Fig. 8.17) (Semmar 2013).

Different terpene families were revealed to be relatively well regulated under different ψ_w ranges: Metabolic regulations of sesquiterpenes tended to be more favored by levels 2 (mainly) and 3 of potential water (Ψ_2, Ψ_3). The most regulated sesquiterpenes by ψ_2 were represented by compounds *15*, *16*, *18*, *20*, *23*, *24*, *26* (Fig. 8.17b) consisting of: β-copaene, aromadendrene, α-muurolene, β-seliene, γ-cadinene, δ-cadinene, α-cadinol (Fig. 8.18c–g, j–k). These compounds showed also high values under ψ_3 after peak observed under Ψ_2 (Fig. 8.19). However, the highest increasing slope was observed between ψ_1 and ψ_2 revealing more significant effect of moderate drought (ψ_2) on regulation of sesquiterpene in *Pinus sylvestris*.

These sesquiterpenes occupied opposite subspace to that of acid resins (diterpenes) in correspondence analysis (Fig. 8.17 b). This opposition was due to the fact that acid resins were relatively more regulated under lower drought stress (C and ψ_1) (Fig. 8.17a, b). The most concerned compounds were *29*, *30*, *32*, *33* (Fig. 8.17b) and corresponded respectively to palustric, levopimaric, abietic and neoabietic acids (Fig. 8.20a–d). All these diterpenes showed a peak at middle water deficit (ψ_1) with

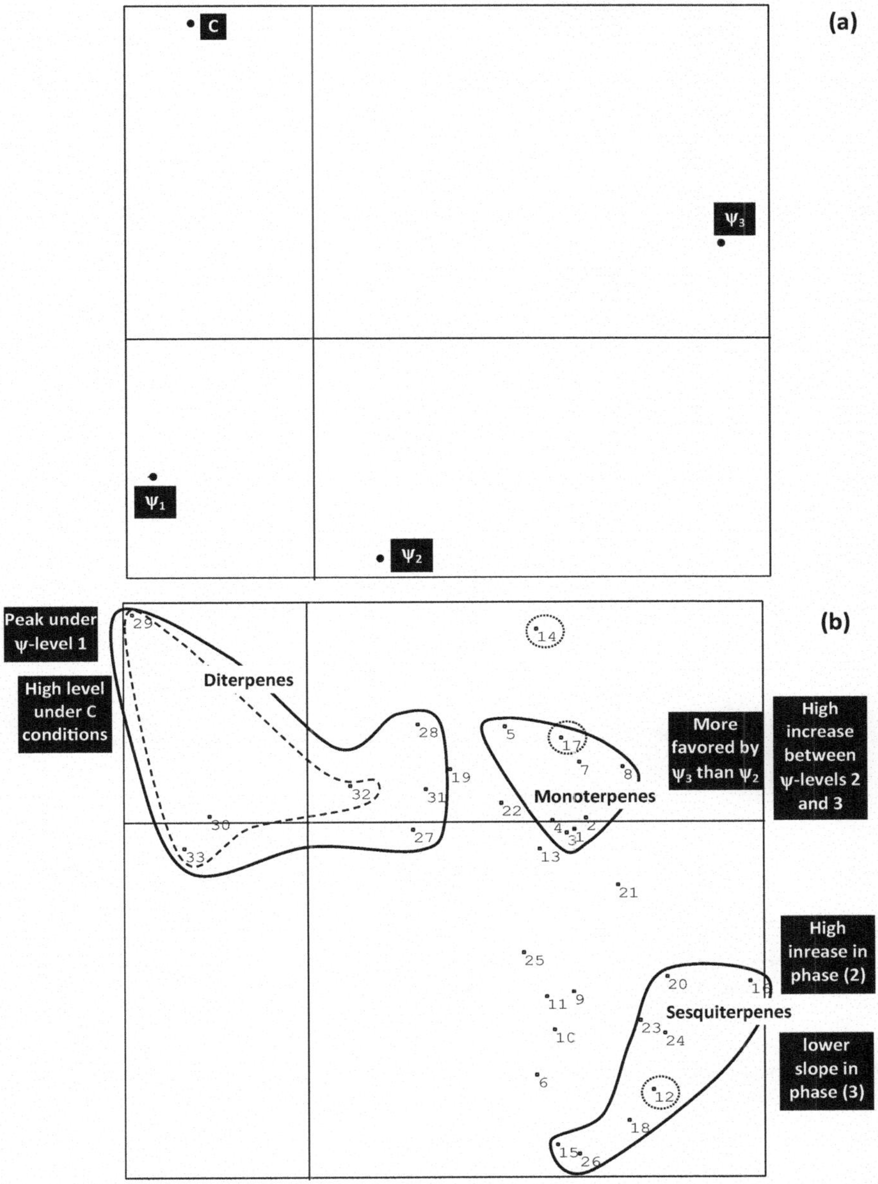

Fig. 8.17 First factorial plot of correspondence analysis applied to the amounts of terpenes (monoterpenes, sesquiterpenes, diterpenes) (**b**) in the needles of *Pinus sylvestris* subjected to different drought stress (**a**). No water deficit (C), middle (ψ_1), moderate (ψ_2), severe water deficit (ψ_3)

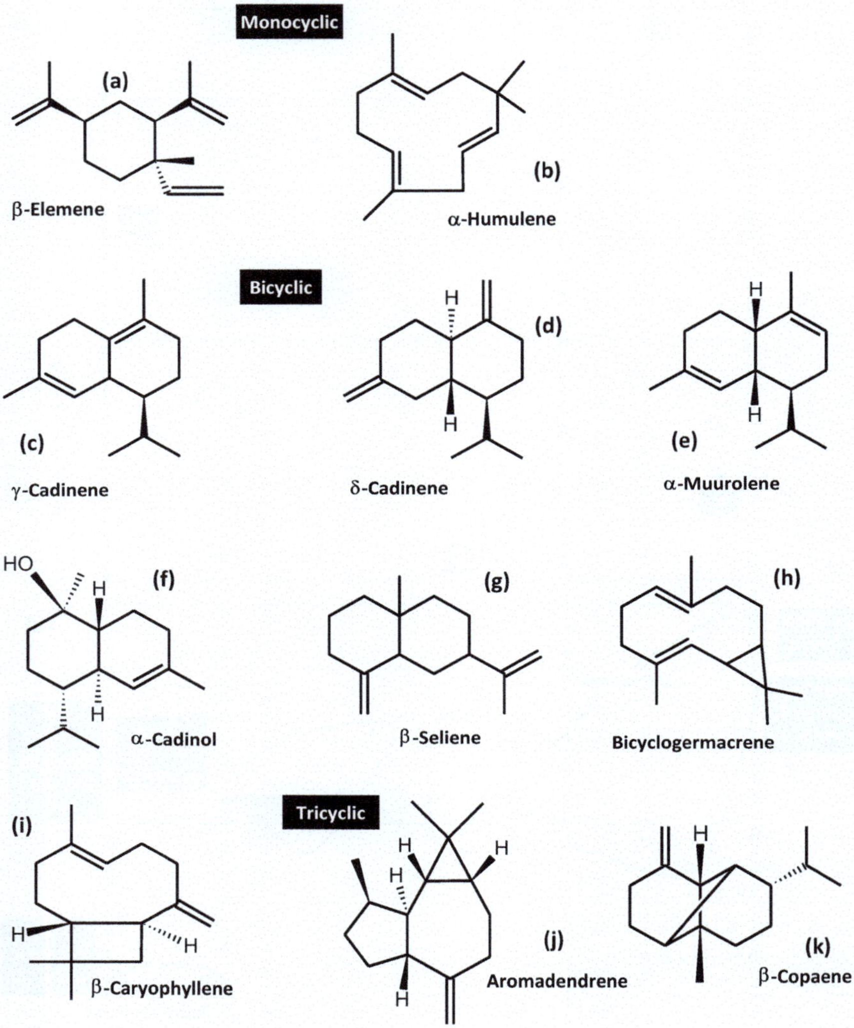

Fig. 8.18 Chemical structures of sesquiterpenes found in the needles of *Pinus sylvestris* during variable drought conditions

relatively high amount under control conditions (C) (Fig. 8.21a–d). The three other acid resins (*27*, *28*, *31*) showed less marked peak under ψ_1 with lower amounts in absence of drought stress (C) (Fig. 8.21e–g). These compounds corresponded to pimaric, isopimaric and dehydroabietic acids (Fig. 8.20e–g). Pimaric and isopimaric acids have well separated lateral double bond; dehydroabietic acid was distinguished by an aromatic cycle.

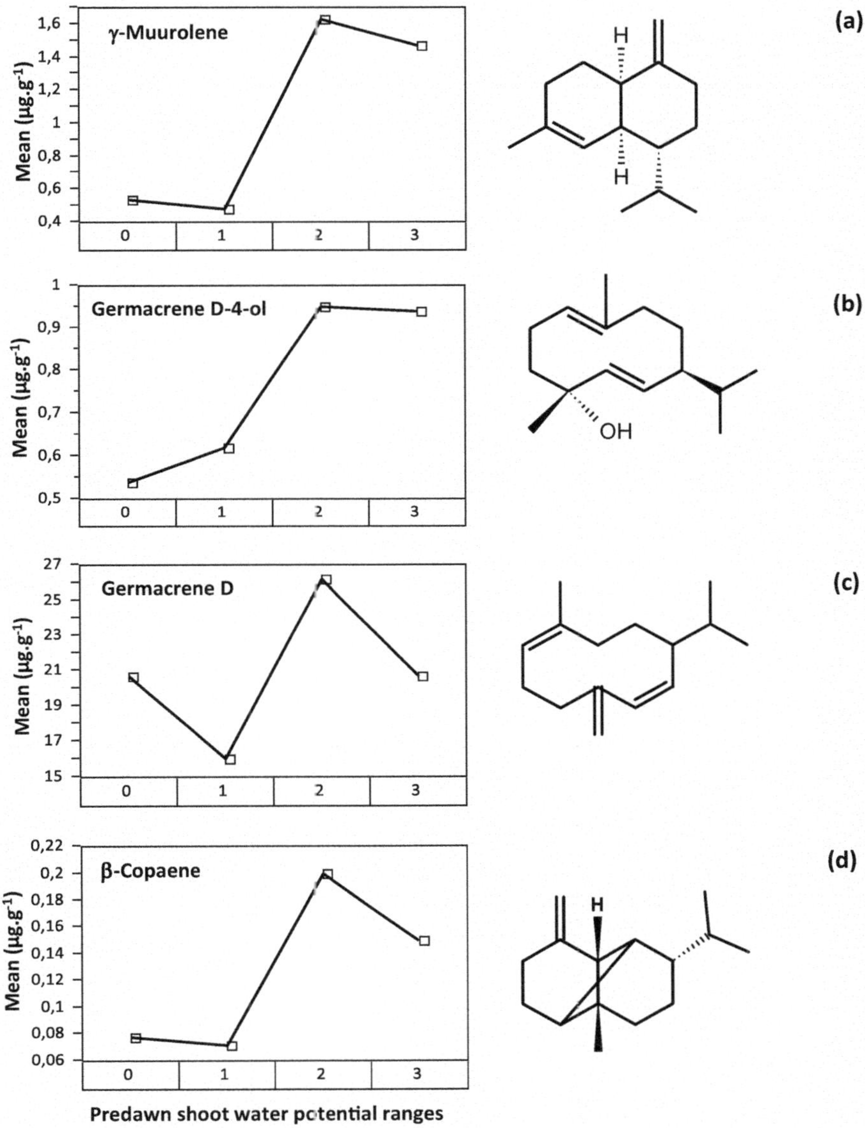

Fig. 8.19 Variations of amounts of sesquiterpenes in *Pinus sylvestris* in relation with different drought levels corresponding to different negative water potentials (ψ_W). 0, no water deficit (control conditions); 1, middle; 2, moderate: 3, severe drought conditions

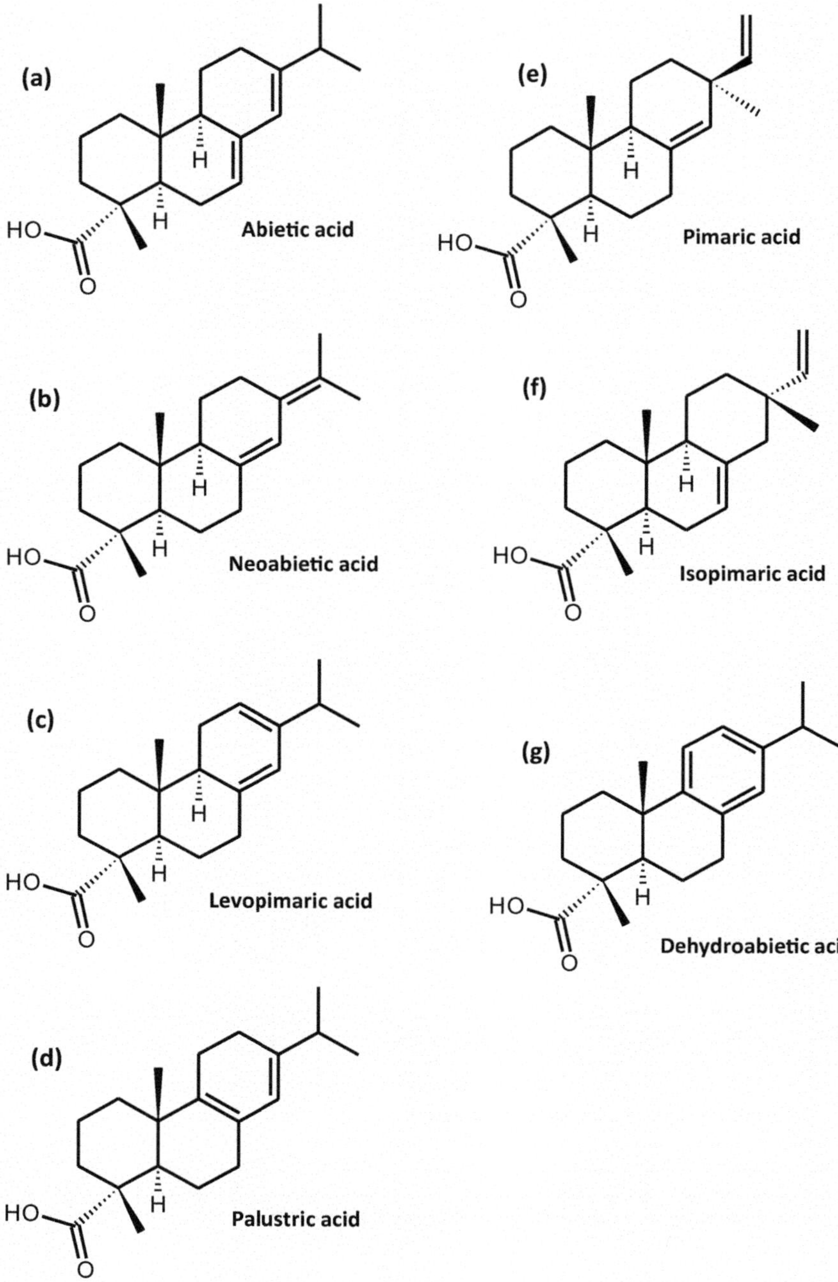

Fig. 8.20 Chemical structures of acid resins (diterpenes) found in the neddles of *Pinus sylvestris* during variable drought conditions

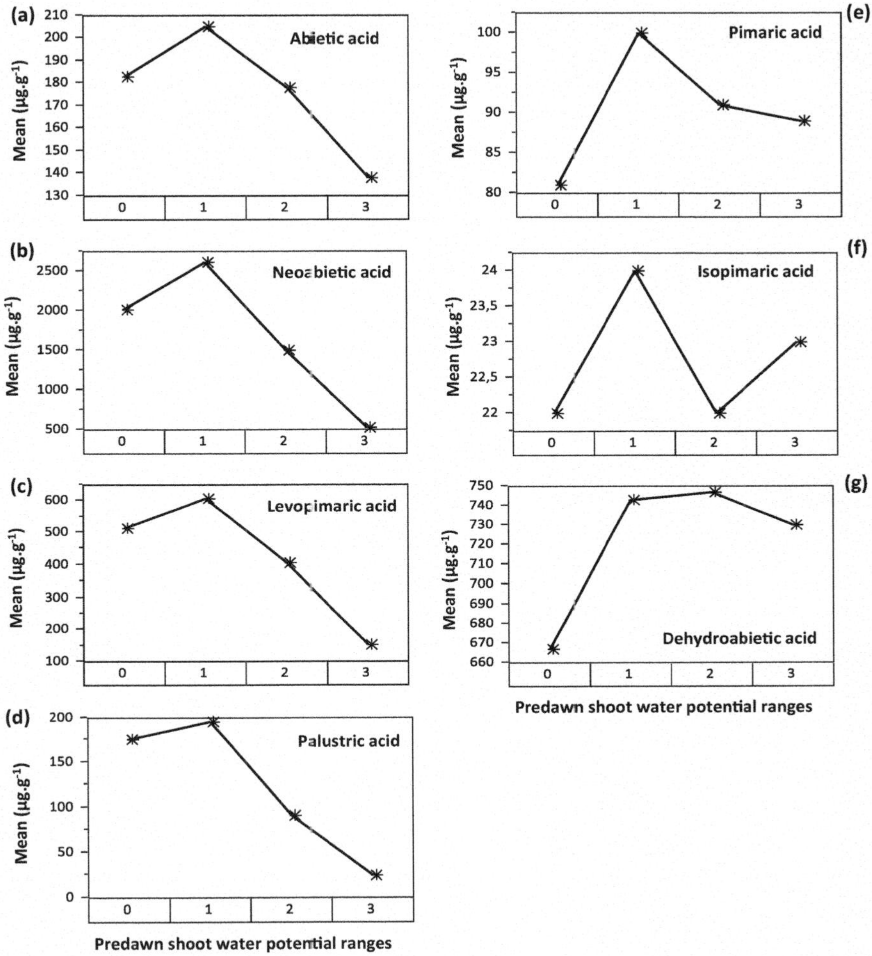

Fig. 8.21 Variations of amounts of acid resins (diterpenes) in *Pinus sylvestris* in relation with different drought levels corresponding to different negative water potentials. 0, no water deficit (control conditions); 1, middle; 2, moderate; 3, severe drought conditions

Finally, monoterpenes tended to be relatively more regulated under severe drought condition (ψ_3) (Fig. 8.17). They essentially concerned compounds *1–5* and *7*, *8* (Fig. 8.17b) which corresponded, respectively, to tricyclene, α-thujene, α-pinene, camphene, sabinene, myrcene and α-phellandrene (Fig. 8.22a, c, d, f–i). These compounds showed continuously increases amounts with a strong slope between ψ_1, ψ_2 and ψ_3 (Fig. 8.23a–g). Other monoterpenes as *6* (β-pinene) (Fig. 8.22e), *9* (limonene), *10* ((Z-)-β-ocimene), *11* ((E-)-β-ocimene) and *12* (terpinolene) showed strong slope reduction between ψ_2 and ψ_3 indicating faster induction by moderate drought followed by cumulative but slower accumulation under severe conditions

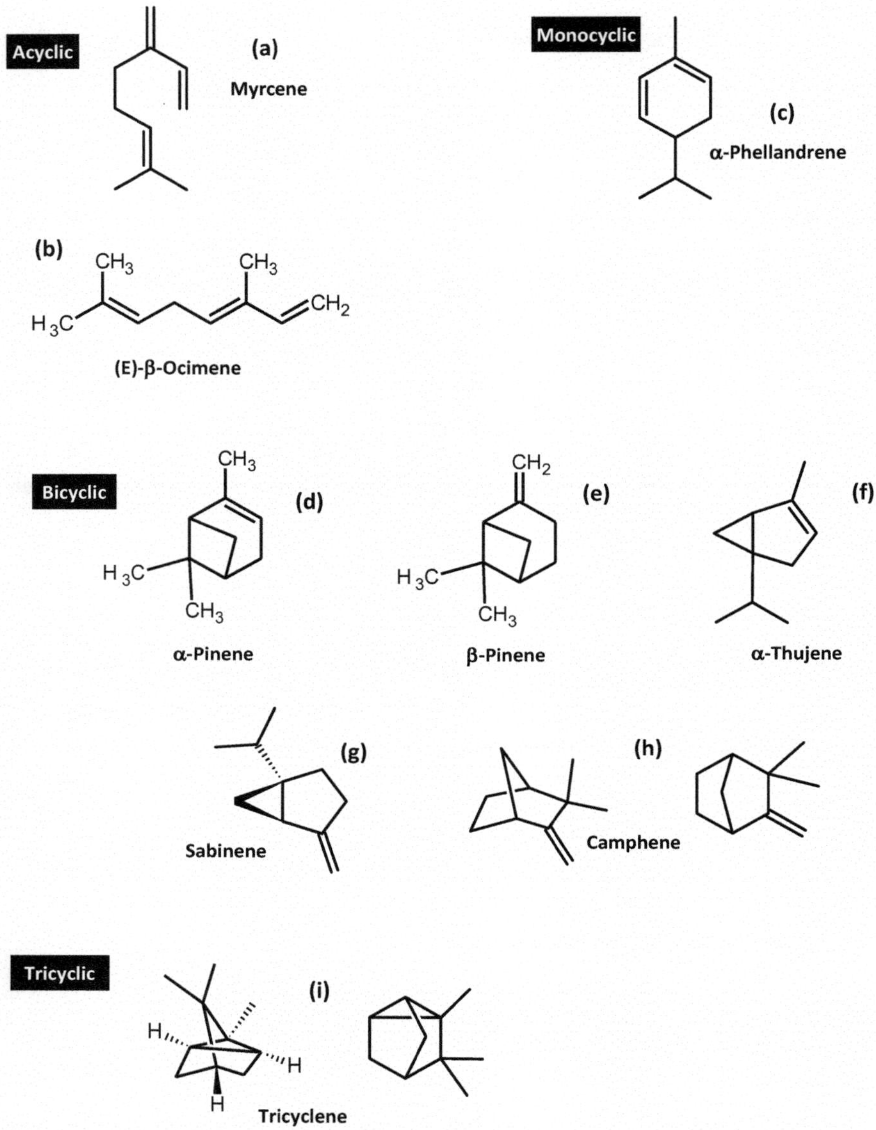

Fig. 8.22 Chemical structures of monoterpenes found in the needled of *Pinus sylvestris* and showing quantitative variation under drought conditions

(Figs. 8.23h, i and 8.24a–c). These monoterpenes projected in the subspace associated with sesquiterpenes which were revealed to be relatively more regulated under moderate drought (Fig. 8.17b).

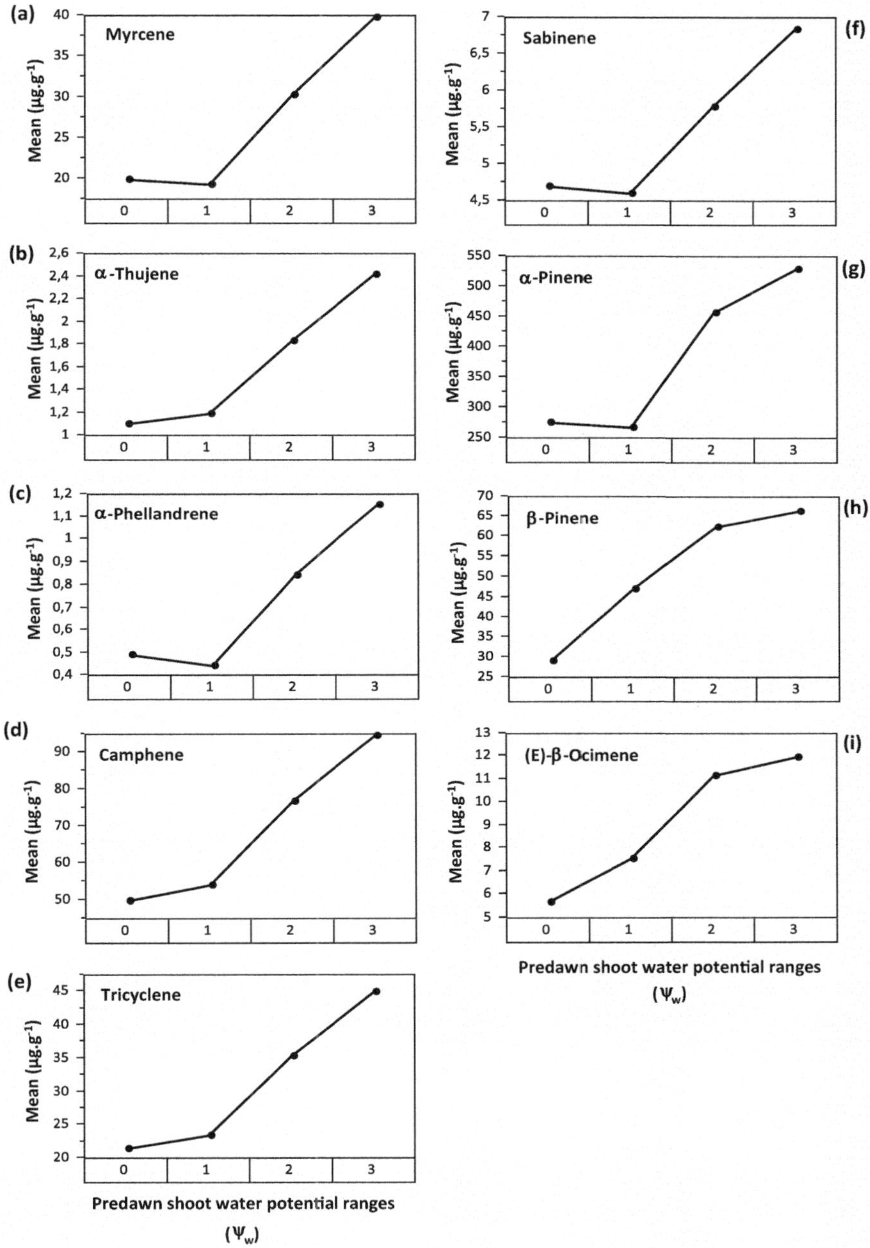

Fig. 8.23 Variations of average amounts (Means (μg.g^{-1} DW)) of monoterpenes in *Pinus sylvestris* in relation with different drought levels corresponding to different negative water potentials. 0, no water deficit (control conditions); 1, middle; 2, moderate; 3, severe drought conditions

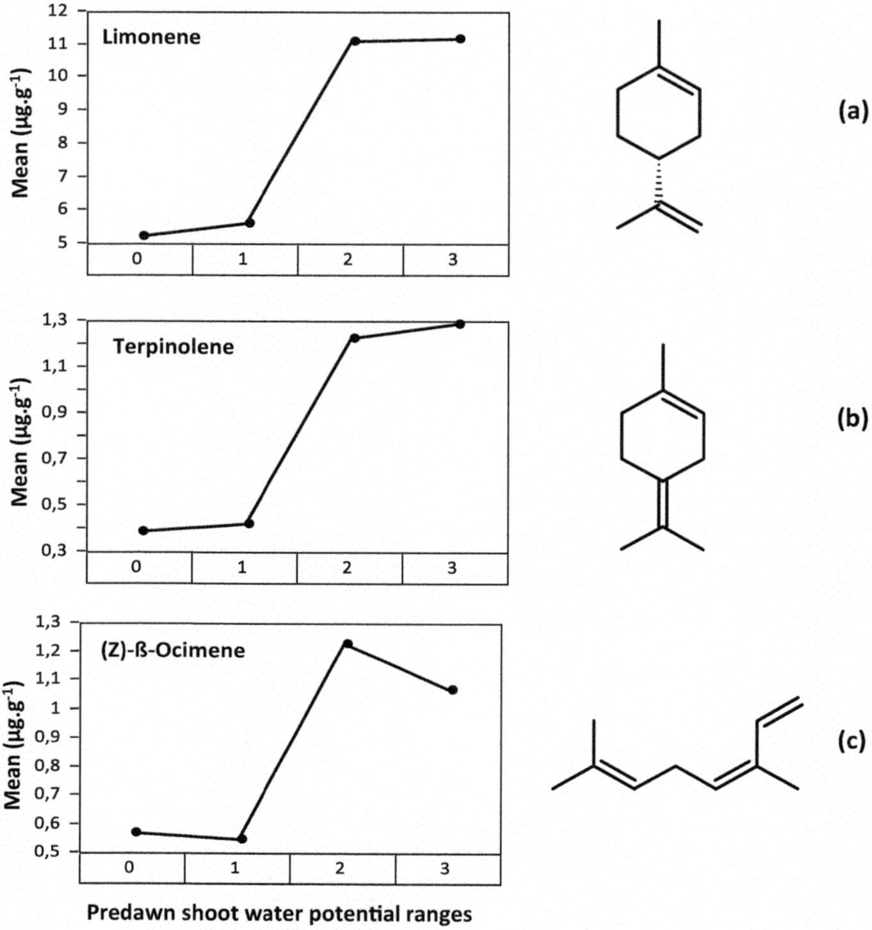

Fig. 8.24 Chemical structures and variations of average amounts (Mean ($\mu g.g^{-1}$ DW)) of monoterpenes in *Pinus sylvestris* in relation with different drought levels corresponding to different negative water potentials. 0, no water deficit (control conditions); 1, middle; 2, moderate; 3, severe drought conditions

References

Ábrahám E, Rigó G, Székely G, Nagy R, Koncz C, Szabados L (2003) Light-dependent induction of proline biosynthesis by abscisic acid and salt stress is inhibited by brassinosteroid in *Arabidopsis*. Plant Mol Biol 51(3):363–372

Anjum SA, Xie X, Wang L, Saleem MF, Man C, Lei W (2011) Morphological, physiological and biochemical responses of plants to drought stress. African J Agric Res 6:2026–2032

Anjum SA, Ashraf U, Tanveer M, Khan I, Hussain S, Shahzad B, Zohaib A, Abbas F, Saleem MF, Ali I et al (2017) Drought induced changes in growth, osmolyte accumulation and antioxidant metabolism of three maize hybrids. Front Plant Sci 8:69

Ashraf MA, Iqbal M, Rasheed R, Hussain I, Perveen S, Mahmood S (2018) Dynamic proline metabolism: importance and regulation in water-limited environments In: Parvaiz A, Ahanger MA, Singh VP, Tripathi DK, Alam P, Alyemeni MN (eds) Plant metabolites and regulation under environmental stress. Academic, London

Asrar A, Elhindi KM (2011) Alleviation of drought stress of marigold (*Tagetes erecta*) plants by using arbuscular mycorrhizal fungi. Saudi J Biol Sci 19:38–46

Bandurska H, Niedziela J, Pietrowska-Borek M, Nuc K, Chadzinikolau T, Radzikowska D (2017) Regulation of proline biosynthesis and resistance to drought stress in two barley (*Hordeum vulgare* L.) genotypes of different origins. Plant Physiol Biochem 118:427–437

Bano A, Ullah F, Nosheen A (2012) Role of abscisic acid and drought stress on the activities of antioxidant enzymes in wheat. Plant Soil Environ 58(4):181–185

Bota J, Flexas J, Medrano H (2004) Is photosynthesis limited by decreased Rubisco activity and RuBP content under progressive water stress? New Phytol 162:671–681

Boudsocq M, Sheen J (2010) Stress signaling II: calcium sensing and signaling. In: Pareek A, Sopory SK, Bohnert HJ, Govindjee (eds) Abiotic stress adaptation in plants: physiological, molecular and genomic foundation, pp 75–90

Brett C, Waldron K (1996) The physiology and biochemistry of the plant cell wall. Chapman and Hall, London

Cardoso AA, Brodribb TJ, Kane CN, DaMatta FM, McAdam SAM (2020a) Osmotic adjustment and hormonal regulation of stomatal responses to vapour pressure deficit in sunflower. AoB PLANTS 12:plaa025

Cardoso AA, Gori A, Da-Silva CJ, Brunetti C (2020b) Abscisic acid biosynthesis and signaling in plants: key targets to improve water use efficiency and drought tolerance. Appl Sci 10(6322):1–15

Cooper M, van Eeuwijk F, Chapman SC, Podlich DW, Löffler C (2006) Genotype-by environment interactions under water-limited conditions. In: Ribaut JM (ed) Drought adaptation in cereals. CRC Press, New York, pp 51–96

Cornic G, Massacci A (1996) Leaf photosynthesis under drought stress. In: Barker NR (ed) Photosynthesis and the environment. Kluwer Academic Publishers, The Netherlands

Davies WJ, Zhang JH (1991) Root Signals and the Regulation of Growth and Development of Plants in Drying Soil. Annu Rev Plant Physiol Plant Mol Biol 42:55–76

de Carvalho K, Kaphan M, de Campos F, Domingues DS, Filipe L, Pereira P, Gonzaga L, Vieira E (2013) The accumulation of endogenous proline induces changes in gene expression of several antioxidant enzymes in leaves of transgenic *Swingle citrumelo*. Mol Biol Rep 40(4):3269

Desclaux D, Roumet P (1996) Impact of drought stress on the phenology of two soybean cultivars. Field Crops Res 46(1–3):61–70

Dietz KJ, Sauter A, Wichert K, Messdaghi D, Hartung W (2000) Extracellular-glucosidase activity in barley involved in the hydrolysis of ABA glucose conjugate in leaves. J Exp Bot 51:937–944

Farahani A, Lebaschi H, Hussein M, Hussein SA, Reza VA, Jahanfar D (2013) Effects of arbuscular mycorrhizal fungi, different levels of phosphorus and drought stress on water use efficiency, relative water content and proline accumulation rate of Coriander (*Coriandrum sativum* L.). J Med Plants Res 2(6):125–131.

Farooq M, Basra SMA, Wahid A, Ahmad N, Saleem BA (2009a) Improving the drought tolerance in rice (*Oryza sativa* L.) by exogenous application of salicylic acid. J Agron Crop Sci 195:237–246

Farooq M, Wahid A, Kobayashi N, Fujita D, Basra SMA (2009b) Plant drought stress: effects, mechanisms and management. Agron Sustain Dev 29:185–212

Farooq M, Hussain M, Wahid A, Siddique KHM (2012) Drought stress in plants: an overview. In: Aroca R (ed) Plant responses to drought stress. Springer, Berlin, pp 1–33

Fukai S (1999) Phenology in rainfed lowland rice. Field Crops Res 64:51–60

Gai Z, Wang Y, Ding Y, Qian W, Qiu C, Xie H, Sun L, Jiang Z, Ma Q, Wang L, Ding Z (2020) Exogenous abscisic acid induces the lipid and favonoid metabolism of tea plants under drought stress. Sci Rep 10:12275

Geerts S, Raes D, Garcia M, Mendoza J, Huanca R (2008) Crop water use indicators to quantify the flexible phenology of quinoa (*Chenopodium quinoa* Willd.) in response to drought stress. Field Crops Res 108:150–156

Ghassemian M, Lutes J, Chang HS, Lange I, Chen W, Zhu T, Wang X, Lange BM (2008) Abscisic acid-induced modulation of metabolic and redox control pathways in *Arabidopsis thaliana*. Phytochemistry 69:2899–2911

Han Y, Watanabe S, Shimada H, Sakamoto A (2020) Dynamics of the leaf endoplasmic reticulum modulate-glucosidase-mediated stress-activated ABA production from its glucosyl ester. J Exp Bot 71:2058–2071

Harb A, Krishnan A, Ambavaram MR, Pereira A (2010) Molecular and physiological analysis of drought stress in *Arabidopsis* reveals early responses leading to acclimation in plant growth. Plant Physiol 154:1254–1271

Hauser F, Waadt R, Schroeder JI (2011) Evolution of abscisic acid synthesis and signaling mechanisms. Curr Biol 21:R346–R355

Hoekstra FA, Golovina EA, Buitink J (2001) Mechanisms of plant desiccation tolerance. Trends Plant Sci 6:431–438

Irigoyen JJ, Einerich DW, Sánchez-Díaz M (1992) Water stress induced changes in concentrations of proline and total soluble sugars in nodulated alfalfa (*Medicago sativa*) plants. Physiol Plant 84(1):55–60

Jiang M, Zhang J (2002) Water stress-induced abscisic acid accumulation triggers the increased generation of reactive oxygen species and up-regulates the activities of antioxidant enzymes in maize leaves. J Exp Bot 53(379):2401–2410

Kavi Kishor PB, Sreenivasulu N (2014) Is proline accumulation per se correlated with stress tolerance or is proline homeostasis a more critical issue? Plant Cell Environ 37(2):300–311

Kempa S, Krasensky J, Dal Santo S, Kopka J, Jonak C (2008) A central role of abscisic acid in stress-regulated carbohydrate metabolism. PLoS ONE 3:e3935

Lee KH, Piao HL, Kim H-Y, Choi SM, Jiang F, Hartung W, Hwang I, Kwak JM, Lee I-J, Hwang I (2006) Activation of glucosidase via stress-induced polymerization rapidly increases active pools of abscisic acid. Cell 126:1109–1120

Li S, Liu J, Liu H, Qiu R, Gao Y, Duan A (2021) Role of hydraulic signal and ABA in decrease of leaf stomatal and mesophyll conductance in soil drought-stressed tomato. Front Plant Sci 12:653186

Li W, Wang Y, Zhang Y, Wang R, Xie Z (2020) Impacts of drought stress on the morphology, physiology, and sugar content of Lanzhou lily (*Lilium davidii* var. *unicolor*). Acta Physiol Plant 42:127.

Liu F, Andersen MN, Jensen CR (2003) Loss of pod set caused by drought stress is associated with water status and ABA content of reproductive structures in soybean. Funct Plant Biol 30:271–280

Loreto F, Tricoli D, Di Marco G (1995) On the relationship between electron transport rate and photosynthesis in leaves of the C4 plant *Sorghum bicolor* exposed to water stress, temperature changes and carbon metabolism inhibition. Aust J Plant Physiol 22:885–892

Mansfield TJ, Atkinson CJ (1990) Stomatal behaviour in water stressed plants. In: Alscher RG, Cumming JR (eds) Stress responses in plants: adaptation and acclimation mechanisms. Wiley-Liss, New York, pp 241–264

McMaster GS, Wilhelm WW (2003) Phenological responses of wheat and barley to water and temperature: improving simulation models. J Agric Sci 141:129–147

Miranda JD, Armas C, Padilla FM, Pugnaire FI (2011) Climatic change and rainfall patterns: Effects on semi-arid plant communities of the Iberian Southeast. J Arid Environ 75:1302–1309

Misra V, Solomon S, Mall AK, Prajapati CP, Hashem A, Abd Allah EF, Ansari MI (2020) Morphological assessment of water stressed sugarcane: a comparison of waterlogged and drought affected crop. Saudi J Biol Sci 27:1228–1236

Mizokami Y, Noguchi KO, Kojima M, Sakakibara H, Terashima I (2015) Mesophyll conductance decreases in the wild type but not in an ABA deficient mutant (aba1) of *Nicotiana plumbaginifolia* under drought conditions. Plant Cell Environ 38:388–398

Nambara E, Marion-Poll A (2005) Abscisic acid biosynthesis and catabolism. Annu Rev Plant Biol 56:165–185

Natarajan SK, Zhu W, Liang X, Zhang L, Demers AJ, Zimmerman MC, Simpson MA, Becker DF (2012) Proline dehydrogenase is essential for proline protection against hydrogen peroxide-induced cell death. Free Radic Biol Med 53(5):1181–1191

Nezhadahmadi A, Prodhan ZH, Faruq G (2013) Drought tolerance in wheat. Sci World J. Article ID 610721, pp 1–12.

Nooden LD (1988) The phenomena of senescence and aging. In: Nooden LD, Leopald AC (eds) Senescence and aging in plants. Academic, USA, pp 1–50

Pamungkas SST, Suwarto S, Farid N (2022) Drought stress: responses and mechanism in plants. Rev Agric Sci 10:168–185

Pantin F, Monnet F, Jannaud D, Costa JM, Renaud J, Muller B, Simonneau T, Genty B (2013) The dual effect of abscisic acid on stomata. New Phytol 197:65–72

Patmi YS, Pitoyo A, Solichatun, Sutarno (2020) Effect of drought stress on morphological, anatomical, and physiological characteristics of Cempo Ireng Cultivar Mutant Rice (*Oryza sativa* l.) strain 51 irradiated by gamma-ray. J Phys Conf Ser 1436:012015.

Raghavendra AS, Gonugunta VK, Christmann A, Grill E (2010) ABA perception and signaling. Trends Plant Sci 15:395–401

Raymond MJ, Smirnoff N (2002) Proline metabolism and transport in maize seedlings at low water potential. Ann Bot 89(7):813–823

Saladin G (2014) From signal perception to gene expression: early responses of plant to drought. Agrobiologia 4:5–10

Sancho-Knapik D, Sanz MA, Peguero-Pina JJ, Niinemets Ü, Gil-Pelegrín E (2017) Changes of secondary metabolites in *Pinus sylvestris* L. needles under increasing soil water deficit. Ann Forest Sci 74(24):1–10

Schachtman DP, Goodger JQD (2008) Chemical root to shoot signaling under drought. Trends Plant Sci 13(6):281–287

Seiler C, Harshavardhan VT, Rajesh K, Reddy PS, Strickert M, Rolletschek H, Scholz U, Wobus U, Sreenivasulu N (2011) ABA biosynthesis and degradation contributing to ABA homeostasis during barley seed development under control and terminal drought-stress conditions. J Exp Bot 62:2615–2632

Semmar N (2013) Two computational simplex approaches to graphical highlighting metabolic phenotypes and their functional origins: correspondence analysis and weighted metabolic profiles analysis. In: Rolin D (ed) Metabolomics coming of age with its technological diversity. Advances in botanical research 67, Academic, Burlington, pp 441–492

Sekmen AH, Ozgur R, Uzilday B, Turkan I (2014) Reactive oxygen species scavenging capacities of cotton (*Gossypium hirsutum*) cultivars under combined drought and heat induced oxidative stress. Environ Exp Bot 99:141–149

Skirycz A, Inzé D (2010) More from less: plant growth under limited water. Curr Opin Biotechnol 21:197–203

Swamy GS (1999) Drought signaling in plants. Resonance 4:34–44

Szabados L, Savouré A (2010) Proline: a multifunctional amino acid. Trends Plant Sci 15(2):89–97

Thalmann M, Pazmino D, Seung D, Horrer D, Nigro A, Meier T, Kölling K, Pfeifhofer HW, Zeeman SC, Santelia D (2016) Regulation of leaf starch degradation by abscisic acid is important for osmotic stress tolerance in plants. Plant Cell 28:1860–1878

Trouverie J, Thévenot C, Rocher JP, Sotta B, Prioul JL (2003) The role of abscisic acid in the response of a specific vacuolar invertase to water stress in the adult maize leaf. J Exp Bot 54(390):2177–2186

Turyagyenda LF, Kizito EB, Ferguson M, Baguma Y, Agaba M, Harvey JJW, Osiru DSO (2013) Physiological and Molecular characterization of drought responses and identification of candidate tolerance genes in Cassava. AoB Plants 5:plt007

Tuteja N (2009) Integrated calcium signaling in plants. In: Baluška F, Mancuso S (eds) Signaling in plants, signaling and communication in plants. Springer, Berlin, pp 29–49

Urano K, Maruyama K, Ogata Y, Morishita Y, Takeda M, Sakurai N, Suzuki H, Saito K, Shibata D, Kobayashi M, Yamaguchi-Shinozaki K, Shinozaki K (2009) Characterization of the ABA-regulated global responses to dehydration in *Arabidopsis* by metabolomics. Plant J 57:1065–1078

Verslues PE, Sharma S (2010) Proline metabolism and its implications for plant-environment interaction. Arabidopsis Book, e0140.

Verslues PE, Agarwal M, Katiyar-Agarwal S, Zhu J, Zhu JK (2006) Methods and concepts in quantifying resistance to drought, salt and freezing, abiotic stresses that affect plant water status. Plant J 45:523–539

Wahid A (2007) Physiological implications of metabolite biosynthesis for net assimilation and heat-stress tolerance of sugarcane (*Saccharum officinarum*) sprouts. J Plant Res 120:219–228

Werner C, Correia O, Beyschlag W (1999) Two different strategies of Mediterranean macchia plants to avoid photoinhibitory damage by excessive radiation levels during summer drought. Acta Oecologica 20:15–23

Willick IR, Lahlali R, Vijayan P, Muir D, Karunakaran C, Tanino KK (2018) Wheat flag leaf epicuticular wax morphology and composition in response to moderate drought stress are revealed by SEM, FTIR ATR and synchrotron X-ray spectroscopy. Physiol Plant 162:316–332

Yaish M (2015) Proline accumulation is a general response to abiotic stress in the date palm tree (*Phoenix dactylifera* L.). Genet Mol Res 14(3):9943–9950.

Yang CW, Wang JW, Kao CH (2000) The relationship between accumulation of abscisic acid and proline in detached rice leaves. Biol Plant 43(2):301–304

Yang X, Lu M, Wang Y, Wang Y, Liu Z, Chen S (2021) Response Mechanism of plants to drought stress. Horticulturae 7:1–50

Yao C, Finlayson SA (2015) Abscisic acid is a general negative regulator of *Arabidopsis* axillary bud growth. Plant Physiol 169:611–626

Zhang C, Wang N, Zhang Y, Feng Q, Yang C, Liu B (2013) DNA methylation involved in proline accumulation in response to osmotic stress in rice (*Oryza sativa*). Genet Mol Res 12(2):1269–1277

Chapter 9
Light Stress

9.1 Light Properties and Effects on Plants

Light is electromagnetic radiation distributed into one visible and two invisible wavelength ranges. Visible wavelength range has a median location extending from 380 to 780 nm, whereas invisible wavelengths are concerned with upper and lower sides including (i) UV (lower side) varying between 100 and 380 nm and (ii) near-infrared and infrared (upper side) varying in the ranges 780–2500 nm and > 2500 nm, respectively (Fig. 9.1).

Wavelengths contain higher energy as they are shorter. Thus, light energy is higher for UV, lower for infra-red and intermediate for visible wavelength. Also, in visible range, energy is relatively higher for blue light, lower for red and intermediate for green, yellow and orange. Visible light has high biological importance because it coincides with the photosynthetically active radiation (PAR) range (400–700 nm) (Fig. 9.1). Solar radiations include visible light, UV and infrared covering the wavelength range 280–2800 nm in which they represent 43%, 4% and 53%, respectively. Solar visible light is needed for plant growth; infrared radiations represent crucial heat energy.

Different wavebands are characterized by specific importance and functions (Fig. 9.1):

- Ultraviolet (UV) radiations range between 100 and 380 nm and consist of three types, UV-A, UV-B and UV-C (Fig. 9.1a).
  - Due to their shorter wavelengths (100–280 nm), UV-C are excessively energetic and cause cell death. They have usefulness in disinfection (e.g. health fields).
  - UV-B (280–320 nm) and UV-A (320–380 nm) cause photomorphogenesis in plants including sun burn and stem elongation. Also, they stimulate

N. Semmar, *Secondary Metabolites in Plant Stress Adaptation*, Signaling and Communication in Plants, https://doi.org/10.1007/978-3-031-52595-7_9

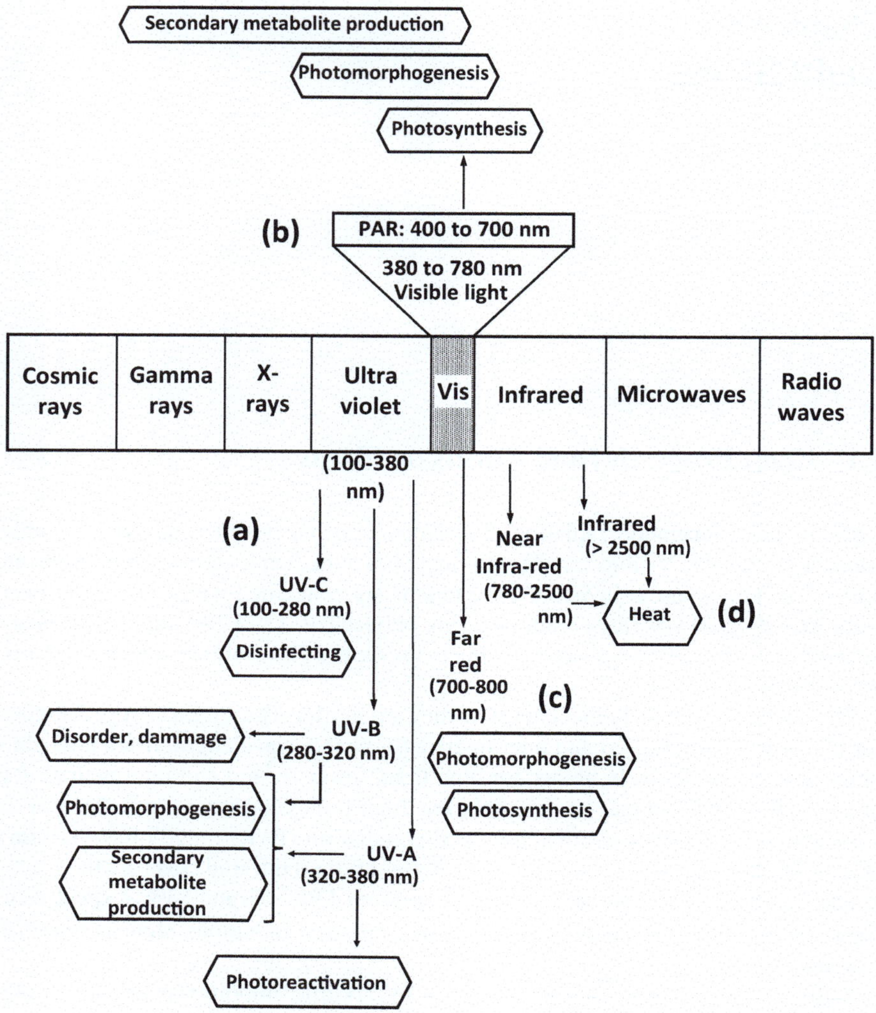

Fig. 9.1 Classification of radiations into different wavelength ranges and their biological effects in plants

productions of secondary metabolites (SMs) including pigments and antioxidant compounds. Due to their shorter wavelengths, UV-B can cause disorders and damages to plants.

– However, UV-A (320–380 nm) are less harmful and can serve as photoreactivation energy source.

- Visible (Vis) radiations range between 380 and 780 nm (Fig. 9.1b); they include photosynthetically active radiation (PAR) range (400–700 nm) and a main part of far red (700–800 nm). PAR is important for photosynthesis and plant growth. Vis-radiations are responsible for several photomorphogenesis processes in plants including seed germination, seedling de-etiolation, shade avoidance response, hypocotyl and root phototropism, circadian oscillation and reproduction response. Moreover, they induce production of secondary metabolites.

- Far red radiations (700–800 nm) are located between the extreme red end of visible region and infrared light (Fig. 9.1c). They contribute to photomorphogenesis processes including seed germination, seedling de-etiolation, shade avoidance response and reproduction response. Also, they are involved in photosynthesis through their excitation of photosystem I.
- Near-infrared (780–2500 nm) and infrared (> 2500 nm) provide heat sources (Fig. 9.1d).

Light represents a multidimensional abiotic factor because its variation involves three aspects (at least) consisting of wavelength (qualitative), intensity and flux ratios (quantitative), and exposure duration or photoperiod (temporal) (Fig. 9.2a–c). Experiences on the effects of light on plant responses tend to simulate complex natural plant-light systems by combining these three aspects with other intrinsic and extrinsic factors (Fig. 9.2). Understanding the effects of light on plants and plant responses to photo-stress requires sequential experiments by (i) varying single wavelengths with their intensities and durations (continuous or by photoperiods). Beyond single radiation, experiments are extended to two (or more) radiations at different supplemented ratios. Experiments can be applied on different (close) species or different cultivars of a same species in order to highlight plant taxon-dependent responses to well-defined photo-stresses (Fig. 9.2d). In more integrative experiments, interactive abiotic factors (e.g. dryness, CO_2) are considered with light variable in studied systems (Fig. 9.2e).

Extension of experiments concerning light effects on plants requires consideration of (i) different development phases of plants and/or (ii) different sampling periods representing well-distinct environment conditions (Fig. 9.2f). Further experiments can be applied in the framework of optimization of secondary metabolite productions by plants under light stress. Among optimization control factors, different radiation types are successively applied according to different temporal sequences through which the best plant responses are subject of researches (Fig. 9.2g).

Quantitatively, received light by plants can be measured by reference to photosynthetic system and by considering different time scales. Apart from the instantaneous light intensity, light intensity of photosynthetic photon flux density (PPFD) refers to the amount of useful light reaching the plant. Also, PPFD represents the number of photosynthetically active photons reaching a given surface per second (expressed in μmol $m^{-2}s^{-1}$). Moreover, daily light integral (DLI) represents the amount of photosynthetic photons received by unit surface (m^2) during one day period; DLI is expressed in mol (light) $m^{-2}d^{-1}$. In plan factory with artificial light (PFAL) where PPFD is kept constant, DLI can be easily calculated from PPFD and photoperiod.

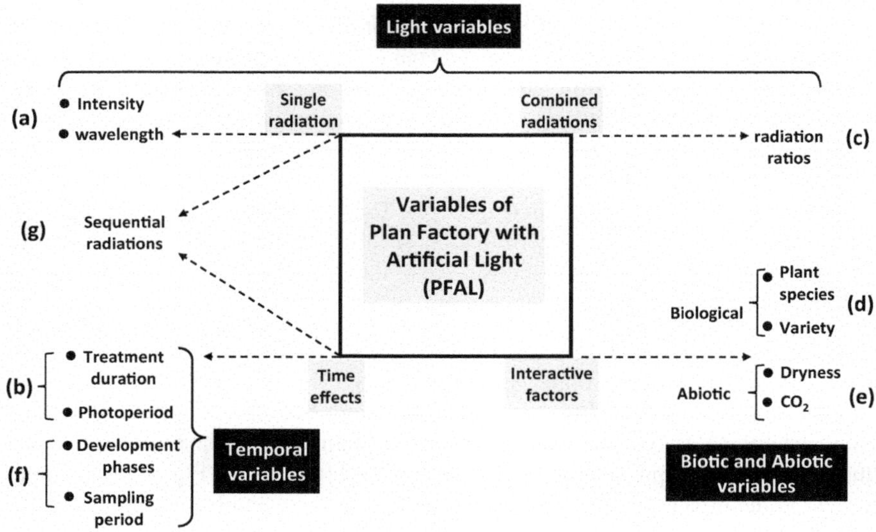

Fig. 9.2 Overview on multifactorial aspect of plan factory with artificial light (PFAL) applied for studying plant responses to radiations and interactions with different factors

Basically, light represents both (i) energy and (ii) information source for plants:

(i) In the first case, plants capture light energy (photons) for photosynthesis and heating. A small part (10%) of captured photons is invested in chemical energy (carbohydrate) production through photosynthesis. The major part of light energy or photons captured by plants is used for heat energy production.

(ii) Apart from energy-energy conversion, light is involved into many signal inductions associated with different types of information that plants perceive to regulate various growth and development processes including photomorphogenesis and photoperiodism. Light signal perceptions in plants are made by photoreceptors which function as light sensors (Fig. 9.3). Subtle changes in light composition of environment are captured by photoreceptors through which plants are informed on the light composition of their growth environment helping them to manifest adaptive physiological responses and morphological behaviors independently from the photosynthetic system.

9.2 Plant Responses to Light Stress

Plants manage received light to be used as energy and signal sources. They use light energy for photosynthesis and heat production on one hand and as signal source on other hand (Fig. 9.3 a). Light is perceived by photoreceptors to induce several metabolic and physiological processes associated with photosynthetic and nonphotosynthetic systems. This implies two types of photoreceptors including photosynthetic

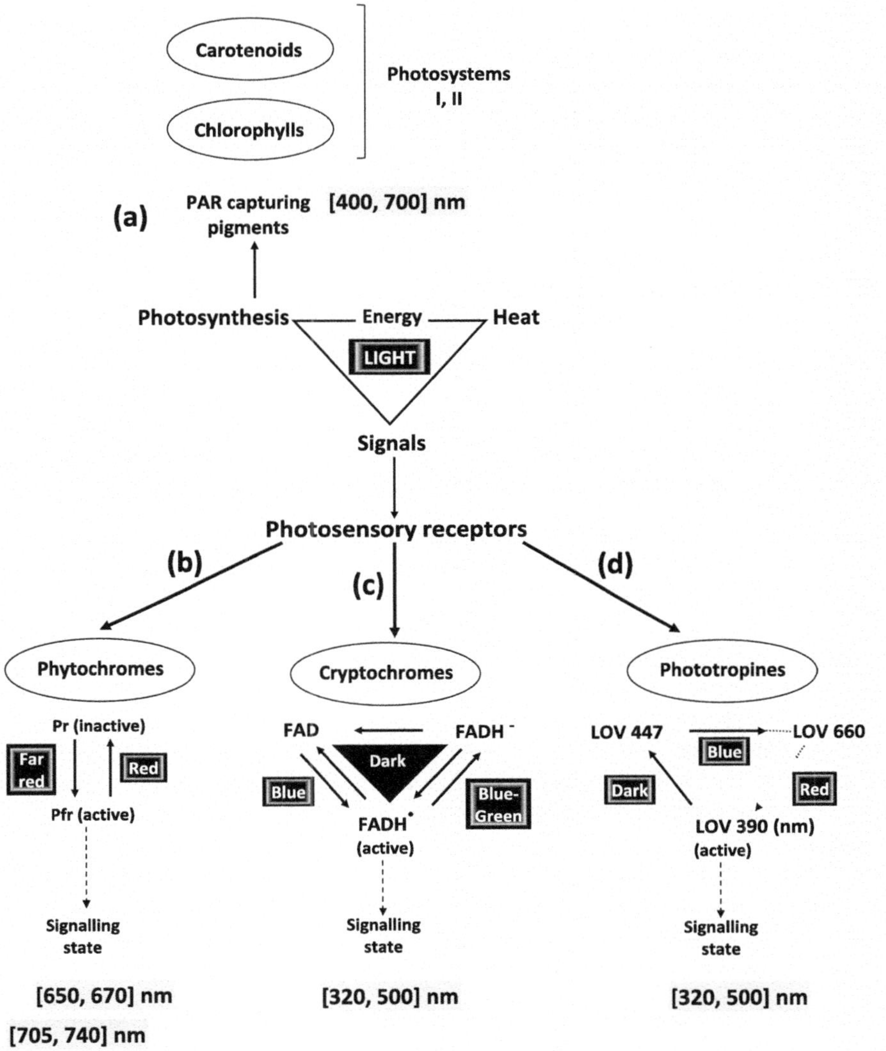

Fig. 9.3 Different types of plant photoreceptors involved in (**a**) photosynthetic and (**b**) nonphotosynthetic processes. LOV, Light-Oxygen-Voltage domain-proteins

pigments (chlorophylls, carotenoids) capturing PAR for photosynthesis (Fig. 9.3 a), and photosensory receptors mediating plant responses to nonphotosynthetic light (Fig. 9.3 b-d). In nonphotosynthetic field, different lights exert different effects on plant behaviours: blue and red lights generally enable plants to orientate appropriately within a canopy, whereas far-red and green-enriched light provide indication on overgrowing vegetation helping for shade avoidance (Casal 2013; Liscum et al.

2014; Wang et al. 2015; Zhang and Folta 2012; Battle and Joens 2020). By combining these light-induced responses, plants tend to optimize their light-gathering capacity.

Photosensory receptors steer the captured light to endogenous signals which induce gene expression and transcription factors associated with growth processes (cell division, enlargement) of different plant tissues (hypocotyl, floral buds, leaf primordia, etc.) (Anpo et al. 2018; Casal 2012).

9.2.1 Photosensory Receptors

Photosensory receptors are classified into five systems including phytochromes, cryptochromes, phototropins, Zeitlupe family members (e.g. F-box proteins), and UV resistance locus 8 (UVR8) (Figs. 9.3 and 9.4) (Franklin and Quail 2010; Rizzini et al. 2011; Kutschera and Briggs 2013). These photoreceptors show different absorption wavelength ranges and are involved in different plant regulation systems:

- Phytochromes are sensitive to red (650–670 nm) and far-red (705–740 nm) (Fig. 9.3 b and 9.4 a); they are involved in regulation of germination, flowering time, shade avoidance response, and de-etiolation.
- Cryptochromes are sensitive to blue/UV-A light (320–500 nm) (Fig. 9.3 c); they are involved in regulation of plant height, flowering time and photomorphogenesis.
- Phototropines are sensitive to blue/UV-A light (320–500 nm) (Fig. 9.3 d); they play key roles in chlorophyll production. They perceive blue light phototropism and are involved in photomovement by controlling the opening of pores (stomas) for water and carbon dioxide exchanges, and regulate the movements of chloroplasts (chloroplast relocation) according to the light intensity (Dou and Niu 2020).
- Zeitlup protein members include F-Box which are sensitive to blue/UV-A light (320–500 nm). Flowering time can be modulated by some of these receptors.
- UVR8 are sensitive to UV-B light (280–320 nm) and play roles in plant defense against these high energy radiations.

Understanding how these photoreceptors are involved in plant response regulations to different light environments could provide deep knowledge on key upstream processes influencing physiological, metabolic and morphological plant systems, and helping for selection of lighting solution in PFALs (Plan Factory with Artificial Light) (Anpo et al. 2018). Three photoreceptor types will be presented in the following sections including phytochromes, cryptochromes and photoropines.

9.2.1.1 Phytochromes

Structurally, phytochromes (*Phy*) consist of linear tetrapyrrolic chromophores (Fig. 9.4a). They are sensitive to red (R, 600–699 nm) and far-red (FR, 700–800 nm)

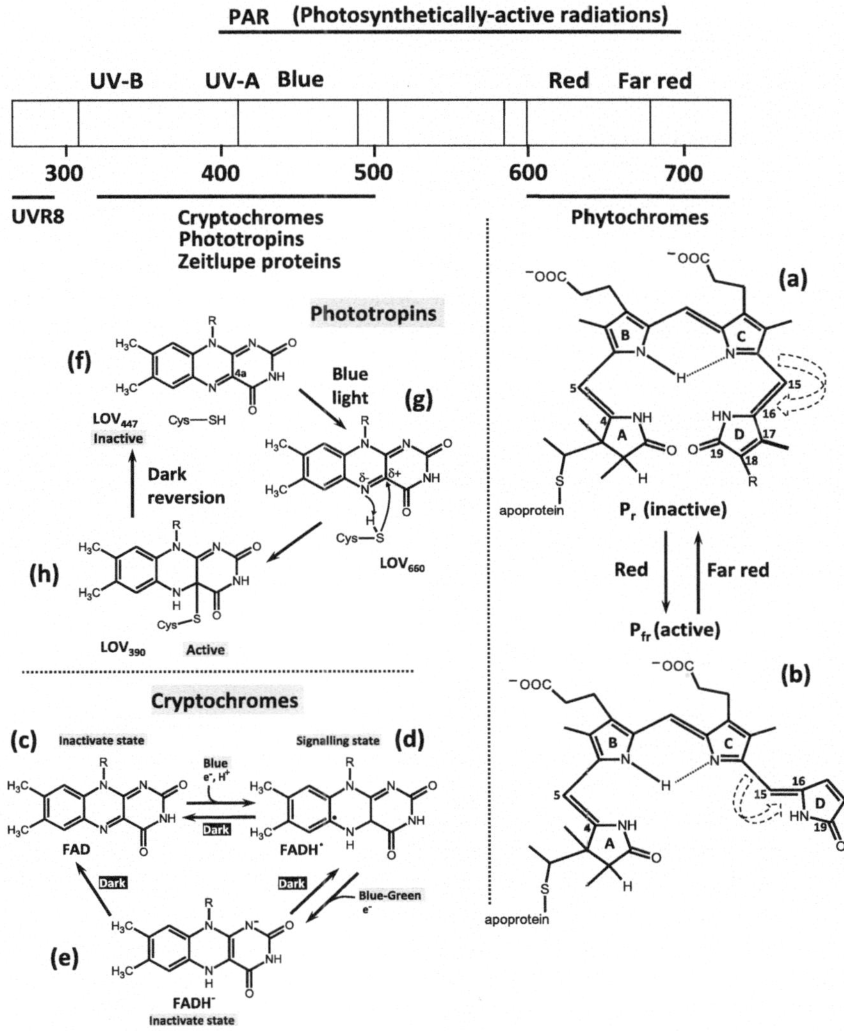

Fig. 9.4 Photocycles and chemical structures of chromophores associated with different plant photoreceptors concerning non-photosynthetic light

due to two reversible isomeric conformations (Pr and Pfr) with absorption peaks at 660 nm and 730 nm, respectively (Fig. 9.3 b and Fig. 9.4 a, b).

Phy synthesis occurs in the Pr form (Fig. 9.4 a). Red light absorption triggers isomerization in the C15-C16 double bond leading to rotation of pyrrolic D-ring (Fig. 9.4b). This rotation occurs with a rearrangement of apoprotein backbone resulting in photoconversion of phytochromes from Pr (inactive) to Pfr (active) (Bae and Choi 2008). The inactive form (Pr) is initially located in the cytoplasm, then biologically active form (Pfr) acts by migrating from cytosol to the nucleus (Van Buskirk et al. 2012; Klose et al. 2015). Therefore, light response is mediated by interaction of Pfr with several regulatory components in the nucleus leading to downstream control of gene expression (Pham et al. 2018). Under saturating light, Pr and Pfr levels reach a photoequilibrium that can be characterized by the proportion of Pfr which depends on (R:FR)-ratio resulting from spectral distribution of light source and *Phy* absorption (Sager et al. 1988; Holmes and Smith 1977).

Back conversion ($P_{fr} \rightarrow P_r$) is favored either by dark or FR absorption (Fig. 9.3 b and 9.4 a, b). The first way is slow and consists of a non-photoinduced reaction (dark reversion). The second way is much faster occurring upon absorption of FR light resulting in P_{fr} excitation and its subsequent relaxation to the P_r form (Mancinelli 1994; Quail 1997; Fankhauser 2001). This reversibility between two configuration forms allows phytochrome to function as a (R:FR)-dependent developmental switch.

Different members of phytochrome family (*PhyA* to *PhyE*) are involved in differential controls of physiological outputs ranging from seed germination to flowering time. Different phytochrome members are excited by different wavelengths and R:FR ratios to exert control of seedling and different vegetative development processes including: seed germination (*PhyA, B, E*), seedling de-etiolation (*PhyA, B, C*), flowering promoting (*Phy A*), shade avoidance (*PhyB, D, E*).

9.2.1.2 Cryptochromes

Cryptochromes (*Crys*) have the structure of photolyases by containing flavin adenine dinucleotide (FAD) (Fig. 9.4c) which is subject to reduction under light excitation leading to the active form of cryptochrome (Fig. 9.4d). Photolyase family includes also DNA repair enzymes catalyzing light-dependent electron transfer to UV-damaged DNA from reduced flavin (FADH-) (Fig. 9.4 e) (Sancar 2003).

Cryptochromes are photoreceptors perceiving blue light in the range of 400–499 nm. Under light, they are subject to a photocycle in which they are converted into different redox states including FAD, $FADH^{\bullet}$ and $FADH^{-}$ (Fig. 9.3 c and 9.4c–e). In dark, *Crys* are in their oxidized redox state corresponding to FAD (inactive form). This molecular form shows a folding of C-terminal domain against the protein; in *Arabidopsis*, it was shown that the C-terminal extension of *Crys* plays key role in protein–protein interaction domain (Lin and Shalitin 2003). Blue light provokes a photoreduction to flavin by electron transfer leading to the $FADH^{\bullet}$ redox form (Fig. 9.4c,d) (Bouly et al. 2007). Moreover, this photoreduction results in a conformation change with unfolded C-terminal domain and makes the activated form ($FADH^{\bullet}$) of the receptor to be ready for signaling partner binding. Further reduction can occur from illumination of the flavin semiquinone ($FADH^{\bullet}$) by an additional photon of either blue or green light (< 600 nm) leading to $FADH^{-}$ representing inactive redox form (Fig. 9.4e). The photocycle is achieved by reoxidation to FAD state (Fig. 9.4e → c) which is spontaneously favored after the return to darkness.

Thus, cryptochrome photocycle induced by blue-illumination generates a continuous cycling between flavin forms showing different redox states responding to darkness adaptation (FAD) and light induction (FADH•) conditions to correlate with downstream biological activities (Fig. 9.3 c). By involving blue light, green light and darkness with different redox states, cryptochrome photocycle provides an efficient perception system that may help plants for sensing the quality of the light environment. *Crys* show increased activity with blue light levels making them to serve as sensors of variable irradiance levels influenced by the degree of shade. Variation of blue/green ratio is perceived by *Crys* resulting in shade avoidance or adaptive responses (Sellaro et al. 2010; Battle and Jones 2020).

Cryptochromes includes three members (*Cry1, 2, 3*) which were identified in *Arabidopsis*, and showing different functions and cell localizations: in *Arabidopsis*, *Cry1* and *Cry2* were localized in nucleus whereas *Cry3* was found in mitochondria and chloroplast (Cashmore et al. 1999; Kleiner et al. 1999).

Cry1 and *Cry2* perform overlapping functions by controlling various aspects of plant growth and development including seedling de-etiolation, elongation growth, flowering initiation and entrainment of circadian clock. The two members show regulation functions of plant growth and development under different light conditions: *Cry1* tends to control seedling de-etiolation regulation under blue light whereas *Cry2* provide flowering control in response to day-length. Unlike *Cry1*, *Cry2* is subject to blue light-dependent ubiquitination and degradation in the nucleus making it preferentially functions under low or limiting light conditions.

Less common functions of *Crys* include activation of transcription activators and modulation of chromatin remodeling (Lin and Shalitin 2003; Valverde et al. 2004; Kumar and Wigge 2010). Also, although phototropins are the main photoreceptors controlling stomate opening, *Crys* were revealed to stimulate this process too.

Apart from light responses, *Crys* were revealed to be implied in regulation processes of chloroplast development and anthocyanin accumulation in plants (Yu et al. 2010). Overexpression of *Cry1* photoreceptor was followed by high accumulation levels of anthocyanins, suggesting a stimulation of anthocyanin accumulation by *Cry1* (Li and Yang 2007).

Cry3 was reported to repair UV-induced damage in single-stranded DNA and in loop structures of double-stranded DNA (Selby and Sancar 2006; Pokorny et al. 2008).

9.2.1.3 Phototropins

Phototropins (Phot) are blue-light receptors consisting of a flavin mononucleotide (FMN) chromophore linked to a protein belonging to the so-called Light-Oxygen-Voltage (LOV) domains. LOV domains constitute a family of ubiquitous protein sensors regulated by external signals including light (L), oxygen (O) or voltage (V), hence the acronym LOV. The structure of LOV domain core is highly conserved in biological world (higher plants, microalgae, fungi, bacteria) on the basis of five-stranded antiparallel β-sheet with ancillary α-helices (α/β scaffold) (Crosson and Moffat 2001; Möglich et al. 2009).

In dark conditions, FMN chromophore is non-covalently bound to the α/β-scaffold resulting in blue-absorbing LOV system with maximum absorption peak at 447 nm (LOV_{447}) (Fig. 9.4f) (Christie et al. 1999, Christie 2007, Celaya and Liscum 2005, Crossan and Moffat 2001, Swartz et al. 2001). Blue radiation drives the apparition of a highly reactive triplet-state flavin that absorbs maximally in the red (660 nm) (Fig. 9.4g). Protonation of this triplet results in the increase of electrophilicity of the C(4a) atom; this promotes nucleophilic attack of the thiol anion at this carbon leading to the formation of the FMN-cysteinyl adduct within microsecond order (Fig. 9.4h) (Christie 2007, Crossan and Moffat 2001, Swartz et al. 2001, Salomon et al. 2000, Swartz et al. 2001). The formed FMN-cysteinyl adduct is characterized by light absorption peak at 390 nm (LOV_{390}) and a reversible covalent bond between the C(4a) carbon of the FMN chromophore and the sulfhydryl side chain of a conserved cysteine residue within the LOV domain (Salomon et al. 2000; Swartz et al. 2001). LOV_{390} is thermodynamically reversible in darkness returning to LOV_{447} state in a time range of seconds to minutes (Salomon et al. 2000; Kasahara et al. 2002). This dark reversion makes the photocycle of phototropin to be achieved. LOV_{390} represents the active signaling state of phototropins (Crosson et al. 2003; Crosson and Moffat 2002; Salomon et al. 2001). By changing cysteine with either serine or alanine, the light-dependent formation of LOV_{390} was suppressed (Swartz et al. 2001; Kottke et al. 2003; Kasahara et al. 2002).

Phototropins play key role in mediating plant phototropism due to correspondence between their absorption properties and the action spectra of phototropism (Christie et al. 1999). In response to blue irradiation, phototropins undergo autophosphorylation.

Phototropins in *Arabidopsis* showed two members (Phot1 and Phot2) which function by overlapping ways to regulate several growth-linked movements including hypocotyl and root phototropism (Sakai et al. 2001), chloroplast accumulation (Kagawa et al. 2001), stomatal opening (Kinoshita et al. 2001), leaf positioning and leaf flattening (Sakamoto and Briggs 2002; Inoue et al. 2008). Under high light, hypocotyl phototropism is particularly regulated by Phot2 (Sakai et al. 2001).

9.2.2 Secondary Metabolite Responses of Plants to Light Stress

Light represents a multidimensional factor inducing plant responses via several combined effects including wavelength, intensity and duration or photoperiod. Phytochemical responses of plant to radiation stress were studied by exposing plants to controlled experiments showing basal conditions (stable for the serials of experiments) and variable factors providing differentiations between experiments (Fig. 9.2). Productions of secondary metabolites (SMs) provide highly variable response profiles of plants resulting from different combined light factors by specific ways to plant species and cultivars. Moreover, phytochemical responses to light stress can be modulated by interactive co-factors such as CO_2.

Three study cases are presented in the next sections to illustrate the variability of SM productions as light-induced responses in plants. Light treatments consisted of supplements of blue-red light or UV-A applied at variables ratios or intensities on same or different cultivated herbaceous species (Vastakaite et al. 2015; Brazaitytė et al. 2015; Dlugozs-Grochowska et al. 2017).

9.2.2.1 Effects of Blue Light Ratio on Phytochemical Profiles and Growth of Microgreen Species/Cultivars

Study case 1: microgreen species/cultivars. Microgreen plants of *Brassica rapa* (Brassicaceae) and *Ocimum basilicium* (basil, Lamiaceae) were exposed to different proportions of red and blue lights by reference to an applied constant PPFD = 302.5 $\mu mol.m^{-2}.s^{-1}$ (Vastakaite et al. 2015). *B. rapa* was represented by two varieties, *chinensis* (or pak choi) and *rosularis* (or rastoi); *O. basilicum* was concerned with a single variety, Sweet Genovese.

The plants were cultivated in growth chamber under stable relative humidity (55 ± 5%) with a day/night photoperiod of 16 h/8 h and 21 °C/17 °C. The total PPFD (302.5 $\mu mol.m^{-2}.s^{-1}$) was distributed between four radiations including red (665 nm, 170 $\mu mol.m^{-2}.s^{-1}$), far red (731 nm, 2.5 $\mu mol.m^{-2}.s^{-1}$) and two variable flux radiations: red (638 nm) and blue (447 nm) sharing the remaining flux (130 $\mu mol.m^{-2}.s^{-1}$) (Fig. 9.5a). Five variable fluxes of these two radiations were applied leading to five light conditions associated with (blue, red) percentages: (43%, 0), (35%, 8%), (26%, 16%), (18%, 25%), (10%, 33%). The plants were harvested after 8 and 18 days of light treatment for *B. rapa* and *O. basilicum*, respectively.

The plants were characterized by their accumulation levels of different phytocompounds including total phenols (*TotPh*), anthocyanins (*Anth*), ascorbic acid (*AsAc*) (amounts, mg/g Fresh Weight) and flavonols (*Fols*) (flavonol index, positively correlated with the content) (Fig. 9.5b–e). Also, antioxidant activity was measured by the DPPH (2,2-diphenyl-1-picrylhydrazyl) radical scavenging activity (μmol/g FW) (Fig. 9.5f–h). Moreover, growth was followed via plant height (*H*) and hypocotyle length (*L*) (cm) (Fig. 9.5i–k).

Ocimum basilicum showed higher *TotPh* production range (2–5 mg/g FW) than *Brassica rapa* (both varieties) (0.5–1.1 mg/g FW) (Fig. 9.5b–d). This highlighted a strong species-dependent levels of *TotPh*. Apart from the quantitative aspect, *TotPh* curves had different shapes providing additional differential key between plants: *TotPh* accumulation increased with the proportion of blue light in *B. rapa* (var. tastoi) and in *O. basilicum* (Fig. 9.5c, d). However, in *B. rapa* (var. pak choi), *TotPh* did not increase continuously, but they showed accumulation peak at 8% followed by a decrease when blue ratio exceeded 8% (Fig. 9.5b). These results showed different light-induced responses in different varieties of a same plant species (*B.*

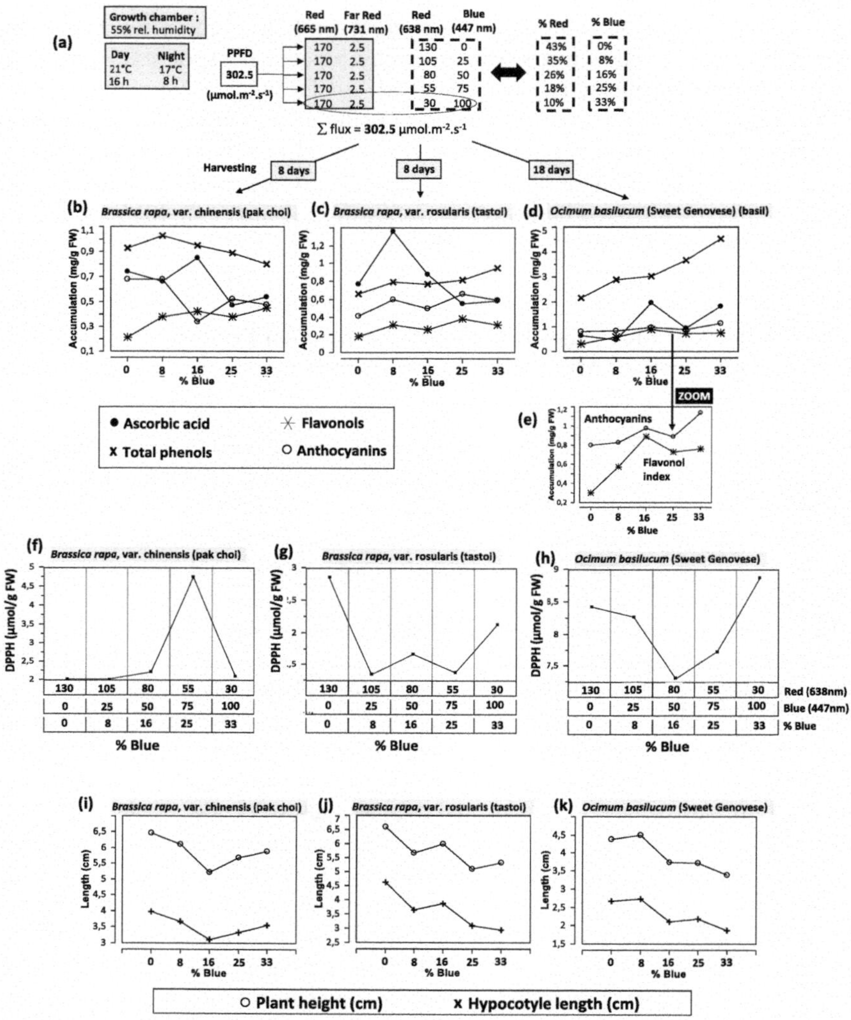

Fig. 9.5 Experimental design for studying irradiance effects of blue-red light ratios (**a**) on antioxidant accumulation (**b–e**), antioxidant activity (**f–h**) and growth level (**i–k**) of *Brassica rapa* (var. pak choi, tastoi) and *Ocimum basilicum* (var Sweet Genovese)

rapa). However, *B. rapa* var. tastoi and *O. basilicum* showed similar trends of *TotPh* (Fig. 9.5c, d); such a sharing of variation curve shapes was confirmed for flavonols (*Fols*) and anthocyanins (*Anth*):

Variation curves of *Fols* and *Anth* showed accumulation peaks that occurred at 25% blue in *B. rapa* (tastoi) and 16% in *O. basilicum* (Fig. 9.5c, d). In basil, accumulation peak associated with 16% blue showed global aspect for flavonols and local aspect for anthocyanins (Fig. 9.5e). The highest anthocyanin content occurred at 33% blue in basil. However, *B. rapa* (var. pak choi) showed a decreasing trend of *Anth* with a marked hole at 16% blue (Fig. 9.5b).

Ascorbic acid (*AsAc*) showed a strong peak at different blue ratios depending on the plant: In *B. rapa* (tastoi), accumulation peak was observed at 8% of blue (Fig. 9.5c). In *B. rapa* (pak choi) and *O. basilicum*, *AsAc*-peak occurred at 16% of blue (Fig. 9.5b, d). This revealed differential phytochemical responses between two cultivars of a same species (*B rapa*).

Beyond the shared *AsAc* peak at 16% in *B. rapa* (pak choi) and basil, differentiation was provided by the global variation trend of accumulation curve (i.e. production way): these curves globally showed increasing and decreasing trends in basil and pak choi, respectively (Fig. 9.5b, c). Also, *B. rapa* (pak choi) showed opposite variations between different metabolic families with increase of blue light ratios (Fig. 9.5b): *Fols* and *TotPh* showed positive and negative trends with the increase of blue ratio, respectively. Moreover, *AsAc* and *Anth* showed maximum and minimum levels at 16% of blue. This highlighted functional negative correlations between productions (accumulations) of different antioxidant compounds in *B. rapa* (pak choi): *TotPh* versus *Fols* and *AsAc* versus *Anth*. However, in *O. basilicum*, the different analyzed antioxidant compounds showed positive trends governed by the increase of blue light ratio (Fig. 9.5d, e). In *B. rapa*, *TotPh* showed distinct variation ranges and opposite trends between the two studied cultivars (Fig. 9.5b, c). These results highlighted the usefulness of considering several phytocompounds for quantitative and qualitative differentiations of light-plant responses at low taxonomical levels.

Individual plant taxa were finally characterized by some production peaks or maxima (Max) of different antioxidants associated with different blue ratios (Table 9.1).

Flavonols were induced at higher blue (B) ratios (25.33% B) in *B. rapa* (both cultivars) than *O. basilicum* (16% B). In *B. rapa* (pak choi), the maximal levels of different antioxidants were associated with more variable blue ratios indicating more diversified production (induction) trends. At specific scale, the maximal levels of antioxidants were induced by different percentages of blue in the two cultivars of *B. rapa*: antioxidant productions in *B. rapa* (pak choi) were more induced at

Table 9.1 Percentages of blue light giving maximal production of different antioxidants in *Brassica rapa* (var. pak choi and tastoi) and *Ocimum basilicum* (var Sweet Genovese). *TotPh*, Total Phenols; *Fols*, Flavonols; *Anth*, Anthocyanins; *AsAc*, Ascorbic Acid

%Blue (B) →	*TotPh*	*Fols*	*Anth*	*AsAc*
B.rapa (pak thoi)	8% B	33% B	0% B	16% B
B.rapa (tatsoi)	33%	25%	25%	8%
O.basilicium	33%	16%	33%	16%

lower and middle blue ratio (0–16% B) except *Fols* (33% B); however, in *B. rapa* (tastoi), antioxidant productions were induced at higher %B (25 and 33%) except *AsAc* (8% B). Finally, the Lamiaceae species, *O. basilicum* showed higher induction of antioxidant production at 16 and 33% of blue.

Variation curves of DPPH showed a peak at 25% B in pak choi vs maximal radical scavenging activities at 0 and 33% in tastoi and basil (Fig. 9.5f–h). These two side aspects could be indicative of two different light stresses due to absence of (needed) blue and blue over dosage.

Finally, plant growth showed negative correlations with antioxidant compounds (Fig. 9.5i–k): in *B. rapa* (tastoi) and *O. basilicum*, plant height and hypocotyl length (Fig. 9.5i–k) showed opposite variations to *TotPh*, *Fols* and *Anth* (Fig. 9.5b–e). Phenolic compounds and growth of plants were revealed to be light-induced via different photoreceptor effects according to plant species (Iwai et al. 2010). In *Brassica napus*, over-expression of *Cry1* (cryptochrome) resulted in the inhibition of cell elongation and more accumulation of anthocyanins (Sharma et al. 2014). In tomato (*Solanum lycopersicum*) growing under white and blue light, cultivar with overexpression of *Cry2* showed reduced hypocotyl length and elevated anthocyanin content compared with wild type plant.

Study case 2: Lamb's lettuce cultivars. Two cultivars of lamb's lettuce *Valerianella locusta* (var. Holländischer and Noordhollandse) were cultivated in greenhouse under stable conditions of temperature and humidity (Dlugosz-Grochowska et al. 2017). After germination, the seedling were subject to 45 days of light treatment with a photoperiod of 16h/8h (day/night) under a constant PPFD of 200 μmol.m^{-2}.s^{-1} shared between blue (440 nm) and red (660 nm) light according to four different ratios (Fig. 9.6a): (Blue:Red) → (0:10), (1:9), (2:8), (3:7). Control conditions consisted of illumination with HPS lamp without blue- and red-supplement (0:0). After harvesting, light-induced plant responses were analyzed from fresh leaves' contents (mg.100g^{-1} Fresh Weight) of phenolic and terpenic structures including contents of chemical families and individual compounds:

- Total phenols (*TotPh*), phenylpropanoids (*PhPr*), flavonols (*Fols*) and anthocyanins (*Anth*).
- *p*-coumaric acid (*p-Coum*), sinapic acid (*Sin*), chlorogenic acid (*ChlrAc*), diosmetin (*Dios*), hesperidin (*Hsp*) and rutin (*Rut*).
- Chlorophylls a and b (*Chl a*, *Chl b*) (in %).
- Total xanthophylls (*Xan*), lutein (*Lut*) and β-carotene (*Car*).

Moreover, radical-scavenging and antioxidant activities were measured by means of HDDP and FRAP (Ferric reducing antioxidant power) methods.

Under global quantitative aspect, chlorogenic acid (*ChlrAc*) was the most accumulated product among the phenolic compounds in both cultivars (Fig. 9.6h). *ChlrAc* reached maxima that corresponded to 2-folds higher levels than control. Chemical structure of *ChlrAc* is characterized by the presence of ortho-di-hydroxyl representing key factor for antioxidative activity.

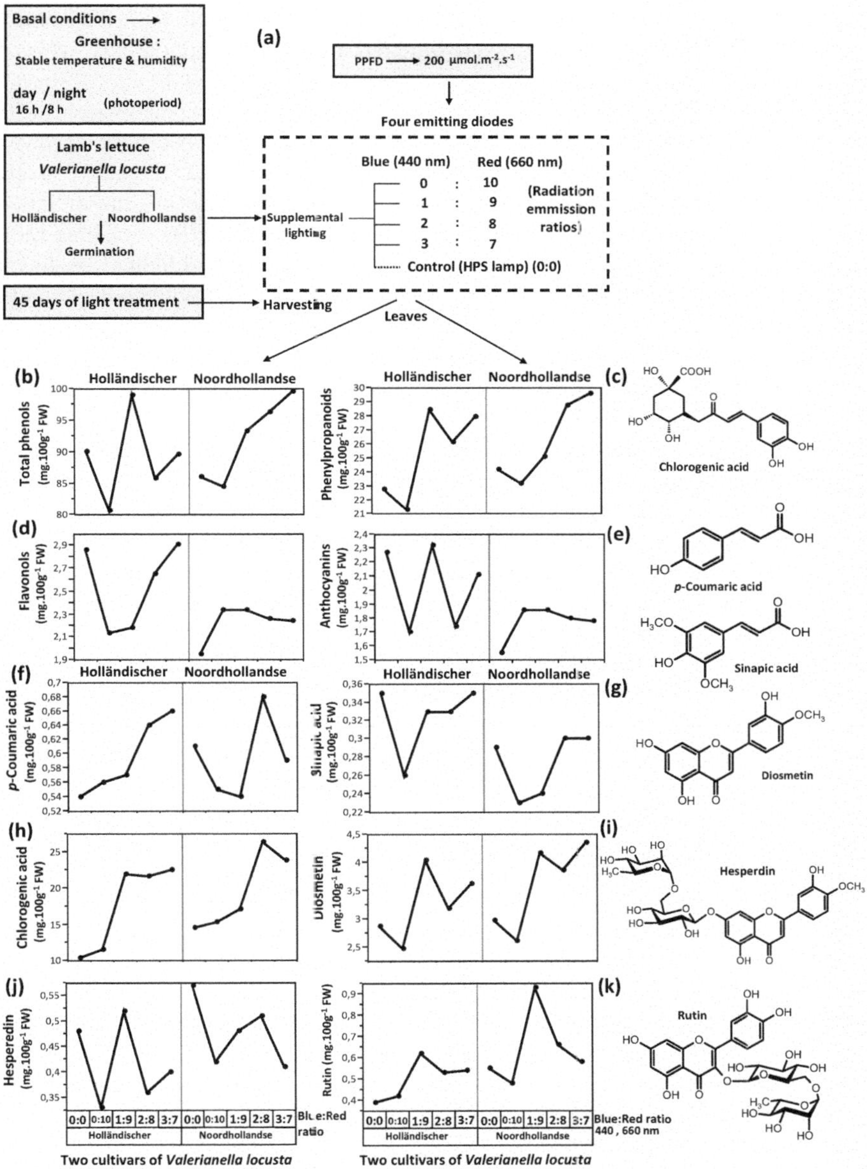

Fig. 9.6 Experimental design for studying irradiance effects of blue-red light ratios (**a**) on phenolic compound contents (**b–k**) in the leaves of two cultivars (Holländischer and Noordhollandse) of the lamb's lettuce *Valerianella locusta*

Phenolic compounds in Noordhollandse cultivar. Compared with the control conditions (HPS lamp), supplemented blue:red light in Noordhollandse (*Noor*) favored the production of several phenolic compounds (Fig. 9.6b–k) except for *Sin* and *Hsp* (Fig. 9.6g, j): *TotPh, PhPr, ChlrAc* and *Dios* showed increasing trend with blue ratio leading to maxima at blue:red ratio (B:R) of 3:7 (Fig. 9.6b, c, h, i). However, *p-Coum* and *Rut* showed production peaks at B:R of 2:8 and 1:9, respectively (Fig. 9.6f, k). In all the cases, these trends highlighted key role of blue radiation in induction of production of phenolic compounds. The importance of blue radiation in phenolic metabolism of Noor cultivar was further highlighted by lower contents observed at B:R ratio 0:10 (absence of blue) compared with control conditions (Fig. 9.6b, c, f–k), except for anthocyanins and flavonols (Fig. 9.6d, e). This could indicate key role of blue radiation in regulation of polyphenol synthesis. Also, HPS lamp lacks radiation in blue wavelengths and resulted in low accumulation levels for most of polyphenols.

Finally, in *Noor*, sinapic acid and hesperedin did not benefit from the applied blue-red illumination conditions: *Sin* showed low amounts at 0:10 and 1:9 ratios (Fig. 9.6g). Its increase with higher blue ratio (2:8 and 3:7) was not sufficient to exceed the control level. Sinapic acid corresponds to dimethylated structure of *p*-coumaric acid which showed a well differentiated peak at 2:8 of B:R ratio (Fig. 9.6f). Methylation (in sinapic acid) seems to influence the shape of such a light induced peak (Fig. 9.6 g).

Concerning the hesperidin (Hsp) in *Noor*, it showed its maximum level under control conditions (0:0) versus a minimum observed at B:R ratio of 3:7, by opposition to its precursor (diosmetin) which reached maximal accumulation at (3:7)-ratio (Fig. 9.6j vs i). *Hsp* represents 7-diglycosylated structure of *Dios* (Fig. 9.6). This could suggest that 7-glycosylation of *Dios* results in less favorable structure (*Hsp*) than the non-substituted compound (*Dios*) under the applied blue-red radiation conditions (wavelengths, ratios).

Phenolic compounds in Holländischer cultivar. In Holländischer (*Hol*) cultivar, phenolic compounds showed different curve shapes from those of *Noor* (Fig. 9.6). *Hol* curves showed several inflexions or inversed variations resulting in less differentiated accumulation levels between B:R ratios and control: *Anth* and *Hsp* showed multiple local variations leading to global invariance of accumulation from control (Fig. 9.6e, j). At B:R ratio = 1:9, they show an indifferent peak with not significantly different level from control. However, several favored compounds showed significant accumulation peaks at B:R = 1:9 (*TotPh, PhPr, ChlrAc, Dios, Rut*) (Fig. 9.6b, c, h, i, k). However, *p-Coum* showed positive trend leading to maximal content favored by higher blue ratio (3:7) (Fig. 9.6f). Finally, *Fols* and *Sin* showed strong minimum at B:R = 0:10 followed by return to similar control state at B:R = 3:7 (Fig. 9.6d, g). Globally, the blue:red ratio of 1:9 was revealed to be the most favorable for phenolic compound accumulation in *Hol* cultivar. Finally, all the phenolic compounds manifested minimum content at B:R = 0:10 (Fig. 9.6b–k). This could provide proof on the need or key role of blue radiation on phenolic metabolism in *Hol* cultivar.

Chlorophylls and tetraterpenes (pigments). Unlike phenolic compounds, chlorophylls and tetraterpenes showed less variability of content curves under applied light conditions (Fig. 9.7). The curves showed a general shape defined by higher and not significantly different levels at 0:0 (HPS lamp control), 2:8 and 3:7 ratios with minimum values at 0:10 and 1:9. These results indicated that the applied irradiation conditions were not favorable for pigment accumulation. Slight accumulation peaks of *Chl a*, *Xan* and *Lut* manifested at 2:8 ratio in *Noor*, but they corresponded to local peaks without significant differences from the control level (Fig. 9.7a, c, d). However, in *Hol* cultivar, control conditions showed significantly higher accumulation levels of pigments than different ratios of supplemental blue-red irradiations (Fig. 9.7a–d). The different shapes of pigment accumulation curves between the two cultivars provided further highlighting on the interest of SMs in functional characterization of plant taxa at the lowest taxonomical levels (variety and cultivar of same species).

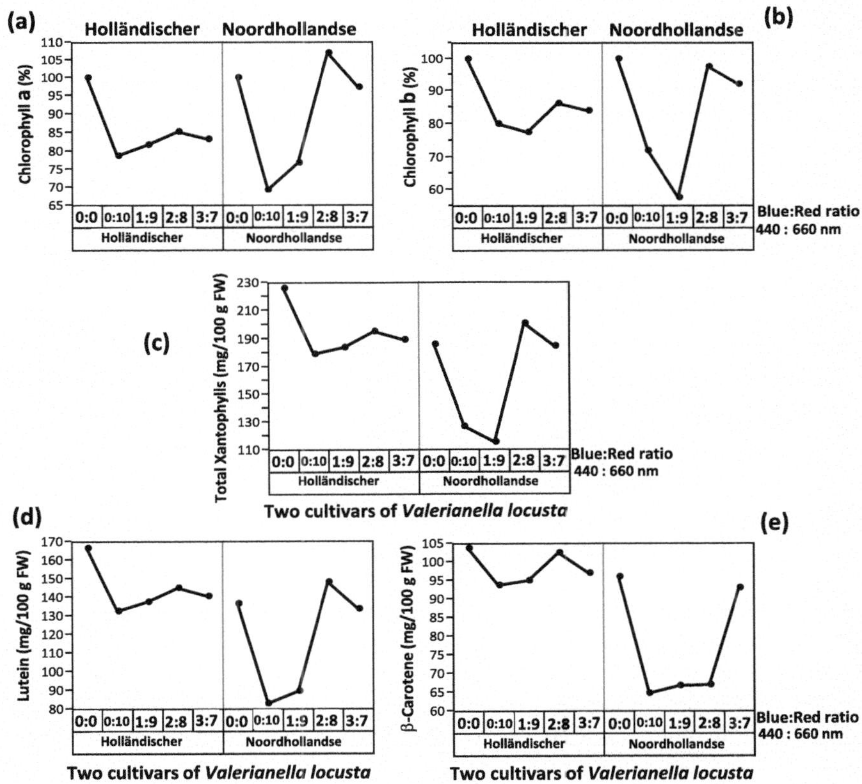

Fig. 9.7 Effects of irradiance by variable blue-red light ratios on accumulation levels of chlorophylls (**a**, **b**) and tetraterpenes (**c**–**e**) in the leaves of two cultivars (Holländischer and Noordhollandse) of the lamb's lettuce *Valerianella locusta*

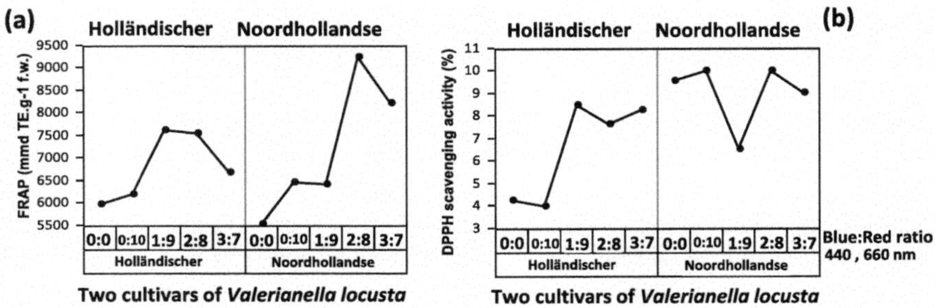

Fig. 9.8 Effects of irradiance by variable blue-red light ratios on antioxidant (**a**) and antiradical (**b**) activities in the leaves of two cultivars (Holländischer and Noordhollandse) of the lamb's lettuce *Valerianella locusta*

Antioxidant and Antiradical Activities. Ferric reducing antioxidant power (FRAP) as well as DPPH radical scavenging activities showed maximal or quasi-maximal levels at B:R ratio of 2:8, in both cultivars (Fig. 9.8). These maximal values were significantly higher than control levels except DPPH in *Noor* cultivar which did not increase significantly. However, DPPH activity was minimal in *Noor* at B:R = 1:9 (Fig. 9.8b).

In conclusion, the (1:9)- and (2:8)-B:R ratios were revealed to be the most characteristic of the studied system:

- B:R ratio = 1:9 was the less favorable for pigments (Fig. 9.7) whereas it favored the most *TotPh*, *PhPr* and *Hsp* (Fig. 9.6b, c, j).
- B:R ratio = 2:8 was globally more favorable for accumulation of several phenolic compounds (Fig. 9.6 d, f, g, h, k), anti-radical activity and slight production trends (or local peaks) of pigments (Fig. 9.7 a-d). These accumulations are linked with antioxidant and antiradical activity provided by phenols and tetraterpenes under light stress.

9.2.2.2 Effects of Supplemental UV-A Flux Density on Phytochemistry and Growth of Microgreen Species/Cultivars

Three microgreen species were cultivated in grown chamber under homogeneous nutrient and humidity conditions: *Beta vulgaris*, *Brassica rapa* (var. *chinensis*) and *Ocimum basilicum* (var. Sweet Genovese) (Fig. 9.9a) (Brazaitytė et al. 2015). Applied photoperiodic conditions were characterized by temperature and light duration with day/night occurring at 21 °C/17 °C for 16 h/8 h, respectively. Four Vis-field radiations were basically applied to all the experiments consisting of blue (447 nm), red (638 nm), deep red (665 nm) and far red (731 nm) with constant flux densities equal to 21, 122, 155, 2.2 μmol m^{-2} s^{-1}, respectively. Therefore, UV-A radiation was supplemented by applying three different wavelengths (366, 390, 402 nm) at two different intensities (6.2 or 12.4 μmol m^{-2} s^{-1}). From each of the six different experiments (3 UV-A wavelengths × 2 intensities), total phenols (*TotPh*), total anthocyanins (*Anth*), tocopherols (*Toc*) and ascorbic acid (*AsAc*) contents (mg g^{-1} FW) and DPPH-scavenging activity (μmol g^{-1} FW) were measured in plant fresh matter.

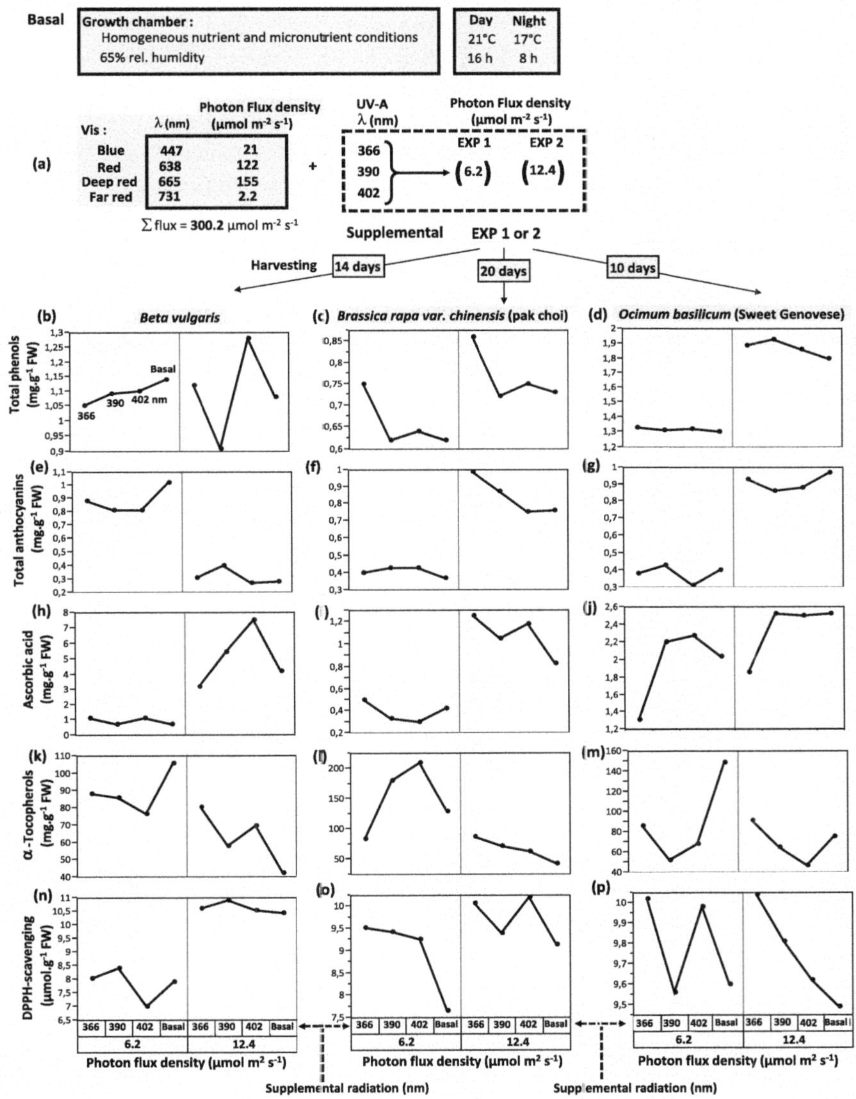

Fig. 9.9 Experimental design for studying effects of UV-A supplemented at three wavelengths under two irradiance intensities (**a**) on antioxidant accumulation (**b–m**) and antioxidant activity (**n–p**) of three microgreen species, *Beta vulgaris*, *Brassica rapa* (var. pak choi, tastoi) and *Ocimum basilicum* (var Sweet Genovese)

Under quantitative aspect, the three species were characterized by different favorite accumulations concerning different phytocompounds: *O. basilicum* showed higher accumulation levels of *TotPh* (Fig. 9.9d) under flux of 12.4 μmol m^{-2} s^{-1}; *Beta vulgaris* showed the highest contents of *AsAc* (peak of 8 mg g^{-1} FW) (Fig. 9.9h); *B. rapa* showed the highest production levels of α-tocopherols (exceeding 200 mg g^{-1} FW) under flux of 6.2 μmol m^{-2} s^{-1} (Fig. 9.9l).

Under functional trend aspect, high phytochemical variations occurred when supplemental UV-A flux density doubled from 6.2 to 12.4 μmol m^{-2} s^{-1} (Fig. 9.9b–p), except for α-tocopherols and DPPH in *O. basilicum* (Fig. 9.9m, p). However, for a given UV-A flux (6.2 or 12.4), variations of chemical contents were of lower order. This showed significant effect of radiation flux resulting in high sensitivity of phytochemical system. Moreover, the high inter-flux variations showed positive trends between accumulation levels of antioxidants and UV-A intensity except for anthocyanins in *Beta vulgaris* (Fig. 9.9e).

Beyond these general trends with outside specific cases, plant species were further characterized by combining UV-A wavelengths with flux density: under higher irradiance, lower wavelength (366 nm), *B. rapa* showed relatively high resulted in more accumulation of *TotPh, Anth, AsAc* and *Toc* (Fig. 9.9c, f, i).

In *Beta vulgaris*, high UV-A irradiance resulted in peak production of *TotPh* and *AsAc* under higher wavelength (402 nm) (Fig. 9.9b, h). Moreover, these two phytochemical responses showed strong content variations through the different wavelengths; this could be indicative of high sensitivity of photoreceptor in this species to UV-A band. However, under lower irradiance, maximal contents of *TotPh, Anth* and *Toc* were obtained under basal illumination, i.e. without UV-A supplements (limited illumination to the four *Vis*-radiations) (Fig. 9.9b, e, k). This indicated flux density-dependent effect of UV-A radiations in *B. vulgaris.* Such conditional effect was further highlighted by anthocyanins which showed higher accumulation levels under lower flux (6.2 μmol m^{-2} s^{-1}) than higher flux (12.4 μmol m^{-2} s^{-1}) (Fig. 9.9e). Moreover, anthocyanin production was more favored by basal illumination conditions under lower flux versus 390 nm under higher flux.

In *O. basilicum*, under high irradiance, *TotPh* and *AsAc* showed maximal contents at 390 nm (Fig. 9.9d, j); however, these maxima were not significantly different from the content levels observed at 402 nm due to low variability of produced amounts. Such a low variability could be indicative of stationarity associated with an optimal radiation band around 390 nm for UV-A induced production of *TotPh* and *AsAc* in *O. basilicum.* Radiation band-linked production of phytocompounds was also observed for *Anth* in *B. vulgaris* (Fig. 9.9e).

Functional phytochemical differentiation between *O. basilicum* and *B. rapa* was provided by *AsAc* production under both high and low UV-A fluxes (Fig. 9.9i, j): the wavelength 366 nm resulted in minimal accumulation levels of *AsAc* in *O. basilicum* by opposition to *B. rapa* showing maximal levels. This functional characterization way shows the importance of relativization of accumulation levels within whole species profiles: although *AsAc* level in *O. basilicum* under high irradiance by 366 nm was higher than corresponding level of *B. rapa*, this later case was associated to a maximum compared to all the *AsAc* content observed in *B. rapa* vs a minimum for *O. basilicum* (Fig. 9.9i, j). Relativizing the content levels of metabolites within and between phytochemical profiles provides a strong methodological way for highlighting specific trends and interspecific differentiations under functional aspects.

Finally, DPPH-scavenging activity was revealed to be high or maximal at 366 nm for the three plant species and both irradiance intensity (Fig. 9.9n–p). These maxima of radical-scavenging activity correlated with maximal or high accumulation of *TotPh* in *B. rapa* and *O. basilicum* (Fig. 9.9 o, p vs 9.9c, d).

9.2.2.3 Effects of UV-A and UV-B and their Interaction with Water Deficit

Field experiments were carried out on *Medicago truncatula* (Fabaceae) by putting light filters above plants of two accessions (F83005-5, French origin; Jemalong A17, Australian origin) (Fig. 9.10a) (Rai et al. 2023). Three separated filters were used for the selection of three treatment light bands (LB): > 290 nm (LB 1), > 350 nm (LB 2), > 400 nm (LB 3) (Fig. 9.10b). These treatments make the assessment of two UV-bands _ (290–350 nm) and (350–400 nm) _ from the differences (LB 1 versus LB 2) and (LB 2 versus LB 3), respectively (Fig. 9.10c). The first UV-band (290–350 nm) includes UV-B radiation (290–315 nm) and short-wave UV-A radiation (315–350 nm). The second UV-band (350–400 nm) represents long-wave UV-A radiation.

The three light bands (LB 1, 2, 3) were combined with two water supply regimes to analyze the effects of radiation (UV-B, -A and visible), water deficit and their interaction on the production of phenolic compounds in each accession of *M. truncatula* (Fig. 9.10d). Plants of both accessions were subject to UV treatment for 37 days after their seedling emergence (Fig. 9.10d1). This light-treatment duration was organized into two successive steps: 29 days in which plants were daily watered (Fig. 9.10d2) followed by 8 days (days 0–7) during which half of plants were subjected to watering cut-off (Fig. 9.10d3); the other moiety continued to be normally watered.

During the eight days of watering cut-off, phytocompound contents were assessed by two ways (Fig. 9.10e):

(i) *In situ* (at day 1 and day 6), using nondestructive spectroscopic techniques which assess phytochemical contents as relative absorbance units. The measures were carried out on the leaves (adaxial side) and concerned epidermal flavonoids, epidermal anthocyanins and mesophyll chlorophylls (Fig. 9.10e1).
(ii) After harvests (at day 0 and day 7), using chromatographical techniques giving amounts in mg.g^{-1} dry weight (DW). They concerned four flavones consisting of apigenin, luteolin, chrysoeriol and tricin (aglycones, glycosylated and acylated forms) (Fig. 9.10e2).

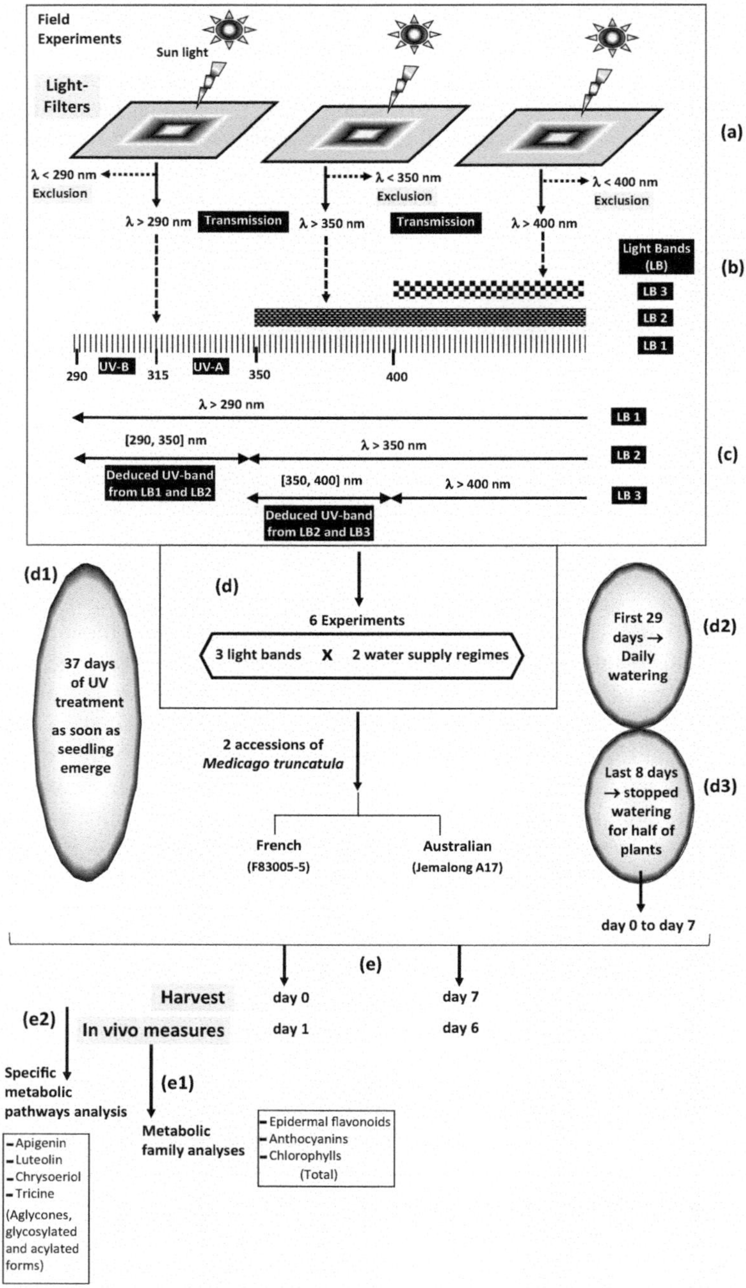

Fig. 9.10 Field experimental design applied to two accessions of *Medicago truncatula* (Fabaceae) to assess the effects of UV radiations (UV-B, UV-A), water deficit and accession (cultivars) on leaves' production of secondary metabolites (phenolic compounds) and chlorophylls

Quantified contents were statistically analyzed by ANOVA leading to assess the significance of light filtering, water deficit and accession effects (separated and combined factors) by considering the duration of water stress (day 0 versus day 7; day 1 versus day 6).

Accumulation levels of epidermal flavonoids, anthocyanins and mesophyll chlorophylls showed more significant variations of contents at day 1 than day 6 (Fig. 9.11). Figure 9.11 reports the *p*-values given by ANOVA applied between accumulation levels of assessed metabolites and the three single and combined factors (accession, water regime, radiation filter). Significant effects ($p < 0.05$) were indicated by circles in the Fig. 9.11a, c, and by bold points in the Fig. 9.11b (zoomed Fig. 9.11a). In both days, the factor "accession" was revealed to be influent on the variations of different metabolic contents.

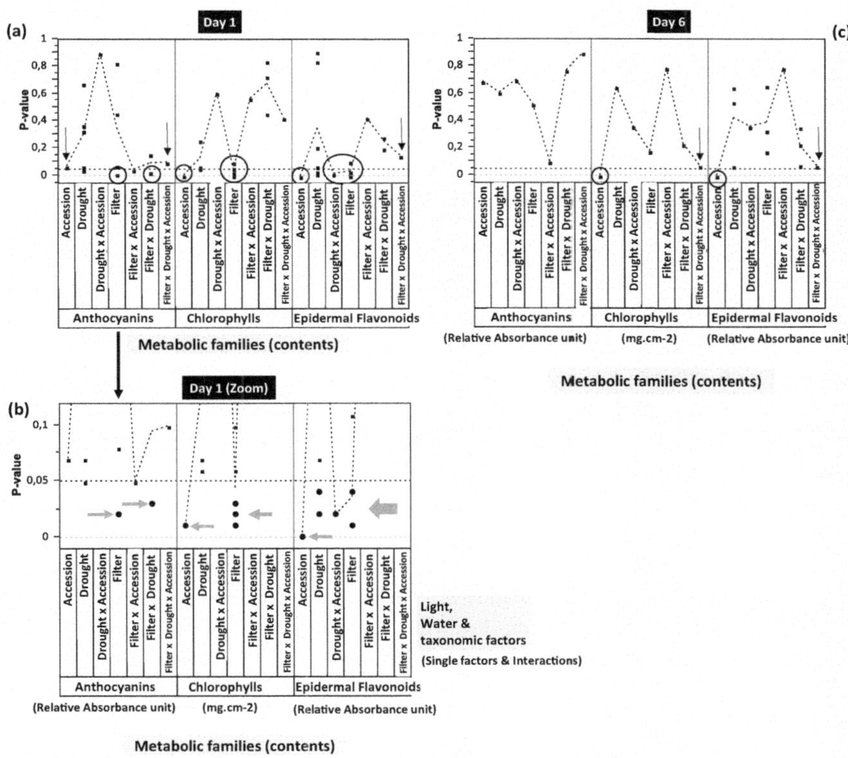

Fig. 9.11 Plots showing the p-values of ANOVA based on the factorial design combining three radiation bands (> 290, > 350, > 400 nm) with two watering mode treatments, applied on two accessions (F83005-5 and Jemalong A17) of *Medicago truncatula* (Fabaceae) from which flavonoids and chlorophyll were assessed at the initial and late period (day 1, day 6) of drought application

On day 1, significant variations of accumulation levels concerned (Fig. 9.11a, c):

- Epidermal flavonoids influenced by accession, filter, drought (water deficit), and accession-drought interaction.
- Mesophyll chlorophyll influenced by accession and filter.
- Epidermal anthocyanins influenced by filter and filter-drought interaction.

On day 6, significant variations were concerned with epidermal flavonoids and chlorophyll under the effect of accession (Fig. 9.11b).

UV-exposition duration resulted in different trends specific to different phyto-compounds: chlorophyll content decreased between day 1 and day 6 whereas anthocyanins showed opposite trend (Fig. 9.12g, h). Effects of accession factor resulted in higher accumulation levels of epidermal flavonoids in F83005-5 than Jemalong A17 (Fig. 9.12a, b) versus opposite trend for chlorophyll (Fig. 9.12e, f). Light filter effects resulted in global accumulation trends of both flavonoids and chlorophyll that were favored by lower UV waves (> 290 nm) compared with higher UV (> 350 nm) and visible light (> 400 nm) (Fig. 9.12a, e). The highest wavelength band (> 400 nm) generally resulted in lower contents of epidermal flavonoids and chlorophyll. These results highlighted induction role of UV-B (short UV waves) toward epidermal flavonoids and chlorophyll during the initial period of water deficit (day 1).

Interaction effects were found for the pairs (accession, drought treatment) and (filter, drought) influencing epidermal flavonoid and anthocyanin contents, respectively (Fig. 9.12a-d): F83005-5 and Jemalong A17 showed higher flavonoid contents under water stress conditions at day 1 and day 6, respectively, compared with watering condition of the same day (Fig. 9.12a, b). This highlights specific response durations in different plant species. For anthocyanins, long waves of UV-A (> 350 nm) combined with drought resulted in lower contents in Jemalong A17 versus relatively high level in F830005-5 (Fig. 9.12c, d). This trend resulted in higher anthocyanin contents in watered than water-stressed Jemalong A17 under the effect of UV-A long waves (> 350 nm) at day 1 (Fig. 9.12c).

Moreover, chlorophyll and anthocyanins varied by opposite ways between day 1 and day 6, irrespective of the accession, UV or drought treatments (Fig. 9.12g, h). Chlorophyll content decreased between day 1 and day 6 (Fig. 9.12g), whereas anthocyanins showed opposite trend (Fig. 9.12h).

Finally, the three factors (accession, filter, drought) tend to interact in direction of effect significance (Fig. 9.11a, c). The *p*-values associated with this trifactorial interaction seem to tend to low levels (close to 0.05) for anthocyanins (and flavonoids) at day 1, and for chlorophyll and flavonoids at day 6.

More specific flavonoid analyses revealed general significant effect of accession on total apigenin, luteolin, chrysoeriol and tricin, both at day 0 and day 7 (p-values $<$ 0.05) (Fig. 9.13a–c). At day 0, filter showed significant effect on luteolin and tricin (Fig. 9.13a). At day 7, significant effects of drought and filter were detected in luteolin (Fig. 9.13c). Structurally, luteolin is characterized by the presence of ortho-dihydroxyl on the C-3′ and C-4′ of lateral phenyl (Fig. 9.14f). This structural aspect (absent in the three other aglycones, Fig. 9.14e, g, h) confers to the molecule higher antioxidant activity.

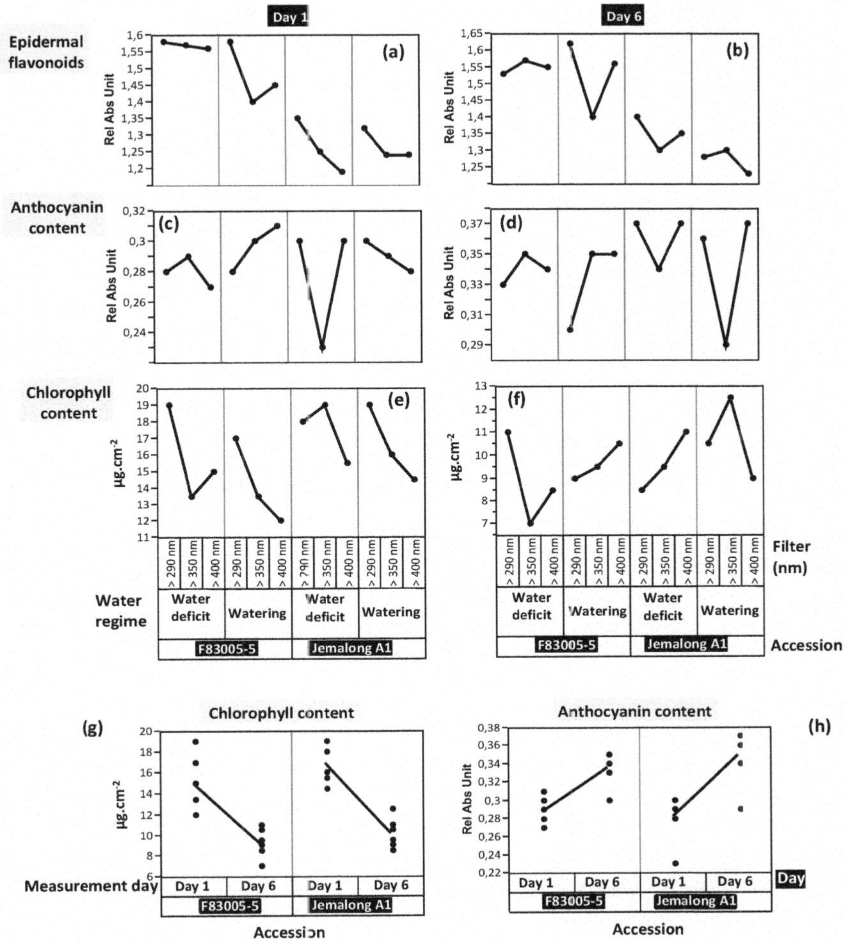

Fig. 9.12 Variations of production levels of different leaf epidermal flavonoids (**a**, **b**), anthocyanins (**c**, **d**, **h**) and chlorophyll (**e**, **f**, **g**) in the leaves of *Medicago truncatula* in relation to plant accession, drought treatment, emitted radiation bands and measure day

Significant effect of accession resulted in significant differences in (apigenin-luteolin) and (chrysoeriol-tricin) pathways between F83005-5 and Jemalong A17 (Fig. 9.14a–d): Jemalong A17 showed higher content of apigenin and luteolin than F83005-5 (Fig. 9.14a, b). However, F83005-5 was characterized by higher production of chrysoeriol and tricin compared to Jemalong A17 (Fig. 9.14c, d).

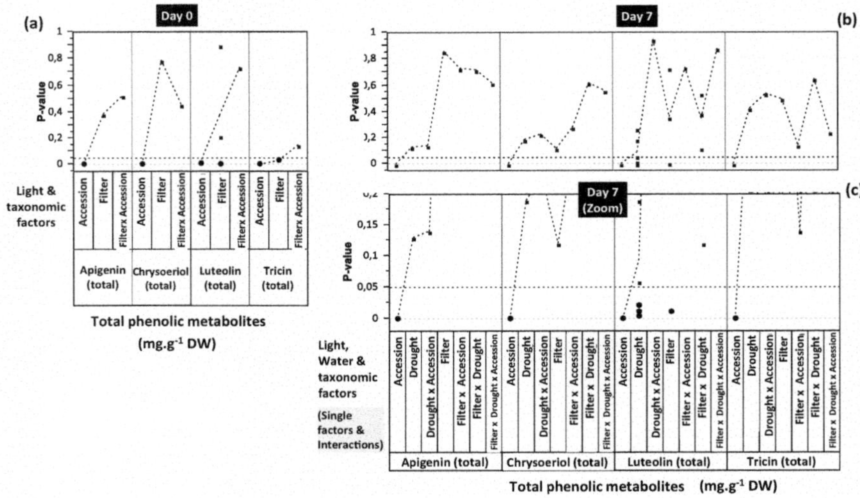

Fig. 9.13 Plots showing the p-values of ANOVA based on the factorial design combining three radiation bands (> 290, > 350, > 400 nm) with two watering mode treatments, applied on two accessions (F83005-5 and Jemalong A17) of *Medicago truncatula* (Fabaceae) from which the total contents of four flavones (apigenin, luteolin, chrysoeriol and tricin) were assessed on day 0 (preceding drought) and on day 7 of drought application

Moreover, within a given accession, flavone contents increased between day 0 and day 7 indicating dynamic plant responses to the (light, water)-stress field conditions (Fig. 9.14a–d). This time-dependent effect showed strong association with water deficit duration for luteolin in both accessions (Fig. 9.14 j): luteolin content was significantly higher in drought treated plants than in normally watered ones. Such a drought-depending trend was also detected for apigenin in Jemalong A17 (Fig. 9.14i). Also, in F83005-5, chrysoeriol and tricin showed slightly higher contents under water deficit conditions (Fig. 9.14k, l).

Significant radiation-effect on luteolin resulted in higher contents favored by UV-B-containing band (> 290 nm) on both days (0, 7) and in both accessions (Fig. 9.15a, b). This effect (UV-B) was revealed to be higher in F83005-5 on day 0 (Fig. 9.15a) and higher in Jemalong A17 on day 7 (290 nm vs 350 and 400 nm) (Fig. 9.15b).

Analysis of individual glycosylated and/or acylated flavones highlighted associations between chemical substitutions and radiation bands (Fig. 9.16): in F83500-5, the 7-glucosyl tricin was revealed to be highly sensitive to the UV-B containing band (> 290 nm, LB1) by showing higher contents under LB1 than LB2 (> 350 nm) or LB3 (> 400 nm), in both watered and droughted conditions (Fig. 9.16a).

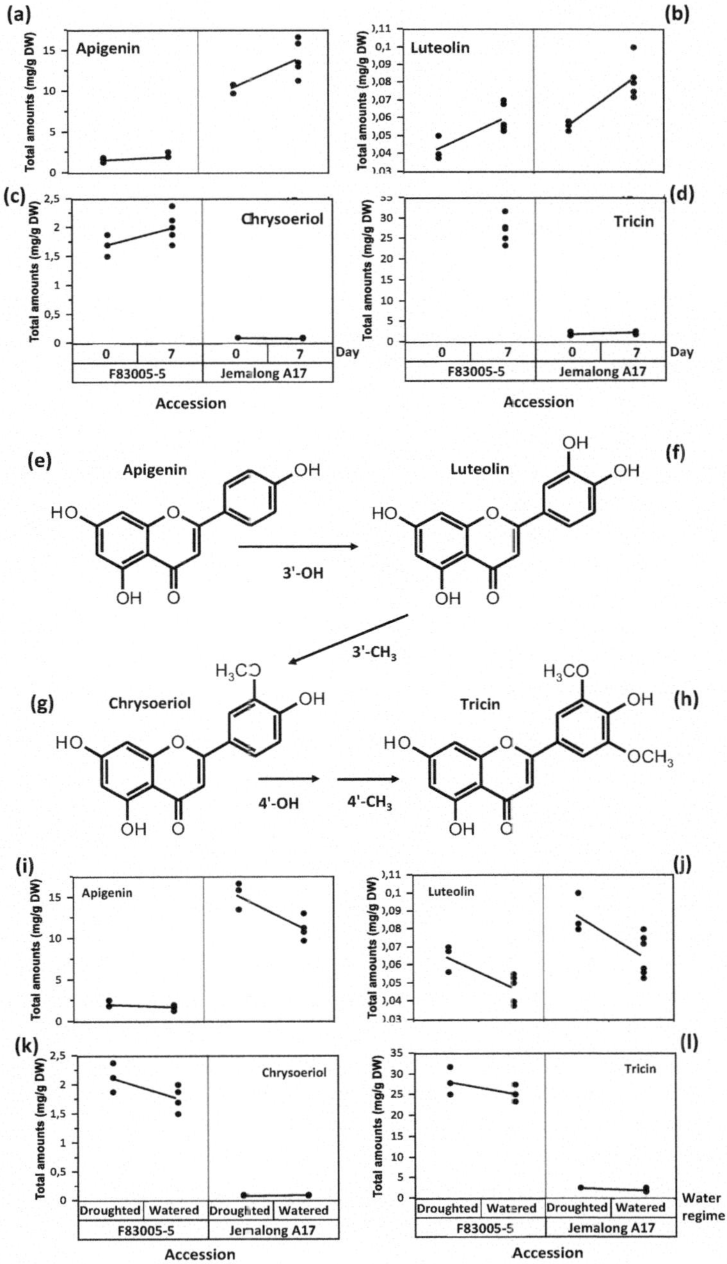

Fig. 9.14 Variations of production levels of total flavones, apigenin (**a**, **i**), luteolin (**b**, **j**), chrysoeriol (**c**, **k**) and tricin (**d**, **l**) in the leaves of two accessions of *Medicago truncatula* harvested on two days (day 0, day 7) (**a-d**) corresponding to different water stress conditions

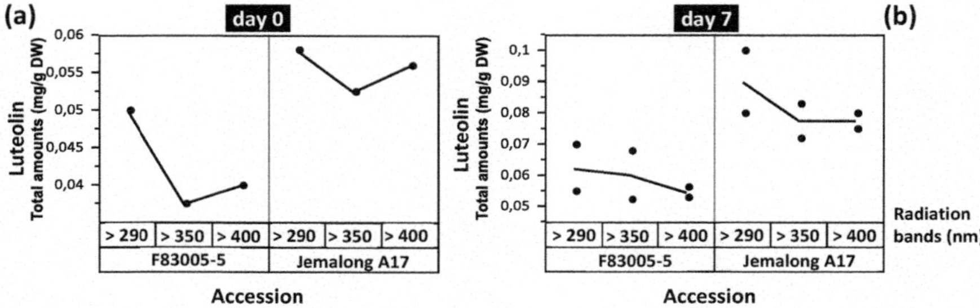

Fig. 9.15 Variations of production levels of total luteolin on day 0 (**a**) and day 7 (**b**) under the effects of three radiation bands (> 290, > 350, > 400 nm) in two accessions (F83005-5 and Jemalong A17) of *Medicago truncatula*

Moreover, LB 1 (> 290 nm) was revealed to be more inductive of other tricin derivatives including isomers of 7-coumaroyl-triglucoside (Fig. 9.16b) and 7-ferruloyl-diglucoside (Fig. 9.16c). However, these two tricin derivatives showed different relative sensitivity toward LB1 according to the (water versus drought)-conditions and duration: LB1 combined with normal watering conditions resulted in maximal content of 7-coumaroyl-triglycosyl tricin on day 0 versus low content on day 7 (Fig. 9.16b); droughted conditions on day 7 gave intermediate content levels. However, 7-ferruloyl-diglucosyl tricin showed opposite trend associated with LB1 resulting in maximal content under watering conditions on day 7 versus low level on day 0 (Fig. 9.16c); droughted conditions on day 7 gave intermediate content levels. These variations should be considered for a same tricin derivative, because the content levels of different metabolites showed different magnitude order and variation scales (e.g. tricin-7-glucoside [10–15 μMol g^{-1} dw] > > tricin-ferruloyl-diglucoside [0.4–1.2 μMol g^{-1} dw] > > tricin-coumaroyl-triglucoside [0.002–0.02 μMol g^{-1} dw]).

In Jemalong A17, 7-ferruloyl-diglucoside and 7-coumaroyl-triglucoside of apigenin were found with content levels varying significantly with the wavelength band (Fig. 9.16d, e). Apigenin-7-coumaroyl-triglucoside showed maximal accumulation under LB1 (> 290 nm) effect in both watered and droughted conditions (Fig. 9.16d). However, apigenin-7-ferruloyl-diglucoside showed strong diversity of accumulation curves influenced by both radiation band and water conditions (Fig. 9.16e): maximal content was reached with LB3 (> 400 nm) in droughted plants on day 7; however, on the same day, LB3 resulted in minimal content of the same compound in watered plants.

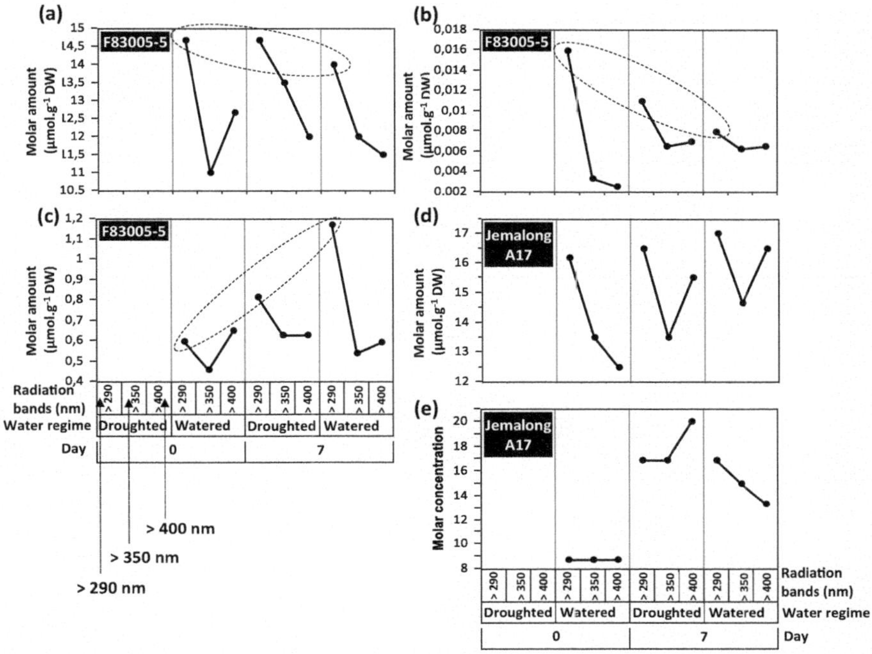

Fig. 9.16 Variations of production levels of different glycosides (acylated and not acylated) of tricin (**a–c**) and apigenin (**d**, **e**) under the effects of three radiation bands (> 290, > 350, > 400 nm), two watering/drought treatment in two accessions (F83005-5 and Jemalong A17) of *Medicago truncatula*. **a** Tricin-7-glucoside; **b** Tricin-7-coumaroyl-glucosyl-glucosyl-glucoside; **c** Tricin-7-ferruloyl-glucosyl-glucoside; **d** Apigenin-7-coumaroyl-glucosyl-glucosyl-glucoside; **e** Apigenin-7-ferruloyl-glucosyl-glucoside

This study revealed that accumulation of different phenolic compounds occurred with different ways depending on the plant accession and favored by separated or combined effects of UV wave bands and watering regime (Fig. 9.17). The UV-B containing band (LB1, > 290 nm) was revealed to be more inductive factors of different total phenols and individual derivative flavones. Water supply regime showed different effects due to watering and water deficit according to the phenolic compound and family. Stress duration showed strong positive and negative effects on anthocyanins and chlorophyll, respectively. Finally, total epidermal flavonoids provided frank separation between the two accessions with higher accumulation levels in F83005-5 than Jemalong A17.

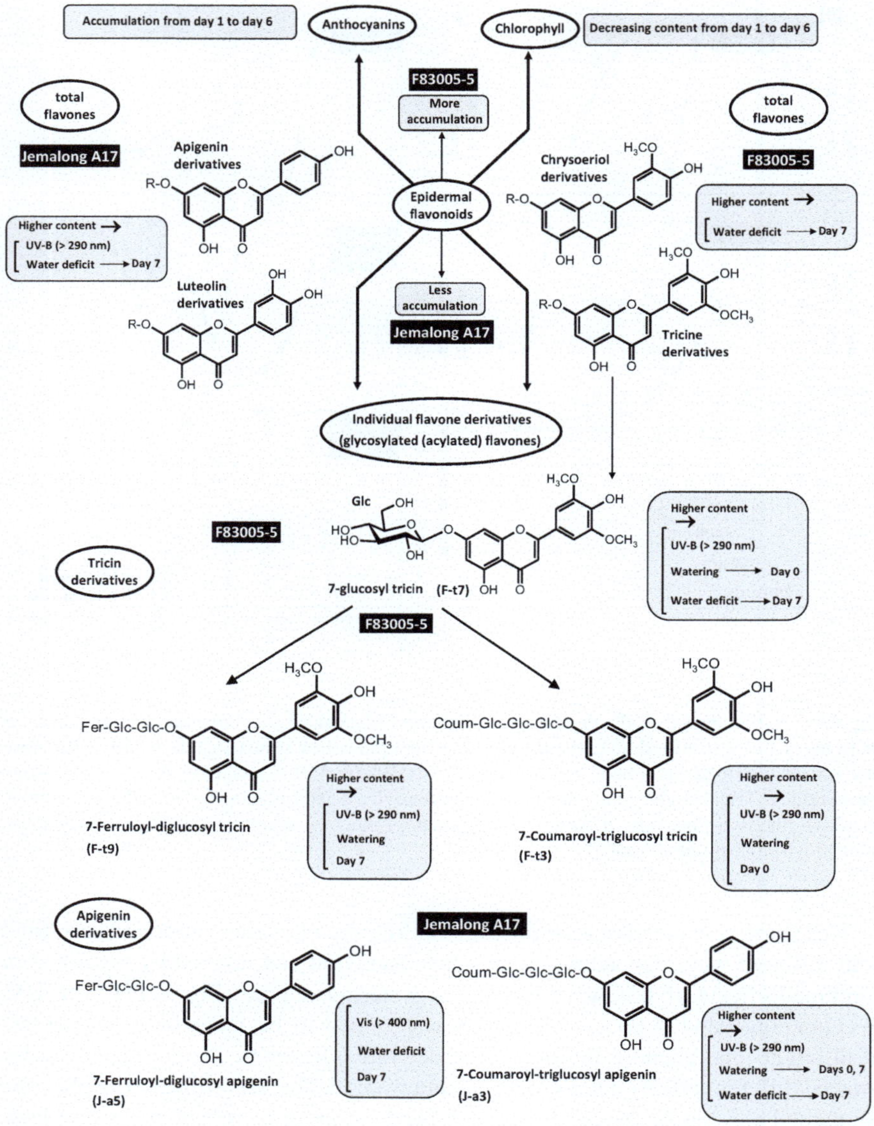

Fig. 9.17 Synthesis scheme on significant effects of radiation bands, water deficit levels, stress duration and plant accessions in *Medicago truncatula* subject to field experiments under filtered sun light and two water supplied regimes. Labels between brackets correspond to those of isomers given in Rai et al. (2023)

References

Anpo M, Fukuda H, Wada T (2018) Plant Factory Using Artificial Light: adapting to Environmental Disruption and Clues to Agricultural Innovation. Elsevier, San Diego, CA

Bae G, Choi G (2008) Decoding of light signals by plant phytochromes and their interacting proteins. Annu Rev Plant Biol 59:281–311

Battle MW, Jones MA (2020) Cryptochromes integrate green light signals into the circadian system. Plant, Cell Environ 43:16–27

Bouly JP, Schleicher E, Dionisio-Sese M, Vandenbussche F, Van Der Straeten D, Bakrim N, Meier S, Batschauer A, Galland P, Bittl R, Ahmad M (2007) Cryptochrome blue light photoreceptors are activated through interconversion of flavin redox states. J Biol Chem 282:9383–9391

Brazaitytė A, Viršilė A, Jankauskienė J, Sakalauskienė S, Samuolienė G, Sirtautas R, Novičkovas A, Dabašinskas L, Miliauskienė J, Vaštakaitė V, Bagdonavičienė A, Duchovskis P (2015) Effect of supplemental UV-A irradiation in solid-state lighting on the growth and phytochemical content of microgreens. Int Agrophys 29(1):13–22

Cashmore AR, Jarillo JA, Wu YJ, Liu D (1999) Cryptochromes: blue light receptors for plants and animals. Science 284:760–765

Casal JJ (2012) Shade Avoidance. Arabidopsis Book 10:e0157

Casal JJ (2013) Photoreceptor signaling networks in plant responses to shade. Annu Rev Plant Biol 64:403–427

Celaya RB, Liscum E (2005) Phototropins and associated signaling: providing the power of movement in higher plants. Photochem Photobiol 81(1):73–80

Christie JM, Salomon M, Nozue K, Wada M, Briggs WR (1999) LOV (light, oxygen, or voltage) domains of the blue-light photoreceptor phototropin (nph1): binding sites for the chromophore flavin mononucleotide. Proc Natl Acad Sci USA 96:8779–8783

Christie JM (2007) Phototropin blue-light receptors. Annu Rev Plant Biol 58:21–45

Crosson S, Moffat K (2001) Structure of a flavin-binding plant photoreceptor domain: insights into light-mediated signal transduction. Proc Natl Acad Sci USA 98:2995–3000

Crosson S, Moffat K (2002) Photoexcited structure of a plant photoreceptor domain reveals a light-driven molecular switch. Plant Cell 14:1–9

Crosson S, Rajagopal S, Moffat K (2003) The LOV domain family: photoresponsive signaling modules coupled to diverse output domains. Biochemistry 42:2–10

Długosz-Grochowska O, Wojciechowska R, Kruczek M et al (2017) Supplemental lighting with LEDs improves the biochemical composition of two *Valerianella locusta* (L.) cultivars. Hortic Environ Biotechnol 58:441–449

Dou H, Niu G (2020) Plant responses to light. Chapter 9. In: Kozai T, Niu G, Takagaki M (eds), Plant factory—an indoor vertical farming system for efficient quality food production. Academic Press, Elsevier, London, UK

Fankhauser C (2001) The phytochromes, a family of red/far-red absorbing photoreceptors. J Biol Chem 276:11453–11456

Franklin KA, Quail PH (2010) Phytochrome functions in Arabidopsis development. J Exp Bot 61:11–24

Holmes MG, Smith H (1977) The function of phytochrome in the natural environment—II. The influence of vegetation canopies on the spectral energy distribution of natural daylight. Photochem Photobiol 25(6):539–545

Inoue S, Kinoshita T, Matsumoto M, Nakayama KI, Doi M, Shimazaki K (2008) Blue light-induced autophosphorylation of phototropin is a primary step for signaling. Proc Natl Acad Sci USA 105:5626–5631

Iwai M, Mari Ohta M, Tsuchiya H, Suzuki T (2010) Enhanced accumulation of caffeic acid, rosmarinic acid and luteolin-glucoside in red perilla cultivated under red diode laser and blue LED illumination followed by UV-A irradiation. J Funct Foods 2:66–70

Kagawa T, Sakai T, Suetsugu N, Oikawa K, Ishiguro S, Kato T, Tabata S, Okada K, Wada M (2001) Arabidopsis NPL1: a phototropin homolog controlling the chloroplast high-light avoidance response. Science 291:2138–2141

Kasahara M, Swartz TE, Olney MA et al (2002) Photochemical properties of the FMN-binding domains of the phototropins from Arabidopsis, rice, and Chlamydomonas. Plant Physiol 129:762–773

Kinoshita T, Doi M, Suetsugu N, Kagawa T, Wada M, Shimazaki K (2001) Phot1 and phot2 mediate blue light regulation of stomatal opening. Nature 414:656–660

Kleiner KW, Raffa KF, Dickson RE (1999) Partitioning of 14C-labeled photosynthate to allelochemicals and primary metabolites in source and sink leaves of aspen: evidence for secondary metabolite turnover. Oecologia 119:8–418

Klose C, Venezia F, Hussong A, Kircher S, Schafer E, Fleck C (2015) Systematic analysis of how phytochrome B dimerization determines its specificity. Nat Plants 1:15090

Kottke T, Heberle J, Hehn D, Dick B, Hegemann P (2003) Phot-LOV1: photocycle of a blue-light receptor domain from the green alga *Chlamydomonas reinhardtii*. Biophys J 84:1192–1201

Kumar SV, Wigge PA (2010) H2A.Z-containing nucleosomes mediate the thermosensory response in Arabidopsis. Cell 140:136–147

Kutschera U, Briggs WR (2013) Seedling development in buckwheat and the discovery of the photomorphogenic shade-avoidance response. Plant Biol 15(6):931–940

Li XY, Yang SF (2007) Influence of loosely bound extracellular polymeric substances (EPS) on the flocculation, sedimentation and dewaterability of activated sludge. Water Res 41(5):1022–1030

Lin C, Shalitin D (2003) Cryptochrome structure and signal transduction. Annu Rev Plant Biol 54:469–496

Liscum E, Askinosie SK, Leuchtman DL, Morrow J, Willenburg KT, Coats DR (2014) Phototropism: growing towards an understanding of plant movement. Plant Cell 26:38–55

Mancinelli AL (1994) The physiology of phytochrome action. In: Kendrick RE, Kronenberg GHM (eds.), Photomorphogenesis in plants. Springer, Dordrecht, pp 211–269

Möglich A, Ayers RA, Moffat K (2009) Design and signaling mechanism of light-regulated histidine kinases. J Mol Biol 385:1433–1444

Pham VN, Kathare PK, Huq E (2018) Phytochromes and phytochrome interacting factors. Plant Physiol 176:1025

Pokorny J, Yanishlieva N, Gordon M (2008) Antioxidants in food. Pract Appl CRC Press 2(1):259–286

Quail PH (1997) An emerging molecular map of the phytochromes. Plant, Cell Environ 20(6):657–665

Rai PK, Song H, Kim KH (2023) Nanoparticles modulate heavy-metal and arsenic stress in food crops: hormesis for food security/safety and public health. Sci Total Environ 902:166064

Rizzini L, Favory JJ, Cloix C, Faggionato D, O'Hara A, Kaiserli E, Baumeister R, Schafer E, Nagy F, Jenkins GI, Ulm R (2011) Perception of UV-B by the Arabidopsis UVR8 protein. Science 332:103–106

Sager JC, Smith WO, Edwards JL, Cyr KL (1988) Photosynthetic efficiency and phytochrome equilibria determination using spectral data. Trans ASAE 31(5):1882–1889

Sakai T, Kagawa T, Kasahara M, Swartz TE, Christie JM, Briggs WR, Wada M, Okada K (2001) Arabidopsis nph1 and npl1: blue light receptors that mediate both phototropism and chloroplast relocation. Proc Natl Acad Sci USA 98:6969–6974

Sakamoto K, Briggs WR (2002) Cellular and subcellular localization of phototropin 1. Plant Cell 14:1723–1735

Salomon M, Christie JM, Knieb E, Lempert U, Briggs WR (2000) Photochemical and mutational analysis of the FMN-binding domains of the plant blue light receptor, phototropin. Biochem 39:9401–9410

Salomon M, Eisenreich W, Durr H et al (2001) An optomechanical transducer in the blue light receptor phototropin from Avena sativa. Proc Natl Acad Sci USA 98:12357–12361

Sancar A (2003) Structure and function of DNA photolyase and cryptochrome blue-light photoreceptors. Chem Rev 103:2203–2237

Selby CP, Sancar A (2006) A cryptochrome/photolyase class of enzymes with single-stranded DNA-specific photolyase activity. Proc Natl Acad Sci USA 103:17696–17700

Sellaro R, Crepy M, Trupkin SA, Karayekov E, Buchovsky AS, Rossi C et al (2010) Cryptochrome as a sensor of the blue/green ratio of natural radiation in arabidopsis. Plant Physiol 154(1):401–409

Sharma A, Bohn K, Jose S, Cropper WP (2014) Converting even-aged plantations to uneven-aged stand conditions: a simulation analysis of silvicultural regimes with slash pine (*Pinus ellioitii Engelm.*). For Sci 60:893–906

Swartz TE, Corchnoy SB, Christie JM, Lewis JW, Szundi I, Briggs WR, Bogomolni RA (2001) The photocycle of a flavin-binding domain of the blue light photoreceptor phototropin. J Biol Chem 276:36493–36500

Valverde F, Mouradov A, Soppe W, Ravenscroft D, Samach A, Coupland G (2004) Photoreceptor regulation of CONSTANS protein in photoperiodic flowering. Sci (new York, NY). 303:1003–1006

Van Buskirk EK, Decker PV, Chen M (2012) Photobodies in light signaling. Plant Physiol 158:52–60

Vaštakaite V, Viršile A, Brazaityte A, Sirtautas R, Novickovas A (2015) The effect of blue light dosage on growth and antioxidant properties of microgreens. SODININKYSTĖ IR DARŽININKYSTĖ. MOKSLO DARBAI. 34:25–36

Wang Y, Zhang T, Folta KM (2015) Green light augments far-red-light-induced shade response. Plant Growth Regul 77:147–155

Yu L, Nie J, Cao C, Jin Y, Yan M, Wang F, Liu J, Xiao Y, Liang Y, Zhang W (2010) Phosphatidic acid mediates salt stress response byregulation of MPK6 in Arabidopsis thaliana. New Phytol 188:762–773

Zhang T, Folta KM (2012) Green light signaling and adaptive response. Plant Signal Behav 7:75–78

Chapter 10
Ozone Stress

10.1 Effects of Ozone on Biological World

Atmospheric ozone (O_3) is present both in stratosphere and troposphere (Figs. 10.1 and 10.2). It has both beneficial (photo-protectant) role and harmful effects which are due to low (undirect effect) and high formation levels, respectively (Figs. 10.1 and 10.3).

10.1.1 Beneficial Role of Ozone

Beneficial effect of ozone consists in its absorbing role of sun-originated UV-C and UV-B (harmful radiations) in the stratosphere leading to a photoprotective role at planetary scale (Fig. 10.1). In stratosphere, O_3 absorbs UV-C (190–280 nm) (99%) and UV-B (280–315 nm) (90%) leading to significantly reduce the penetration of these harmful radiations in the troposphere (Figs. 10.1a and 10.2a). In stratosphere, O_3 is naturally formed from O_2 and free O (Fig. 10.2a). Monoatomic oxygen (O) is released by photodissociation of O_2 (Fig. 10.2b). Dioxygen molecule (O_2) is dissociated by UV-wavelengths < 240 nm corresponding to photon energy > 5 electron-volts (Fig. 10.2b):

$$O_2 + \text{UV}(E > 5\text{ eV}) \rightarrow O + O$$

When sun light penetrates in atmosphere, the UV radiations will be gradually absorbed but remain enough available down while O_2 tends to increase. The best conditions for O_3 formation occur in the lower part of stratosphere containing the ozone layer (Fig. 10.1a). Collision between O_2 and O (issued from photodissociation of O_2) results in the formation of O_3:

N. Semmar, *Secondary Metabolites in Plant Stress Adaptation*, Signaling and Communication in Plants, https://doi.org/10.1007/978-3-031-52595-7_10

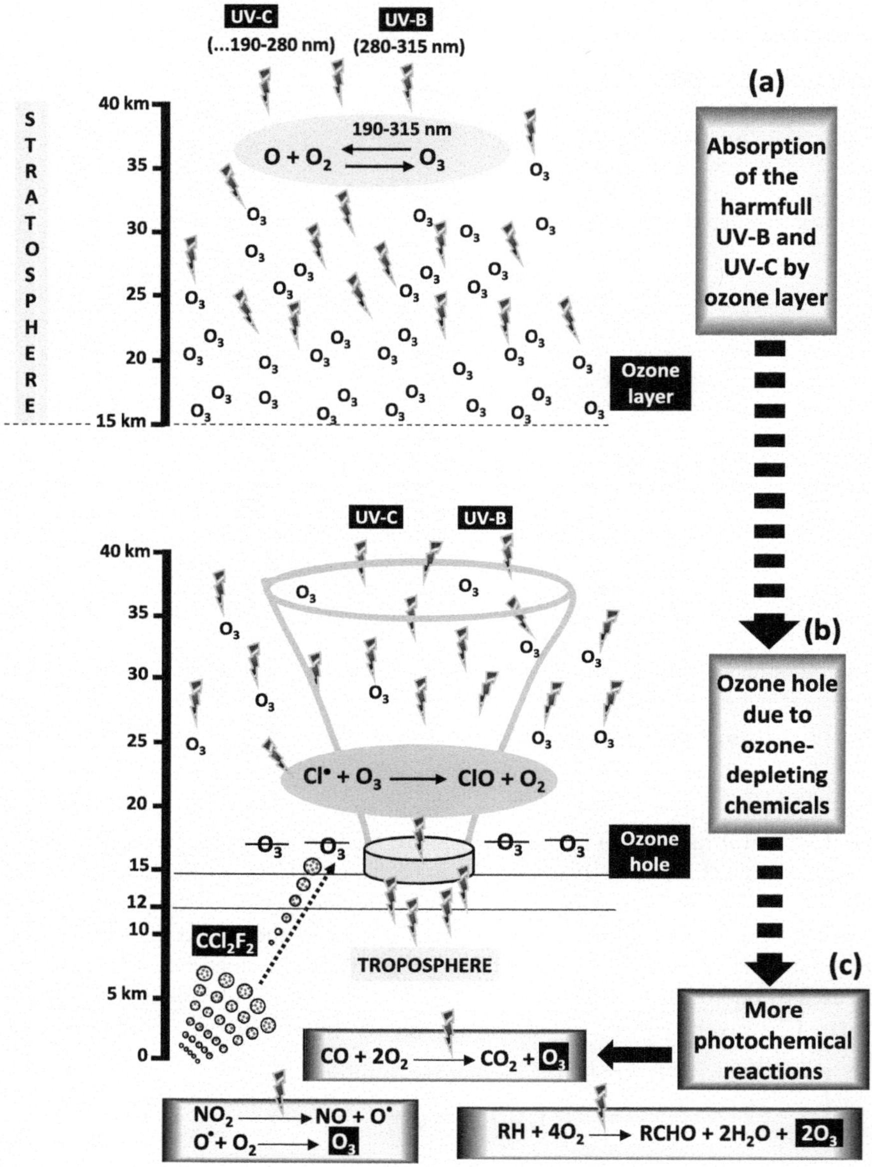

Fig. 10.1 **a** Photoprotective role of ozone (O_3) against harmful sun-originating UV (UV-C, UV-B) through the stratospheric ozone layer formed from collision between O_2 and O. **b** Apparition of ozone hole subsequently to the destruction of ozone by radicals (e.g. chlorin) originated from anthropogenic emitted compounds including mainly the chlorofluorocarbons (CFC)

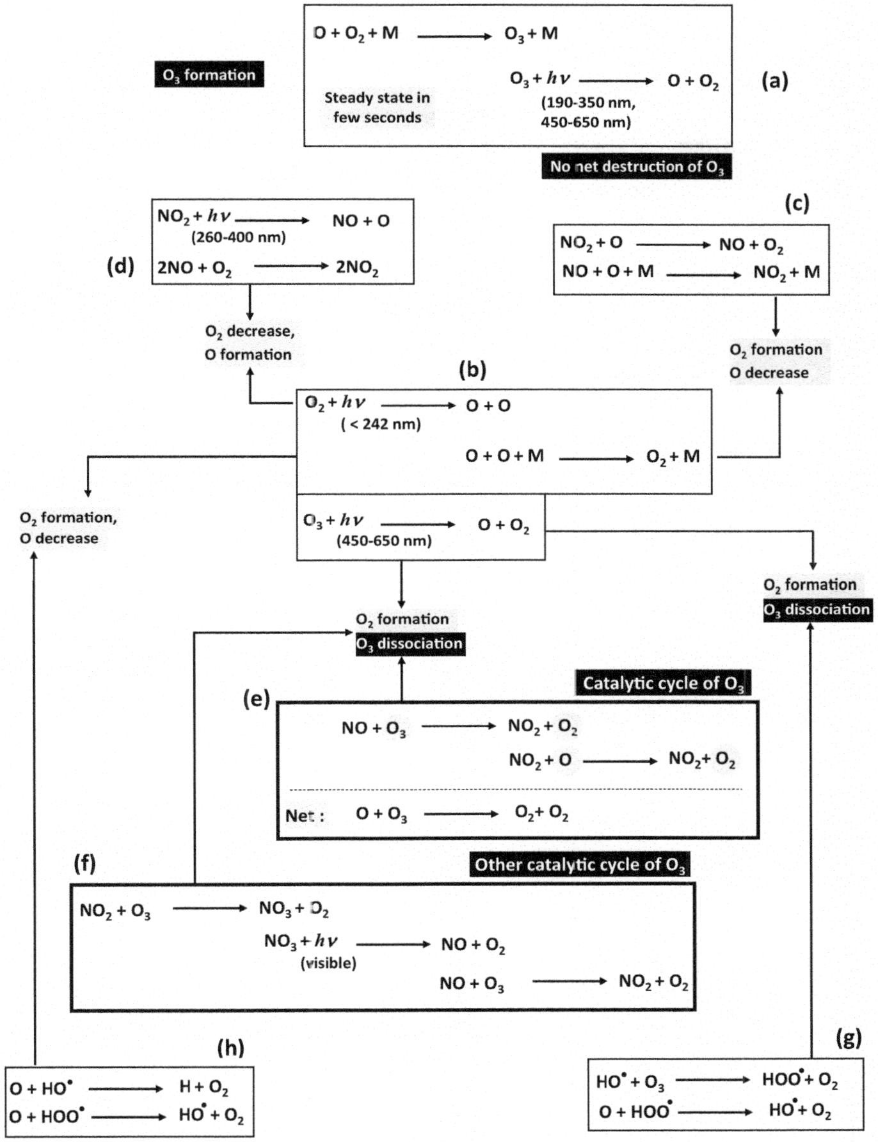

Fig. 10.2 Different photoreactions leading to dissociation and formation of ozone and involving radical reactions from nitrogen oxides or water. M: gas

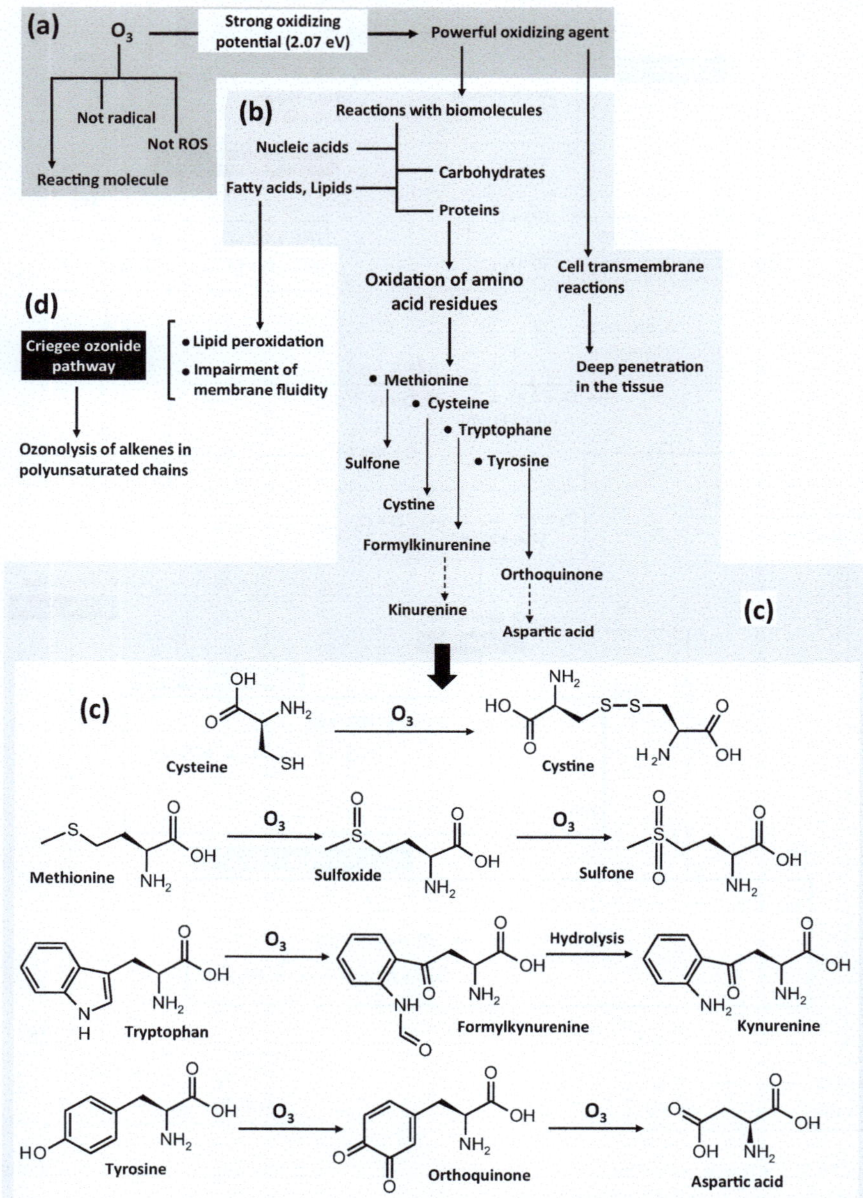

Fig. 10.3 Different harmful processes of ozone on biochemical systems due to its strong oxidizing potential against nucleic acids, protein, amino acids and alkenes (unsaturated hydrocarbon chains)

$$O_2 + O \rightarrow O_3$$

Thus, two molecules of ozone (O_3) are formed from three molecules of dioxygen (O_2) in presence of sun UV photon having enough energy to dissociate one molecule O_2. Production flux of stratospheric O_3 depends on UV photon flux which shows nictemeral (null at night), seasonal and latitudinal variations. This natural formation-depletion of O_3 involving dissociation-regeneration of O_2 doesn't show net destruction of O_3 (Fig. 10.2a), by opposition to destructive catalytic cycles due to halogen-containing organic compounds (Fig. 10.1b) and nitrogen oxides (Fig. 10.2e–f).

10.1.2 Indirect Harmful Effects of Ozone (Linked to Ozone Hole)

Indirect harmful effects of ozone are due to the formation of stratospheric ozone hole making UV radiations to enter more in troposphere. Ozone hole is due to O_3 destruction by reactive radicals originating from anthropogenic emitted polluting compounds (Crutzen and Lelieveld 2001):

Stratospheric O_3 level decreased due to reactions with several anthropogenic polluting emissions containing halogen atoms (Cl, F, Br) which release chlorine, fluorine or bromine radicals by photoexcitation. These radicals react with stratospheric ozone causing its depletion. Chlorine radical apparition increased subsequently to anthropogenic emissions of chlorofluorocarbons (CFC). CFC are stable organic compounds and consequently are able to reach the stratosphere where they react with O_3; they are referred as ozone-depleting substances (ODS) (Fig. 10.1b). They are transported into stratosphere by turbulent flow where they are photo-dissociated to release chlorine radical ($Cl^{\cdot}$) which catalyze the breakdown of O_3 into O_2 through a serial of chlorine radical regeneration:

$$CFCl_3 + \text{light} \rightarrow Cl^{\cdot} + {}^{\cdot}CFCl_2 \quad \text{(Chlorine radical release from CFC)}$$

Chlorine radical removes an oxygen to ozone leading to the formation of ClO which is also reactive with O_3:

$$Cl^{\cdot} + O_3 \rightarrow ClO + O_2$$

The ClO reacts with O_3 by removing O and by regenerating the chlorine radical:

$$ClO + O_3 \rightarrow Cl^{\cdot} + 2O_2$$

Apart from halogen radicals, nitric oxide radical ($NO^{\bullet}$) is highly active against O_3 causing its depletion. $NO^{\bullet}$ radical is released by nitrogen oxides (NO_x: mainly NO and NO_2) which play strong catalytic effects on O_3 either by directly dissociating O_3 or by reacting with (decreasing) O and O_2 (Fig. 10.2c–f). A same NO_x molecule can destroy a large number of O_3 molecules through a catalytic cycle that can be repeated over and over indefinitely (Fig. 10.2) (Johnston 1971):

$$O + NO_2 \rightarrow O_2 + NO$$

$$O_3 + NO \rightarrow O_2 + NO_2$$

By this catalytic way, NO_2 reacts with O to give NO which reacts with O_3 to dissociate it followed by NO_2 regeneration. In net, two O_3 are repeatedly transformed into three O_2.

Excess of nitrogen oxides in stratosphere was originated from emissions by planes, more particularly supersonic transport (Johnston 1971). Resulting excess of nitrogen oxides increased the rate of O_3 depletion.

Ozone depletion occurs also by the effect of water originated radicals including hydroxyl ($HO^{\cdot}$) and hydroperoxyl ($HOO^{\cdot}$), either through direct dissociation (Fig. 10.2g) or via O-consumption (Fig. 10.2h).

10.1.3 Harmful Oxidative Effect of Ozone Due to Excessive Formation

In troposphere, ozone naturally exerts a strong oxidative effect on atmospheric alkene-structures (unsaturated hydrocarbons) among which isoprene represents the most abundant target molecule. Isoprene (2-methyl-1,3-butadiene) is the most abundant nonmethane volatile organic compound in troposphere with emissions approaching 6×10^{11} kg per year (Sindelarova et al. 2014). More than 90% of tropospheric isoprene are issued from plant volatile releases. Its removal cycle is governed by several oxidative processes initiated by ozone, hydroxyl radicals ($^{\cdot}OH$) and nitrogen oxides (NO_x) (Atkinson and Arey 2003).

However, tropospheric O_3 level was subject to increasing trend due to anthropogenic emissions of greenhouse effect compounds including carbon oxides, nitrogen oxides, sulfur oxides and hydrocarbons. These primary polluting compounds are involved into photochemical reactions favoring additional (excessive) formation of ozone (considered as secondary polluting) (Fig. 10.1c). Thus, decrease vs increase of ozone in stratosphere and troposphere, respectively, provide a strong evaluating pattern of chemical composition alteration of atmosphere by anthropic emissions.

In living systems, ozone causes detrimental oxidative effects to biological tissues and metabolism that depend on both O_3 concentration and exposure duration (Fig. 10.3). It is highly reactive oxidant with strong oxidative potential (2.07 eV) whereas it is neither radical nor ROS (Fig. 10.3a). Chlorosis and necrotic lesions were observed in leaves of plants exposed to acute high ozone concentrations vs accelerated leaf senescence under chronic exposition to low O_3 levels (Baier et al. 2005; Rao et al. 2000, 2002).

O_3 reacts with biomolecules causing oxidative damage to nucleic acids, structural modifications and functional alterations of proteins (Fig. 10.3b). Amino acid residues, mainly methionine (Met), cysteine (Cys), tryptophane (Trp) and tyrosine (Tyr) are the most sensitive to ozone (Fig. 10.3b, c) (Cataldo 2003; Kotiaho et al. 2000; Sharma and Graham 2010). These amino acids are characterized by either aromatic cycle or the presence of sulfhydryl (SH) group (Fig. 10.3c). O_3-induced oxidation of Cys leads to the formation of disulphide bridges and dimerization of cysteine into cystine. Met is oxidized into sulfoxide then sulfone. Trp (the most sensitive amino acid) is degraded into formylkynurenine by oxidation of the pyrrole-ring double-bond; it can be followed by kynurenine issued from hydrolysis. Tyr is oxidized into ortho-quinone that can be further oxidized into aspartic acid.

Moreover, ozone is a powerful oxidizing agent inducing lipid peroxidation reactions that can result in impairment of cell membrane fluidity. O_3-induced lipid peroxidation occurs according to multiple reaction ways known as Criegee mechanism (Figs. 10.3d, 10.4 a) (Criegee 1957). This involves ozonolysis of alkene (cleavage of alkene by O_3) in polyunsaturated chains by electrophilic addition of O_3 across the C–C double bonds leading to the formation of ozonide (Criegee ozonide) (Fig. 10.4a–d). Also, primary ozonide decomposes to give organic radicals, aldehydes and peroxides (Fig. 10.4b, c, e, f).

10.2 Multiway Oxidative Mechanisms of Ozone with Unsaturated Hydrocarbons

Reaction between alkene (e.g. isoprene, fatty acid, etc.) and O_3 initially provides 1,2,3-trioxolane which represents primary ozonide (POZ) (Fig. 10.4a, b) (Pryor et al. 1991). This reaction consists of a cycloaddition of ozone across a given double bond (C=C) leading to a given POZ conformation. In the case of isoprene (containing two double bonds), two POZ conformations are produced (Fig. 10.5). POZ rapidly decomposes into carbonyl oxide (called Criegee intermediate, CI) and an aldehyde (Fig. 10.4c). Criegee intermediates show zwitterion character resulting from the ozonolysis of unsaturated hydrocarbons. They can strongly interact with hydrogen-bonding molecules like H_2O, CH_3OH, NH_3, etc. Carbonyl oxide can react with aldehyde or with water leading to the formation of Criegee ozonide (Fig. 10.4d) and hydroxy hydroperoxide (Fig. 10.4e), respectively. Reaction with water is faster than

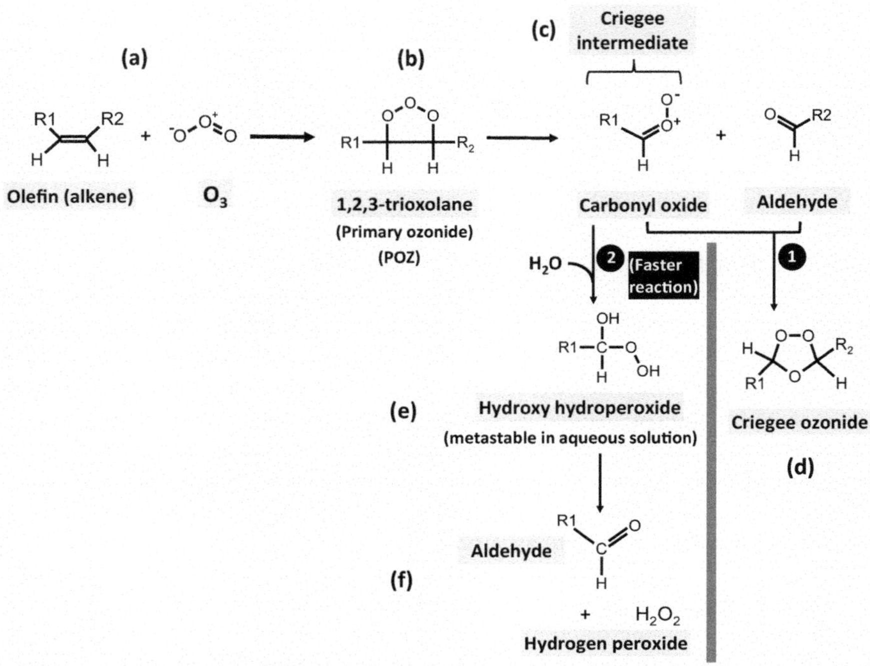

Fig. 10.4 Ozonolysis of alkene structures (**a**) resulting from oxidative mechanisms of ozone (O_3) in absence or presence of water leading to the formation of ozonide (**d**) or hydroxy hydroperoxide (**e**), respectively, via primary ozonide (**b**) and reactive Criegee intermediate(s) (**c**)

that with aldehyde. In aqueous middle, hydroxy hydroperoxide is metastable and it hydrolyzes to give hydrogen peroxide and an aldehyde (Fig. 10.4f).

Studies at high water concentrations (high humidity) revealed that reaction of carbonyl oxide (CH_2OO) with water is extremely rapid making it represents the major atmospheric sink for CH_2OO (Lewis et al. 2015).

Ozonolysis of Earth's atmospheric alkenes results in the formation of carbonyl oxide. For a given oxidized alkene, several conformational forms of carbonyl oxide can be generated (called Criegee intermediates, CIs). Multiple conformational forms of CIs are linked to the double bonds at which ozonolysis occurred. Illustrative example of multiple reactional pathways associated with multi-conformation aspect of generated CIs is presented for O_3-oxidation of tropospheric isoprene (Fig. 10.5): ozonolysis of isoprene can occur at the C1=C2 or C3=C4 bonds resulting in POZ that generates methacrolein oxide (MACR-oxide) (Fig. 10.5a, c) or methyl vinyl ketone oxide (MVK-oxide) (Fig. 10.5b, d). Proportions of CIs issued from ozonolysis of isoprene were estimated to 58% (CH_2OO), 23% (MVK-oxide) and 19% (MACR-oxide) (Nguyen et al. 2016). Ozonolysis reaction pathways are further extended via the formation of two conformation types in each of MVK-oxide and MACR-oxide groups (called *syn* and *anti*) that are distinguished by the orientation of the

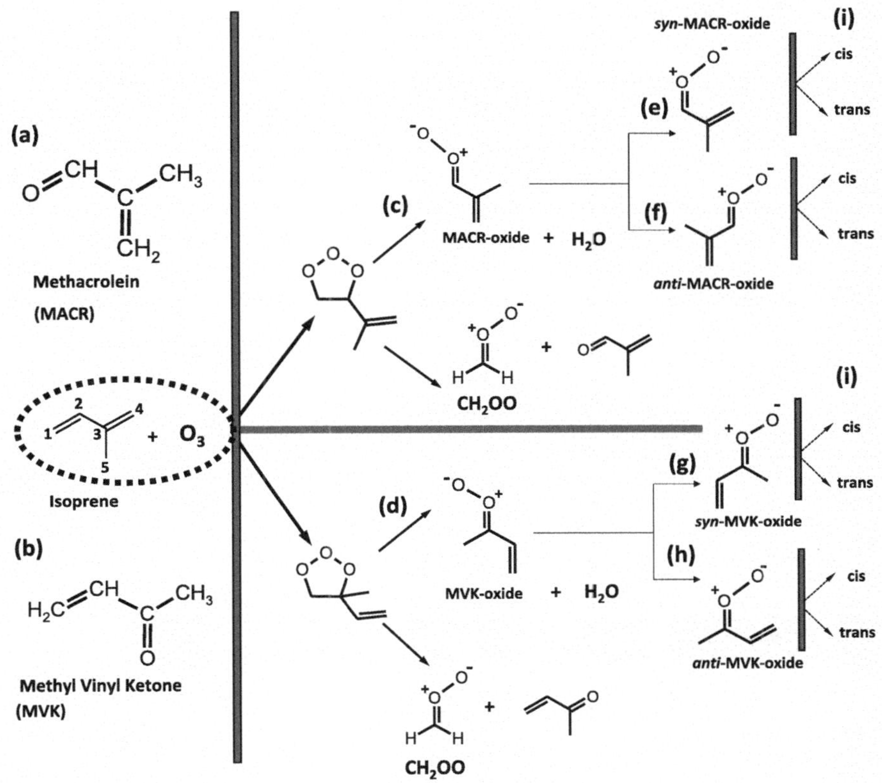

Fig. 10.5 Isoprene-based illustration of different ozonolysis pathways through different conformation types (MVK vs MACR, *syn* or *anti*) of Criegee intermediates formed by reaction of O_3 with different double bonds of the target alkene (isoprene)

terminal oxygen with respect to the vinyl group (CH_2=CH) (Fig. 10.5e–h). This results in four conformations consisting of *syn*-MVK, *anti*-MVK, *syn*-MACR, *anti*-MACR. Moreover, each of *syn*- and *anti*-conformation can be under cis or trans forms that correspond to different relative orientations of the C=C and C=O bonds. This results in eight conformations combining (MACR vs. MVK) with (*syn* vs. *anti*) and (cis vs. trans) (Fig. 10.5i) (i.e. 2^3). Under atmospheric conditions, cis and trans conformations within each group show rapid interconversions through rotation about the C–C bond (Caravan et al. 2020). However, the distinct conformers of MVK and MACR remarkably differ by their decay process rates impacting significantly the tropospheric composition.

Apart from the isoprene (the most abundant alkene in the atmosphere), ozonolysis is responsible for removing (regeneration) of different volatile unsaturated hydrocarbons including monoterpenes and sesquiterpenes (Nguyen et al. 2016).

10.3 Plant Responses to Ozone Oxidative Stress

Ozone induces several oxidative signals in plants preceding transcription processes of antioxidative defensive systems. Acute O_3 exposition results in rapid changes in protein activities (and gene expression); such changes can occur in few minutes from the onset of exposure (Vainonen and Kangasjarvi 2015). O_3-induced oxidative signals sequentially occur in different cell compartments from apoplast to nucleus via plasma membrane and cytosol (Fig. 10.6). Plant responses to O_3-induced oxidative stress consist of a complex cascade system including (i) entrance control through stomata, (ii) detoxification mechanisms in the apoplast, (iii) endogenous ROS production and

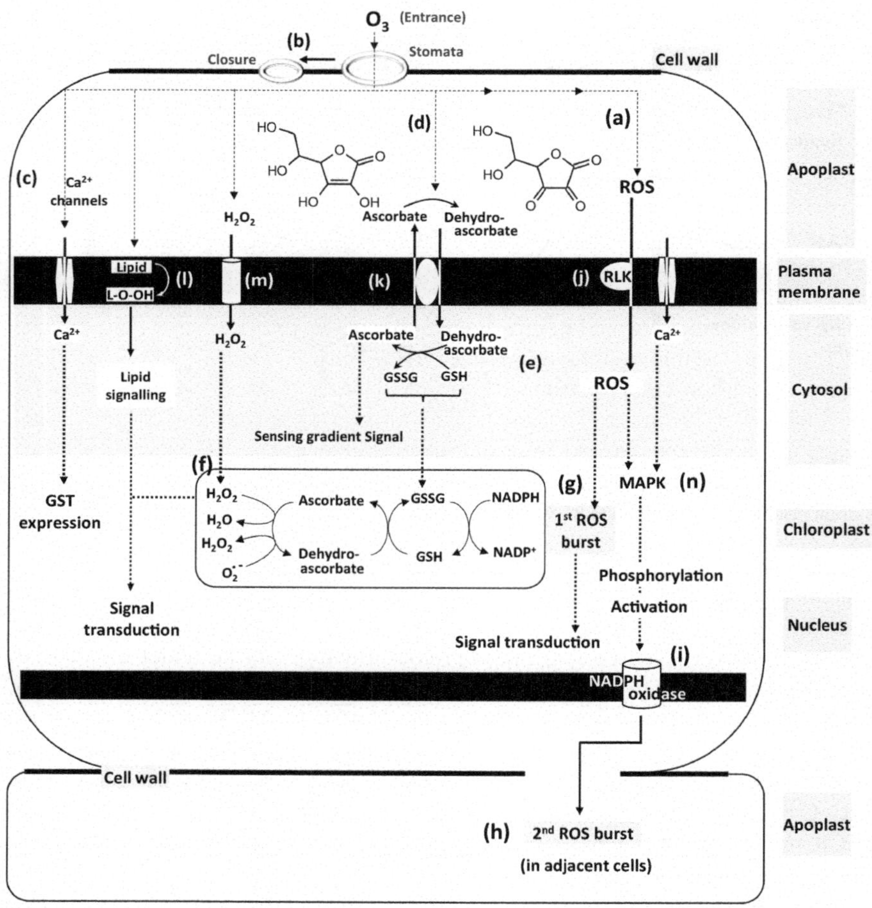

Fig. 10.6 Multiple plant defensive chemical responses and signaling pathways induced by ozone exposure

(iv) sensing, (v) activation of signaling cascades of protein kinases (e.g. MAPKs), (vi) ethylene biosynthesis and nitric oxide (NO) production.

Initially, O_3 enters by stomata and is rapidly degraded into ROS in the apoplast (Fig. 10.6a). (i) Control of stromata aperture tends to restrict ozone flux in leaves, and represents the first defense line against O_3 (Fig. 10.6b). This occurs via the activation of different types of guard cell ion channels, required for stomatal closure. Crucial control of channels' activities occurs by protein phosphorylation, a reversible process mediated by protein kinases and phosphatases (Hedrich 2012).

Elevated atmospheric O_3 induces oxidative activation of Ca^{2+} channels leading to specific increase in intracellular Ca^{2+} concentration (Fig. 10.6c) (Clayton et al. 1999). Under O_3 exposure, calcium shows rapid increase in cytosol of guard cells (within seconds after onset of exposure) (Evans et al. 2005). This causes rapid changes in the protein phosphorylation pattern which plays key role in downstream signaling process (Agrawal et al. 2002a, b, c). Interestingly, experiments on *Arabidopsis* showed O_3-specific responses that differ from other stress-induced responses (Evans et al. 2005; Short et al. 2012): fumigation by increasing O_3 levels resulted in two sequential peaks of cytosolic Ca^{2+} in *Arabidopsis*. This biphasic process represents a specific signal associated with O_3-induced oxidative stress in plant. The second O_3-induced elevation was revealed to be required for successful coupling between O_3 perception and glutathione-S-transferase (GST) expression (antioxidant defense enzyme).

(ii) Under O_3 oxidative stress, plants respond by detoxifying ROS that are immediately formed after O_3 enters the apoplast (Fig. 10.6a). For detoxification, ascorbic acid (*Asc*) is initially involved according to a biphasic mechanism: it initially accumulates in apoplast followed by its activity for ROS-scavenging (Fig. 10.6d) (Noctor and Foyer 1998; Plöchl et al. 2000; Foyer & Noctor 2009). Therefore, *Asc* is oxidized leading to the formation of dehydroascorbate (DHA). This reaction generates a need for *Asc* to maintain apoplastic antioxidant activity. This extracellular need results in a perturbation of the intracellular redox balance. Therefore, the loss of apoplastic *Asc* is compensated by a flux of intracellular *Asc* generated from the reduction of cellular DHA by DHA-reductase; the formation of *Asc* occurs at the expense of cellular glutathione (GSH) which oxidizes into glutathione disulfide GSSG (Fig. 10.6e) (Foyer and Halliwell 1976; Foyer and Noctor 2009, 2011). This *Asc* recycling process undertakes the cellular *Asc* redox state by involving a time-dependent relationship between the oxidation of extracellular *Asc* and subsequent oxidation of intracellular gluthatione (Luwe et al. 1993). Ascorbate (*Asc*) accumulation level and its redox state in leaf tissue have been shown to correlate with ozone tolerance in many plant species (Conklin et al. 1996; Conklin and Barth 2004; Feng et al. 2010; Cheng et al. 2007). A genetic proof of this defensive trend was provided by isolated *Asc*-biosynthetic mutants showing increased sensitivity to O_3 fumigation (Conklin et al. 1999; Yoshida et al. 2006). O_3-oxidative stress induces also the formation of hydrogen peroxide (H_2O_2) by oxidative reactions in aqueous middle (apoplast). Formed H_2O_2 can be transformed in H_2O by ascorbic acid reaction (Fig. 10.6d). Such a detoxification can also occur in intra-cellular compartments:

(iii) Apoplast serves as the first space for ROS detoxification after ozone induced oxidative stress. However, when stress level increase, the antioxidant capacity of apoplast tends to be exceeded leading to propagation of ROS accumulation to intracellular compartment followed by ROS burst process (Fig. 10.6f–h). H_2O_2 can also diffuse through aquaporins of plasma membrane to reach cytosol and organelles where it is reduced to water by the ascorbate–glutathione cycle in chloroplast (Fig. 10.6f).

ROS signals are firstly mediated from the apoplast to the cytosol by means of ROS-sensitive transmembrane proteins called receptor-like-kinases (RLKs) (Fig. 10.6j). RLKs have extracellular signal-sensing domain and an intracellular protein kinase domain. Among the RLKs, cysteine-rich ones are strong candidates for mediating ROS signals from apoplast to cytosol because of the sensitivity of cysteine residues to redox regulation (Fig. 10.3c) (Wrzaczek et al. 2010). Oxidative modification of RLKs by modifying critical amino acid residues tends to activate MAPK proteins (mitogen activated protein kinases) (Fig. 10.6m) involved in next ROS burst.

Increased levels of early formed ROS are followed by ROS burst which represents a key process for systemic signaling and spread of ROS signal (Fig. 10.6g, h) (Miller et al. 2009; Dubiella et al. 2013). ROS burst is mediated by two types of enzymes:

- Mainly, NADPH oxidases localized in plasma membrane that generates superoxide (Fig. 10.6i) (Torres et al. 2002; Foreman et al. 2003). NADPH oxidases (also known as Respiratory Burst Oxidase Homologs (RBOHs) act under high control based on phosphorylation by protein kinases (MAPK) which are activated by calcium (Sierla et al. 2013; Wrzaczek et al. 2013).
- Type III cell wall peroxidases that produces H_2O_2 (Daudi et al. 2012; O'Brien et al. 2012).

ROS burst shows a biphasic process or bimodal kinetics associated with a first and second burst (Joo et al. 2005): the first ROS peak occurs in chloroplast (Fig. 10.6g); this suggests a tight connection between apoplast and chloroplast in ROS signaling. Chloroplast peak is followed by a second ROS peak manifesting in the apoplasts of adjacent cells to the affected ones (Fig. 10.6h) (Krasensky et al. 2017).

(iv) ROS issued from O_3 degradation or subsequent oxidative burst must be sensed to be followed by appropriate responses:

- Ascorbate gradient across the plasma membrane and the resulting disturbance of the intracellular redox balance represent signal source that could be sensed by redox-regulated transcription factors (Fig. 10.6d, e, k) (Foyer and Noctor 2009; Munné-Bosch et al. 2013).
- Lipid oxidation of the plasma membrane results in a change in membrane redox state (Fig. 10.6l) (Mueller and Berger 2009).
- Plasma membrane crossing by apoplastic H_2O_2 through aquaporins (Fig. 10.6m) (Dynowski et al. 2008; Bienert and Chaumont 2014).
- Chlroplastic ROS burst could participate in the propagation of the signal from the apoplast (Shapiguzov et al. 2012; Sierla et al. 2013) (Fig. 10.6g). The signal from the chloroplast is transmitted to the nucleus through multiple alternative pathways,

leading to changes in the nuclear gene expression (Leister 2012; Estavillo et al. 2013).

- Cystein-rich RLKs (CRKs) show transcriptional activation in response to ozone attack; this provides strong indication on their involvement in sensing or mediating ROS signaling from the apoplast (Fig. 10.6j).

(v) RLKs, in presence of calcium, activate MAPK cascade signaling mechanisms involving sequential phosphorylation of different protein kinases (Fig. 10.6n). Therefore, protein kinases phosphorylate a wide range of substrates (from enzymes to transcription factors). The multi-aspect phosphorylation activity of MAPK provides a network of biochemical activation ways in favor of specific responses to specific stresses (Colcombet and Hirt 2008; Pitzschke et al. 2009; Rodriguez et al. 2010).

In O_3-stressed *Arabidopsis*, two MAPKs (MPK3, MPK6) showed rapid and transient (0.5–2h) activation (Ahlfors et al. 2004). Subsequently to their activation, the proteins translocate to the nucleus for interaction with transcription factors and gene expression regulations. MPK3 and MPK6 were revealed to be reciprocally regulated: the suppression of one protein led to more intense activation of the other one (Miles et al. 2005). Both genotypes were ozone sensitive. After MPK3/6 reach the nucleus for transcription and gene expression, they are dephosphorylated by MAPK phosphatase (MKP2) (Lee and Ellis 2007). This feedback process is needed for balancing proper ozone response in favor of enhanced ozone sensitivity associated with variable strength and/or time of activation (Meskiene et al. 2003).

Some experiments revealed the existence of alternative ways to ascorbate for ROS detoxification: in *Trifolium repens* L., an ozone sensitive clone showed more reduced ascorbate in the apoplast than a tolerant clone (D'Haese et al. 2005). Alternative detoxifying molecules include phenolic glycosides that showed induced accumulation in the apoplast under O_3 exposure in favor of protective antioxidative role (Mahalingam et al. 2005; Tosti et al. 2006; Booker et al. 2012). O_3-stimulated transcription of several phenylpropanoid pathway genes, increasing the apoplastic concentration of sinapoyl malate.

Ozone-induced responses, including oxidative burst and hypersensitive responses (HR) (e.g. programmed cell death) showed similarities with plant defense responses to pathogen infections (Kangasjärvi et al. 1994, 2005; Baier et al. 2005). It has been shown that both O_3 exposure and pathogen attack induce salicylic acid (SA) synthesis within some hours following plant-O_3 or plant-pathogen contact (Örvar et al. 1997; Overmyer et al. 2005).

10.4 Variations of Secondary Metabolites in O_3-Stressed Plants

10.4.1 Variations of Phenylpropanoids in Leaves and Stem Wood of Populus tremula

Populus tremula x *alba* was subject to O_3-fumigation for 53 days by either light period (daylight) or continuous (light and dark) exposition (200 ± 20 nL L^{-1}) (Richet et al. 2012). Plants under control conditions were exposed to charcoal-filtered air. After O_3-fumigation, transcript levels and activities of enzymes involved in phenylpropanoid and lignin biosynthesis were measured in the leaves and wood tissue consisting of two compartments: opposite wood representing the lower part of step and tension wood representing the upper part. Wood represents the main lignin producing tissue. Lignins are produced by polymerisation processes of phenylpropanoid derivatives (Fig. 10.7). After fumigation, ten types of transcript levels concerning biosynthesis enzymes of phenylpropanoids were measured in the different tissues. Results revealed different metabolic regulation levels of phenylpropanoids associated with the different tissues with globally positive trend in the leaves vs negative trend in the wood (Fig. 10.8). At lower scale, different lignin pathways showed different induction levels in the leaves.

Transcript abundances showed higher levels in the leaves than the wood (Fig. 10.8a–c vs. d–g). This difference reflects the fact that leaves are directly exposed to air-O_3. Moreover, the two tissues showed opposite variation trends: in leaves (Fig. 10.8a–c), transcript levels of phenylpropanoid enzymes where higher after ozone-exposition than control vs opposite trend in wood (Fig. 10.8d–g). These two opposite variation trends between tissues could be indicative of readjustment of carbon availability for phenylpropanoid biosynthesis with higher carbon pool transfer in the leaves (directly exposed to O_3) versus reduction in the stem wood (Richet et al. 2012). Therefore, the decrease of lignin synthesis in wood seemed to be a consequence of reduction of carbon availability in this tissue after plant fumigation by O_3.

In the leaves, the highest transcript levels concerned four enzymes (Fig. 10.8a–c): 4CL, CAD, CCR and PAL. These four enzymes are involved in substitution mechanism concerning the lateral propane chain. The succession of these four enzymes forms a whole metabolic pathway associated with monohydroxylated aromatic ring and favoring the biosynthesis of H-lignin (Fig. 10.7). However, HCT, H3C, CCoAOMT, Cald5H and COMT transcript levels were relatively lower than the former (Fig. 10.8). As these enzymes are involved in the metabolism of S- and G-lignin, these two lignin types could have been less induced than H-lignin (Fig. 10.7). This showed tight and target metabolic regulations of phenylpropanoid and monolignol under O_3-stress effect. Stimulation of phenylpropanoid and lignin metabolic pathways in leaves under ozone was associated with possible defensive process against ROS propagation (Cabané et al. 2004; Betz et al. 2009). This hypothesis

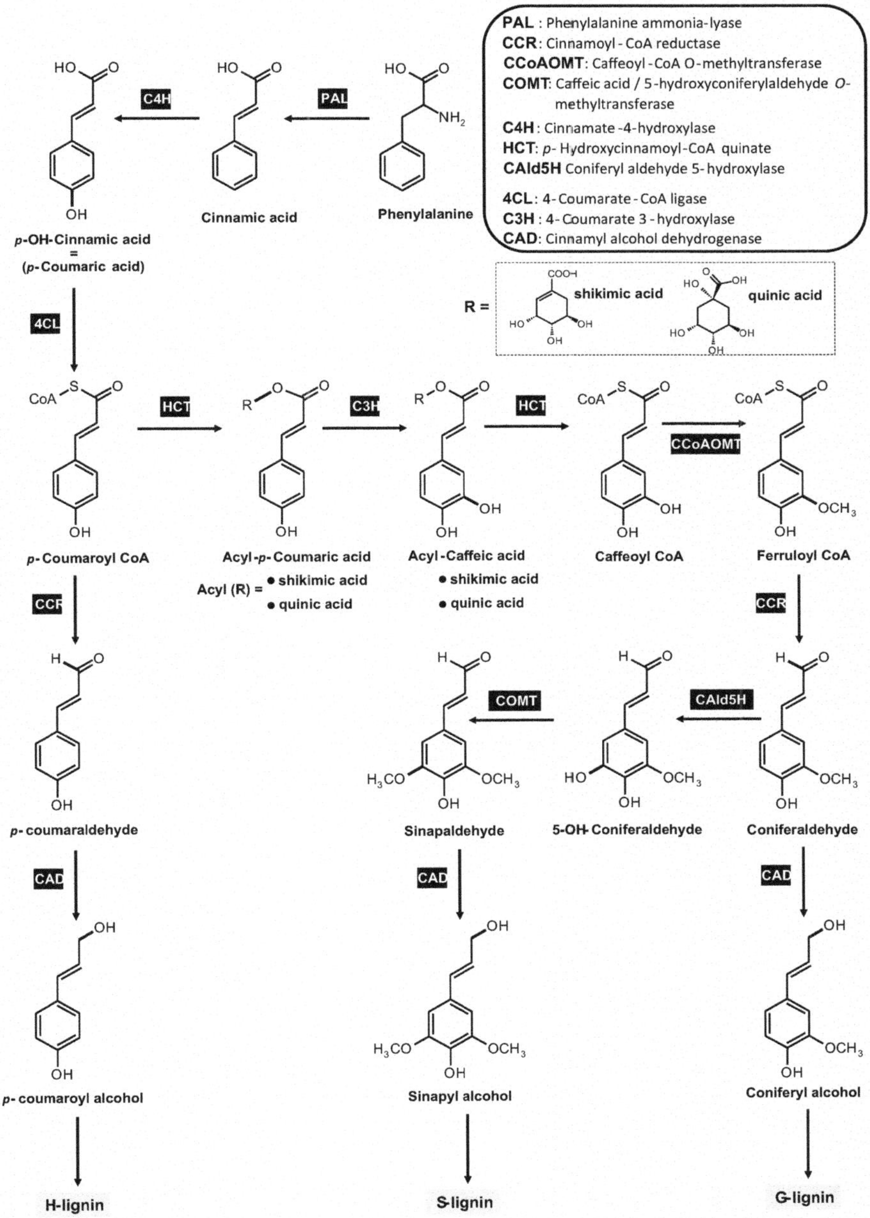

Fig. 10.7 Biosynthesis pathways of three types of lignin (monolignol) from different phenylpropanoid derivatives and regulation enzymes

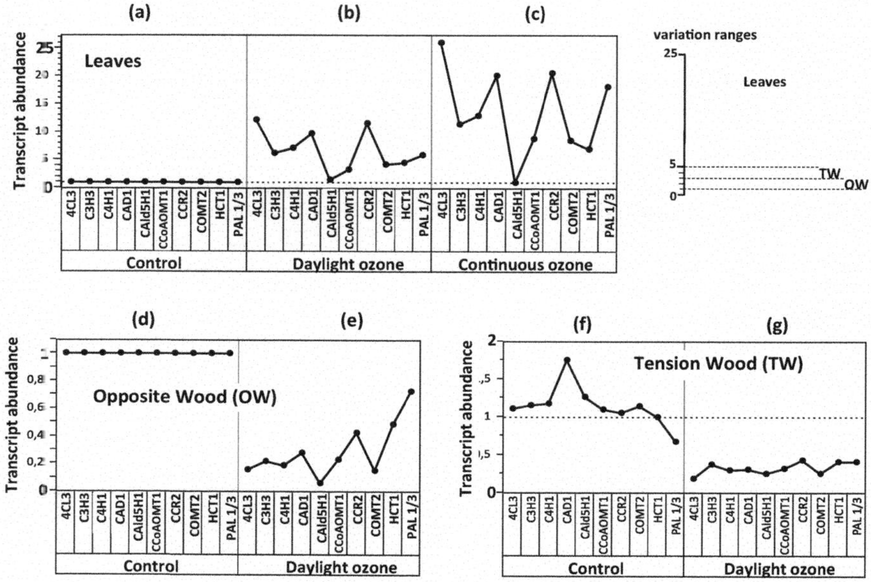

Fig. 10.8 Variation of transcript levels of different regulation enzymes involved in phenylpropanoid and lignin synthesis in the leaves (**a–c**) and stem wood (opposite wood (**d**, **e**), tension wood (**f**, **g**)) of *Populus tremula* after 53 days of exposition to ozone fumigation (200 nM L^{-1}): opposite wood (**d**, **e**) tension wood (**f**, **g**). Transcript levels were expressed in terms of folds by reference to the control conditions corresponding to the absence of ozone treatment (**a**, **d**, **f**). Ozone treatments were applied either under daylight (**b**, **e**, **g**) or continuously (in the leaves, **c**). Daylight and continuous exposition to ozone gave similar trends in stem wood; therefore, only daylight results are presented (**e**, **g**)

agreed with the fact that continuous fumigation by O_3 resulted in higher induction levels of phenylpropanoids than daily light-limited treatments (Fig. 10.8c vs. b). Moreover, PAL and CAD activities showed more rapid increase in the first 18 days under continuous than daylight fumigation (Fig. 10.9b vs. c). Also, transcript levels of genes encoding HCT1, C3H3, CCoAOMT1 and COMT2 were higher under continuous fumigation compared with daylight treatment (Fig. 10.8c vs. b). These intermediate enzymes catalyze hydroxylation-methylation of aromatic ring at C-3 and C-5 positions (Fig. 10.7). Under continuous ozone exposure, the highest stimulation of RNA levels in the leaves concerned enzymes which are involved in initial (PAL1, 4CL3) and final (CCR2, CAD1) steps of monolignol biosynthesis pathway, with amplification factors reaching 17 to 26-fold compared to their levels under control conditions (Fig. 10.8c vs. a).

Kinetic analysis of enzyme expression was carried out for shikimate dehydrogenase (SHDH), PAL and CAD activities (Fig. 10.9). These enzymes are involved in early step of phenolic metabolism (SHDH), phenylpropanoid formation (PAL) and catalysis of monolignol (lignin) production (CAD). In leaves, the three enzymes

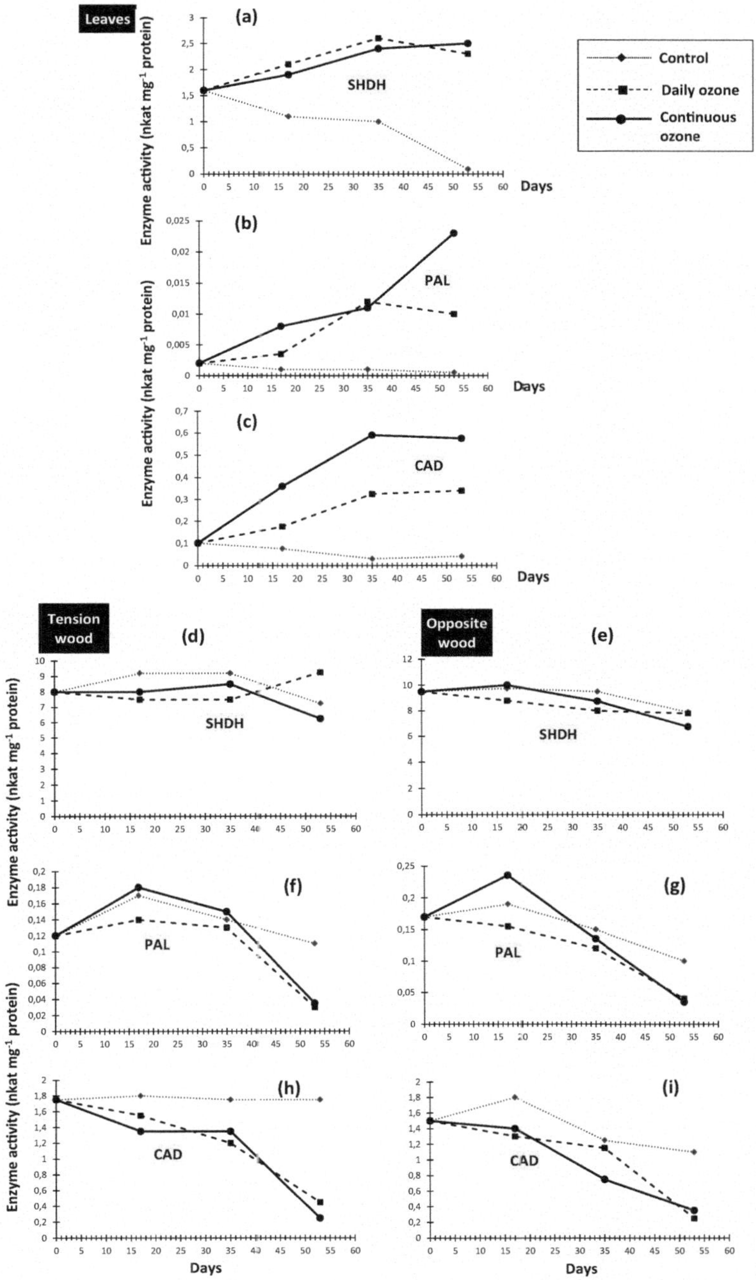

Fig. 10.9 Variation of three enzyme activities (SHDH, PAL, CAD) involved in phenylpropanoid and lignin metabolism under the effect of ozone fumigation and duration treatment in leaves (**a–c**), tension wood (**d**, **f**, **h**) and opposition wood (**e**, **g**, **i**) of *Populus tremula* x *alba* exposed to ozone for 53 days either in daylight or by continuous way

showed increased levels induced by ozone treatment compared to control, with amplification effect of treatment duration (Fig. 10.9a–c).

Ozone-induction of shikimate pathway in leaves of poplar was reported by other works (Cabané et al. 2004; Guidi et al. 2005). However, wood tissue (tension and opposite woods) showed opposite trends with decreased activities of the three enzymes (SHDH, PAL, CAD) resulting from ozone treatment and fumigation duration (Fig. 10.9d–i). These variation trends associated with enzyme activities agreed with those concerning transcript abundances (Fig. 10.8), and provided additional proof to the hypothesis of negative adjustment of phenylpropanoid and lignin metabolism in wood vs positive trend (accumulation) in leaves (directly exposed to O_3). Ozone stimulation of phenylpropanoid enzymes has been reported in various plant species (Sandermann et al. 1998; Moura et al. 2010).

References

Agrawal GK, Rakwal R, Iwahashi H (2002a) Isolation of novel rice (*Oryza sativa* L.) multiple stress responsive MAP kinase gene, OsMSRMK2, whose mRNA accumulates rapidly in response to environmental cues. Biochem Biophys Res Commun 294:1009–1016

Agrawal GK, Rakwal R, Yonekura M, Saji H (2002b) Rapid induction of defense/stress-related proteins in leaves of rice (*Oryza sativa*) seedlings exposed to ozone is preceded by newly phosphorylated proteins and changes in a 66-kDa ERK-type MAPK. J Plant Physiol 159:361–369

Agrawal GK, Rakwal R, Yonekura M, Kubo A, Saji H (2002c) Proteome analysis of differentially displayed proteins as a tool for investigating ozone stress in rice (*Oryza sativa* L.) seedlings. Proteomics 2:947–959

Ahlfors R, Macioszek V, Rudd J, Brosché M, Schlichting R, Scheel D, Kangasjärvi J (2004) Stress hormone-independent activation and nuclear translocation of mitogen-activated protein kinases in Arabidopsis thaliana during ozone exposure. Plant J 40:512–522

Atkinson R, Arey J (2003) Atmospheric degradation of volatile organic compounds. Chem Rev 103:4605–4638

Baier M, Kandlbinder A, Golldack D, Dietz KJ (2005) Oxidative stress and ozone: perception, signalling and response. Plant Cell Environ 28:1012–1020

Betz GA, Knappe C, Lapierre C, Olbrich M, Welzl G, Langebartels C, Heller W, Sandermann H, Ernst D (2009) Ozone affects shikimate pathway transcripts and monomeric lignin composition in European beech (*Fagus sylvatica* L.). Eur J for Res 128:109–116

Bienert GP, Chaumont F (2014) Aquaporin-facilitated transmembrane diffusion of hydrogen peroxide. Biochimica et Biophysica Acta 1840(5):1596–1604

Booker FL, Burkey KO, Jones AM (2012) Re-evaluating the role of ascorbic acid and phenolic glycosides in ozone scavenging in the leaf apoplast of *Arabidopsis thaliana* L. Plant Cell Environ 35:1456–1466

Cabané M, Pireaux JC, Leger E, Weber E, Dizengremel P, Pollet B, Lapierre C (2004) Condensed lignins are synthesized in poplar leaves exposed to ozone. Plant Physiol 134:586–594

Caravan RL, Vansco MF, Au K, Khan MAH, Li YL, Winiberg FAF, Kristen Zuraski Lin YH, Chao W, Trongsiriwat N, Walsh PJ, Osborn DL, Percival CJ, Lin JJM, Shallcross D, Sheps L, Klippenstein SJ, Taatjes CA, Lester MI (2020) Direct kinetic measurements and theoretical predictions of an isoprene-derived Criegee intermediate. PNAS 117(18):9733–9740

Cataldo F (2003) On the action of ozone on proteins. Polym Degrad Stab 82:105–114

Cheng FY, Burkey KO, Robinson JM, Booker FL (2007) Leaf extracellular ascorbate in relation to O_3 tolerance of two soybean cultivars. Environ Pollut 150:355–362

Clayton H, Knight MR, Knight H, McAinsh MR, Hetherington AM (1999) Dissection of the ozone-induced calcium signature. Plant J 17(5):575–579

Colcombet J, Hirt H (2008) *Arabidopsis* MAPKs: a complex signalling network involved in multiple biological processes. Biochem J 413:217–226

Conklin PL, Barth C (2004) Ascorbic acid, a familiar small molecule intertwined in the response of plants to ozone, pathogens and the onset of senescence. Plant, Cell Environ 27:959–970

Conklin PL, Williams EH, Last RL (1996) Environmental stress sensitivity of an ascorbic acid-deficient Arabidopsis mutant. Proc Natl Acad Sci USA 93:9970–9974

Conklin PL, Norris SR, Wheeler GL, Williams EH, Smirnoff N, Last RL (1999) Genetic evidence for the role of GDP-mannose in plant ascorbic acid (vitamin C) biosynthesis. Proc Natl Acad Sci USA 96:4198–4203

Criegee R (1957) The course of ozonation of unsaturated compounds. Record Chem Progr 18:110–120

Crutzen PJ, Lelieveld J (2001) Human impacts on atmospheric chemistry. Ann Rev Earth Planet Sci 29:17–45

D'Haese D, Vandermeiren K, Asard H, Horemans N (2005) Other factors than apoplastic ascorbate contribute to the differential ozone tolerance of two clones of *Trifolium repens* L. Plant, Cell Environ 28:623–632

Daudi A, Cheng Z, O'Brien JA, Mammarella N, Khan S, Ausubel FM, Bolwell GP (2012) The apoplastic oxidative burst peroxidase in *Arabidopsis* is a major component of pattern-triggered immunity. Plant Cell 24:275–287

Dubiella U, Seybold H, Durian G, Komander E, Lassig R, Witte CP, Romeis T (2013) Calcium-dependent protein kinase/NADPH oxidase activation circuit is required for rapid defense signal propagation. Proc Natl Acad Sci USA 110:8744–8749

Dynowski M, Schaaf G, Loque D, Moran O, Ludewig U (2008) Plant plasma membrane water channels conduct the signalling molecule H_2O_2. Biochem J 414:53–61

Estavillo GM, Chan KX, Phua SY, Pogson BJ (2013) Reconsidering the nature and mode of action of metabolite retrograde signaling from the chloroplast. Front Plant Sci 3:300

Evans NH, McAinsh MR, Hetherington AM, Knight MR (2005) ROS perception in *Arabidopsis thaliana*: the ozone-induced calcium response. Plant J 41:615–626

Feng Z, Pang J, Nouchi I, Kobayashi K, Yamakawa T, Zhu J (2010) Apoplastic ascorbate contributes to the differential ozone sensitivity in two varieties of winter wheat under fully open-air field conditions. Environ Pollut 158:3539–3545

Foreman J, Demidchik V, Bothwell JH, Mylona P, Miedema H, Torres MA, Dolan L (2003) Reactive oxygen species produced by NADPH oxidase regulate plant cell growth. Nature 422:442–446

Foyer CH, Halliwell B (1976) The presence of glutathione and glutathione reductase in chloroplasts: a proposed role in ascorbic acid metabolism. Planta 133:21–25

Foyer CH, Noctor G (2009) Redox regulation in photosynthetic organisms: signaling, acclimation, and practical implications. Antioxid Redox Signal 11:861–905

Foyer CH, Noctor G (2011) Ascorbate and glutathione: the heart of the redox hub. Plant Physiol 155:2–18

Guidi L, Degl'Innocenti E, Genovesi S, Soldatini GF (2005) Photosynthetic process and activities of enzymes involved in the phenylpropanoid pathway in resistant and sensitive genotypes of *Lycopersicon esculentum* L. exposed to ozone. Plant Sci 168:153–160

Hedrich R (2012) Ion channels in plants. Physiol Rev 92:1777–1811

Johnston H (1971) Reduction of stratospheric ozone by nitrogen oxide catalysts from supersonic transport exhaust. Science 173:517–522

Joo JH, Wang S, Chen JG, Jones AM, Fedoroff NV (2005) Different signaling and cell death roles of heterotrimeric G protein alpha and beta subunits in the *Arabidopsis* oxidative stress response to ozone. Plant Cell 17:957–970

Kangasjärvi J, Talvinen J, Utriainen M, Karjalainen R (1994) Plant defense systems induced by ozone. Plant Cell Environ 17:783–794

Kangasjärvi J, Jaspers P, Kollist H (2005) Signalling and cell death in ozone-exposed plants. Plant Cell Environ 28:1021–1036

Kotiaho T, Eberlin MN, Vainiotalo P, Kostiainen R (2000) Electrospray mass and tandem mass spectrometry identification of ozone oxidation products of amino acids and small peptides. J Amer Soc Mass Spectrom 11:526–535

Krasensky J, Carmody M, Sierla M, Kangasjärvi J (2017) Ozone and reactive oxygen species. In: eLS. Wiley, Chichester. https://doi.org/10.1002/9780470015902.a0001299.pub

Lee JS, Ellis BE (2007) *Arabidopsis* MAPK phosphatase 2 (MKP2) positively regulates oxidative stress tolerance and inactivates the MPK3 and MPK6 MAPKs. J Biol Chem 282:25020–25029

Leister D (2012) Retrograde signaling in plants: from simple to complex scenarios. Front Plant Sci 3:135

Lewis TR, Blitz MA, Heard DE, Seakins PW (2015) Direct evidence for a substantive reaction between the Criegee intermediate, CH_2OO, and the water vapour dimer. Phys Chem Chem Phys 17:4859–4863

Luwe MWF, Takahama U, Heber U (1993) Role of ascorbate in detoxifying ozone in the apoplast of spinach (*Spinacia oleracea* L.). Plant Physiol 101:969–976

Mahalingam R, Shah N, Scrymgeour A, Fedoroff N (2005) Temporal evolution of the Arabidopsis oxidative stress response. Plant Mol Biol 57:709–730

Meskiene I, Baudouin E, Schweighofer A, Liwosz A, Jonak C, Rodriguez PL, Jelinek H, Hirt H (2003) Stress-induced protein phosphatase 2C is a negative regulator of a mitogen activated protein kinase. J Biol Chem 278:18945–18952

Miles GP, Samuel MA, Zhang Y, Ellis BE (2005) RNA interference based (RNAi) suppression of AtMPK6, an *Arabidopsis* mitogen-activated protein kinase, results in hypersensitivity to ozone and misregulation of AtMPK3. Environ Pollut 138:230–237

Miller G, Schlauch K, Tam R, Cortes D, Torres MA, Shulaev V, Mittler R (2009) The plant NADPH oxidase RBOHD mediates rapid systemic signaling in response to diverse stimuli. Sci Signal 2:ra45

Moura JCMS, Bonine CAV, de Oliveira Fernandes Viana J, Dornelas MC, Mazzafera P (2010) Abiotic and biotic stresses and changes in the lignin content and composition in plants. J Integr Plant Biol 52:360–376

Mueller MJ, Berger S (2009) Reactive electrophilic oxylipins: pattern recognition and signalling. Phytochemistry 70:1511–1521

Munné-Bosch S, Queval G, Foyer CH (2013) The impact of global change factors on redox signaling underpinning stress tolerance. Plant Physiol 161:5–19

Nguyen TB, Tyndall GS, Crounse JD, Teng AP, Bates KH, Schwantes RH, Coggon MM, Zhang L, Feiner P, Milller DO, Skog KM, Rivera-Rios JC, Dorris M, Olson KF, Koss A, Wild RJ, Brown SS, Goldstein AH, de Gouw JA, Brune WH, Keutsch FN, Seinfeld JH, Wennberg PO (2016) Atmospheric fates of Criegee intermediates in the ozonolysis of isoprene. Phys Chem Chem Phys 18:10241–10254

Noctor G, Foyer CH (1998) Ascorbate and glutathione: keeping active oxygen under control. Ann Rev Plant Physiol Plant Mol Biol 49:249–279

O'Brien JA, Daudi A, Finch P, Butt VS, Whitelegge JP, Souda P, Ausubell FM, Bolwell GP (2012) A peroxidase-dependent apoplastic oxidative burst in cultured *Arabidopsis* cells functions in MAMP-elicited defense. Plant Physiol 158(4):2013–2027

Örvar BL, McPherson J, Ellis BE (1997) Pre-activating wounding response in tobacco prior to high-level ozone exposure prevents necrotic injury. Plant J 11:203–212

Overmyer K, Brosché M, Pellinen R, Kuittinen T, Tuominen H, Ahlfors R, Keinänen M, Saarma M, Scheel D, Kangasjärvi J (2005) Ozone-induced programmed cell death in the Arabidopsis radical-induced cell death1 Mutant. Plant Physiol 137(3):1092–1104

Pitzschke A, Schikora A, Hirt H (2009) MAPK cascade signalling networks in plant defence. Curr Opin Plant Biol 12:421–426

Plöchl M, Lyons T, Ollerenshaw J, Barnes J (2000) Simulating ozone detoxification in the leaf apoplast through the direct reaction with ascorbate. Planta 210:454–467

Pryor WA, Das B, Church DF (1991) The ozonation of unsaturated fatty acids: aldehydes and hydrogen peroxide as products and possible mediators of ozone toxicity.

Rao MV, Koch JR, Davis KR (2000) Ozone: a tool for probing programmed cell death in plants. Plant Mol Biol 44:345–358

Rao MV, Lee H, Davis KR (2002) Ozone-induced ethylene production is dependent on salicylic acid, and both salicylic acid and ethylene act in concert to regulate ozone-induced cell death. Plant J 32:447–456

Richet N, Tozo K, Afif D, Banvoy J, Legay S, Dizengremel P, Cabane M (2012) The response to daylight or continuous ozone of phenylpropanoid and lignin biosynthesis pathways in poplar differs between leaves and wood. Planta 236:727–737

Rodriguez MC, Petersen M, Mundy J (2010) Mitogen-activated protein kinase signaling in plants. Annu Rev Plant Biol 61:621–649

Sandermann HJ, Ernst D, Heller W, Langebartels C (1998) Ozone: an abiotic elicitor of plant defence reactions. Trends Plant Sci 3:47–50

Shapiguzov A, Vainonen JP, Wrzaczek M, Kangasjärvi J (2012) ROS-talk—how the apoplast, the chloroplast, and the nucleus get the message through. Front Plant Sci 3:292

Sharma VK, Graham NJD (2010) Oxidation of amino acids, peptides and proteins by ozone: a review. Ozone Sci Eng 32:81–90.

Short EF, North KA, Roberts MR, Hetherington AM, Shirras AD, McAinsh MR (2012) A stress-specific calcium signature regulating an ozone-responsive gene expression network in *Arabidopsis*. Plant J 71:948–961

Sierla M, Rahikainen M, Salojärvi J, Kangasjärvi J, Kangasjärvi S (2013) Apoplastic and chloroplastic redox signaling networks in plant stress responses. Antioxid Redox Signal 18:2220–2239

Sindelarova K, Granier C, Bouarar I, Guenther A, Tilmes S, Stavrakou T, Müller JF, Kuhn U, Stefani P, Knorr W (2014) Global data set of biogenic VOC emissions calculated by the MEGAN model over the last 30 years. Atmos Chem Phys 14:9317–9341

Torres MA, Dangl JL, Jones JD (2002) *Arabidopsis* gp91phox homologues AtrbohD and AtrbohF are required for accumulation of reactive oxygen intermediates in the plant defense response. Proc Natl Acad Sci USA 99:517–522

Tosti N, Pasqualini S, Borgogni A, Ederli L, Falistocco E, Crispi S, Paolocci F (2006) Gene expression profiles of O_3-treated Arabidopsis plants. Plant, Cell Environ 29:1686–1702

Vainonen JP, Kangasjärvi J (2015) Plant signalling in acute ozone exposure. Plant Cell Environ 38:240–252

Wrzaczek M, Brosché M, Salojärvi J et al (2010) Transcriptional regulation of the CRK/DUF26 group of receptor-like protein kinases by ozone and plant hormones in *Arabidopsis*. BMC Plant Biol 10:95

Wrzaczek M, Brosché M, Kangasjärvi J (2013) ROS signaling loops—production, perception, regulation. Curr Opin Plant Biol 16:575–582

Yoshida S, Tamaoki M, Shikano T, Nakajima N, Ogawa D, Ioki M, Aono M, Kubo A, Kamada H, Inoue Y, Saji H (2006) Cytosolic dehydroascorbate reductase is important for ozone tolerance in *Arabidopsis thaliana*. Plant Cell Physiol 47:304–308

Part IV
Secondary Metabolites as Plant Responses to Biotic Interactions

Chapter 11
Overview on Different Biotic Interactions Involving Phytochemical Signals

Production of SMs by plants results in a wide phytochemical diversity which plays key roles in protective barriers and target defensive ways against biological enemies (Fig. 11.1a). Also, involvements of SMs in biological interactions include communicative chemical signals with partners such as pollinators, symbionts and assistance organisms.

In individual plants, diversified production of SMs provides multiple ways for cell/tissue protection against different types, levels and apparition rates (rhythms) of environment stresses (Yeshi et al. 2022). At population scale (in a given species), plants with higher phytochemical diversity tend to be more resistant to environmental stresses. Moreover, phytochemical diversity results in different sensitivities of plant consumers toward differentiated plant taxa (Espinosa-García 2022).

11.1 Overview on Interactions Between Plant and Enemies (Herbivores, Parasites)

Plants confront different attacks from herbivore and pathogenic organisms. They overcome these wounding situations by producing different types of SMs playing sequential and complementary roles in defenses against outside attacks. Phytochemical defense strategies in plants include (Fig. 11.1a):

(i) Anticipated chemical barriers provided by pre-synthesized and stored SMs that manifest astringency, repulsive odors, irritant or toxic effects in response to expected animal or microbial attacks (Fig. 11.1a1).

N. Semmar, *Secondary Metabolites in Plant Stress Adaptation*, Signaling and Communication in Plants, https://doi.org/10.1007/978-3-031-52595-7_11

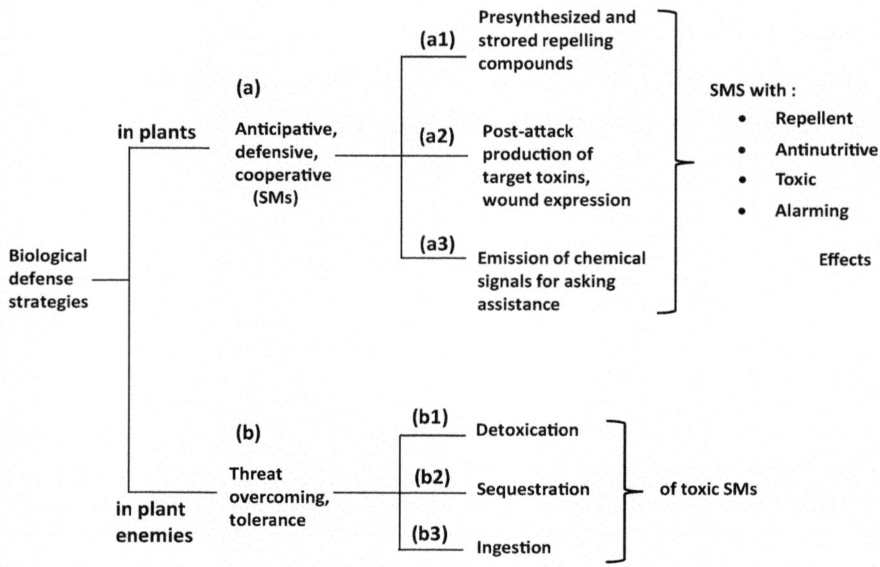

Fig. 11.1 Different protective/defensive strategies between plants and their enemies based on different effects and fates of secondary metabolites (SMs)

(ii) Specialized or target toxins that are synthesized after attack to respond to unexpected or sudden disturbances and to express wounding internal states (Fig. 11.1a2).

(iii) Emitted volatile SMs for asking assistance. Emitted SMs are mixtures with specific smells attracting the enemies of plant enemies (e.g. predator of herbivore) (Fig. 11.1a3).

Beyond balanced confrontation strategies based on repulsive and toxic phyto-compounds, plants developped assistance strategies based on chemical attractions of predators of plants' enemies using released (volatile) SMs with specific smells. Attractive released volatile phytocompounds can be mixed to result in specifically attractant scents of called predator assistant.

Production by host plants of potentially harmful chemical compounds against parasites and herbivores result in the development of different tolerance capacities within enemies' populations (Hadacek 2002). In plants' enemies (parasites, herbivores), toxicity of SMs was confronted through different adaptive strategies including detoxication and sequestration of reactive compounds for their subsequent use. In animal world, tolerance strategies were developed to overcome repelling and toxic effects of plant SMs. Such strategies include (Fig. 11.1b):

(i) Detoxication by antagonistic catalytic reactions facilitating elimination of exogenous compounds (Shapiro 1991; Brattsten 1992) (Fig. 11.1b1).

(ii) Sequestration of SMs for their reutilization (Shapiro 1991; Hartmann 1992) (Fig. 11.1b2).

(iii) Efficient ingestion with trophic assistance of endosymbiotic bacteria (Fig. 11.1b3). Endosymbiotic bacteria of gut have shorter generation cycles and more rapid adaptation than their host animals (insects). Such a microbial strategy played crucial role in the coevolution of plants and herbivore insects (Spiteller et al. 2000).

Tolerance strategies favored co-evolution between plants and animal or microbial species leading to high biodiversity and shared (mutual) benefits in ecosystems.

Interactions between plants and other species represent a strong functional source for generation and maintaining phytochemical diversity (Espinosa-García 2022). Individuals and species which potentially benefit from phytochemical diversity are naturally selected leading to coevolution processes. Pairwise coevolution (e.g. plant–insect) is governed by reciprocal gene-for-gene interactions (Janzen 1980). Insect species which are closely related to plants with divergent SM chemistry tend to specialize on such plants (Jermy 1984; Beran et al. 2019). Co-evolved traits are continuously mixed through gene-flows between populations (Thompson 1993, 1999).

Apart from the offensive-defensive metabolic ways between plants and their enemies, phytocompounds are involved in cooperative strategies between plants and partners including pollinators and symbionts. Pigments as anthocyanins and tetraterpenes are responsible for different colors of plant tissues (mainly flowers) which exert selective or specific attraction of plant partners. Apart from the colors and scents, attractive chemical signals are exerted by released SMs which are enzymatically recognized by specific plant partners: typical case is provided by symbiose between Leguminosae and atmospheric nitrogen-fixing soil bacteria in nitrogen poor soils (Franche et al. 2009; Dong and Song 2020).

A synthesis on interactions of plant SMs with biotic (and abiotic) factors is given in Fig. 11.2. This book mainly focuses on the defensive aspect of SMs against biotic stresses including pathogen infections and herbivore attacks (see Chaps. 12 and 13).

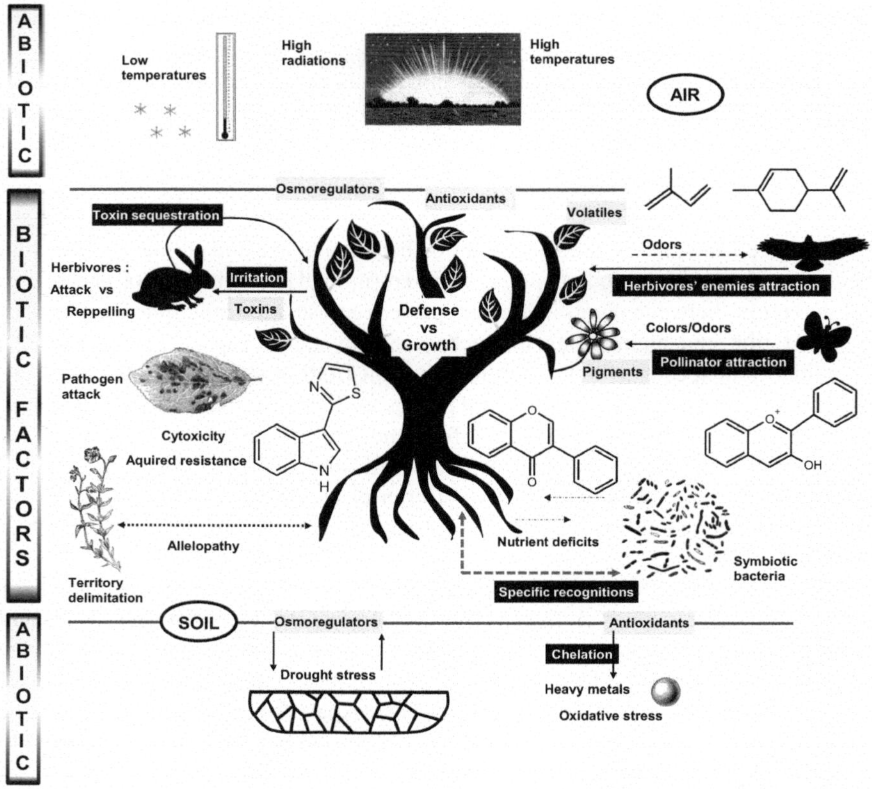

Fig. 11.2 A synthesis on interactions of plant SMs with biotic (and abiotic) factors

References

Beran F, Kollner TG, Gershenzon J, Tholl D (2019) Chemical convergence between plants and insects: biosynthetic origins and functions of common secondary metabolites. New Phytol 223:52–67

Brattsten LB (1992) Metabolic defenses against plant allelochemicals. In: Rosenthal GA, Berenbaum MR (eds) Herbivores. Their interaction with secondary plant metabolites. Ecological and evolutionary processes, 2nd edn. Academic, San Diego, pp 175–242

Dong W, Song Y (2020) The significance of flavonoids in the process of biological nitrogen fixation. Int J Mol Sci 21:5926

Espinosa-García FJ (2022) The role of phytochemical diversity in the management of agroecosystems. Bot Sci 100(Special):S245–S262.

Franche C, Lindström K, Elmerich C (2009) Nitrogen-fixing bacteria associated with leguminous and non-leguminous plants. Plant Soil 321:35–59

Hadacek F (2002) Secondary metabolites as plant traits: current assessment and future perspectives. Crit Rev Plant Sci 21(4):273–322

Hartmann T (1992) Alkaloids. In: Rosenthal GA, Berenbaum MR (eds) Herbivores. Their interaction with secondary plant metabolites. The chemical participants, 2nd edn. Academic, San Diego, pp 79–121

Janzen DH (1980) When is it coevolution? Evolution 20:611–612

Jermy T (1984) Evolution of insect/host plant relationships. Am Nat 124:609–630

Shapiro JP (1991) Phytochemicals at the plant–insect interface. Arch Insect Biochem Physiol 17:191–200

Spiteller D, Dettner K, Boland W (2000) Gut bacteria may be involved in interactions between plants, herbivores and their predators: microbial biosynthesis of N-acylglutamine surfactants as elicitors of plant volatiles. Biol Chem 381:755–762

Thompson JN (1993) Preference hierarchies and the origin of geographical speciation in host use in swallowtail butterflies. Evolution 47:1585–1594

Thompson JN (1999) Specific hypothesis on geographic mosaic of coevolution. Am Nat 153 (Suppl.):S1–S14

Yeshi K, Crayn D, Ritmejeryte E, Wangchuk P (2022) Plant Secondary metabolites produced in response to abiotic stresses has potential application in pharmaceutical product development. Molecules 27:313

Chapter 12
Plant Defense Against Pathogen Attacks

12.1 Effects of Pathogen Infections on Plants

Biotic stresses can originate from biological organisms that use plants as nutritive sources or from neighbor plants that compete for territory delimitation. Consumer organisms of plants include pathogenic agents (fungi, bacteria, viruses, nematodes) and herbivores.

Pathogenic agents cause several diseases to different plant parts (roots, seedling, leaves, stem, flowers, fruits, tuber, etc.) and can be identified from different symptoms. Symptoms correspond to visible reactions manifested by infected plants and associated with different causal agents. They are of different types including damping off, wilts, leaf spots, cankers, blights, galls, root knots, mosaics, storage rots, etc. More details are given in Fig. 12.1 (Byther et al. 2000; Pscheidt et al. 2019; Bell and Guggenheim 2019).

Correct diagnosis requires the location of damage on plants, identification of its types and subsequent symptoms.

Occurrence and development of pathogenic infections require coincidence between appropriate states of three factors associated with plant host, pathogen and environment (Fig. 12.2) (Moore et al. 2019):

- Plant host should be susceptible to pathogen agent with appropriate stage of development (Fig. 12.2a).
- Pathogen agent should be virulent with sufficient number (called inoculum potential) (Fig. 12.2b). Appropriate vectors or agents of dispersal could be revealed to be necessary such as wind, water, insects, etc.
- Environment conditions should be favorable for development of pathogenic infections (Fig. 12.2c). They concern several factors including temperature, humidity, light, wind, soil texture, pH, density of planting and nutritional conditions (mineral deficiency or excess).

N. Semmar, *Secondary Metabolites in Plant Stress Adaptation*, Signaling and Communication in Plants, https://doi.org/10.1007/978-3-031-52595-7_12

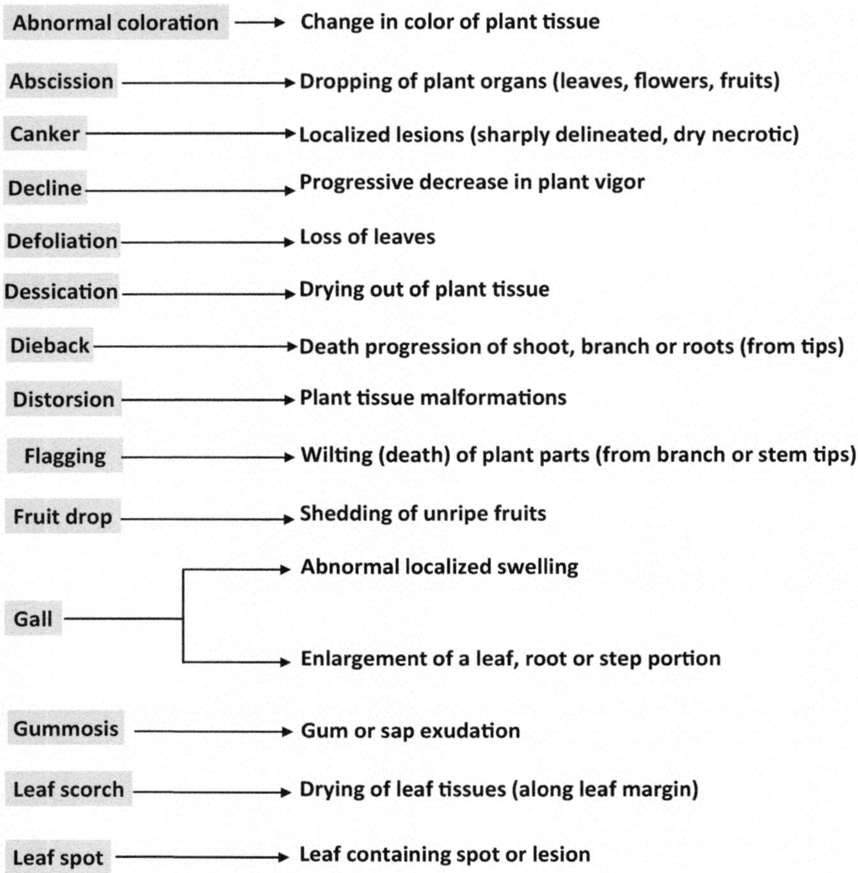

Fig. 12.1 Different symptoms visually manifested by plants infected by pathogenic agents

When plant host, pathogen agent and environment conditions coincide, pathogenic infection is developed into several steps including successively: (i) contact between susceptible plant and virulent pathogen (inoculation step) followed by (ii) the penetration of pathogen into the plant tissue leading to the parasitic relationship or infection (Fig. 12.2d). Penetration can occur either mechanically (as in fungi and nematodes) or passively through wounds, stomata or other natural opening (Fig. 12.2e). Therefore, pathogen may release enzymes causing symptoms in infected plant tissues (Fig. 12.2f). Infection and symptom manifestation are separated by a more or less long period (incubation period) that can take some days or extend to years (Fig. 12.2g). The infection stage is characterized by the reproduction of many pathogens favoring their dissemination to other plants (Fig. 12.2h).

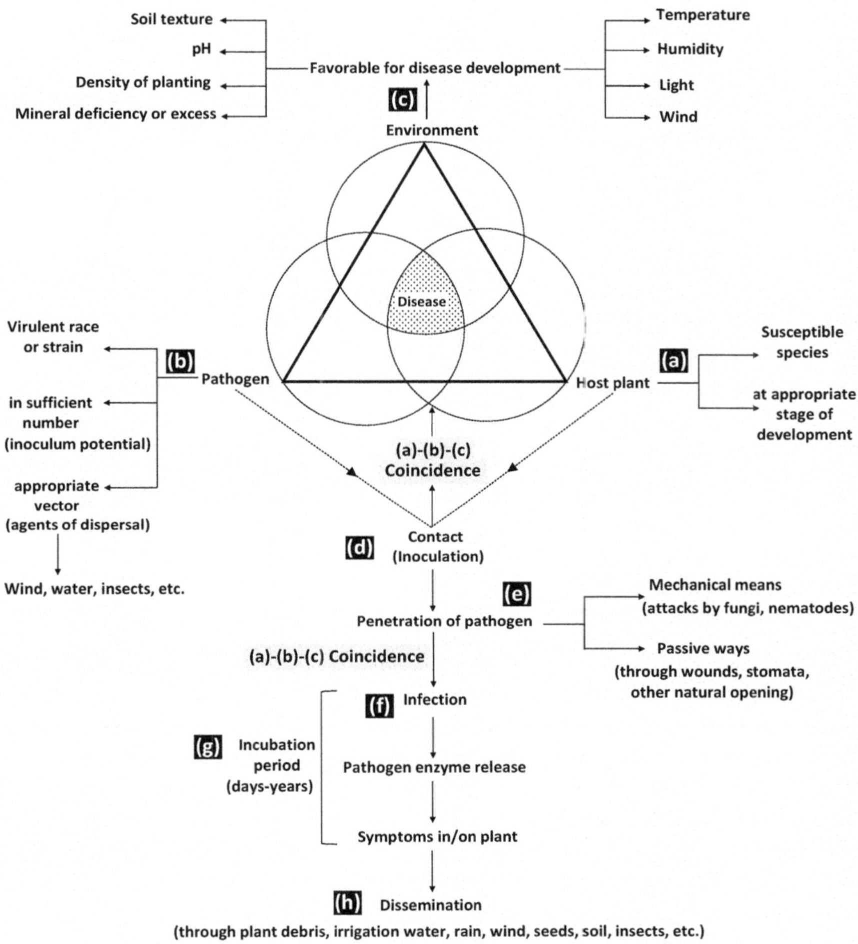

Fig. 12.2 Different conditions associated with three interactive factors involved in pathogenic infections of plants (**a–c**) according to sequential processes including inoculation, pathogen penetration, infection, incubation and dissemination (**d–h**)

Among fungi, phytopathogenic species were isolated from the genera *Alternaria*, *Botrytis*, *Colletotrichum*, *Fusarium*, *Helminthosporium*, and *Phoma*. These pathogens produce highly diversified SMs showing toxic activities against host plants. Fungal phytotoxic SMs include polyketides, phenolic compounds, terpenoids and nitrogen-containing metabolites. Chemical structures, biological sources, phytotoxic activities and structure–activity relationships of such SMs can be extensively consulted in the review of Xu et al. (2021).

Alternaria brassicicola and *Alternaria brassicae* among other *Alternaria* species are fungal pathogens causing diseases to *Brassica* vegetables including black spot or leaf spot disease. Also, visible symptoms of infections include necrotic and chlorotic lesions on the leaf, inflorescence, petiole, stem and seed. Several fungal pathogens of Brassicaceae were shown to be able of phytoalexin detoxication by means of selective enzymes and simple chemical reactions including oxidation, reduction, hydrolysis, and in one case, conjugation with glucose (Pedras and Ahiahonu 2005).

12.2 Plant Responses to Pathogen Infections

Plants protect themselves against pathogens and herbivores through different chemical, biochemical and physiological ways. These defense ways should be optimally controlled to:

(i) Trigger their manifestations at the appropriate times (stress conditions) against sudden or anticipated risks.
(ii) Guarantee their efficient protective effects through appropriate (required) production levels.
(iii) Avoid or decrease their harmful effects on plants (self-poisoning avoidance) while their toxicity is needed for stopping disturbances from enemies.

Triggering of defense processes are well-controlled through inducible resistance involving *de novo* synthesis of phytocompounds to face sudden wounding (external attack) event on one hand. These biotic attack-inducible phytocompounds are known as phytoalexins (Müller and Börger 1940). On the other hand, anticipated (expected) threats are managed by anticipative resistance involving activation of constitutive phytocompounds that play the role of a prepared protective chemical barrier against biotic attacks. These compounds are known as phytoanticipins and are previously stored in plant tissues prior to infection (VanEtten et al. 1994). They are stored and separated from their respective activating enzymes via different cellular compartments. Storage or sequestration of such toxic compounds in plants is favored by their glycosylation resulting in stabilization of the parent compounds before their activation by hydrolysis and deglycosylation triggered by infection processes. Moreover, a plant defense can be effective or not according to the type of enemy leading to the need of diversification of protection ways (phytochemical polymorphism) on one hand, and self-poisoning avoidance on other hand.

Plant defenses against pathogen attacks involve sequential mechanisms including pathogen recognition (Fig. 12.3a), downstream signaling (Fig. 12.3b) and transcriptional activation of defense-related genes (Fig. 12.3c) leading to final defense responses that restrict infection propagation, stop pathogen virulence (Fig. 12.3d) and improve immunity against next infections (Fig. 12.3e). These mechanisms are correlated with programmed cell death process in damaged tissues.

For instance, *Pseudomonas synringae* (gram-negative bacterium) activates plant immune response by translocating effector proteins (called virulence factors) into

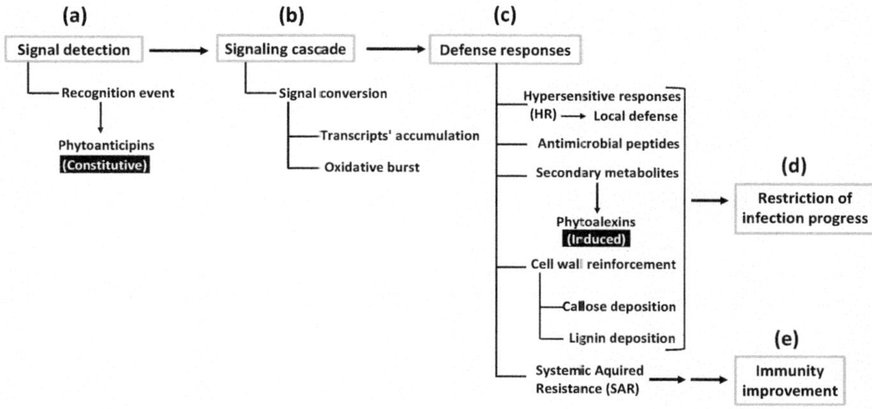

Fig. 12.3 Sequential processes involving signal detection of infection followed by cascade signal conversion leading to defensive reactions including hypersensitive response (HR) and systemic acquired resistance (SAR) against pathogen

host plant cells through a type III secretion system (Nimchuk et al. 2000). Type III effectors are detected by resistance proteins encoded by R genes. This recognition event activates a signaling cascade that is converted into rapid and localized programmed cell death corresponding to hypersensitive response (HR). HR leads to restriction of pathogen development at infection site. Moreover, it contributes to the development of a systemic response mediated by salicylic acid (SA) known as systemic acquired resistance (SAR). SAR makes noninfected leaves to acquire resistance against a wide range of pathogens (bacteria, fungi).

Efficient phytochemical responses against pathogens require appropriate (optimal) spatiotemporal distribution of secondary metabolites (the bioactive end products). This involves the accumulation of these products at appropriate time and location with sufficient amounts to satisfy their functions. Well targeted localization of required bioactive (toxic) SMs represents a crucial factor to avoid detrimental toxication of plant by its self-products. Moreover, after infection, activating enzymes and their substrates tend to be dynamically transferred to the plant-pathogen interface leading to efficient defensive local accumulations of phytocompounds. This relates to hypersensitive responses playing key role in local surrounding of infection to avoid its propagation in intact cells (Fig. 12.3c). Bioactive phytochemicals provide immunity aspects to plants through different ways including disintegration of cell membranes and inhibition of protein activities of pathogenic organisms. Pathogen infection can be stopped or braked by plants by reinforcing the cell wall via deposition of lignin and callose (Fig. 12.3c) (Maroti et al. 2011; Bednarek 2012a; Miedes et al. 2014; Voigt 2014). Moreover, SMs significantly contribute to plant immunity by either pre-synthesis (storage) or pathogen-induction mechanisms. Phytochemicals associated with these two complementary defense mechanisms correspond to phytoanticipins (Fig. 12.3a) and phytoalexins (Fig. 12.3c), respectively.

12.2.1 Phytoanticipins

Pre-synthesized and stored compounds in plant tissues correspond to phytoanticipins (VanEtten et al. 1994). They are pre-formed (pre-invasive) compounds that provide preventive aspect against attempted or sudden biological attacks. Phytoanticipins (toxic) are initially separated within storage compartments to protect plant against their toxicities. In addition to compartmentation, self-toxicity protection in plants is chemically brought by glycosylation of phytoanticipins. Under biological attack or infection, phytoanticipins are released from their storage compartments then activated by the cleavage (hydrolysis) of glycosidic bond by means of glycosidase (Halkier and Gershenzon 2006; Morant et al. 2008; Frey et al. 2009). Under other aspect, the amphiphilic character found in some phytoanticipins as saponins and steroid alkaloids favors their targeted interactions with cell membranes and subsequently their antibiotic activities. Also, preventive or pre-invasive compounds (phytoanticipins) should be secreted outside plant cells to potentially stop invasive growth of parasites. Under this aspect, many volatile terpenes are released as defensive compounds (Dudareva et al. 2006).

When biological stress results in loss of cellular integrity, previously compartmentalized glucosinolates are released and become in contact with hydrolase-type enzymes (β-thioglucoside glucohydrolases, TGGs, also known as myrosinases). This leads to hydrolysis and subsequent release of isothiocyanates (ITCs) which are highly reactive and potentially toxic compounds against pathogens (Fig. 12.4a) (Bones and Rossiter 2006; Wittstock and Burow 2010). Antimicrobial properties of ITCs result in restriction of the growth of microbial pathogens (Mithen et al. 1986; Brader et al. 2001; Tierens et al. 2001). ITCs are predominant and spontaneously formed products from glucosinolate degradation.

Apart from ITCs, glucosinolate degradation can give other possible products that involve enzyme action and/or appropriate pH (Fig. 12.4):

- Thiocyanate formation is enforced by special enzymes (thiocyanate forming proteins, TFPs) that catalyze appropriate molecular rearrangement to yield thiocyanates (Burow et al. 2006) (Fig. 12.4b).
- Also, nitrile formation from glucosinolate degradation is favored by low pH values, promoted by ferrous ions and catalyzed by epithiospecifier protein (ESP) (Larsen 1981; Selmar 2010) (Fig. 12.4c).

The central carbon of the isothiocyanate group (–N=C=S) is strongly electrophile and it plays a key role in the high chemical reactivity of ITCs through rapid addition reactions with nucleophilic centers (e.g. amino and thiol groups) (Brown and Hampton 2011). This provides direct reactions with:

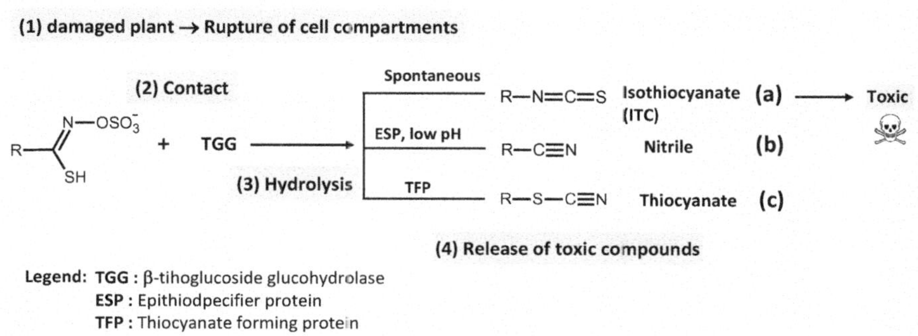

Fig. 12.4 Release of isothiocyanates from glucosinolates under the hydrolytic effect of β-thioglucoside glucohydrolases (TGGs) (or myrosinases)

- Cysteine, selenocysteine or amine residues leading to negative effects on enzyme activities.
- Thiol group of glutathione leading to disturbing balance between reduced and oxidized glutathione affecting subsequently cell redox homeostasis (Kirlin et al. 1999; Brown and Hampton 2011). Depletion of reduced glutathione has repercussion on the generation of ROS subsequently impacting programmed cell death and stomata closure (Noctor et al. 2011; Andersson et al. 2015).

Cyanogenic glucosides (CGs) display their biological function after hydrolysis reactions catalyzed by associated cyanogenic β-glucosidases (Morant et al. 2008). CG-hydrolysis releases unstable α-hydroxynitriles which decompose to hydrogen cyanide (HCN) either spontaneously or by enzymatic effect α-hydroxynitrile lyases (Fig. 12.5). Cyanide causes the inhibition of respiratory oxidation leading to stopping cellular energy production (Morant et al. 2008). Moreover, decomposition of α-hydroxynitriles releases aldehyde or ketone according to the parental CG structure (Møller 2010; Gleadow and Møller 2014).

Plants self-protect against cyanogenic glucosides by separating them from their respective glucosidases through compartmentation mechanisms at tissue and cellular scales: for instance, CGs and associated glucosidases are selectively accumulated in epidermal cells and neighboring mesophyll cells, respectively (Heraud et al. 2018; Thayer & Conn 1981). Within cells, CGs are stored in vacuole

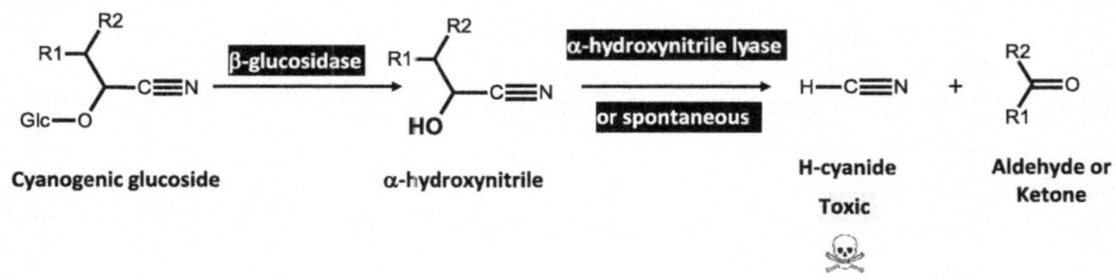

Fig. 12.5 Release of reactive and toxic hydrogen cyanide from hydrolysis of cyanogenic glucosides

12.2.2 Phytoalexins

SMs that are synthesized de novo in response to pathogen challenge were defined as phytoalexins (Müller and Börger 1940; Paxton 1981). Phytoalexins represent a heterogeneous group of pathogen damage-induced metabolites providing (Shinbo et al. 2006; Schmelz et al. 2011; Huffaker et al. 2011):

- Curative responses against biotic stress (e.g. pathogen infections).
- Molecular markers of disease resistance.

Phytoalexins tend to exert harmful effects against cell membrane of pathogen agent through various mechanisms. The roles of phytoalexins in defense against plant pathogens were studied by means of different ways including *in silico*, *in vitro* and bioengineering approaches. Some illustrative examples are given including:

- Camalexin, a phytoalexin of *Arabidopsis thaliana* (Brassicacea) showing anti-*Pseudomonas synringae* activity in *in vitro* cultures (Fig. 12.6a–c) (Rogers et al. 1996).
- Camelexin, brassilexin, rutalexin and spirobrassinin studied by docking with different pathogen proteins (*in silico* approach) (Fig. 12.16) (Pathak et al. 2016). Docking served for analysis the abilities of phytoalexins of *Brassica* to neutralize proteins of *Alternaria* spp.

In vitro essays showed that camalexin (500 μg/ml) (Fig. 12.6a) caused several damage types to membranes of *Pseudomonas synringae* pv. *maculicola* ES4326 (*Psm* ES4326) consisting of increases in ion leakage, protein leakage and supernatant serine deaminase activity (Fig. 12.6a) (Rogers et al. 1996). These effects were similar to those associated with toluene which is known to be a membrane disruptive agent (De Smet et al. 1978; Jackson and DeMoss 1965). Moreover, camalexin rapidly reduced proline uptake and stopped it within 3 min of its adding in cell culture middle (Fig. 12.6b). 14C proline was added at 0 min in *in vitro* cultures of *Psm* ES4326, then either 5% of dimethyl sulfoxide (DMSO) or camalexin-DMSO (500 μg/ml–5%, vol/vol) was added at 2 min (indicated by arrow in Fig. 12.6b). Proline uptake was measured by means of counts per minute which clearly indicated a strong negative effect of camalexin. Further confirmation on such negative effect was provided by comparing proline uptake between *Psm* ES4326 and two partially resistant mutants to camalexin (Fig. 12.6c): mutants showed more ability to take up proline than wild-type *Psm* ES4326 in presence of camalexin (Rogers et al. 1996). However, in all the cases, higher supplements of camalexin resulted in relatively lower uptake of proline by *Pseudomonas syringae*.

Docking (*in silico* approach) consists in computationally identifying protein cavity(ies) favoring binding interaction(s) with ligand (metabolite), and evaluating subsequent ligand–protein binding affinity by system energetic scores (Wei et al.

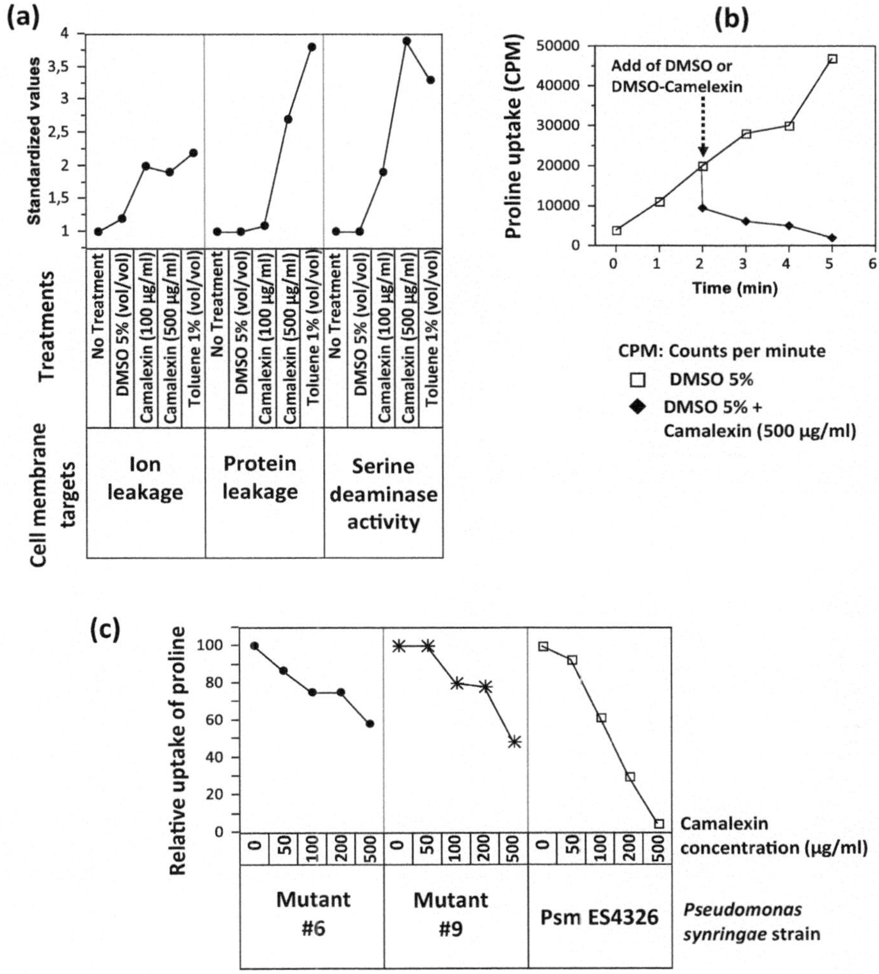

Fig. 12.6 Negative effects of camalexin (phytoalexin) on cell membrane of *Pseudomonas synringae* pv. *maculicola* ES4326 (*Psm* ES4326) in *in vitro* cultures (Rogers et al. 1996). **a** Effect on ion leakage, protein leakage and serine deaminase activity; **b** effect on proline uptake; **c** compared effect on proline uptake between wild and mutant strain of *Psm* ES4326

2002). More negative docking energy is associated with stronger ligand–protein interaction (more stable docking system). Molecular docking was applied between different phytoalexins of *Brassica* plants and proteins of *Alternaria* spp. pathogen (Pathak et al. 2016). Phytoalexins included camalexin, brassilexin, rutalexin and spirobrassinin (Fig. 12.16a). Pathogenic proteins included *Alternaria* ABC transporter, Amr1, Beta-tubulin, cutinase, fusicoccadine synthase and glutathione transferase. By reference to the most negative docking energies, the highest binding

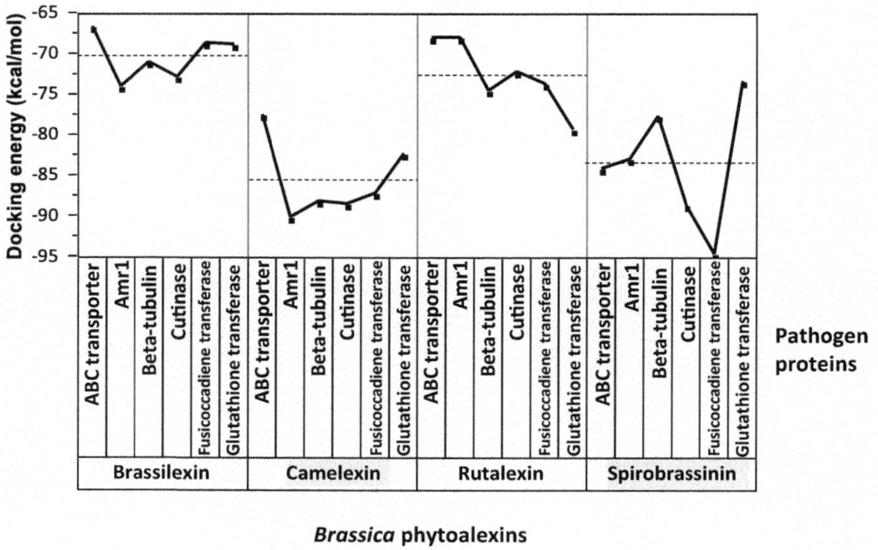

Fig. 12.7 Variation of docking energies associated with different interactions between *Brassica* phytoalexins and *Alternaria* (pathogen) proteins

affinity toward *Alternaria* pathogenic proteins was revealed by spirobrassinin with values ranging from − 73.09 to − 94.46 kcal/mol (Fig. 12.7). It was followed by camalexin with docking energies varying between − 77.34 and − 90.06 kcal/mol. However, camalexin showed less significant interactions because of absence of hydrogen bonds toward amino acids of fusicoccadiene synthase and glutathione synthase (Fig. 12.8c).

High binding affinities of spirobrassinin toward pathogen proteins were linked to strong H-bond interactions with some amino acid residues including (Fig. 12.8d):

- Asn712 and Tyr626 of ABC transporter.
- Pro48 of Amr1.
- Arg12 of beta-tubulin.
- Ser40 and His187 of cutinase.
- Gln596 of fusicoccadine synthase.
- Gln4 of glutathione transferase.

In infected leaves of *Brassica juncea*, spirobrassinin was the major produced phytoalexin (Pedras et al. 2009). Lower docking energies of spirobrassinin compared with the other phytoalexins revealed that this metabolite tends to be strong candidate model of phytoalexin for designing agriculturally protective molecules of *Brassica* against *Alternaria* spp. infections (Abdel-Farid et al. 2006).

By considering the most potent phytoalexin (spirobrassinin) for pathogen protein inhibition, different derivative metabolites were subjected to new docking after structural change of chemical substitution group R (Fig. 12.9). Such a variation aimed

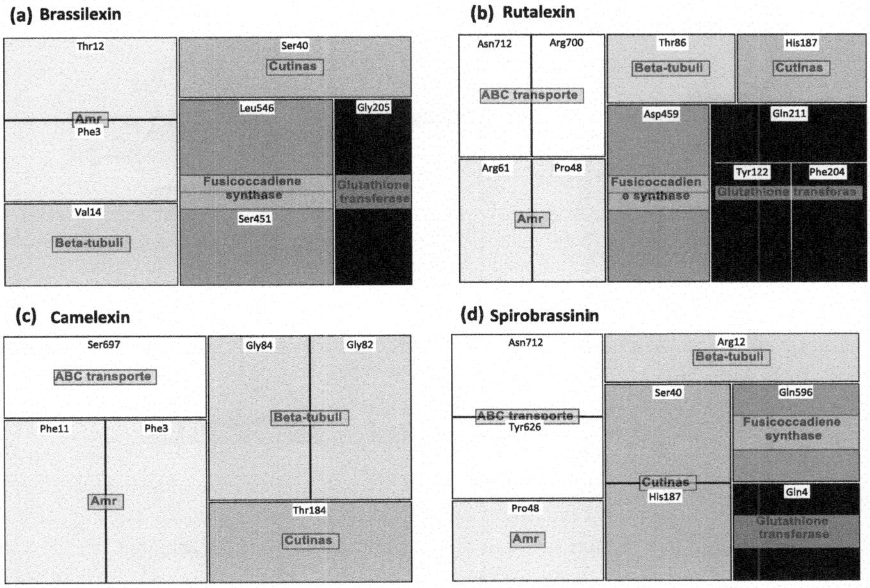

Fig. 12.8 Different localized amino acid residues of *Alternaria* (pathogen) proteins involved in H-bonds for docking with different phytoalexins of *Brassica* plants. Pathogen proteins: ABC transporter, Amr1, Beta-tubulin, cutinase, fusicoccadine synthase and glutathione transferase. Amino acid legend: Arg, arginin; Asn, asparagin; Gly, glycin; His, histidin; Leu, leucin; Phe, phenylalanine; Pro, prolin; Ser, serine; Thr, threonin; Val, valine

to modify the affinity of phytoalexin toward pathogenic proteins by conserving the molecular parts which are involved in ligand–protein interactions (pharmacophore parts) responsible for pathogenic protein inhibition. The R group list included S-OH, OH, CH_3, O–CH_3 and S–O–CH_3 corresponding to spirobrassinin derivatives 1–5, respectively. Docking of these metabolic structures with different pathogenic proteins revealed some advantages of spirobrassinins 2 and 5 toward the protein glutathione transferase due to their lower docking energies compared with that of spirobrassinin (Fig. 12.10). Interactive amino-acid residues of gluthatione transferase were mainly concerned with Ser and Gln (Fig. 12.11a) with more or less maintained locations (Fig. 12.11b): Gln4 in spirobrassinin and spirobrassinin 2; Ser78 in spirobrassinins 1, 3, 4, 5; Ser59 and Gln211 in spirobrassinins 1 and 2, respectively. Moreover, spirobrassinin 2 showed Asn3 as a third interactive amino acid residue (Fig. 12.11b).

From other experiment types, genetic engineering is commonly applied to highlight difference between wild and resistant or deficient strains leading to conclude about the phytoalexin role of considered phytocompound: for instance, isoflavonoid phytoalexins have been shown to play key role in programmed cell death in soybean tissues damaged by pathogen infection (Graham et al. 2007): RNA interference-mediated silencing of isoflavone synthase or chalcone reductase resulted in reduced production of isoflavonoid phytoalexins in soybean which consequently manifested a loss of hypersensitive response toward the oomycete pathogen *Phytophthora sojae*.

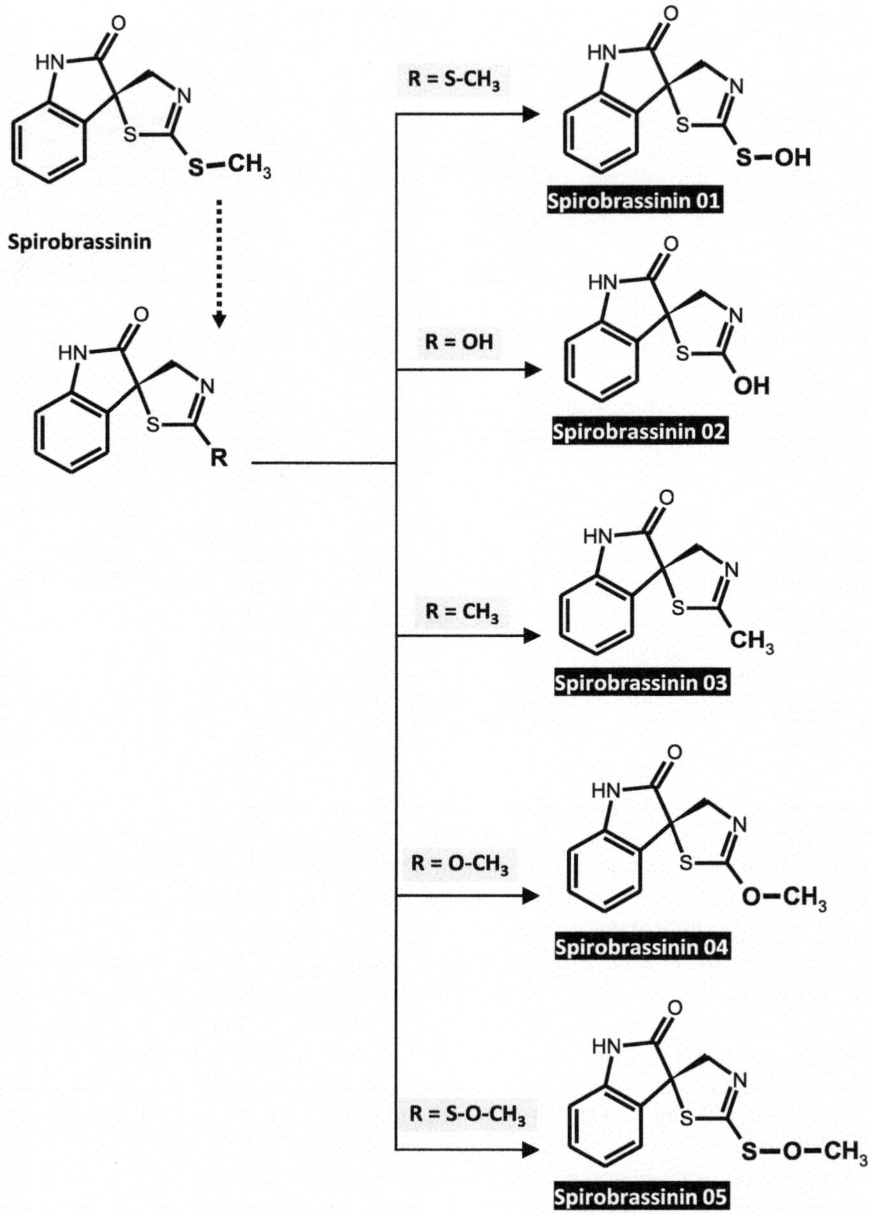

Fig. 12.9 Chemical structures of different metabolites based on spirobrassinin scaffold (potent phytoalexin)

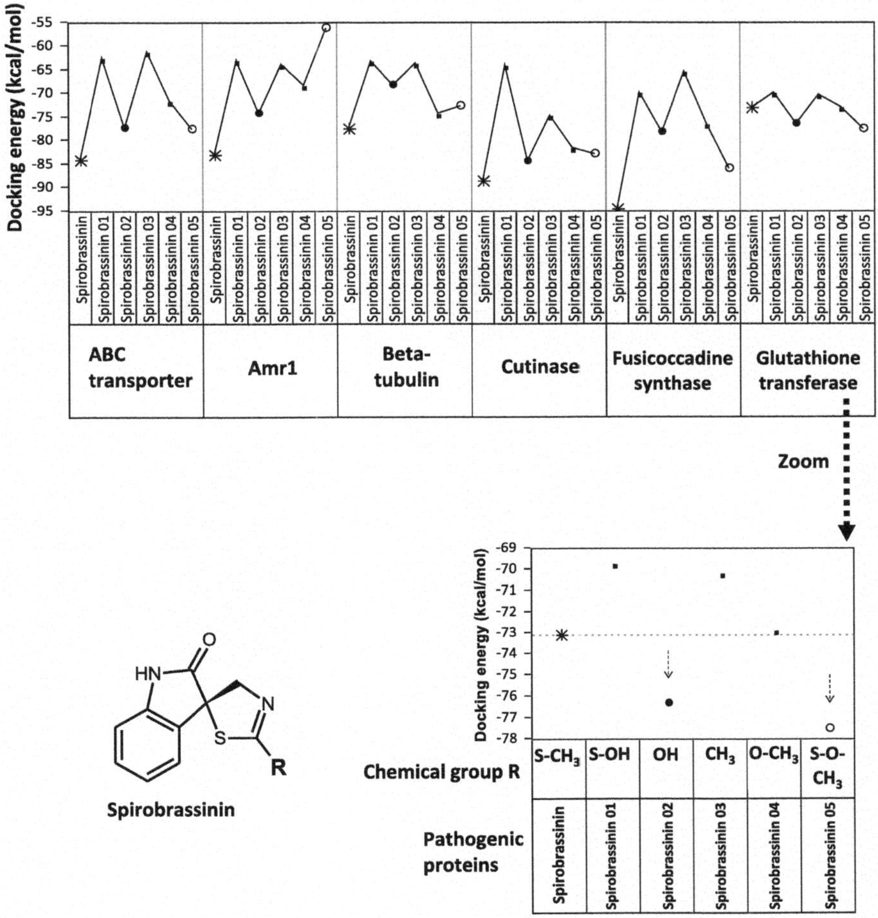

Fig. 12.10 Variation of docking energy for different spirobrassinin derivatives (phytoalexins) of *Brassica* plants with different pathogenic proteins of *Alternaria* showing stronger interactions for spirobrassinin 2 (black circle) and spirobrassinin 5 (white circle) toward glutathione transferase compared with results obtained with the mother molecule, the spirobrassinin (in asterisk)

12.3 Phytochemical Patterns Specific to Different Plant Species Under Pathogen Infections

12.3.1 Phytoanticipins

Phytoanticipins include saponins, steroidal glycoalkaloids, glucosinolates and cyanogenic glucosides (Figs. 12.12, 12.13 and 12.14).

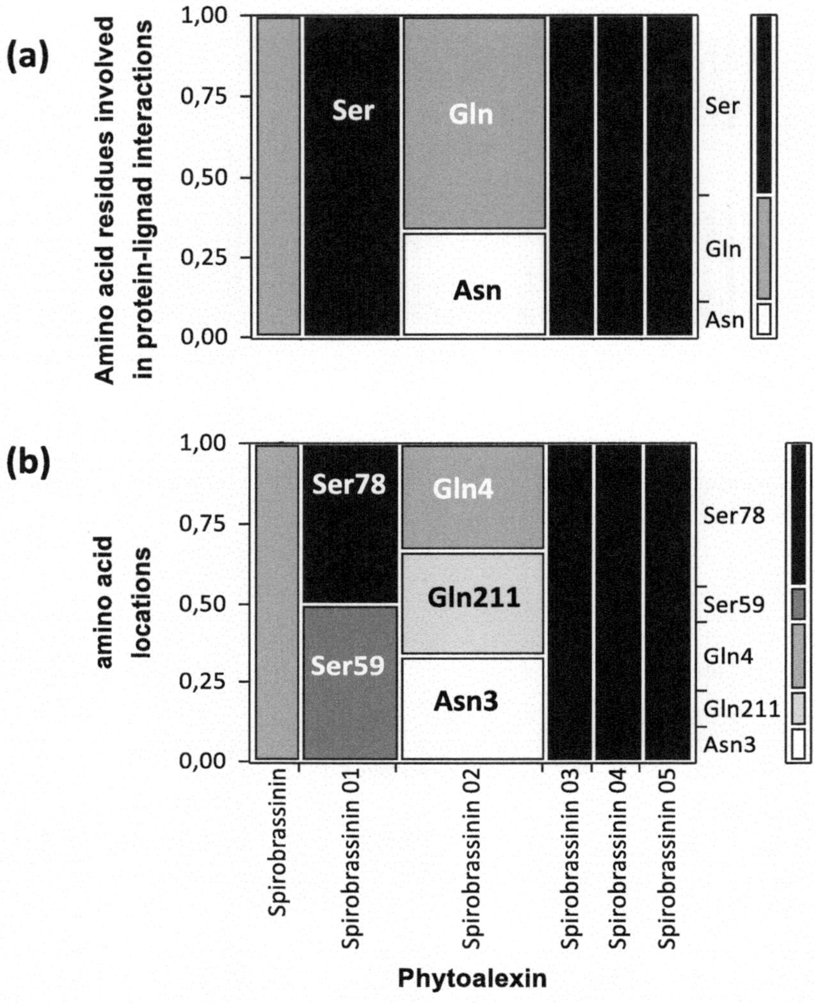

Fig. 12.11 Different types (**a**) and locations (**b**) of amino acid residues of glutathione transferase interacting in H-bond with different spirobrassinin derivatives (spirobrassinins 1–5)

12.3.1.1 Saponins and Steroidal Glycoalkaloids

Saponins and steroidal glycoalkaloids showed *in vitro* membranolytic activity attributing them potential antibiotic effects and revealing their key roles in plant immunity (Ahuja et al. 2012; Bednarek 2012a). Saponin-type phytoanticipins include avenacins (A1, A2, B1, B2) and avenacosides (A, B) accumulated in the roots and

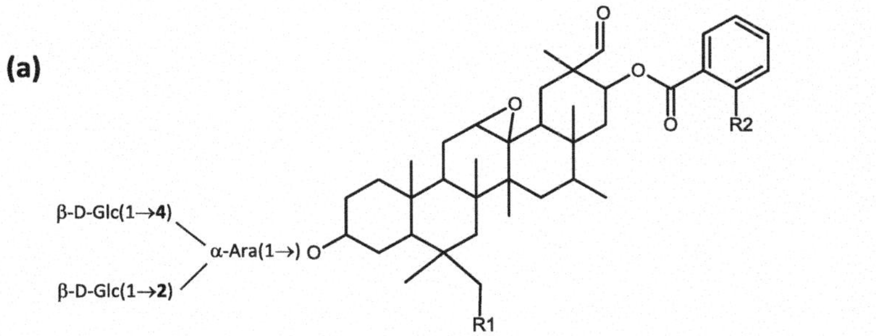

	Avenacin A-1	Avenacin A-2	Avenacin B-1	Avenacin B-2
R1	OH	H	OH	H
R2	$NHCH_3$	$NHCH_3$	H	H

in roots of *Avena* spp

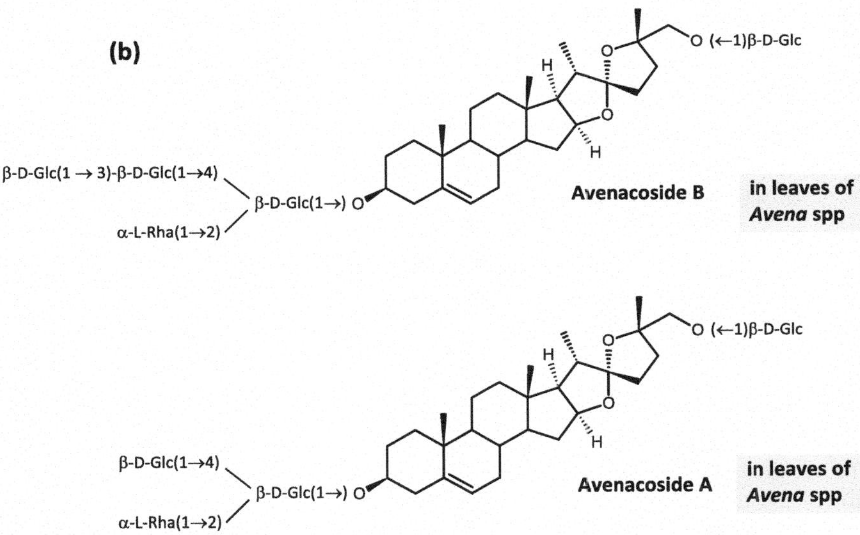

Fig. 12.12 Avenacins (**a**) and avenacosides (**b**) representing saponin and steroid saponin phytoanticipins found in the root and the leaves of oat species (*Avena* spp.), respectively

leaves of oat species (*Avena* spp.), respectively (Fig. 12.12). Steroidal glycoalkaloids include α-tomatine which is produced by all the organs of tomato (*Lycopersicon esculentum*) (Fig. 12.13) (Arneson and Durbin 1968). *In vitro* studies showed that saponins as avenacin A-1 and α-tomatine can disrupt lipid bilayers (Steel and Drysdale 1988; Keukens et al. 1995; Armah et al. 1999). Due to its lipophilic character, aglycone moiety of these molecules can interact with membrane-resident sterol moieties; thereafter, glycosidic chains of these molecules interact with each other leading to subsequent reorganization of sterols in the microbes' membranes

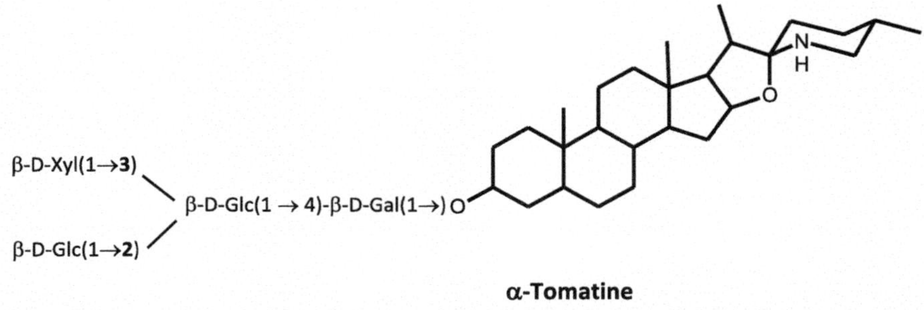

Fig. 12.13 Structure of α-tomatine, a major steroidal glycoalkaloid phytoanticipin produced by all the organs of tomato *Lycopersicon esculentum*

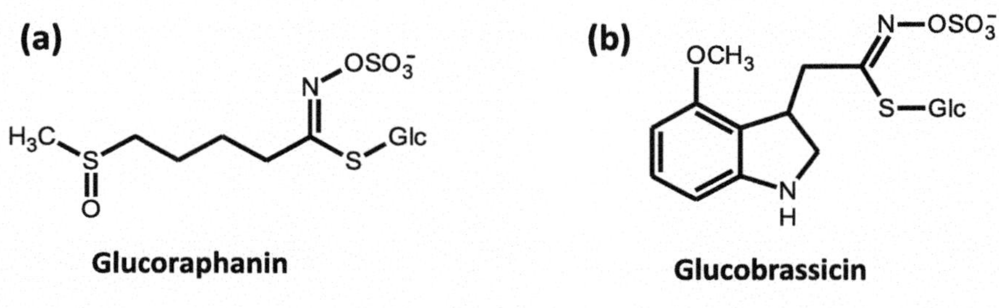

Fig. 12.14 Illustrative structures of aliphatic and indolic glucosinolates found in *Arabidopsis*

followed by pore formation (Keukens et al. 1995; Armah et al. 1999; Osbourn 1996; Osbourn et al. 2011; Morant et al. 2008). However, fungi adapted on saponin- or steroidal glycoalkaloid-producing plants exert their pathogenicity by using enzymatic cleavage of sugar chain of respective aglycone (Bowyer et al. 1995; Bouarab et al. 2002; Ökmen et al. 2013).

12.3.1.2 Glucosinolates

Glucosinolate-type phytoanticipins include aliphatic and indolic structures (Fig. 12.14). Illustration is given by glucoraphanin (aliphatic) and glucobrassicin (indolic) found in *Arabidopsis* and derived from methionine and tryptophane, respectively. *Arabidopsis* produces Met-derived aliphatic glucosinolates and Try-derived indolic glucosinolates with smaller production of Phe-derived benzyl glucosinolates.

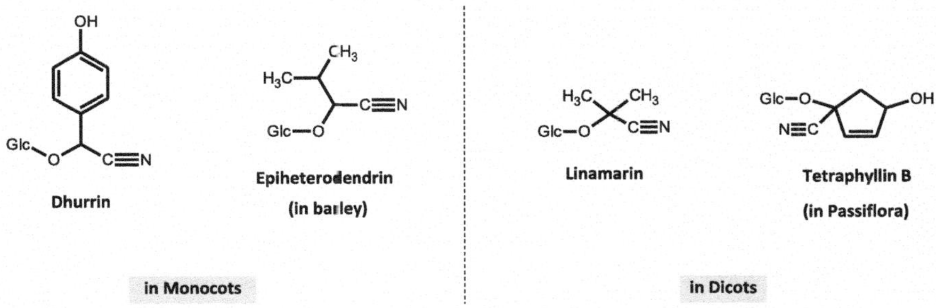

Fig. 12.15 Some structures of cyanogenic glucosides found in some plant taxa, issued from different amino acids and showing phytoanticipin properties

12.3.1.3 Cyanogenic Glucosides

Cyanogenic glucosides (CGs) are amino acid-derived phytoanticipins associated with tyrosine, phenylalanine, valine, isoleucine or leucine (Fig. 12.15). CGs are associated with different plant taxa including dhurrin and epiheterodendrin in monocots, and linamarin and tetraphillin B in dicots.

12.3.2 Phytoalexins

Phytoalexins include many secondary metabolites associated with different metabolic families including alkaloids, phenolics and terpenes (Ahuja et al. 2012). Different plant families are distinguished by different phytoalexin types (Fig. 12.16).

12.3.2.1 Alkaloid-Derived Phytoalexins

In Brassicaceae, phytoalexins include sulfur-containing tryptophan-derived alkaloids which are synthesized and accumulated in response to pathogenic infections (Fig. 12.16a) (Pedras et al. 2011; Bednarek 2012b). Among these compounds, camalexin is indole phytoalexin showing major defensive role in *Arabidopsis* and more generally in the Camelinae tribe (Browne et al. 1991; Tsuji et al. 1992; Bednarek et al. 2011). Camalexin is well produced by *Arabidopsis thaliana* and *Camelina sativa* (Pedras and Yaya 2014). Sulfur-containing indole alkaloid phytoalexins in *Arabidopsis* include also rapalexins (Fig. 12.16a) (Pedras and Adio 2008).

Rapalexin A and rapalexin B (5-hydroxyl rapalexin A) are characterized by indolyl isothiocyanate structures (Fig. 12.17b, c) (Pedras et al. 2007). Rapalexins A and B were found to be synthesized in the leaves of canola (*Brassica rapa*) infected with the white rust oomycete, *Albugo candida* (Pedras et al. 2007). They were absent in non-stressed plants by biotic infections or UV radiations. Moreover, following

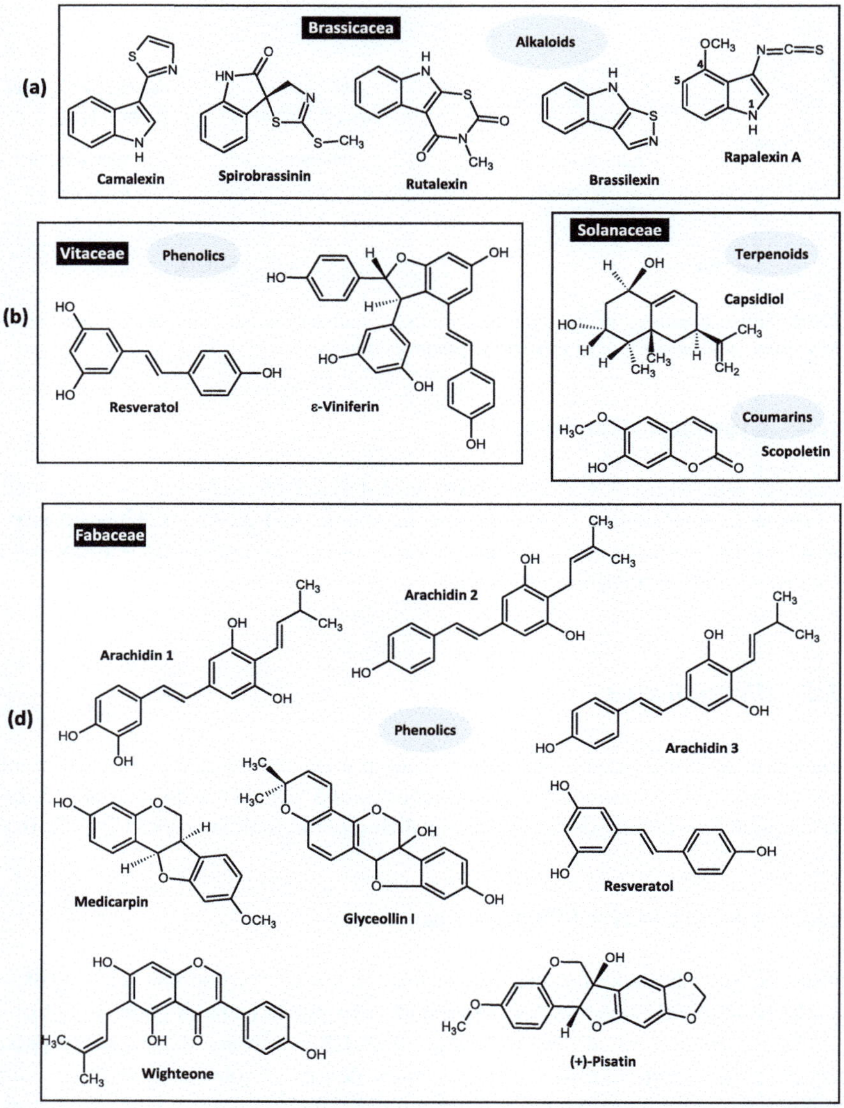

Fig. 12.16 Different phytoalexins of alkaloid-, phenolic and terpene-types characterizing different plant families (Brassicaceae, Vitaceae, Solanaceae, Fabaceae)

Albugo candida infection, rapalexins A and B showed different biosynthesis levels and kinetics (Fig. 12.17a–c) (Pedras et al. 2007). Rapalexin A manifested from day 8 of infection with amounts 10 folds less than those of rapalexin B which manifested from day 5. Rapalexin A showed more potent effect than rapalexin B against zoospore germination as revealed from a bioassay with impregnated cellulose membranes

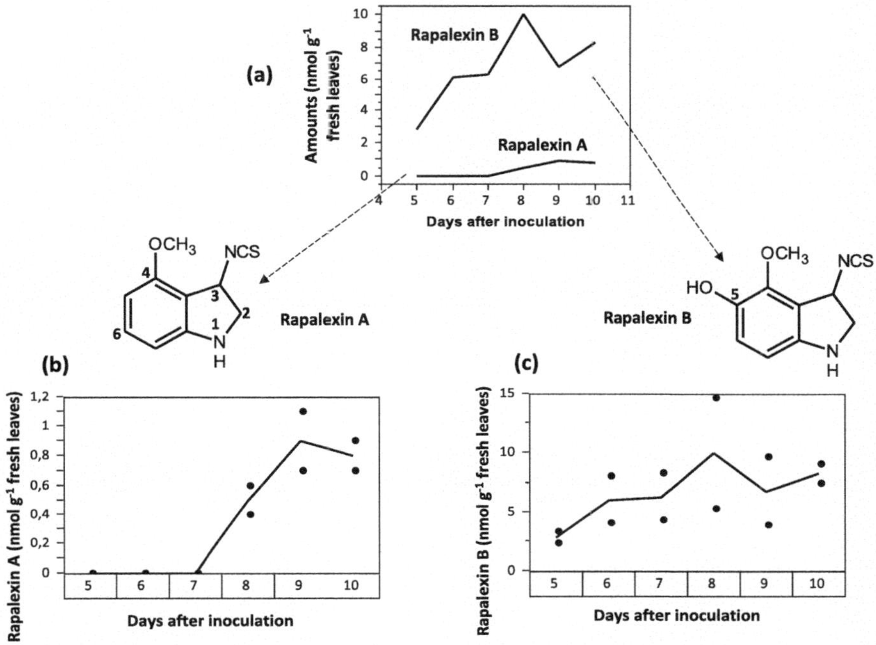

Fig. 12.17 Kinetic variation of production levels of rapalexins A and B in the leaves of *Brassica rapa* (canola) infected with the white rust oomycete *Albugo candula*

(von Röpenack et al. 1998). This could agree with lower amounts of rapalexin A needed for anti-pathogen effect.

12.3.2.2 Phenolic Phytoalexins

Among phenolic phytoalexins, resveratrol showed accumulation in infected grapevine (*Vitis vinifera*) (Vitaceae) and peanut (*Arachis hypogaea*) (Fabaceae) (Fig. 12.16b, d) (Langcake and Pryce 1976; Sobolev et al. 1995). The presence of this stilbene phytoalexin in phylogenetically unrelated species could be linked to its simple metabolism involving only one enzyme (stilbene synthase, STS) and ubiquitous precursors in plant kingdom (p-coumaroyl-coenzyme A and malonyl-CoA) (Fig. 12.18; Fig. 3.2e, Part I; Fig. 4.2g, Part II) (Tropf et al. 1994). Resveratrol was successively expressed in tobacco (*Nicotiana tabacum*) by transferring grapevine STS gene VTS1 leading to pathogen-inducible biosynthesis of resveratrol and better disease resistance to the necrotrophic fungus *Botrytis cinerea* (Hain et al. 1993). This provided further proofs on significance of resveratrol in plant immunity.

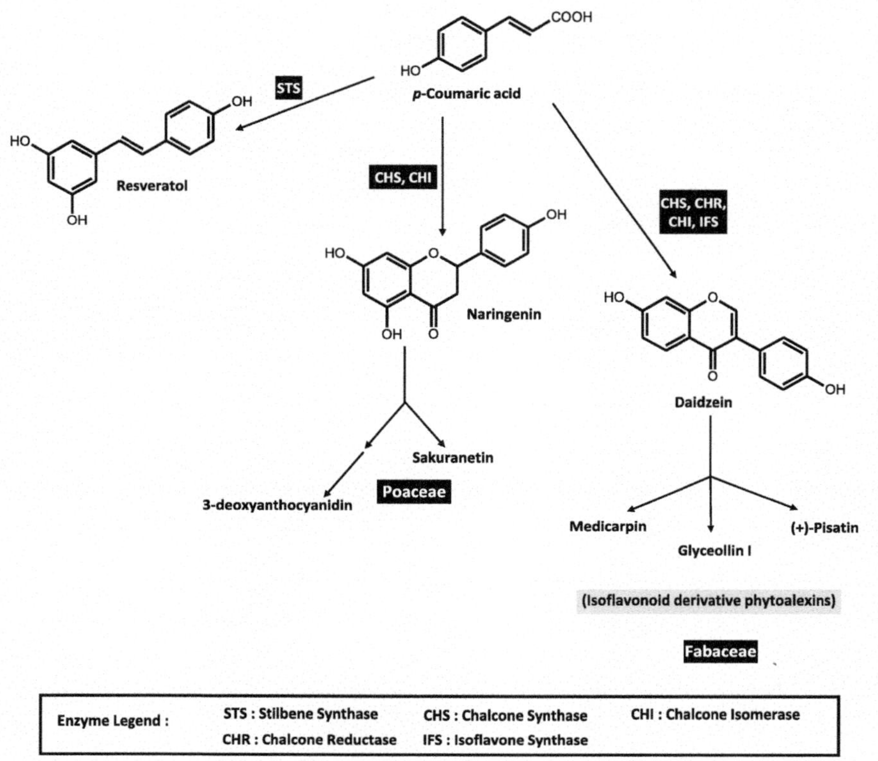

Fig. 12.18 Biosynthesis pathways and involved enzymes of some phytoalexins in some plant families. Resveratrol shows a short and unique enzyme-based pathway deriving directly from p-coumaric acid

Phenolic phytoalexins include also isoflavonoids which are commonly produced by Fabaceae. Different isoflavonoid-derived phytoalexins were found in different plant species including (Fig. 12.16d): pisatin in pea (*Pisum sativum*) (Perrin and Bottomley 1961), medicarpin in alfalfa (*Medicago sativa*) (He and Dixon 2000) and glyceollins in soybean (*Glycine max*) (Ebel et al. 1976). These compounds are synthesized from daidzein (Fig. 12.18).

In soybean, silencing of isoflavone synthase or chalcone reductase (mediated by RNA interference, RNAi) resulted in strong suppression of isoflavonoids (including glyceollins, Fig. 12.16d) leading to loss of plant resistance to the oomycete pathogen *Phytophtora sojae* (Graham et al. 2007). Beyond oomycete pathogen, defense effect of isoflavonoid phytoalexins in soybean was revealed to extend to bacteria (Morgan et al. 2010).

Moreover, prenylated isoflavonoids were characterized by inducible aspects in Papilionoideae (Fabaceae) (Ibrahim 2005; Wang et al. 2001; Botta et al. 2005; Yazaki et al. 2009). Also, prenylated flavanones were found to play key defensive roles in plants.

Apart from isoflavonoids in Fabaceae, phenolic phytoalexins were also concerned with flavonoids in Poaceae (Fig. 12.19):

Sakuranetin (a flavanone, Fig. 12.19c) tends to accumulate in rice (*Oryza sativa*) infected by *Magnaporthe oryzae* (ascomycete fungus causing rice browning) (Duan et al. 2014).

Apigeninidin and luteolinidin (3-deoxyanthocyanidins) (Fig. 12.19c) tend to accumulate in sorghum after pathogenic infection (Snyder & Nicholson 1990). In Sorghum, biosynthesis of 3-deoxyanthocyanidins and phlobaphenes are regulated by a MYB transcription factor which is in turn encoded by the gene Y1 (yellow seed1). Loss-of-function y1 alleles in sorghum resulted in non-production of 3-deoxyanthocyanidins upon pathogen challenge making the plant to be more susceptible to *Colletotrichum sublineolum* (anthracnose causing-fungus in rice and sorghum) (Ibraheem et al. 2010). This provided proofs on the key role of 3-deoxyanthocyanidins in combating anthracnose fungi.

Targeted localization of 3-deoxyanthocyanidin phytoalexins under fungal infection was well highlighted in sorghum; the red coloration of this plant species allowed macroscopic study of subcellular localization of these phytoalexin pigments:

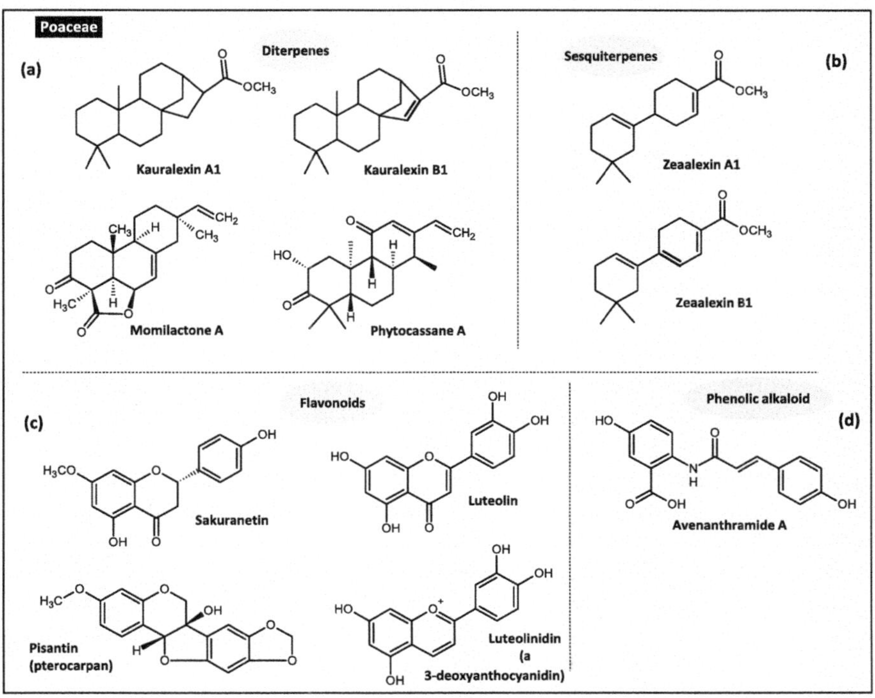

Fig. 12.19 Different phytoalexins of phenolic and terpene types associated with Poaceae family

when epidermal cells of sorghum were attacked by the ascomycete fungus *Colletotrichum graminicola*, 3-deoxyanthocyanidin phytoalexins accumulated in cytoplasmic vesicle-like inclusion bodies (Snyder and Nicholson 1990). Moreover, accumulation focally occurred beneath the fungal appressorium resulting in high local concentrations of 3-deoxyanthocyanidin at pathogen entry sites (Snyder and Nicholson 1990; Nielsen et al. 2004).

12.3.2.3 Terpenoid-Type Phytoalexins

Rice produces diterpenoid phytoalexins based on labdane type-scaffold (Fig. 12.19a) (Schmelz et al. 2011, 2014). These defense compounds include momilactones A and B, oryzalexins and phytocassanes (Fig. 12.19a) (Cartwright et al. 1977; Peters 2006; Schmelz et al. 2014). Maize shows, however, production of diterpene and sesquiterpene-type phytoalexins including kauralexins and zealexins, respectively (Fig. 12.19a, b) (Huffaker et al. 2011; Schmelz et al. 2011).

The (E)-β-caryophyllene is among the most commonly emitted sesquiterpenes by plants within floral volatiles occurring in more than half of angiosperm families. In *Arabidopsis*, mutant plants (tps21) which were deficient in (E)-β-caryophyllene showed more colonized stigmas by the bacterial pathogen *Pseudomonas syringae* DC3000 than stigmas of wild plants (WT) (Huang et al. 2012).

12.3.3 Involvement of Phytoalexins in Environment Management

Phytoalexins and bioengineered derivative molecules can be used as substituents to fungicides in agrosystems leading to the avoidance of environmental damage caused by synthetic pesticides.

Several pathogenic fungi of Brassicaceae developed detoxication ability of phytoalexins via selective enzymes and simple reactions including oxidation, reduction, hydrolysis and glycosylation (Fig. 12.20) (Pedras and Abdoli 2017; Pedras and Ahiahonu 2005). *Leptosphaeria maculans* (canola virulent isolates) detoxifies erucalexin by reduction reaction leading to the formation of dihydroerucalexin (Fig. 12.20a). Camalexin is detoxified by several pathogenic fungi including *Alternaria brassicicola*, *Botrytis cinerea*, *Rhizoctonia solani*, and *Sclerotinia sclerotiorum*. These pathogens have different detoxication ways including hydrolysis, oxidation and glycosylation. Detoxification by oxidation and glycosylation is illustrated by *Sclerotinia sclerotiorum* case (Fig. 12.20b). However, *Alternaria brassicae or Leptosphaeria maculans* (canola virulent isolates) did not show detoxication toward camalexin. Spirobrassinin is detoxified by *Leptosphaeria maculans*

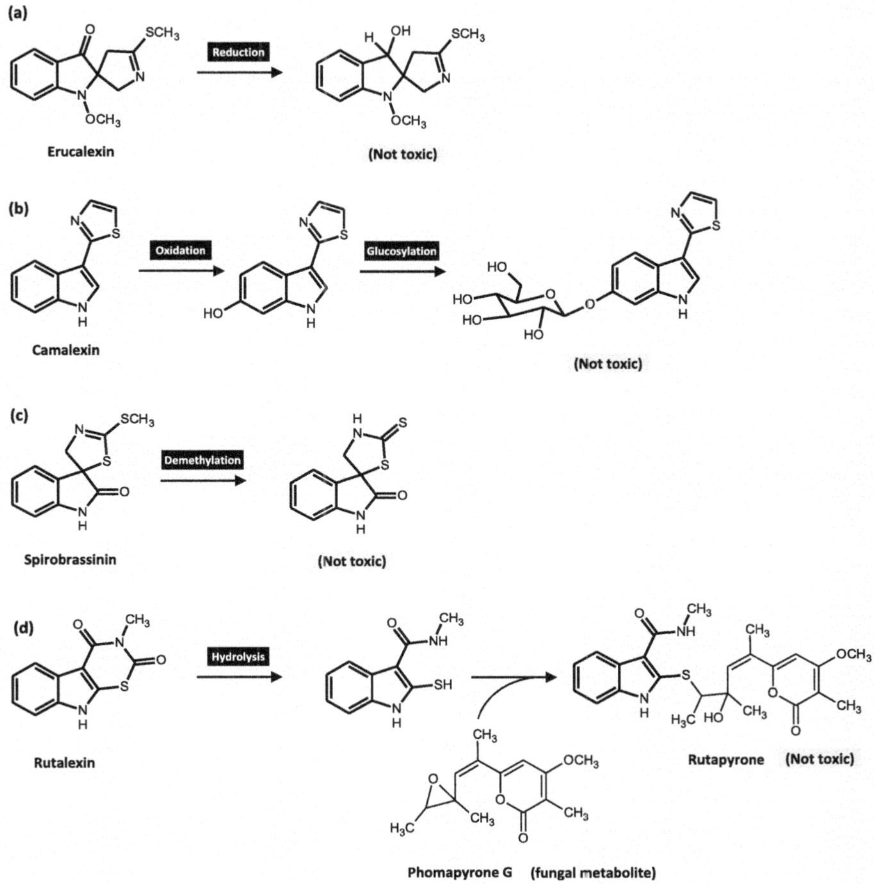

Fig. 12.20 Detoxification mechanisms of some Brassicaceae phytoalexins by fungal pathogens

(brown mustard virulent isolates) through a demethylation mechanism (Fig. 12.20c). Rutalexin is detoxified by the pathogen fungus *Alternaria brassicicola* via sequential reactions consisting of a hydrolysis followed by the addition of fungal metabolite (phomapyrone G) leading to the formation of rutapyrone (Fig. 12.20d).

Potential protective way against pathogenic detoxications was provided by synthetic phytoalexin derivatives called paldoxins (**p**hyto**al**exin **detox**ication inhibitor**s**) which showed inhibition abilities of pathogen detoxifying enzymes (Pedras et al. 2003). Subsequently, paldoxins could be used as alternative to pesticides providing a potentially sustainable strategy with minimal impacts on the environment.

References

Abdel-Farid IB, Choi YH, Kim HK, Hondel CVD, Meijden E, Verpoorte R (2006) The role of secondary metabolites in *Arabidopsis* and *Brassica* in the interaction with fungi. Curr Topics Plant Bio 7:47–73

Ahuja I, Kissen R, Bones AM (2012) Phytoalexins in defense against pathogens. Trends Plant Sci 17:73–90

Andersson MX, Nilsson AK, Johansson ON, Boztas G, Adolfsson LE, Pinosa F, Garcia C, Aronsson H, Mackey D, Tor M, Hamberg M, Ellerström M (2015) Involvement of the electrophilic isothiocyanate sulforaphane in Arabidopsis local defense responses. Plant Physiol 167:251–261

Armah CN, Mackie AR, Roy C, Price K, Osbourn AE, Bowyer P, Ladha S (1999) The membrane-permeabilizing effect of avenacin A-1 involves the reorganization of bilayer cholesterol. Biophys J 76:281–290

Arneson PA, Durbin RD (1968) Sensitivity of fungi to a-tomatine. Phytopathology 58:536–537

Bednarek P (2012a) Chemical warfare or modulators of defence responses—the function of secondary metabolites in plant immunity. Curr Opin Plant Biol 15:407–414

Bednarek P (2012b) Sulfur-containing secondary metabolites from *Arabidopsis thaliana* and other Brassicaceae with function in plant immunity. ChemBioChem 13:1846–1859

Bednarek P, Pislewska-Bednarek M, Loren V, van Themaat E, Maddula RK, Svatos A, Schulze-Lefert P (2011) Conservation and clade-specific diversification of pathogen-inducible tryptophan and indole glucosinolate metabolism in *Arabidopsis thaliana* relatives. New Phytol 192:713–726

Bell S, Guggenheim C (2019) Plant disease diagnosing and management. In: West A, Jensen J, Bell SM, Ellis L (eds) Idaho master gardener handbook, 19th edn

Bones AM, Rossiter JT (2006) The enzymic and chemically induced decomposition of glucosinolates. Phytochemistry 67:1053–1067

Botta B, Vitali A, Menendez P et al (2005) Prenylated flavonoids: pharmacology and biotechnology. Curr Med Chem 12:717–739

Bouarab K, Melton R, Peart J, Baulcombe D, Osbourn A (2002) A saponin detoxifying enzyme mediates suppression of plant defences. Nature 418:889–892

Bowyer P, Clarke BR, Lunness P, Daniels MJ, Osbourn AE (1995) Host range of a plant pathogenic fungus determined by a saponin detoxifying enzyme. Science 267:371–374

Brader G, Tas E, Palva ET (2001) Jasmonate-dependent induction of indole glucosinolates in *Arabidopsis* by culture filtrates of the nonspecific pathogen *Erwinia carotovora*. Plant Physiol 126:849–860

Brown KK, Hampton MB (2011) Biological targets of isothiocyanates. Biochim Biophys Acta Gen Subj 1810:888–894

Browne LM, Conn KL, Ayer WA, Tewari JP (1991) The camalexins: new phytoalexins produced in the leaves of *Camelina sativa* (Cruciferae). Tetrahedron 47:3909–3914

Burow M, Markert J, Gershenzon J, Wittstock U (2006) Comparative biochemical characterization of nitrile-forming proteins from plants and insects that alter myrosinase-catalysed hydrolysis of glucosinolates. FEBS J 273(111):2432–2446

Byther RS, Foss CR, Antonelli AL, Maleike RR, Bobbitt VM (2000) Landscape plant problems: a pictorial diagnostic manual. Washington State University Extension, MISC0194

Cartwright D, Langcake P, Pryce RJ, Leworthy DP, Ride JP (1977) Chemical activation of host defence mechanisms as a basis for crop protection. Nature 267:511–513

Duan L, Liu H, Li X, Xiao J, Wang S (2014) Multiple phytohormones and phytoalexins are involved in disease resistance to *Magnaporthe oryzae* invaded from roots in rice. Physiol Plant 152:486–500

Dudareva N, Negre F, Nagegowda DA, Orlova I (2006) Plant volatiles: recent advances and future perspectives. Crit Rev Plant Sci 25:417–440

Ebel J, Ayers AR, Albersheim P (1976) Host–pathogen interactions: XII. Response of suspension-cultured soybean cells to the elicitor isolated from *Phytophthora megasperma* var. *sojae*, a fungal pathogen of soybeans. Plant Physiol 57:775–779

Frey M, Schullehner K, Dick R, Fiesselmann A, Gierl A (2009) Benzoxazinoid biosynthesis, a model for evolution of secondary metabolic pathways in plants. Phytochemistry 70:1645–1651

Gleadow RM, Møller BL (2014) Cyanogenic glycosides: synthesis, physiology, and phenotypic plasticity. Annu Rev Plant Biol 65:155–185

Graham TL, Graham MY, Subramanian S, Yu O (2007) RNAi silencing of genes for elicitation or biosynthesis of 5-deoxyisoflavonoids suppresses race-specific resistance and hypersensitive cell death in *Phytophthora sojae* infected tissues. Plant Physiol 144:728–740

Hain R, Reif HJ, Krause E, Langebartels R, Kindl H, Vornam B, Wiese W, Schmelzer E, Schreier PH, Stocker RH, Stenzel K (1993) Disease resistance results from foreign phytoalexin expression in a novel plant. Nature 361:153–156

Halkier BA, Gershenzon J (2006) Biology and biochemistry of glucosinolates. Annu Rev Plant Biol 57:303–333

He XZ, Dixon RA (2000) Genetic manipulation of isoflavone 7-omethyltransferase enhances biosynthesis of 4′-O-methylated isoflavonoid phytoalexins and disease resistance in alfalfa. Plant Cell 12:1689–1702

Heraud P, Cowan MF, Marzec KM, Møller BL, Blomstedt CK, Gleadow R (2018) Label-free Raman hyperspectral imaging analysis localizes the cyanogenic glucoside dhurrin to the cytoplasm in sorghum cells. Sci Rep 8:2691

Huang M, Sanchez-Moreiras AM, Abel C, Sohrabi R, Lee S, Gershenzon J, Tholl D (2012) The major volatile organic compound emitted from *Arabidopsis thaliana* flowers, the sesquiterpene (E)-β-caryophyllene, is a defense against a bacterial pathogen. New Phytol 193:997–1008

Huffaker A, Kaplan F, Vaughan MM, Dafoe NJ, Ni X, Rocca JR, Alborn HT, Teal PEA, Schmelz EA (2011) Novel acidic sesquiterpenoids constitute a dominant class of pathogen-induced phytoalexins in maize. Plant Physiol 156:2082–2097

Ibraheem F, Gaffoor I, Chopra S (2010) Flavonoid phytoalexin-dependent resistance to anthracnose leaf blight requires a functional yellow seed1 in *Sorghum bicolor*. Genetics 184:915–926

Ibrahim RK (2005) A forty-year journey in plant research: original contributions to flavonoid biochemistry. Can J Bot 83:433–450

Jackson RW, DeMoss JA (1965) Effects of toluene on *Escherichia coli*. J Bacteriol 90:1420–1425

Keukens EAJ, de Vrije T, van den Boom C, de Waard P, Plasman HH, Thiel F, Chupin V, Jongen WMF, de Kruijff B (1995) Molecular basis of glycoalkaloid induced membrane disruption. Biochimica et Biophysica Acta (BBA)—Biomembranes 1240:216–228

Kirlin WG, Cai J, Thompson SA, Diaz D, Kavanagh TJ, Jones DP (1999) Glutathione redox potential in response to differentiation and enzyme inducers. Free Radical Biol Med 27:1208–1218

Langcake P, Pryce RJ (1976) Production of resveratrol by *Vitis vinifera* and other members of Vitaceae as a response to infection or injury. Physiol Plant Pathol 9:77–86

Larsen PO (1981) Glucosinolates. In: Stumpf PK, Conn EE (eds) The biochemistry of plants, vol 7, secondary plant products. Academic, New York, pp 501–525

Maroti G, Kereszt A, Kondorosi E, Mergaert P (2011) Natural roles of antimicrobial peptides in microbes, plants and animals. Res Microbiol 162:363–374

Miedes E, Vanholme R, Boerjan W, Molina A (2014) The role of the secondary cell walls in plant resistance to pathogens. Front Plant Sci 5:358

Mithen RF, Lewis BG, Fenwick GR (1986) *In vitro* activity of glucosinolates and their products against *Leptosphaeria maculans*. Trans Br Mycol Soc 87:433–440

Møller BL (2010) Functional diversifications of cyanogenic glucosides. Curr Opin Plant Biol 13:337–346

Moore D, Robson GD, Trinci APJ (2019) 14.9 plant disease basics: the disease triangle. Twenty-First Century Guidebook to Fungi (online version), 2nd edn. Cambridge University Press, New York. http://www.davidmoore.org.uk/21st_Century_Guidebook_to_Fungi_PLATINUM/Ch14_09.htm

Morant AV, Jørgensen K, Jørgensen C, Paquette SM, Sanchez-Perez R, Møller BL Bak S (2008) β-Glucosidases as detonators of plant chemical defense. Phytochemistry 69:1795–1813

Morgan RL, Zhou H, Lehto E, Nguyen N, Bains A, Wang X, Ma W (2010) Catalytic domain of the diversified *Pseudomonas syringae* type III effector HopZ1 determines the allelic specificity in plant hosts. Mol Microbiol 76:437–455

Müller KO, Börger H (1940) Experimental studies on the Phytopthora resistance of potatoes. A contribution to the problem of acquired resistance in the plants. Arbeitender Biologischen Reichsanstalt Für Land Und Forstwirtschaft 23:189–231

Nielsen KA, Gotfredsen CH, Buch-Pedersen MJ, Ammitzboll H, Mattsson O, Duus JO, Nicholson RL (2004) Inclusions of flavonoid 3-deoxyanthocyanidins in *Sorghum bicolor* self-organize into spherical structures. Physiol Mol Plant Pathol 65:187–196

Nimchuk Z, Marois E, Kjemtrup S, Leister RT, Katagiri F, Dangl JL (2000) Eukaryotic fatty acylation drives plasma membrane targeting and enhances function of several type III effector proteins from *Pseudomonas syringae*. Cell 101:353–363

Noctor G, Queval G, Mhamdi A, Chaouch S, Foyer CH (2011) Glutathione. Arabidopsis Book 9:e0142

Ökmen B, Etalo DW, Joosten MHAJ, Bouwmeester HJ, de Vos RCH, Collemare J, de Wit PJ (2013) Detoxification of a-tomatine by *Cladosporium fulvum* is required for full virulence on tomato. New Phytol 198:1203–1214

Osbourn A (1996) Saponins and plant defence—a soap story. Trends Plant Sci 1:4–9

Osbourn A, Goss RJM, Field RA (2011) The saponins—polar isoprenoids with important and diverse biological activities. Nat Prod Rep 28:1261–1268

Pathak RK, Taj G, Pandey D, Kasana VK, Baunthiyal M, Kumar A (2016) Molecular modeling and docking studies of phytoalexin(s) with pathogenic protein(s) as molecular targets for designing the derivatives with anti-fungal action on *Alternaria* spp. of *Brassica*. Plant Omics J 9(3):172–182

Paxton JD (1981) Phytoalexins—a working redefinition. Phytopathol Z 101:106–109

Pedras MSC, Abdoli A (2017) Pathogen inactivation of cruciferous phytoalexins: detoxification reactions, enzymes and inhibitors. RSC Adv 7:23633–23646

Pedras MSC, Adio AM (2008) Phytoalexins and phytoanticipins from the wild crucifers *Thellungiella halophila* and *Arabidopsis thaliana*: rapalexin A, wasalexins and camalexin. Phytochemistry 69:889–893

Pedras MSC, Ahiahonu WK (2005) Metabolism and detoxification of phytoalexins and analogs by phytopathogenic fungi. Phytochemistry 66:391–422

Pedras MSC, Yaya EE (2014) Tenualexin, other phytoalexins and indole glucosinolates from wild cruciferous species. Chem Biodivers 11(6):910–918

Pedras MSC, Jha M, Ahiahonu WK (2003) The synthesis and biosynthesis of phytoalexins produced by cruciferous plants. Curr Org Chem 7:1635–1647

Pedras MSC, Zheng Q-A, Gadagi RS (2007) The first naturally occurring aromatic isothiocyanates, rapalexins A and B, are cruciferous phytoalexins. Chem Commun 2007:368–370

Pedras MSC, Chumala PB, Jin W, Islam MS, Hauck DW (2009) The phytopathogenic fungus *Alternaria brassicicola*: phytotoxin production and phytoalexin elicitation. Phytochemistry 70:394–402

Pedras MSC, Yaya EE, Glawischnig E (2011) The phytoalexins from cultivated and wild crucifers: chemistry and biology. Nat Prod Rep 28:1381–1405

Perrin DR, Bottomley W (1961) Pisatin—antifungal substance from *Pisum sativum* L. Nature 191:76–77

Peters RJ (2006) Uncovering the complex metabolic network underlying diterpenoid phytoalexin biosynthesis in rice and other cereal crop plants. Phytochemistry 67:2307–2317

Pscheidt JW, Ocamb CM (eds) (2019) Pacific Northwest plant disease management handbook. Oregon State University, Corvallis

Rogers EE, Glazebrook J, Ausubel FM (1996) Mode of action of the *Arabidopsis thaliana* phytoalexin camalexin and its role in *Arabidopsis*-pathogen interactions. Mol Plant Microbe Interact 9:748–757

Schmelz EA, Kaplan F, Huffaker A, Dafoe NJ, Vaughan MM, Ni X, Rocca JR, Alborn HT, Teal PE (2011) Identity, regulation, and activity of inducible diterpenoid phytoalexins in maize. Proc Natl Acad Sci USA 108:5455–5460

Schmelz EA, Huffaker A, Sims JW, Christensen SA, Lu X, Okada K, Peters RJ (2014) Biosynthesis, elicitation and roles of monocot terpenoid phytoalexins. Plant J 79:659–678

Selmar D (2010) Biosynthesis of cyanogenic glycosides, glucosinolates and non-protein amino-acids. Ann Plant Rev 40:92–181

Shinbo Y, Nakamura Y, Altaf-Ul-Amin M, Asahi H, Kurokawa K, Arita M, Saito K, Ohta D, Shibata D, Kanaya S (2006) KNApSAcK: a comprehensive species-metabolite relationship database. In: Saito K, Dixon RA, Willmitzer L (eds) Plant metabolomics. Springer, Berlin, pp 165–181

De Smet MJ, Kingma J, Witholt B (1978) The effect of toluene on the structure and permeability of the outer and cytoplasmic membranes of *Escherichia coli*. Biochim Biophys Acta 4 506(1):64–80

Snyder BA, Nicholson RL (1990) Synthesis of phytoalexins in Sorghum as a site-specific response to fungal ingress. Science 248:1637–1639

Sobolev VS, Cole RJ, Dorner JW, Yagen B (1995) Isolation, purification, and liquid chromatographic determination of stilbene phytoalexins in peanuts. J AOAC Int 78:1177–1182

Steel CC, Drysdale RB (1988) Electrolyte leakage from plant and fungal tissues and disruption of liposomemembranes by α-tomatine. Phytochemistry 27:1025–1030

Thayer SS, Conn EE (1981) Subcellular localization of dhurrin β-glucosidase and hydroxynitrile lyase in the mesophyll cells of Sorghum leaf blades. Plant Physiol 67:617–622

Tierens KFMJ, Thomma BPHJ, Brouwer M, Schmidt J, Kistner K, Porzel A, Mauch-Mani B, Cammue BPA, Broekaert WF (2001) Study of the role of antimicrobial glucosinolate-derived isothiocyanates in resistance of *Arabidopsis* to microbial pathogens. Plant Physiol 125:1688–1699

Tropf S, Lanz T, Rensing SA, Schroder J, Schroder G (1994) Evidence that stilbene synthases have developed from chalcone synthases several times in the course of evolution. J Mol Evol 38:610–618

Tsuji J, Jackson EP, Gage DA, Hammerschmidt R, Somerville SC (1992) Phytoalexin accumulation in *Arabidopsis thaliana* during the hypersensitive reaction to *Pseudomonas syringae* pv *syringae*. Plant Physiol 98:1304–1309

VanEtten HD, Mansfield JW, Bailey JA, Farmer EE (1994) Two classes of plant antibiotics—phytoalexins versus "phytoanticipins." Plant Cell 6:1191–1192

Voigt CA (2014) Callose-mediated resistance to pathogenic intruders in plant defense-related papillae. Front Plant Sci 5:168

von Röpenack E, Parr A, Schulze-Lefert P (1998) Structural Analyses and dynamics of soluble and cell wall-bound phenolics in a broad spectrum resistance to the powdery mildew fungus in barley. J Biol Chem 273:9013–9022

Wang Y, Tan W, Li WZ, Li Y (2001) A facile synthetic approach to prenylated flavanones: first total syntheses of (±)-bonannione A and (±)-sophoraflavanone A. J Nat Prod 64:196–199

Wei BQ, Baase WA, Weaver LH, Matthews BW, Shoichet BK (2002) A model binding site for testing scoring functions in molecular docking. J Mol Biol 322:339–355

Wittstock U, Burow M (2010) Glucosinolate breakdown in *Arabidopsis*: mechanism, regulation and biological significance. Arabidopsis Book 8:e0134

Xu D, Xue M, Shen Z, Jia X, Hou X, Lai D, Zhou L (2021) Phytotoxic secondary metabolites from fungi. Toxins 13:261

Yazaki K, Sasaki K, Tsurumaru Y (2009) Prenylation of aromatic compounds, a key diversification of plant secondary metabolites. Phytochemistry 70:1739–1745

Chapter 13
Plant Defense Against Herbivore Attacks

13.1 Organization Aspects of Phytochemical and Phytohormonal Defenses

Phytochemical defenses against herbivore attacks are either constitutive or induced (Fig. 13.1). Constitutive defense secondary metabolites (SMs) are synthesized before attack and are stored into specialized cell or tissue compartments such as xylemic resin canals used for terpene storage in coniferous trees or cell wall accumulating polyphenols in cereals (Fig. 13.1a) (Sampedro et al. 2011; Santiago et al. 2018). Induced SMs are synthesized following herbivore attacks by a cascade of signaling processes involving upstream biochemical effectors such as herbivore elicitors, plant or herbivore enzymes and/or phytohormones (Fig. 13.1b). Induced SMs include the production of herbivore-induced plant volatiles (HIPVs) which exert negative (defensive/offensive) effects against plant enemies. HIPVs consist of various SMs including volatile terpenoids, benzenoids, phenylpropanoids, volatile fatty acid derivatives and volatile amino acid derivatives (Turlings et al. 2018).

Phytochemical defense induction represents a complex system involving cascade processes between several sequential signaling cues combining SMs, transcription factors and phytohormones (Fig. 13.2). Volatile SMs can act as alerting signal inducing phytohormonal defense system (Fig. 13.2a) (Bouwmeester et al. 2019). Under herbivore attack, induced phytohormonal signaling network is organized by investing in defense way at the expense of growth one (Fig. 13.2b) (Divekar et al. 2022). Plant resistance and growth processes are increased and decreased, respectively, by the action of transcription factor (WRKY70) (Fig. 13.2c) resulting in positive effect on defense hormone (jasmonic acid) and negative effect on growth hormone (gibberellin) (Fig. 13.2d). Such a phytohormone-based survival strategy was highlighted in different plant-herbivore systems including the attack of rice by *Chilo suppressalis* (chewing herbivore).

N. Semmar, *Secondary Metabolites in Plant Stress Adaptation*, Signaling and Communication in Plants, https://doi.org/10.1007/978-3-031-52595-7_13

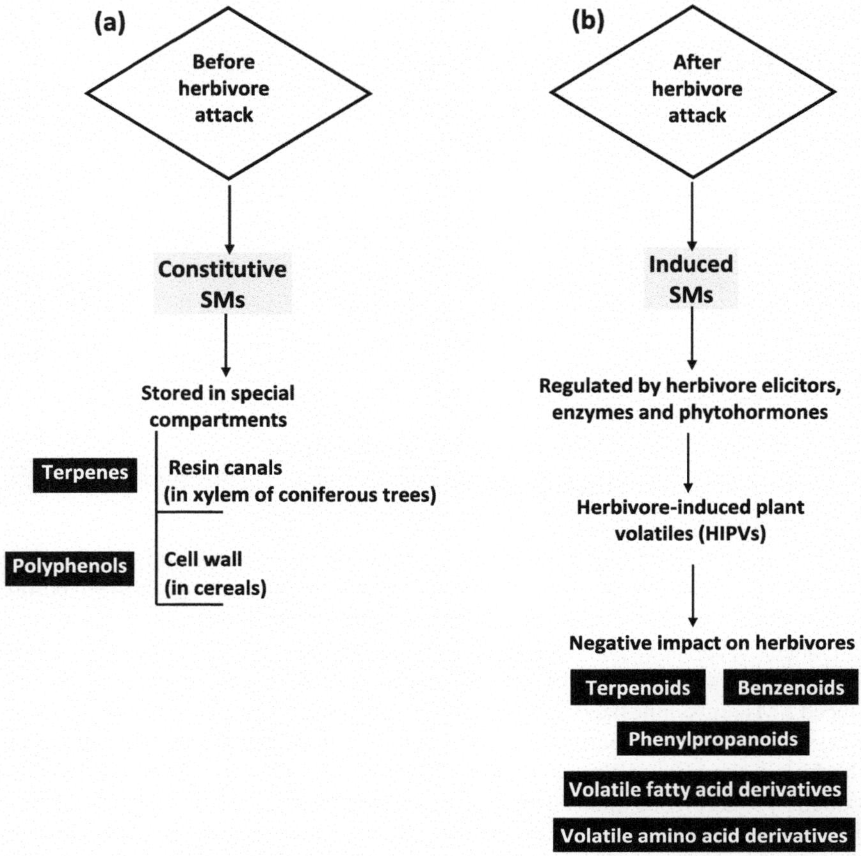

Fig. 13.1 General characteristics of constitutive and induced phytochemical defenses

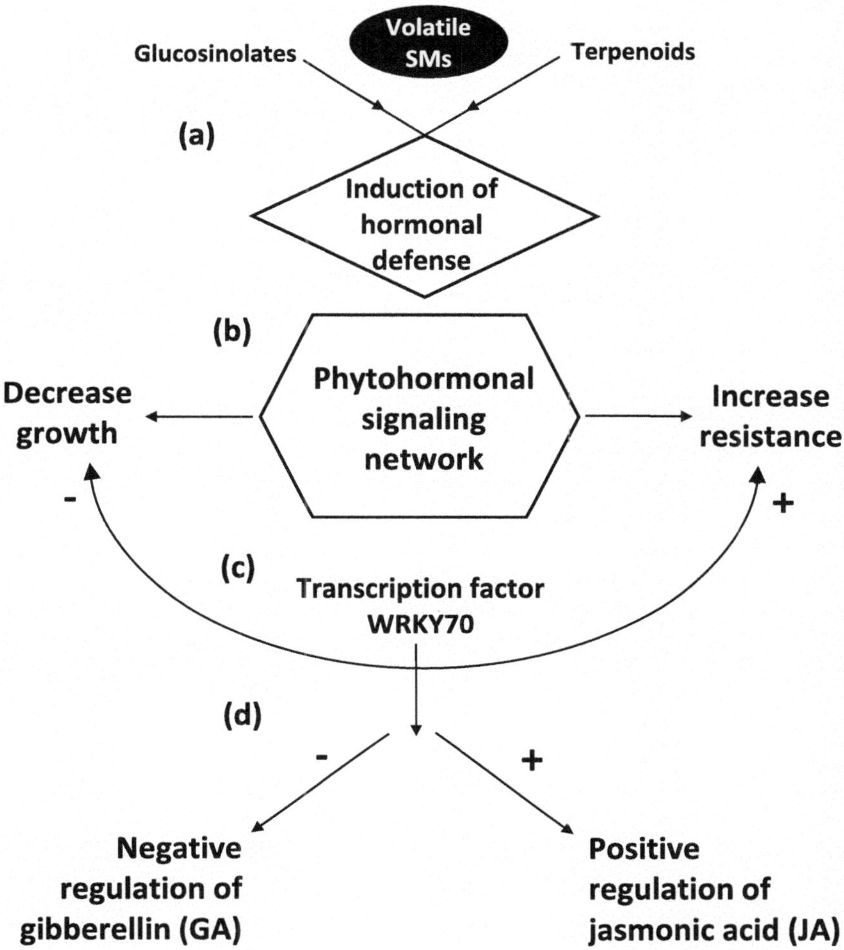

Fig. 13.2 Simplistic overview on the involvement of secondary metabolites (SMs), transcription factor(s) and phytohormones in plant defense against herbivore attacks

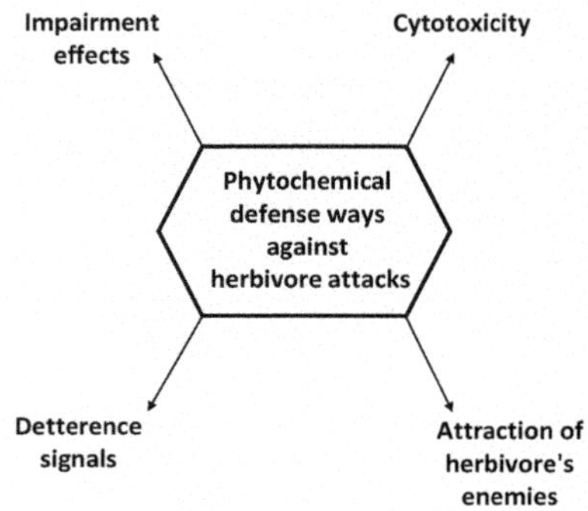

Fig. 13.3 Different phytochemical strategies used by plants for direct or indirect defenses against herbivores

13.2 Various Phytochemical Defense Ways Against Herbivore Attacks

Apart from phytohormonal regulations, attacked plants by herbivores proceed to their defenses by direct or indirect phytochemical ways that manifest with more or less severe levels (Fig. 13.3): phytochemical defense strategies include various impairment effects, cytotoxicity, direct repelling of herbivores and indirect way based on attraction of herbivores' enemies (via volatile compounds) (Dötterl and Gershenzon 2023).

13.2.1 Impairment Effects of Phytocompounds to Herbivores

Attacked plants by herbivores oppose to their enemies by producing different SMs causing various impairments at molecular, cellular and physiological levels (Fig. 13.4). Impairments can be concerned with blocking or disturbance of growth processes (cell division, protein synthesis), movements, respiration and behavior (neural dysfunction) in animal. Impairment of growth can be caused by interruption of cell mitotic division by alkaloids such as colchicine and sanguinarine (Fig. 13.4a) (Schmeller et al. 1997; Wink and Schimmer 2010). Also, alkaloids, as lycorine and emetine, cause alterations of protein synthesis in ribosomes (Fig. 13.4b) (Wink and Schimmer 2010). Impairment effects of plant alkaloids to herbivores include also affectation of neural signal transduction resulting in adverse effects at physiological and behavioral levels (Fig. 13.4c) (Wink 2012). Such effects can be caused by nicotine, caffeine, muscarine, theophyllijne and tubocuranine. Moreover, caffeine and nicotine showed inhibitory effects toward phosphodiesterase activity resulting in movement disorders and paralysis in herbivore (Fig. 13.4d) (Hollingsworth et al. 2002; Pavela 2016). These effects advantageously attribute high sustainable interest to these alkaloids as potential biopesticides. Phytochemical defense can also manifest through the inhibition of respiratory chain in animal cells (Fig. 13.4e). This effect

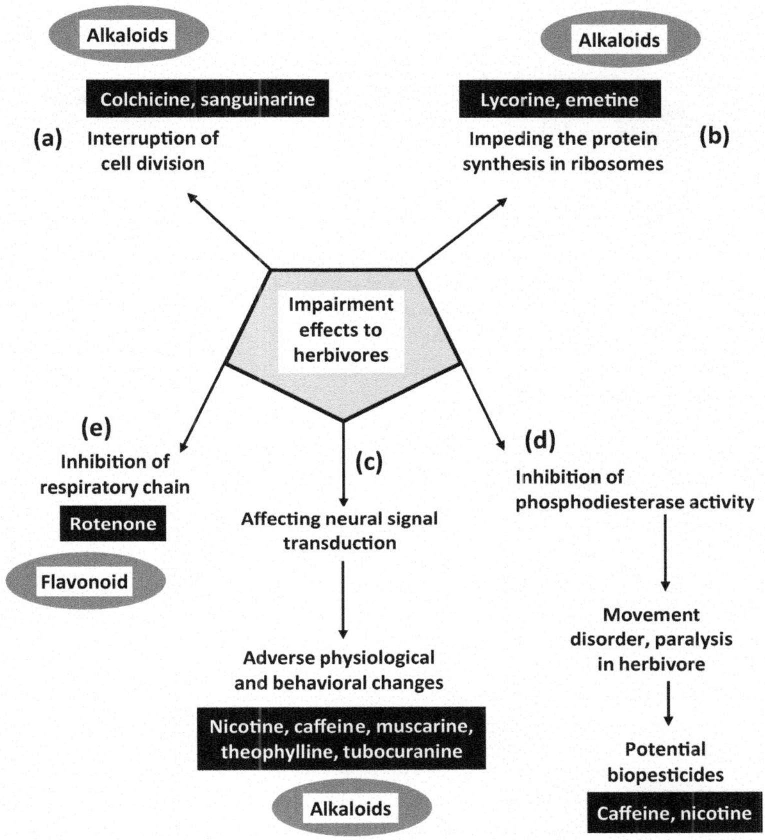

Fig. 13.4 Impairment effects caused to herbivore by different plant secondary metabolites at molecular, cell and physiological levels

was revealed by rotenone, a highly active isoflavonoid with allelochemical aspect (Kobayashi et al. 2021; Wink and Schimmer 2010; Lindahl and Öberg 1961). This metabolite is naturally produced by the roots and stems of some tropical plants such as *Derris elliptica* (Fabaceae).

Impairment effects associated with phytocompounds include also the alteration of protein functions, and cell membrane structures (Fig. 13.5): cell membrane structures can be altered by lipophilic SMs including mainly terpenes (C10, C15, C20, C30) (Fig. 13.5a) (Divekar et al. 2022). Moreover, these compounds are able to alter the fluidity and permeability properties of biomembranes. However, phenolic compounds (containing hydroxyl groups) can interact with proteins via H- and ionic bonds leading to modification of tridimensional structures and functional properties of these macromolecules (Fig. 13.5b) (Wink 2012). SMs-protein interactions can

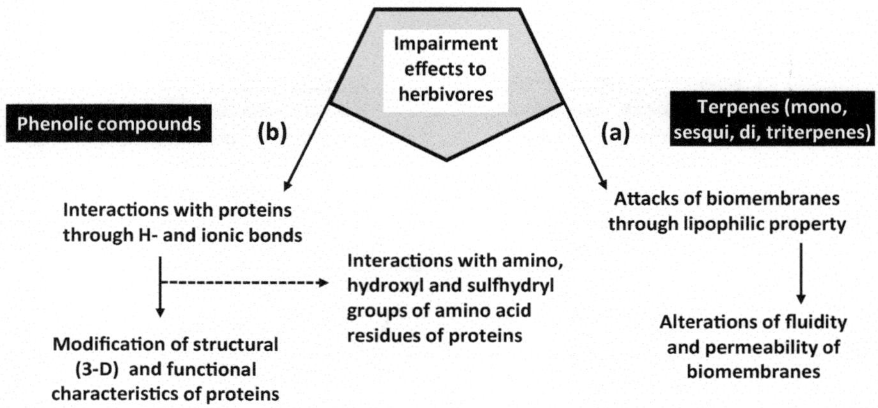

Fig. 13.5 Structural and functional impairments of plant secondary metabolites caused to herbivore body through their interactions with cell membranes and proteins

occur through different chemical groups of protein amino acid residues including hydroxyl, amino and sulfhydryl.

13.2.2 Cytotoxic Effects

Some SMs can exert cytotoxic effects due to their intercalation in DNA (Fig. 13.6). The intercalation occurs between members of nucleotide pairs such as glycine-cytosine pairs. This process averts the activities of helicases and subsequently impairs DNA replication, DNA repair and transcription leading to mutation ways and cell death (Wink 2007). DNA intercalating metabolites include several alkaloids such as sanguinarine, berberine, quinine, emetine and furanoquinoline alkaloids (e.g. dictamnine), as well as furanocoumarins and anthraquinones (Schmeller et al. 1997; Wink 2003, 2010).

13.2.3 Repelling and Deterrence of Herbivore

Plants can use various mixtures of volatile compounds to produce some repelling odors toward specific herbivores (Fig. 13.7). For example, in tomato, repelling olfactory cues combine different sesquiterpenes (zingiberene, curcumene) and monoterpenes (*p*-cymene, α-terpinene, α-phellandrene) leading to significant decrease of visitations by the whitefly (*Bemisia tabaci*) (Bleeker et al. 2009). Sensible detection of these deterrent volatile compounds by whitefly occur by means of its antennae.

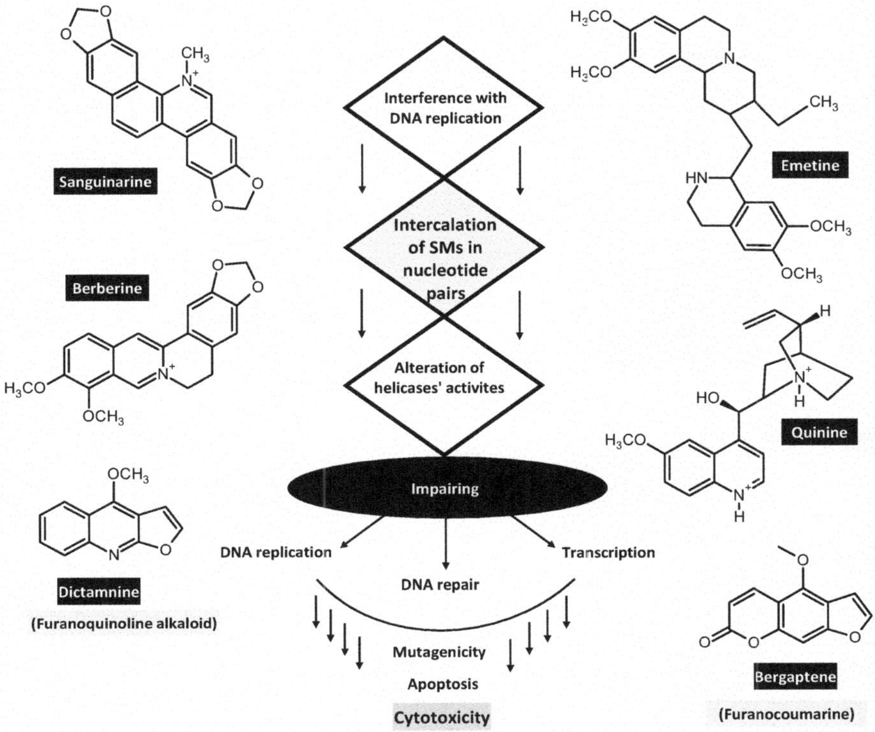

Fig. 13.6 Overview on cytotoxic effects of secondary metabolites (SMs) via their intercalation with DNA

Neuronal olfactory detection by insect's antennae occurs on the basis of ratios of compounds for both host and nonhost (e.g. partner) identification (Webster et al. 2008; Nottingham et al. 1991). After the perception of a repelling mixture of volatile compounds, the insect refrains from probing and subsequently goes away from the repelling source (Pickett et al. 2008).

In Brassicaceae, direct deterrent effects on insect herbivores are provided by glucosinolate-deriving volatiles (e.g. isothiocyanate, nitrile, thiocyanate) which are released by enzymatic action of myrosinase (β-glucosidase) (Fig. 13.8): glucosinolates are initially stored in vacuole and separated from their hydrolytic enzyme (myrosinase) (Fig. 13.8a). Then, mechanical damages from herbivores result in plant tissue breakage and cell disruption making different cell compartments to release

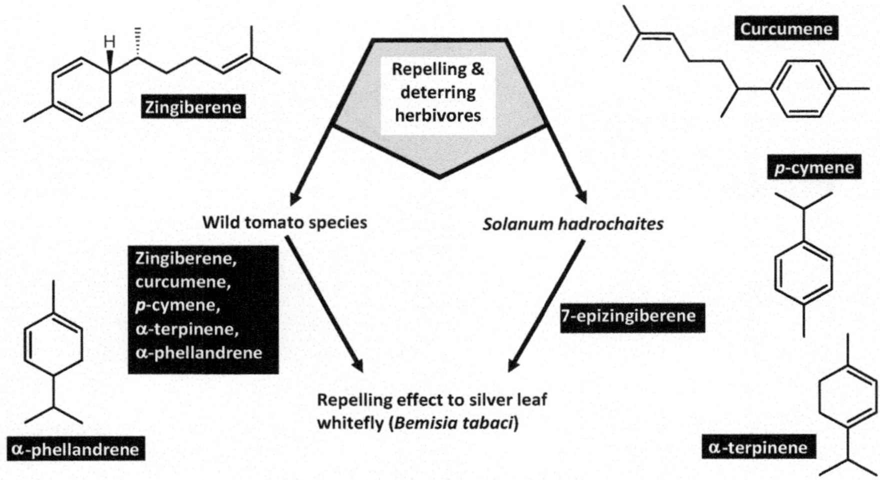

Fig. 13.7 Plant repelling and deterring ways of herbivores based on appropriate mixtures of deterrent volatile compounds (e.g. sesquiterpenes, monoterpenes). Illustrative example is given for tomato against whitefly

their storage contents (glucosinolates, myrosinases) (Fig. 13.8b). Therefore, glucosinolates will be in direct contact with β-glucosidases leading to their hydrolysis and subsequent formation of glucosinolate-deriving volatiles such as isothiocyanates, nitriles and thiocyanates (Fig. 13.8c) (Zhou and Jander 2022; Halkier and Gershenzon 2006). These breakdown products of glucosinolates have strong bioinsecticide effects by repelling herbivorous insects from concerned plant sources (Fig. 13.8d) (Li et al. 2000).

13.2.4 Attraction of Natural Enemies of Herbivores

Plants can defend against herbivores by indirect way consisting of chemical attraction of natural enemies of herbivores. This interspecific assistance occurs by the release of plant volatile compounds that play guidance attractive role making predator or parasitoid to come in by finding its prey beside the attacked host plant (Fig. 13.9):

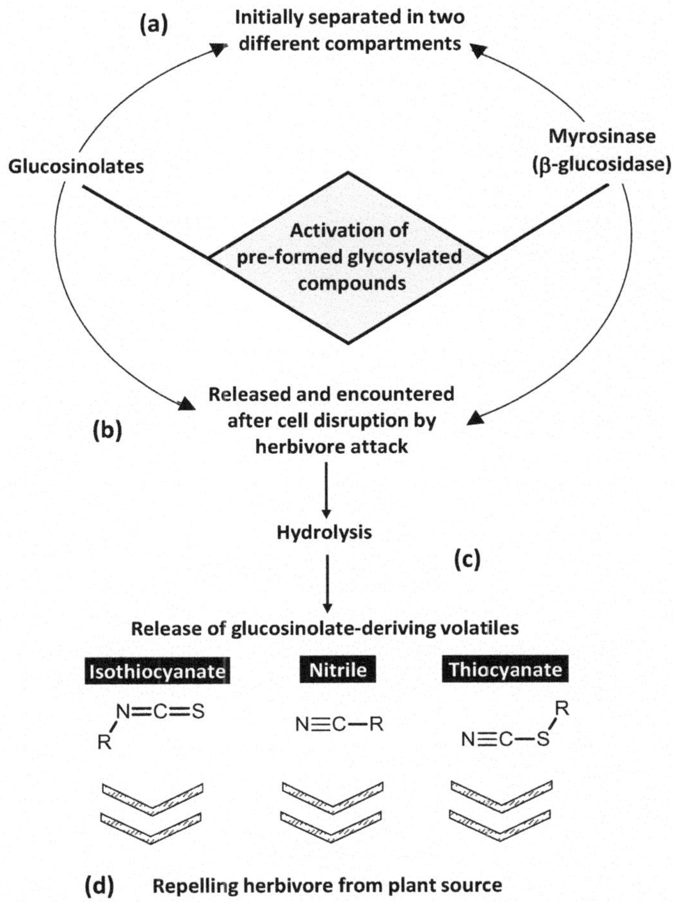

Fig. 13.8 Repelling effect of glucosinolate-deriving volatiles on herbivores

Herbivores contain elicitors in their oral secretions or oviposition fluids influencing the release of volatile organic compounds by plants as defensive/alarming response signals (Bonaventure 2018; Zunjarrao et al. 2020; Aljbory and Chen 2018; Whitman and Eller 1990). Elicitors of oral secretions (of herbivore insects) include enzymes, fatty acid-amino acid conjugates (FACs) and peptides (Fig. 13.9a). Released elicitors in oviposition fluids include esters and benzyl cyanide (Fig. 13.9b). Released elicitors by herbivores represent specific active substances that are recognized by host plants which will manifest resistance to herbivores through downstream signaling cascades that are primarily mediated by increased levels of phytohormone secretions (Fig. 13.9c, d) (Bonaventure 2018). Defensive phytochemical system leads to the release of a series of volatiles that attract natural enemies (parasitoids, predators) of herbivores (Fig. 13.9e, f).

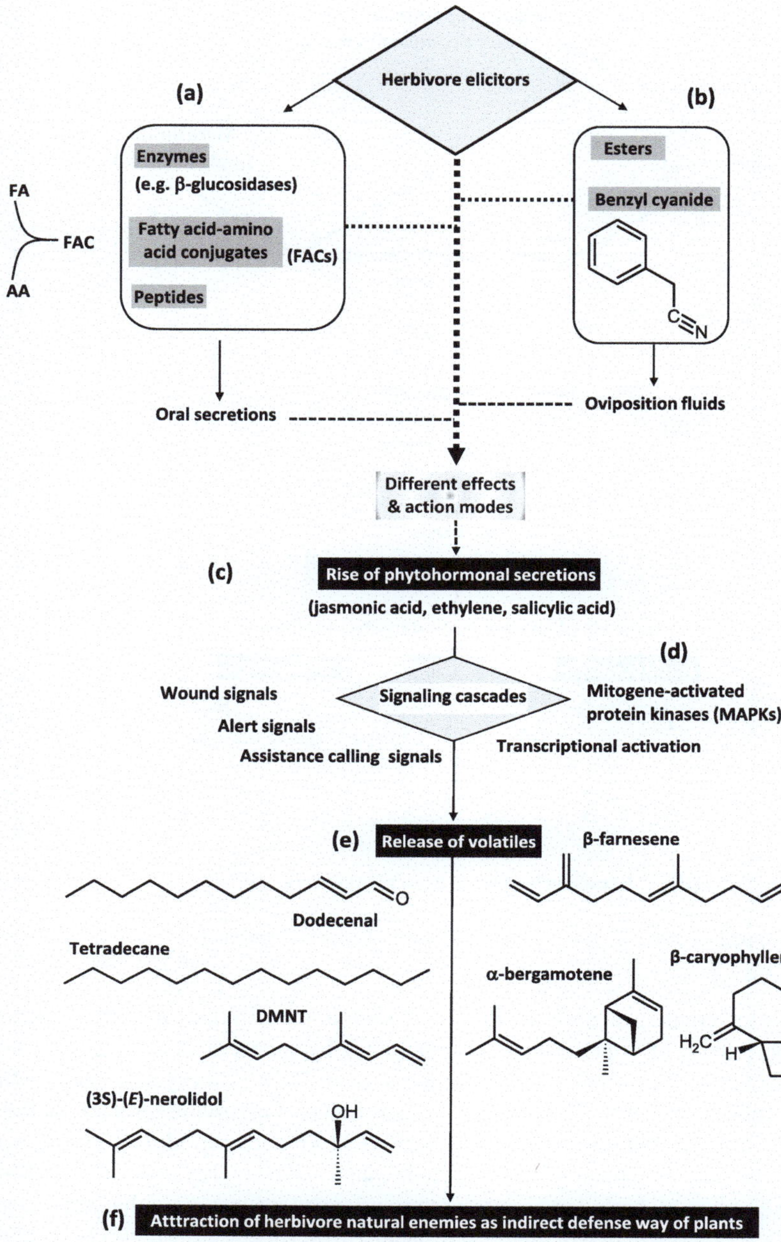

Fig. 13.9 Sequential processes involved in the release of plant volatile compounds playing key role in attraction of natural enemies of herbivores. AA—amino acid; FA—fatty acid

Parasitoid attractiveness was enhanced in tomato, rice and soybean after treating crops with jasmonate resulting in induced production of volatile organic compounds (VOCs) (Thaler 1999; Lou et al. 2005; Moraes et al. 2009).

Positive effects of terpenoids on parasitoid attractiveness were demonstrated by bioengineering in Laboratory and greenhouse experiments. In *Arabidopsis*, single gene modifications resulted in production of novel terpenoid emitting plants able to attract natural enemies of herbivores:

- Bioengineering provided productive *Arabidopsis* of (3S)-(E)-nerolidol (sesquiterpene) and (3E)-4,8-dimethyl-1,3,7-nonatriene (DMNT) (homoterpene, C11) (Kappers et al. 2005). This increased the attractiveness trend of the predatory mites, *Phytoseiulus persimilis*, by *Arabidopsis.*
- Production of (E)-β-farnesene (sesquiterpene) and (E)-α-bergamotene (sesquiterpene), was increased in bioengineered *Arabidopsis* with the maize TPS10 terpene synthase (Schnee et al. 2006). This resulted in more attractive plants for *Cotesia marginiventris* through produced odors helping this parasitoid wasp to locate its lepidopteran hosts.
- *Arabidopsis* was also bioengineered with a gene from *Mentha* x *piperita* (peppermint) for production of (E)-β-farnesene (EβF) (Beale et al. 2006). This sesquiterpene plays the role of alarm pheromone in aphid making natural enemy to locate the herbivore prey. Consequently, more productive *Arabidopsis* strain of EβF was revealed to be more attractive for the parasitoid wasp *Diaeretiella rapae* which is cued by EβF to find its aphid host *Myzus persicae.*

The (E)-β-farnesene production was also increased by engineering in *Triticum aestivum* (wheat) resulting in attractive effect for the parasitoid wasp of aphids, *Aphidius ervi*, in laboratory setting (Bruce et al. 2015). Under other (alternative) aspect, higher EβF production by the wheat strain showed deterrent effects on three aphid species, *Sitobion avenae* (English grain aphid), *Rhopalosiphum padi* (bird cherryoat aphid), and *Metopolophium dirhodum* (rose-grain aphid).

Genetic modification approaches have been also applied to tomato, rice, and cotton (Bleeker et al. 2012; Li et al. 2018; Zhang et al. 2020). These experiments provided highlights on attractive effects of volatile compounds (essentially terpenes) toward natural enemies of herbivores. However, these results should be further validated in field experiments where the natural conditions are more complex than the controlled laboratory and greenhouse conditions. Natural environment is characterized by many interactions, competition and adaptive processes occurring between biological species involved in tritrophic systems, under the effects of several local conditions.

Interestingly, field experiments showed that the larvae of pest species, *Diabrotica virgifera* (corn rootworm), were fivefold more infected by nematodes in the maize variety which produced (E)-β-caryophyllene compared with that does not (Rasmann et al. 2005).

13.2.5 Combined Effects of Herbivore and Atmospheric Chemistry on Volatile Terpenoid Emissions and Parasitoid Attraction

Agrosystems are managed against pest insects by using plants containing transgenically incorporated insect resistance. Among the most insect-resistance ways, thiringiensine (Bt) is a natural toxin with bioinsecticide properties produced by the positive gram bacteria *Bacillus thuringiensis*. Bt can be produced by plants in which genomic fragments of *B. thuringiensis* was introduced.

A key question around the transgenic way consists in asking about the effects of allocated Bt production by plants on their terpenoid emissions induced by herbivore attacks and involved in multitrophic communication (e.g. parasitoid attraction). Beyond this biological aspect, volatile organic compounds (VOCs) are also influenced by climate change conditions including increase of tropospheric CO_2 and O_3. Himanen et al. (2009) carried out an integrative study on the variations of terpenoid patterns of the oilseed rape, *Brassica napus* (Brassicaceae), by considering the effects of (i) allocated Bt production, (ii) herbivore-caused damages and (iii, iv) air concentrations of CO_2 and O_3. The study was carried out on early vegetative-stage plants characterized by younger leaves with strongly active photosynthesis and typically higher VOC synthesis levels and induction capabilities (Takabayashi et al. 1994; Gouinguené and Turlings 2002).

Two (non-Bt and Bt) *Brassica napus* ssp. *oleifera* L. were cultivated in computer-controlled growth chambers with the following light, temperature and water conditions: photoperiod → 16h (light): 8h (dark); temperature → 20 °C (day):16 °C (night); daily watering. CO_2 and O_3 effects were separately studied in different experiments with two levels: control versus doubled concentration for CO_2 (360 $\mu l\ l^{-1}$ vs. 720$\mu l\ l^{-1}$; 24 h daily), and control (filtered air) versus 100 ppb (8 h daily) for O_3. A subset of plants was exposed to the herbivorous diamondback moth (*Plutella xylostella*) leading to damaged versus not damaged plant groups. In responses to this multifactorial experiment design, several terpenoid emissions by plants and olfactometric orientation of introduced *Cotesia vestalis* (endoparasitoid of *P. xylostella*), were assessed under the corresponding CO_2 and O_3 concentrations.

Total terpenoid emissions (TTE) showed higher level under CO_2 exposure than O_3 (Fig. 13.10). The high induction of terpenoid emissions favored by CO_2 would be in agreement with its main role in photosynthetic and non-photosynthetic biosynthesis pathways. Moreover, this hypothesis is in agreement with the increase of TTE under elevated CO_2 compared with control (Fig. 13.10a vs. c). Under each CO_2 level, herbivore damaged plants showed significantly higher terpenoid emissions compared with intact plants. This highlighted strong induction effect of herbivore attacks on emitted levels of plant terpenoids. The increase of TTE under herbivore effect was observed both in Bt and non-Bt plants (Bt + Hbv vs. Bt; Non Bt + Hbv vs. Non Bt). Moreover, among the herbivore-damage plants, non-Bt plants showed higher TTE than Bt ones. This provides highlighting on the key defensive role of terpenoid emissions in non-Bt plants which have not Bt toxin. Finally, the highest TTE levels were

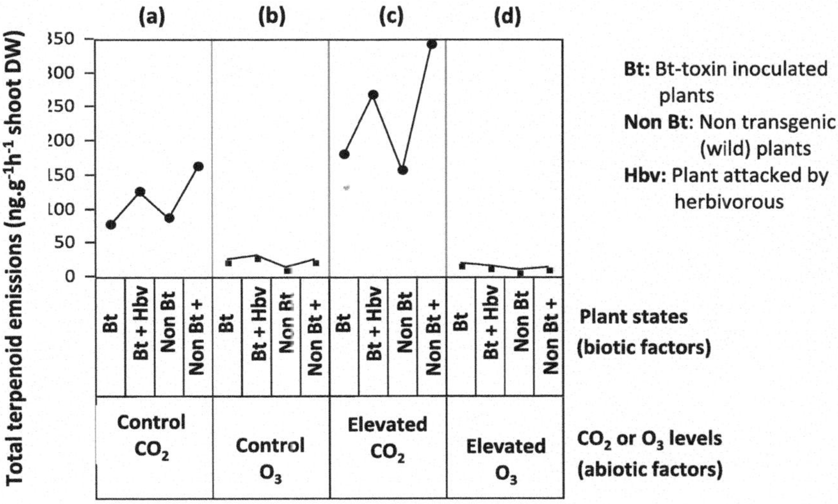

Fig. 13.10 Effects of single and combined abiotic (CO_2, O_3) and biotic (Bt-toxin allocation, herbivore attack) factors on total terpenoid emissions in *Brassica napus* ssp. *oleifera* L. (**a**, **b**) Control conditions of air CO_2 and O_3 concentration; **c**, **d** elevated air CO_2 and O_3 concentrations. Bt and Non Bt → Bt toxin-allocated and not allocated plants, respectively; Hbv, herbivorous damaged plants

observed in non-Bt plants growing under higher air CO_2 concentration and exposed to herbivore attacks (Fig. 13.10c). This shows that elevated CO_2 could positively interact with herbivore-induced defenses to further increase terpene emission levels at vegetative stages of *B. napus*, and more particularly in non-transgenic plants.

In CO_2 variation-based experiments, terpenoids were more emitted by herbivores-attacked plants than by intact ones (Figs. 13.11 and 13.12): the 1,8-cineol, limonene and sabinene showed the highest concentrations which reached their maximum levels under elevated CO_2 (vs. control) and herbivore attack (vs. intact plants) (Fig. 13.12c). Other terpenoids with lower variation scales were more emitted by attacked plants: under control CO_2, herbivore induced an increase in β-elemene, β-myrcene, α-thujene and (E)-DMNT emission by plants (Fig. 13.12a). Under elevated CO_2, herbivorous attacks induced further increase of the aforementioned terpenes in addition to α-pinene and δ-3-carene (Fig. 13.12c vs 13.12a).

Further sensitive phytochemical responses can be highlighted by considering variation ratios of terpenoids between different experiment conditions such as herbivorous damaged plants relatively to intact ones (Figs. 13.13 and 13.14):

Thus, the ratios of each terpenoid between the two situations of damaged and intact plants showed maximal levels for (*E*)-DMNT and β-elemene in non-Bt plants under both elevated and control CO_2 (Fig. 13.14b, d). These results highlight sensitive

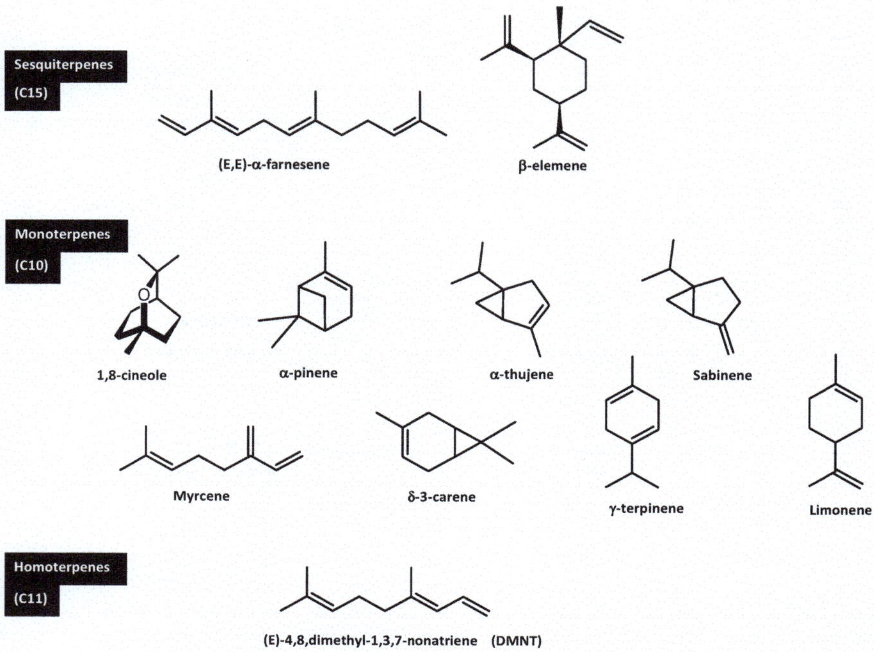

Fig. 13.11 Chemical structures of monoterpenes, sesquiterpenes and alkatriene emitted with variable levels by *Brassica napus* (Brassicaceae) growing under different abiotic and biotic conditions combining (i) two genetic strains _ thiringiensine (Bt)-transgenic and nontransgenic_ with (ii) elevated or control air-CO_2 levels and (iii) occurrence or absence of herbivorous attack (diamondback moth, *Plutella xylostella*)

variations of individual volatile terpenes in herbivorous damaged non-Bt plants. Also, these results highlight strong positive effect of wild plant variety on terpene production levels. Moreover, the highest ratios were observed under elevated CO_2 growing conditions (Fig. 13.14b). By this way, regulation ratios of emitted specific terpenes provide useful indications on defensive reactions of different plant strains against herbivore attacks under well-defined abiotic conditions.

Further associations between high or low metabolic regulation ratios and abiotic/biotic conditions can be highlighted by correspondence analysis (CA) (Fig. 13.15):

CA highlighted strong hierarchical effects concerning biotic and abiotic factors on different terpenoid emissions: the experiments associated with herbivore effects showed topological separation from those concerning non attacked plants (non-herbivore) (Fig. 13.15a). Factorial plot of terpenoids highlighted three metabolites which were sensitively regulated in presence of herbivore attack (Fig. 13.15b): (E)-DMNT, β-elemene and δ-3-carene (projected in same subspace than Hbv points). Under herbivore attack, these three terpenes showed significant regulation variations between Bt and non-Bt plants: (E)-DMNT and β-elemene were highly regulated under herbivore attack in non-Bt plants (Fig. 13.15b). They projected in the

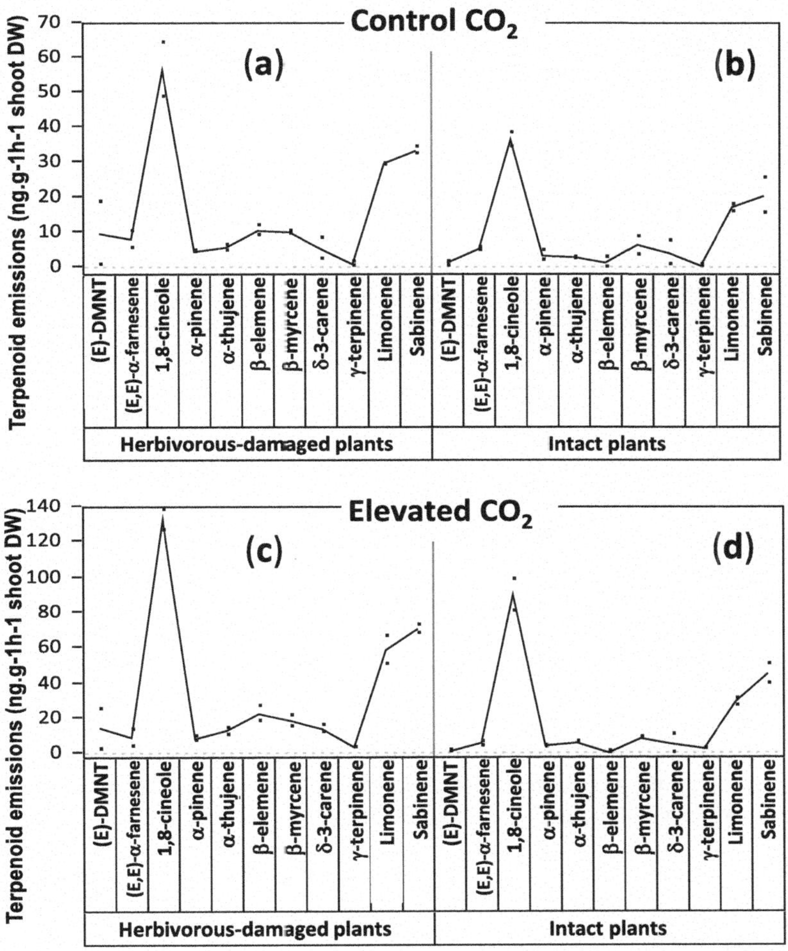

Fig. 13.12 Global terpenoid emission patterns of *Brassica napus* exposed to control (**a**, **b**) and elevated (**c**, **d**) CO_2 conditions combined with herbivorous attack (**a**, **c**) or not (**b**, **d**)

upper-right and lower-right quadrants corresponding to those of Ctr.(0Bt, Hbv) and CO_2.(0BT, HBv), respectively (Fig. 13.15a, b). This indicated that emission levels of (E)-DMNT and β-elemene in (herbivorous damaged, non-Bt)-plants were relatively more favored by control and elevated CO_2, respectively. These positive trends agreed with the strong ratios previously shown for emitted terpenes by damaged plants relatively to intact plants. Under herbivore attacks:

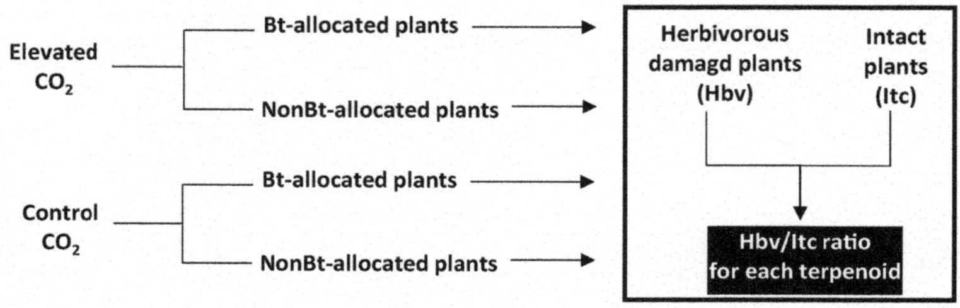

Fig. 13.13 Applied design for calculation of variation ratios of emitted terpenes between herbivorous damaged (Hbv) and intact (Itc) plants belonging to either Bt or nonBt transgenic strains of *Brassica napus* and growing under elevated or control CO_2 conditions

- (E)-DMNT ratio was maximal in non-Bt plants under CO_2 control (Figs. 13.14b, 13.15b).
- β-elemene ratio strongly increased under elevated CO_2 condition to reach comparable level to (E)-DMNT (Figs. 13.14d, 13.15b).

For both terpenes, elevated CO_2 conditions favored an increase of emission levels in damaged non Bt plants, but the β-elemene was frankly more advantaged by elevated CO_2 after its induction by herbivore attack (in non Bt plants). However, in Bt damaged plants, the emissions levels of these two terpenes showed remarkable relative reduction (Fig. 13.14a, c). This could be linked to some alternative defensive ways based on Bt toxin. This negative effect of Bt toxin allocation on terpene emission by damaged plants was highlighted in CA through opposition between the projection space of CO_2. (**Bt**, Hbv), Ctr.(**Bt**, Hbv) points and that of (E)-DMNT, β-elemene along the first factorial axis F1 (Fig. 13.15b vs. a). Bt-concerned points generally projected along the left side of the first principal component (F1) whereas (E)-DMNT and β-elemene projected on the right side of F1. This indicated the key defensive/communicative roles of these two terpenes in damaged **non Bt** plants.

Another effect of Bt allocated toxin was concerned with γ-terpinene which projected in the upper left quadrant of CA (Fig. 13.15b) by sharing left side projection with (CO_2.Bt) and (CO_2.0Bt) points (Fig. 13.15a). This highlighted less sensitive terpene to herbivore attacks on one hand versus more efficient indicative role of elevated CO_2 on other hand (Fig. 13.17). Also, combined positive effects of elevated CO_2 and Bt allocation resulted in higher relative emission level of γ-terpinene. In general, the ratios of γ-terpinene calculated for elevated CO_2 relatively to CO_2 control conditions (Fig. 13.16) was high in both damaged and intact Bt plants (Fig. 13.17a, b). Moreover, elevated CO_2 resulted in higher emissions of most of the volatile terpenes as indicated by ratios > 1 (Fig. 13.17). The highest ratios were observed in damaged non-Bt plants exposed to elevated CO_2 levels, whereas lesser increase of such ratios occurred in intact non-Bt plants (Fig. 13.17c, d). This highlighted significant interaction between elevated CO_2 and herbivory in the induction of volatile terpenes in wild (non-transgenic) plants.

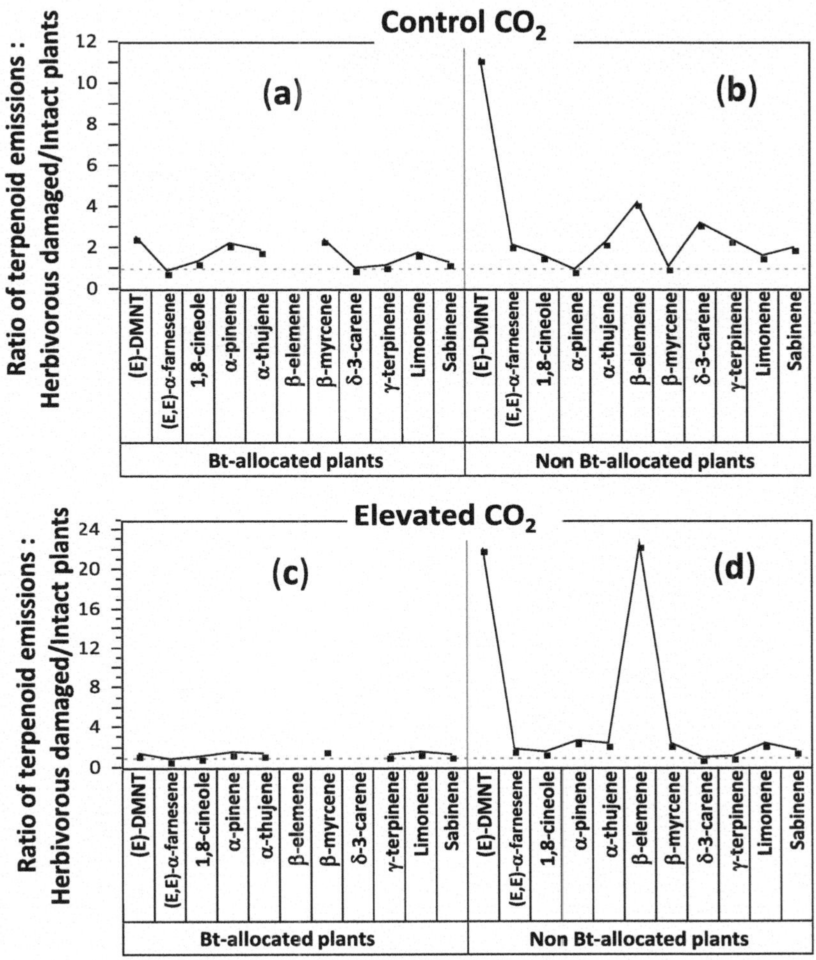

Fig. 13.14 Variation ratios of emitted levels of individual terpenoids calculated between attacked and intact plants under control (**a**, **b**) versus elevated CO_2 conditions (**c**, **d**), and in Bt-transgenic (**a**, **c**) versus not transgenic (**b**, **d**) plants

However, δ-3-carene showed minimal variation ratio in intact Bt plant growing under elevated CO_2 relatively to control conditions (Fig. 13.17b). This highlighted opposite trend to γ-terpinene which showed the maximal level in Bt (damaged and intact) plants. The δ-3-carene was particularly distinguished by strongly flexible

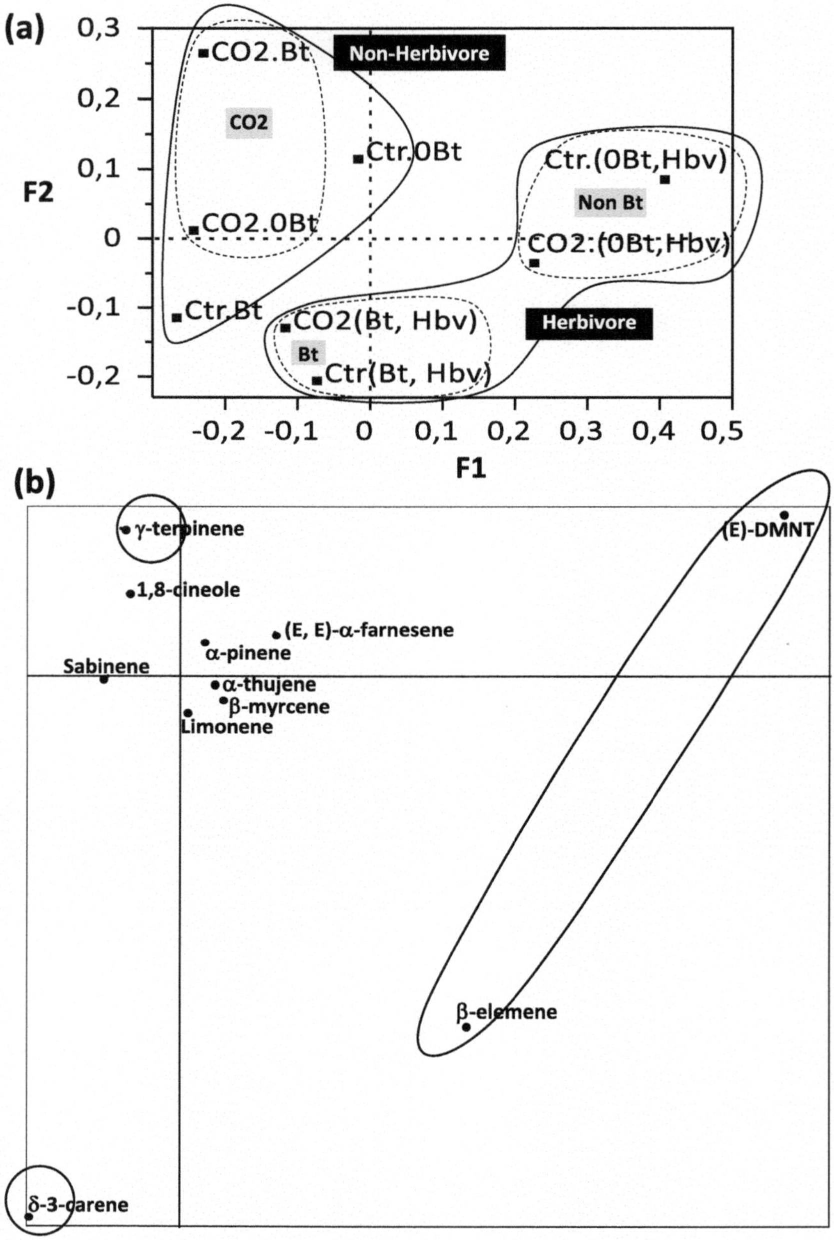

Fig. 13.15 First factorial plot (F1F2) of correspondence analysis showing association and variation trends between terpenoid emissions in *Brassica napus* and its growth conditions combining air CO_2 concentration, Bt toxin allocation and herbivorous damage manifestation. Hbv, herbivorous attacked plants; Ctr and CO_2, control and elevated CO_2, respectively; Bt and 0Bt, allocated and not allocated Bt plants, respectively

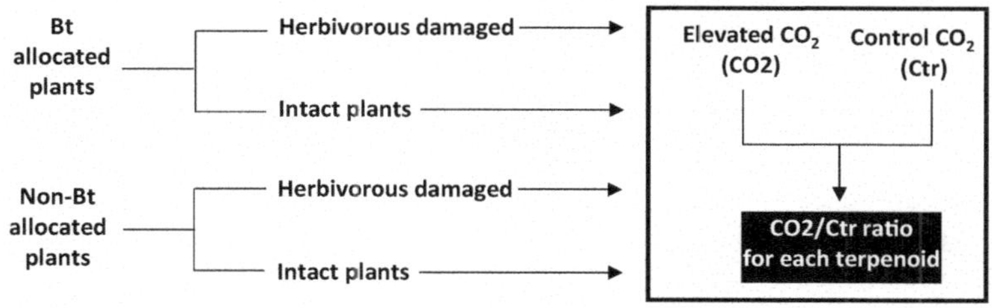

Fig. 13.16 Applied design for calculation of variation ratios of emitted terpenes between elevated and control CO_2 conditions in herbivorous damaged or intact plants and belonging to either Bt or non-Bt transgenic strains

variation due to absolute maximal ratio in non-Bt (Fig. 13.17d) versus absolute minimal ratio in Bt plants (Fig. 13.17b).

In CA, this opposition between δ-3-carene and γ-terpinene was highlighted along the second principal component (F2) where δ-3-carene opposed also to (CO_2.Bt) points indicating its low relative emission (contrary to γ-terpinene) in intact Bt plants exposed to elevated CO_2 levels (Fig. 13.15b vs. a). However, the low variation ratio of δ-3-carene was improved in damaged plants (Fig. 13.17a) indicating some sensitivity of δ-3-carene to herbivorous attack in Bt plants exposed to elevated CO_2. This was highlighted in CA by the projection of CO_2(Bt, Hbv) point close to δ-3-carene (Fig. 13.15a vs. b). The δ-3-carene was also characterized by multi-directional (flexible) variations due to its response to multiple factor states: relative emission levels of δ-3-carene showed also positive trend in non-Bt plants under elevated CO_2 (Fig. 13.17c, d) leading to maximal variation ratios of δ-3-carene between elevated CO_2 and control conditions. This positive trend of relative emission was highlighted in CA by the projection of both δ-3-carene and CO_2.0Bt (non-Bt plants exposed to elevated CO_2) along the left side of F1 (Fig. 13.15a, b).

Beyond emission levels of terpenoids following herbivorous attack, plant defense was analyzed through the relative attraction of *Cotesia vestalis*, the endoparasitoid of *Plutella xylostella* (herbivore of *Brassica napus*) (Fig. 13.18). The results showed that attraction of parasitoids was highly associated with host-damaged plants compared with intact plants whatever CO_2 levels in air and plant genetic strain (Bt and non-Bt) (Fig. 13.18a): the percentages of attracted parasitoids by host-damaged plants varied between 71.4 and 76.5% versus 23.5–28.6% for intact plants. This highlighted a strong hierarchical effect of herbivore attack on parasitoid attraction by volatile terpenes of damaged plants.

In plants growing under variable air-O_3 concentration, the effects of herbivore and subsequent odor(s) of emitted terpenoids on parasitoid attraction were also observed expect in (Bt plants, elevated O_3) case where C. *versalis* was less oriented to damaged plants (Fig. 13.18b): 36.8% vs. 63.2% in damaged and intact plants, respectively. This could be indicative of some negative interaction between high O_3 level and plant

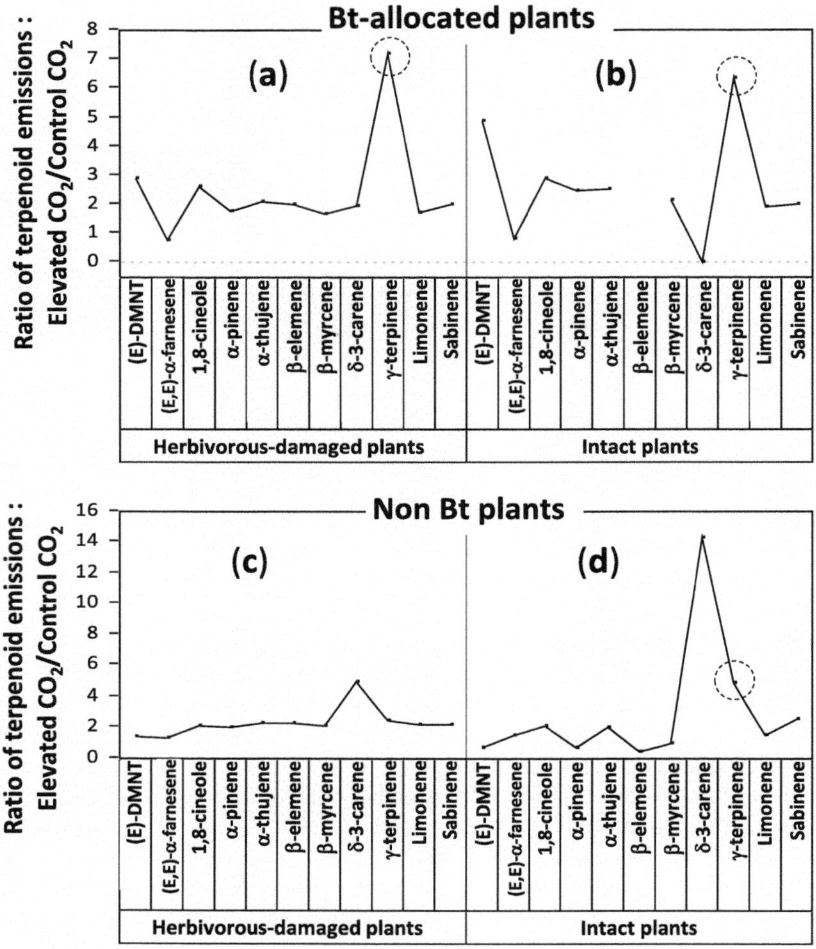

Fig. 13.17 Variation ratios of emitted levels of individual terpenoids between elevated CO_2 relatively to control conditions in Bt (**a**, **b**) and non-Bt plants (**c**, **d**), and in herbivorous attacked (**a**, **c**) versus non attacked (**b**, **d**) plants

strain for parasitoid attraction by emitted terpenes (e.g. variation of olfactory stimuli resulting from variations of different terpene regulation levels). Compared with CO_2 which directly influences the photosynthesis activity and indirectly SM biosynthesis, O_3 shows relatively contrasted effects by activating defense signaling and defense protein synthesis on one hand, and by damaging tissues and O_3-sensitive proteins on other hand (Fiscus et al. 2005). Moreover, O_3 and VOC can react leading to the formation of atmospheric aerosols and a subsequent loss of VOCs (McFrederick

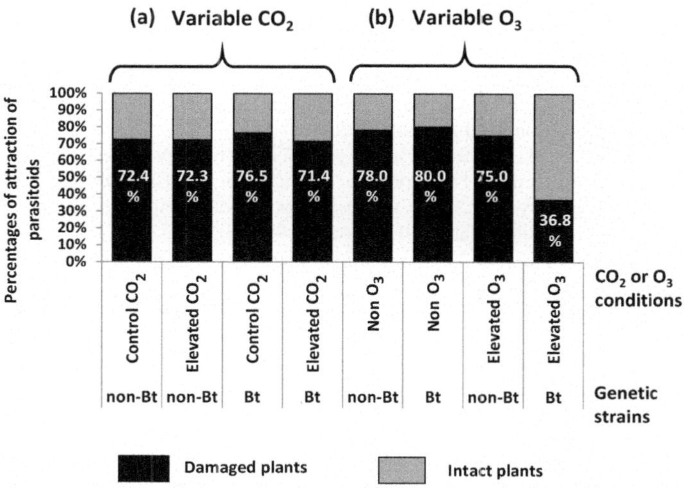

Fig. 13.18 Percentages of attraction of parasitoids of herbivore resulting from terpenoid emissions by damaged (in black) and intact (in grey) plants *Brassica napus* associated with Bt and non-Bt allocated strains and growing under variable CO_2 or O_3 levels in air

et al. 2008). By this way, elevated O_3 could have negative impact on tritrophic signaling (Pinto et al. 2007a, b, c). Moreover, reactivity between O_3 and VOCs in the atmosphere is compound-specific leading to multiple variation ways affecting the initially emitted terpene mixtures. Moreover, O_3 reactivity is characterized by sudden and rapid aspects that could result in severe confuse orientation of parasitoids (Himanen et al. 2009).

References

Aljbory Z, Chen M-S (2018) Indirect plant defense against insect herbivores: a review. Insect Sci 25:2–23

Beale MH, Birkett MA, Bruce TJA et al (2006) Aphid alarm pheromone produced by transgenic plants affects aphid and parasitoid behavior. Proc Natl Acad Sci USA 103:10509–10513

Bleeker PM, Diergaarde PJ, Ament K, Guerra J, Weidner M, Schütz S, de Both MT, Haring MA, Schuurink RC (2009) The role of specific tomato volatiles in tomato-whitefly interaction. Plant Physiol 151:925–935

Bleeker PM, Mirabella R, Diergaarde PJ, VanDoorn A, Tissier A, Kant MR, Prins M, de Vos M, Haring MA, Schuurink RC (2012) Improved herbivore resistance in cultivated tomato with the sesquiterpene biosynthetic pathway from a wild relative. Proc Natl Acad Sci USA 109:20124–20129

Bonaventure G (2018) Plants recognize herbivorous insects by complex signalling networks. Annu Plant Rev 1–35

Bouwmeester H, Schuurink RC, Bleeker PM, Schiestl F (2019) The role of volatiles in plant communication. Plant J 100:892–907

Bruce TJA, Aradottir GI, Smart LE et al (2015) The first crop plant genetically engineered to release an insect pheromone for defence. Sci Rep 5:11183

Divekar PA, Narayana S, Divekar BA, Kumar R, Gadratagi BG, Ray A, Singh AK, Rani V, Singh V, Singh AK et al (2022) Plant secondary metabolites as defense tools against herbivores for sustainable crop protection. Int J Mol Sci 23:2690

Dötterl S, Gershenzon J (2023) Chemistry, biosynthesis and biology of floral volatiles: roles in pollination and other functions. Nat Prod Rep. https://doi.org/10.1039/D3NP00024A

Fiscus EL, Booker FL, Burkey KO (2005) Crop responses to ozone: uptake, modes of action, carbon assimilation and partitioning. Plant Cell Environ 28:997–1011

Gouinguené SP, Turlings TCJ (2002) The effects of abiotic factors on induced volatile emissions in corn plants. Plant Physiol 129:1296–1307

Halkier BA, Gershenzon J (2006) Biology and biochemistry of glucosinolates. Annu Rev Plant Biol 57:303–333

Himanen SJ, Nerg AM, Nissinen A, Pinto DM, Stewart NC Jr, Poppy GM, Holopainen JK (2009) Effects of elevated carbon dioxide and ozone on volatile terpenoid emissions and multitrophic communication of transgenic insecticidal oilseed rape (*Brassica napus*). New Phytol 181:174–186

Hollingsworth RG, Armstrong JW, Campbell E (2002) Caffeine as a repellent for slugs and snails. Nature 417:915–916

Kappers IF, Aharoni A, van Herpen TW, Luckerhoff LL, Dicke M, Bouwmeester HJ (2005) Genetic engineering of terpenoid metabolism attracts bodyguards to *Arabidopsis*. Science 309:2070–2072

Kobayashi K, Sasamoto H, Sasamoto Y, Sugiyama A, Fujii Y (2021) Evaluation of isoflavones as allelochemicals with strong allelopathic activities of kudzu using protoplast co-culture method with digital image analysis. Am J Plant Sci 12:376–393

Li Q, Eigenbrode S, Stringam GR, Thiagarajah MR (2000) Feeding and growth of *Plutella xylostella* and *Spodoptera eridania* on *Brassica juncea* with varying glucosinolate concentrations and myrosinase activities. J Chem Ecol 26:2401–2419

Li F, Li W, Lin YJ, Pickett JA, Birkett MA, Wu K, Wang G, Zhou JJ (2018) Expression of lima bean terpene synthases in rice enhances recruitment of a beneficial enemy of a major rice pest. Plant Cell Environ 41:111–120

Lindahl PE, Öberg KE (1961) The effect of rotenone on respiration and its point of attack. Exp Cell Res 23:228–237

Lou YG, Du MH, Turlings TC, Cheng JA, Shan WF (2005) Exogenous application of jasmonic acid induces volatile emissions in rice and enhances parasitism of *Nilaparvata lugens* eggs by the parasitoid *Anagrus nilaparvatae.* J Chem Ecol 31:1985–2002

McFrederick QS, Kathilankal JC, Fuentes JD (2008) Air pollution modifies floral scent trails. Atmos Environ 42:2336–2348

Moraes MC, Laumann RA, Pareja M, Sereno FT, Michereff MF, Birkett MA, Pickett JA, Borges M (2009) Attraction of the stink bug egg parasitoid *Telenomus podisi* to defence signals from soybean activated by treatment with cis-jasmone. Entomol Exp Appl 131:178–188

Nottingham SF et al (1991) Behavioral and electrophysiological responses of aphids to host and non-host plant volatiles. J Chem Ecol 17:1231–1242

Pavela R (2016) History, presence and perspective of using plant extracts as commercial botanical insecticides and farm products for protection against insects—a review. Plant Prot Sci 52:229–241

Pickett JA, Birkett MA, Logan JG (2008) DEET repels ORNery mosquitoes. Proc Natl Acad Sci USA 105:13195–13196

Pinto DM, Tiiva P, Miettinen P, Joutsensaari J, Kokkola H, Nerg AM, Laaksonen A, Holopainen JK (2007a) The effects of increasing atmospheric ozone on biogenic monoterpene profiles and the formation of secondary aerosols. Atmos Environ 41:4877–4887

Pinto DM, Blande JD, Nykänen R, Dong WX, Nerg AM, Holopainen JK (2007b) Ozone degrades common herbivore-induced plant volatiles: does this affect herbivore prey location by predators and parasitoids? J Chem Ecol 33:683–694

Pinto DM, Nerg AM, Holopainen JK (2007c) The role of ozone-reactive compounds, terpenes, and green leaf volatiles (GLVs), in the orientation of *Cotesia plutellae*. J Chem Ecol 33:2218–2228

Rasmann S, Köllner TG, Degenhardt J, Hiltpold I, Toepfer S, Kuhlmann U, Gershenzon J, Turlings TCJ (2005) Recruitment of entomopathogenic nematodes by insect-damaged maize roots. Nature 434:732–737

Sampedro L, Moreira X, Zas R (2011) Costs of constitutive and herbivore-induced chemical defences in pine trees emerge only under low nutrient availability. J Ecol 99:818–827

Santiago R, López-Malvar A, Souto C, Barros-Ríos J (2018) Methods for determining cell wall-bound phenolics in maize stem tissues. J Agric Food Chem 66:1279–1284

Schmeller T, Latz-Brüning B, Wink M (1997) Biochemical activities of berberine, palmatine and sanguinarine mediating chemical defence against microorganisms and herbivores. Phytochemistry 44:257–266

Schnee C, Köllner TG, Held M, Turlings TC, Gershenzon J, Degenhardt J (2006) The products of a single maize sesquiterpene synthase form a volatile defense signal that attracts natural enemies of maize herbivores. Proc Natl Acad Sci USA 103:1129–1134

Takabayashi J, Dicke M, Takahashi S, Posthumus MA, Van Beek TA (1994) Leaf age affects composition of herbivore-induced synomones and attraction of predatory mites. J Chem Ecol 20:373–386

Thaler JS (1999) Jasmonate-inducible plant defences cause increased parasitism of herbivores. Nature 399:686–688

Turlings TCJ, Erb M (2018) Tritrophic interactions mediated by herbivore-induced plant volatiles: mechanisms, ecological relevance, and application potential. Annu Rev Entomol 63:433–452

Webster B, Bruce T, Dufour S, Birkemeyer C, Birkett M, Hardie J, Pickett J (2008) Identification of volatile compounds used in host location by the black bean aphid, *Aphis fabae*. J Chem Ecol 34:1153–1161

Whitman DW, Eller FJ (1990) Parasitic wasps orient to green leaf volatiles. Chemoecology 1:69–76

Wink M (2003) Evolution of secondary metabolites from an ecological and molecular phylogenetic perspective. Phytochemistry 64:3–19

Wink M (2012) Medicinal plants: a source of anti-parasitic secondary metabolites. Molecules 17:12771–12791

Wink M, Schimmer O (2010) Molecular modes of action of defensive secondary metabolites. Funct Biotechnol Plant Second Metab 39:21–161

Wink M (2007) Molecular Modes of action of cytotoxic alkaloids: from DNA intercalation, spindle poisoning, topoisomerase inhibition to apoptosis and multiple drug resistance. Alkaloids Chem Biol 64:1–47

Zhang L, Lu G, Huang X, Guo H, Su X, Han L, Zhang Y, Qi Z, Xiao Y, Cheng H (2020) Overexpression of the caryophyllene synthase gene GhTPS1 in cotton negatively affects multiple pests while attracting parasitoids. Pest Manag Sci 76:1722–1730

Zhou S, Jander G (2022) Molecular ecology of plant volatiles in interactions with insect herbivores. J Exp Bot 73(2):449–462

Zunjarrao SS, Tellis MB, Joshi SN, Joshi RS (2020) Plant-Insect interaction: the saga of molecular coevolution. In: Reference series in phytochemistry. Springer Science and Business Media LLC, Berlin, pp 19–45

Index

N. Semmar, *Secondary Metabolites in Plant Stress Adaptation*, Signaling and Communication in Plants, https://doi.org/10.1007/978-3-031-52595-7

Q

R

S

The manufacturer's authorised representative in the EU is Springer Nature Customer Service Centre GmbH, Europaplatz 3, 69115 Heidelberg, Germany. If you have any concerns regarding our products, please contact ProductSafety@springernature.com

Printed and bound by CPI Group (UK) Ltd, Croydon, CR0 4YY
05/10/2026
02228137-0002